Smart Grid and Energy Storage

Smart Grid and Energy Storage

Guest Editor

Adel Merabet

MDPI

Basel • Beijing • Wuhan • Barcelona • Belgrade • Novi Sad • Cluj • Manchester

Guest Editor
Adel Merabet
Division of Engineering
Saint Mary's University
Halifax
Canada

Editorial Office
MDPI AG
Grosspeteranlage 5
4052 Basel, Switzerland

This is a reprint of the Special Issue, published open access by the journal *Energies* (ISSN 1996-1073), freely accessible at: https://www.mdpi.com/journal/energies/special_issues/H43287V2KK.

For citation purposes, cite each article independently as indicated on the article page online and as indicated below:

Lastname, A.A.; Lastname, B.B. Article Title. *Journal Name* **Year**, *Volume Number*, Page Range.

ISBN 978-3-7258-7226-8 (Hbk)
ISBN 978-3-7258-7227-5 (PDF)
https://doi.org/10.3390/books978-3-7258-7227-5

Contents

About the Editor

Adel Merabet

Adel Merabet is a full professor at the Division of Engineering, Faculty of Science, Saint Mary's University, Halifax, NS, Canada. He was a Visiting Academic at the Department of Electrical Engineering, Khalifa University, Abu Dhabi, United Arab Emirates, in 2024, as well as at the Department of Sustainable and Renewable Energy Engineering, University of Sharjah, Sharjah, United Arab Emirates, in 2016-2017. In 2019, he was awarded the Tan Chin Tuan Exchange Fellowship in Engineering from Nanyang Technological University, Singapore, and held a limited-term appointment as a Research Fellow in its School of Electrical and Electronic Engineering. He was included in the "World's Top 2% Scientists List" every year from 2021 to 2025. His research interests include renewable energy systems, energy storage, energy management, optimization, control, and AI applications.

energies

MDPI

Editorial

Smart Grid and Energy Storage: Innovations, Challenges, and Applications

Adel Merabet

Division of Engineering, Saint Mary's University, Halifax, NS B3H 3C3, Canada; adel.merabet@smu.ca;
Tel.: +1-902-420-5712

Smart grids are complex systems that require efficient solutions integrating demand response, distributed generation, and energy storage to achieve optimal performance [1]. Furthermore, energy storage plays an increasingly important role in stabilizing and supporting power systems while also providing auxiliary services [2]. In smart grid-connected hybrid renewable energy systems, energy management is essential to achieve the benefits of reliability, efficiency, cost-effectiveness, and high power quality [3]. For these reasons, research activity is expanding in terms of both smart grid applications and energy storage technologies, reflecting their critical role in modern power systems. Artificial intelligence is also being used to modernize energy production and distribution systems, resulting in a revolutionary transformation [4,5].

This Special Issue of *Energies*, titled "Smart Grid and Energy Storage", aims to collect and publish research on recent advances in smart grids, including integration of renewable energy resources, enhancement of energy efficiency, decentralization of electric energy generation, and distribution through small- to medium-scale electric infrastructures. Considering the intermittency of renewable energy resources and the stochastic nature of energy demand, energy storage systems are essential for enhancing energy resilience and enabling the achievement of operational objectives, despite existing technical and economic constraints. Smart grid architectures integrated with energy storage comprise multiple heterogeneous and interconnected energy systems that require advanced optimization, coordination, and control strategies to ensure their efficient operation while satisfying multi-objective performance criteria, including economic, social, and health-related factors. Our aim is to explore innovative solutions and publish original research related to smart grids and energy storage.

In this Special Issue, twelve (12) research articles and two (2) review articles have been published to address challenging topics related to smart grids and energy storage. Rasheed et al. (1) present a probabilistic visualization technique using polar histograms to identify the dynamic influence zones of distributed generation units. This technique provides an enhanced reflection-based methodology that accounts for reverse power flows and concurrent real/reactive power to improve grid visibility and stability. Stanchev et al. (2) evaluate the physical and cyber-physical resilience of PMSG-dominated smart grids using a unified simulation framework that compares five distinct control strategies. A hybrid architecture is proposed to optimize frequency and voltage stability while mitigating the impacts of cyber disturbances like communication delays. Barua et al. (3) introduce a hybrid metaheuristic optimization, the Jackal Arithmetic Algorithm, for parametric identification of the solar photovoltaic model. Compared to standalone optimization methods, the proposed dual-strategy framework balances exploration and exploitation and achieves superior estimation accuracy and convergence stability compared to standalone optimization methods.

Gharbi et al. (4) propose an adaptive multi-layer heuristic framework within a unified real-time loop to simultaneously optimize demand-side management, dynamic pricing, communication infrastructure, and AC load-flow feasibility. This proxy-gated heuristic process ensures physical feasibility and grid stability while achieving high computational efficiency for residential energy systems. Bradło et al. (5) present a feasibility study evaluating the achievement of the net-zero energy system in Poland by 2060. The study models a transition toward weather-dependent renewables, nuclear power, and biomethane-fueled generation through a framework that uses centralized electrolyzers as dispatchable load units to stabilize the grid by absorbing renewable energy surpluses for industrial hydrogen production. Zhang et al. (6) propose a coordinated scheduling strategy using 5G base station dormancy, load transfer, and many-to-many energy sharing to optimize flexible load resources. The proposed optimization framework reduces grid load variance and operational cost. Avila Saccol et al. (7) propose an AI-assisted methodology using deep neural network surrogate models to automate the selection of optimal DC/AC power converter topologies. It is a fast, accurate estimation framework for efficiency and thermal performance that eliminates the need for time-consuming, resource-intensive physical simulations. Martínez-Nolasco et al. (8) present a low-cost controller hardware-in-the-loop (CHL) platform for validating current-sharing and hot-swap algorithms in islanded DC microgrids. This validation tool can accurately emulate parallel photovoltaic systems to ensure grid stability and reliability. Jlidi et al. (9) propose a hybrid methodology combining artificial neural networks for weather forecasting with minimal risk problem-sliding mode control to optimize photovoltaic panel performance. This framework transforms stochastic weather uncertainties into deterministic risk problems, enabling highly accurate and robust maximum power point tracking (MPPT). Maiwald et al. (10) use agent-based modeling to analyze the correlation between prosumer numbers, flexible pricing, demand, and generation balance within cellular electricity systems, showing that economic incentives require integrated technical solutions to effectively manage seasonal supply volatility and finite consumer willingness to pay. Zhang et al. (11) study the integration of micrometeorology into power systems to manage renewable energy variability and grid stability throughout the operational lifecycle, showing that wind power forecasting accuracy can be improved using a high-resolution reanalysis dataset and localized meteorological data. Long et al. (12) propose a high-concurrency "cloud, edge, and end collaboration" architecture based on edge computing for digital substation operation support systems. This framework integrates cloud decision-making with localized edge intelligence, which helps in improving real-time monitoring and grid operational efficiency. The review by Nazir et al. (13) evaluates the security, privacy, and operational balance of smart grids by analyzing diverse architectures, cyber threats, and demand–supply management strategies. Their synthesized framework, which includes privacy-preserving techniques and storage technologies, identifies research gaps for advancing resilient and sustainable power networks. Finally, the review by Ishfaq et al. (14) provides information about cybersecurity challenges and AI-driven solutions across different layers of modern smart grids. The holistic synthesis of cyber defense and energy-routing protocols offers a unified framework to enhance the resilience and efficiency of decentralized energy networks.

We intend for this Special Issue to advance the study of energy storage systems and smart grid technologies and assist in the identification of future research avenues to support sustainable, dependable, and efficient power networks.

Acknowledgments: The author would like to thank the editorial board of *Energies* for the opportunity to serve as the Guest Editor of this Special Issue. Many thanks to all of the researchers who submitted articles to this Special Issue for their excellent contributions. Thank you to all of the reviewers for their comments and suggestions to improve the quality of the contributions. Finally, congratulations to the *Energies* editorial office team on their success in managing this Special Issue.

Conflicts of Interest: The author declares no conflicts of interest.

List of Contributions:

1. Rasheed, Y.N.A.; Agalgaonkar, A.P.; Muttaqi, K. Probabilistic Visualisation Approach Using Polar Histograms to Examine the Influence of Networked Distributed Generation. *Energies* **2026**, *19*, 799.
2. Stanchev, P.; Hinov, N. Smart Grids and Sustainability in the Age of PMSG-Dominated Renewable Energy Generation. *Energies* **2026**, *19*, 772.
3. Barua, S.; Paul, S.; Merabet, A. Solar Photovoltaic Model Parameter Identification with Improved Metaheuristic Algorithm Based on Balanced Search Strategies. *Energies* **2026**, *19*, 315.
4. Gharbi, A.; Ayari, M.; Albalawi, N.; Alshammari, A.; Ben Halima, N.; Klai, Z. An Adaptive Multi-Layer Heuristic Framework for Real-Time Energy Optimization in Smart Grids. *Energies* **2026**, *19*, 307.
5. Bradło, D.; Żukowski, W.; Porzuczek, J.; Olek, M.; Berkowicz-Płatek, G. Towards Net Zero in Poland: A Novel Approach to Power Grid Balance with Centralized Hydrogen Production Units. *Energies* **2025**, *18*, 1576.
6. Zhang, N.; Liu, D.; Liu, T.; Zhang, X.; Guo, J.; Lan, F.; Li, Q.; Lu, W.; Yang, X. A Win-Win Coordinated Scheduling Strategy Between Flexible Load Resource Operators and Smart Grid in 5G Era. *Energies* **2025**, *18*, 1510.
7. Avila Saccol, G.; Bui, V.-H.; Su, W. A Deep-Neural-Network-Based Surrogate Model for DC/AC Converter Topology Selection Using Multi-Domain Simulations. *Energies* **2024**, *17*, 6467.
8. Martínez-Nolasco, J.; Sámano-Ortega, V.; Rodriguez-Estrada, H.; Santoyo-Mora, M.; Rodriguez-Segura, E.; Zavala-Villalpando, J. Controller Hardware in the Loop Platform for Evaluating Current-Sharing and Hot-Swap in Microgrids. *Energies* **2024**, *17*, 3803.
9. Jlidi, M.; Barambones, O.; Hamidi, F.; Aoun, M. ANN for Temperature and Irradiation Prediction and Maximum Power Point Tracking Using MRP-SMC. *Energies* **2024**, *17*, 2802.
10. Maiwald, J.; Schütte, T. Prosumer Impact on Cellular Power Systems. *Energies* **2024**, *17*, 2195.
11. Zhang, H.; Zhang, M.; Yi, R.; Liu, Y.; Wen, Q.H.; Meng, X. Growing Importance of Micro-Meteorology in the New Power System: Review, Analysis and Case Study. *Energies* **2024**, *17*, 1365.
12. Long, Y.; Bao, Y.; Zeng, L. Research on Edge-Computing-Based High Concurrency and Availability "Cloud, Edge, and End Collaboration" Substation Operation Support System and Applications. *Energies* **2024**, *17*, 194.
13. Nazir, I.; Mushtaq, N.; Amin, W. Smart Grid Systems: Addressing Privacy Threats, Security Vulnerabilities, and Demand–Supply Balance (A Review). *Energies* **2025**, *18*, 5076.
14. Ishfaq, H.; Kanwal, S.; Anwar, S.; Abdussalam, M.; Amin, W. Enhancing Smart Grid Security and Efficiency: AI, Energy Routing, and T&D Innovations (A Review). *Energies* **2025**, *18*, 4747.

References

1. Hua, L.-G.; Bilal, M.; Hafeez, G.; Ali, S.; Alghamdi, B.; Alsafran, A.S.; Kraiem, H. An energy-efficient system with demand response, distributed generation, and storage batteries for energy optimization in smart grids. *J. Energy Storage* **2025**, *117*, 115491. [CrossRef]
2. Deng, L.; Huan, J.; Wang, W.; Zhang, W.; Xie, L.; Dong, L.; Guo, J.; Li, Z.; Huang, Y.; Xiang, Y. Market operation of energy storage system in smart grid: A review. *Energy Eng.* **2024**, *121*, 1403–1437. [CrossRef]
3. Senthil, S.S.R.; Balamurugan, R. Hybrid approach of energy management and power quality enhancement in smart grid-connected hybrid renewable energy system. *Sustain. Energy Technol. Assess.* **2026**, *85*, 104763.

4. Mastoi, M.S.; Wang, D.; Ma, N.; Hassan, M.; Shafiullah, M.; Bashir, T.; Hassan, A.; Flah, A. AI-driven control and optimization for renewable energy integration in smart grids: Challenges, applications, and future research directions. *Energy Strategy Rev.* **2026**, *64*, 102049. [CrossRef]
5. Anis, G.; Naz, N.S.; Ghazal, T.M.; Farooq, M.S.; Saleem, M.; Yeun, C.Y.; Ahmad, M.; Adnan, K.M. Smart and transparent grid stability prediction for efficient energy management using explainable AI. *Energy Strategy Rev.* **2026**, *64*, 102083. [CrossRef]

https://doi.org/10.3390/en19051251

energies

MDPI

Article

Probabilistic Visualisation Approach Using Polar Histograms to Examine the Influence of Networked Distributed Generation†

Yasmin Nigar Abdul Rasheed [1,2,*], Ashish P. Agalgaonkar [1,2,*] and Kashem Muttaqi [1,2]

1 The Australian Power and Energy Research Institute, University of Wollongong, Wollongong, NSW 2522, Australia; kashem_muttaqi@uow.edu.au
2 School of Electrical, Computer & Telecommunications Engineering, University of Wollongong, Wollongong, NSW 2522, Australia
* Correspondence: yasminnigarchazath@yahoo.com (Y.N.A.R.); ashish@uow.edu.au (A.P.A.)

† This paper is an extended version of our paper published in 2023 IEEE International Conference on Energy Technologies for Future Grids (ETFG), Wollongong, Australia, 3–6 December 2023; pp. 1–6.

Abstract

The variability of renewable energy sources, coupled with the decentralised configuration of distributed generation (DG), significantly complicates grid management, necessitating sophisticated visual analytics to enhance power system performance and energy distribution. This paper presents a probabilistic visualisation technique based on polar histograms to identify the dynamic influence zones of DG units by analysing line current flows. The proposed framework explicitly accounts for the probabilistic representation of reverse power flows, which provides an overall view of DG impacts on distribution networks. Quasi-dynamic simulations are conducted on a 33-bus distribution system using DIgSILENT PowerFactory 2020, MATLAB R2020, and Python 3.8. The results demonstrate that the polar histogram approach provides intuitive insights into DG influence, revealing zones of grid-dominated, DG-dominated, and shared interactions. These findings act as a potential practical tool for voltage management, demand balancing, and secure integration of renewable DG units into modern power grids.

Keywords: dataanalytics; sphere of influence; polar histogram; visual analytics; probabilistic visualisation; consumer energy resources; modern power grid

1. Introduction

The pursuit of the net-zero target is prominent worldwide. In quest of achieving the net-zero target, electric networks are undergoing a paradigm shift towards consumer energy resources or distributed generation (DG). Integrating DG enables a significant reduction in greenhouse gas emissions by incorporating renewable sources such as solar, wind, biomass, and hydropower. Locating DG units closer to end-users offers several operational advantages, including voltage support, loss reduction, deferred infrastructure upgrades, and improved system reliability, all of which collectively contribute to lowering the overall carbon footprint [1,2].

Furthermore, the built-in scalability of DG increases the generation in the network, which makes it easier to adapt to fluctuating energy demands and evolving technological advancements. DG systems such as solar, wind and hydrogenerators can be implemented in varying sizes, from small, residential-scale installations to large farm-scale installations. These size variations aid in gradually increasing energy capacity as demand increases or additional funding becomes available. For instance, a community starting with a modest solar farm can add more panels over time as energy consumption grows, which reduces dependency on the conventional grid.

Moreover, DG plays a pivotal role in enhancing grid resilience and stability. By distributing generation across multiple smaller sources, power networks are less vulnerable to single points of failure. This enhances the security and stability of the system, ensuring that if one source in the electric line goes offline, the impact on the overall system is minimised, and other sources can potentially increase output to compensate. Additionally, as DGs are closer to the load of consumption, they reduce potential loss and stress on the grid infrastructure, further enhancing the network stability.

Though DG brings about advantages in improving system performance and sustainability, its incorporation into power grids poses challenges. In network scenarios, DG may unexpectedly result in issues such as voltage instability, grid congestion, and added complexity. Additionally, renewable power generation patterns and time-varying load demands are stochastic in nature. Furthermore, different DG generators have different reactive power capabilities. The non-consideration of generation pattern, time-varying load, and reactive power capability of DG type can have different impacts on the power system, affecting DG's size. To study and analyse the network, DNSPs must understand and observe the ongoing effects happening within the network when generation and loads are varying with time. Recognising the level of certainty within the network is crucial to properly assess and manage its inherent uncertainties.

These factors emphasise that a novel methodology must be developed to optimally allocate DG in the distribution network. Furthermore, the outcome of the methodology produces a vast amount of data, which is difficult to interpret and to analyse how the integration of DG affects the rest of the system [3–5]. Hence, a visualisation technique is to be developed that identifies the system condition lucidly and identifies the sphere of influence of DG. The identification of the influence of DG has practical benefits for operations: it aids in reactive power management during voltage fluctuation, locates and focuses on protection settings within the affected area, improves grid performance instantly after disruptive events and amplifies grid visibility to identify precursors of extreme events.

By employing visualisation, the influence of DG connected in the network can be clearly delineated, thereby enabling DNSPs to take necessary actions to mitigate the effect of DG. For example, electric vehicles acting as both loads during charging and generators during vehicle-to-grid operation can create bidirectional and time-varying stresses on distribution feeders [6]. The ability of the visualisation to highlight probabilistic influence zones under such scenarios provides the network operators with early warning of potential vulnerabilities, thereby enabling the development of proactive mitigation strategies. With this approach, growing penetration of DG can be accommodated while maintaining grid reliability, strengthening the grid security and resilience of modern distribution networks. Hence, this research paper proposes a novel approach considering both real and reactive powers of DG units concurrently and by introducing a probabilistic visualisation technique based on polar histograms or wind rose.

1.1. Techniques in Determining DG Sphere

This section delves into the existing literature review of determining the sphere of influence of DG in the network. The sphere of influence is defined as the segment of the distribution network where a DG unit exerts a dominant influence on current flows and voltage behaviour under varying conditions. Unlike power flow, which provides instantaneous current or voltage values at a specific location, it describes a region of DG impact across different nodes and line sections.

https://doi.org/10.3390/en19030799

Historically, three main techniques, namely sensitivity analysis, probabilistic analysis, and visualisation techniques, have been employed to analyse the sphere of influence of DG in the distribution networks. Authors in [7–12] applied the sensitivity analysis method to quantify the changes in network parameters such as voltage profiles, power flows, and stability margins under DG integration. These methods identify critical locations and the relative influence of DG units. However, they are deterministic and assume DG is constant and neglects the stochastic nature of variable generation. Furthermore, it approximates that DG is supplying either real or reactive power and does not take into consideration of both real and reactive power simultaneously.

Recognising these limitations, several studies [13–15] have applied probabilistic approaches to characterise the DG influence zones. The statistical evaluation of network response under varying conditions is determined by modelling the stochastic nature of DG and using Monte Carlo Simulation (MCS), while such frameworks provide a broader representation of DG variability, they typically build on sensitivity analysis and continue to evaluate real and reactive power separately. Furthermore, translating these complex probabilistic results into actionable insights remains challenging task without effective visualisation techniques.

In parallel, visualisation techniques have gained momentum in illustrating DG influence zones. Methods such as quiver plots [16], contour plots [17], Sankey diagrams [18], and connectivity diagrams [19] have been employed to offer intuitive spatial representations of DG impacts on distribution networks. Although these visualisations provide valuable visual insights on the DG sphere, the stochastic aspects of DG variability and its impacts on neighbouring buses are not incorporated.

To address these gaps, an approach proposed in [20] concurrently considers both real and reactive power contributions of DG units and introduces a probabilistic visualisation framework based on polar histograms to capture dynamic influence zones, while this approach incorporates real-time load data and acknowledges the stochastic behaviour of DG units, it does not explicitly account for the visual representation of reverse power flows, which remains critical limitation in existing polar histogram techniques. Hence, this paper proposes an enhanced polar histogram-based visualisation approach to determine dynamic spheres of influence, explicitly accounting for reverse power flows within the network. Such an advancement empowers grid operators to evaluate how often and to what extent DGs influence power flows, thereby identifying critical areas where voltage and network stability issues may arise. In addition, it facilitates proactive measures to mitigate associated challenges and enables better decision-making processes, including dispatch by DG units, grid operation, and planning, thereby improving grid reliability, stability, and efficiency.

1.2. Organisation of the Paper

The remainder of this paper is organised as follows: Section 2 illustrates the probabilistic modelling of DG and load. Section 3 details the application of probabilistic approaches in power systems. Section 4 depicts the proposed polar histogram approach. Section 5 describes the methodology of polar histogram to determine the DG spheres. Section 6 illustrates the results and simulation, and Section 7 presents the concluding remarks.

2. Probabilistic Modelling of DG and Load

This section details probabilistic modelling of DG and load. Probabilistic modelling is essential to analyse the variability and uncertainty associated with renewable DG and time-varying load demand.

https://doi.org/10.3390/en19030799

2.1. Modelling Solar Photovoltaic Generation

The solar power is uncertain in nature and exhibits variability, which primarily arises from cloud cover, time of the day, wind speed, humidity, seasonal changes, and geographical locations. The power output generated by solar photovoltaic (PV) is dependent on solar irradiance, solar panel efficiency, and system losses, and can be formulated as follows [21]:

$$P_{pv} = \eta_{pv} \cdot I_{pv} \cdot A_{pv} \cdot (1 - L_{pv}) \tag{1}$$

where P_{pv} is the power generated by solar PV, η_{pv} is the solar panel efficiency, I_{pv} is the solar irradiance which can be modelled from several probabilistic distributions, A_{pv} is the total area of the PV module and L_{pv} is the lossess due to PV cell temperature, i.e., temperature coefficient. Assuming solar irradiance follows normal distribution, the model for I_{pv} is given by

$$I_{pv} \sim \mathcal{N}(\mu_{I_{pv}}, \sigma^2_{I_{pv}}) \tag{2}$$

where $\mu_{I_{pv}}$ is the mean irradiance and $\sigma_{I_{pv}}$ is the standard deviation of the irradiance. As power generated by solar PV is directly proportional to solar irradiance, the output power follows normal distribution which is given as

$$P_{pv} \sim \mathcal{N}(\eta \cdot \mu_{I_{pv}} \cdot A \cdot (1 - L), (\eta \cdot A \cdot (1 - L))^2 \cdot \sigma^2_{I_{pv}}) \tag{3}$$

There are also several probabilistic distribution functions (PDFs), such as Weibull, beta, lognormal, logistic, and normal, that can be applied to model solar irradiance [22].

2.2. Load Modelling

The active and reactive power components of load demand can be modelled from real time load data or can be expressed as various PDFs. Load demand that follows normal distribution can be expressed as follows [23]:

$$f(P_{ap}) = \frac{1}{\sqrt{2\pi\sigma^2}} \exp\left(-\frac{(P_{ap} - \mu_{ap})^2}{2\sigma^2_{ap}}\right) \tag{4}$$

$$f(P_{rp}) = \frac{1}{\sqrt{2\pi\sigma^2}} \exp\left(-\frac{(P_{rp} - \mu_{rp})^2}{2\sigma^2_{rp}}\right) \tag{5}$$

where P_{ap} and P_{rp} are the active and reactive powers, μ_{ap} and μ_{rp} are their mean values, and σ_{ap} and σ_{rp} are their standard deviations, respectively.

The other method for load modelling is by obtaining hourly load factor data. The load factor is defined as the ratio of actual load to the maximum load during a specific time period. It provides valuable insights into how the load varies hourly, daily, or seasonally. The load factor for each hour is calculated by dividing the actual load by the maximum load observed in the dataset or within specific periods, such as daily, monthly, and annually:

$$\text{Load Factor} = \frac{\text{Actual Load}}{\text{Maximum Load}} \tag{6}$$

Equation (6) normalises the load data, which enhances the analysis and comparison of electrical load data across different times or seasons.

3. Application of Probabilistic Approaches in Power Systems

This section delves into application of probabilistic approaches in power system. Probabilistic approaches are crucial in managing and understanding the inherent uncertainties associated with load demand, DG output, sphere of influence, and network configuration.

https://doi.org/10.3390/en19030799

3.1. Probabilistic Approaches in Power System

Probabilistic approaches encompass a variety of techniques to account for unpredictability and fluctuation in power systems. One of the fundamental methods is an MCS [24,25], which is used to model the behaviour of complex systems by generating a large number of random samples. In power systems, an MCS is often employed to simulate different scenarios of power generation, load demand, and system parameters. By running simulations with random inputs, it provides probabilistic distributions of outcomes, such as voltage levels, power flows, and system reliability.

Probabilistic load flow analysis (PLF) [22,26] is another essential technique that extends traditional load flow analysis by considering uncertainties from existing load demand, variable generation, and network configuration. It incorporates multiple load flow scenarios with the probability density function of the generator and load as input parameters in the system. The output from the PLF generates a probability density function of output variables such as bus voltage, line current, and power flow to compute the probabilistic behaviour of the power network under different operating conditions.

Statistical methods [23,27–29] are integral to probabilistic approaches in power systems. Techniques such as descriptive statistics, regression analysis, probability distributions, hypothesis testing, and time series analysis are employed to model uncertainty and variability in power system analysis. Descriptive statistics provides data profiling and aids in understanding system operating conditions. Probabilistic distributions such as normal, log normal, beta, and Weibull distributions model the variability of load demand and DG units. Hypothesis testing and regression analysis are applied to infer relationships and make predictions based on historical data. Time series analysis helps in forecasting load and generation demand over time.

Probabilistic visualisation [13,20] effectively represents probabilistic data in power systems. Several visualisation techniques such as heatmap, contour plots, choropleth maps, and box plots are deployed in the power system to show trends, patterns, probability densities, confidence intervals, and distribution. However, a polar histogram is a highly effective probabilistic visualisation tool that analyses where directional data and probability distributions must be exhaustively represented. Here is a detailed exploration of why polar histograms excel in this role:

1. Directional data representation: polar histograms are designed to represent directional data, making them ideal for scenarios where understanding directional influences is critical.
2. Intuitive interpretation the bin and circular format of polar histograms provides intuitive interpretation of directional data and probability distributions. Users can quickly identify predominant directions or patterns in the data.
3. Incorporation of probability distributions: polar histograms can easily incorporate probability distributions, providing a quick representation of the likelihood of different outcomes, which enables stakeholders to assess uncertainty effectively.
4. Comparative analysis: polar histograms easily compare different generating sources or scenarios. Also, multiple histograms can be overlaid on a same network, facilitating comparative analysis and aiding decision-making processes.
5. Dynamic visualisation: polar histograms represent dynamic data that changes over time, enabling stakeholders to explore probabilistic data interactively. This dynamic visualisation helps users understand complex relationships and data flows that are difficult to interpret using snapshot charts.

https://doi.org/10.3390/en19030799

3.2. Key Roles of Probabilistic Visualisation

Probabilistic visualisation plays a vital role in power systems by translating complex probabilistic data into intrinsic, transparent visual formats, particularly with the increasing incorporation of DG from renewable sources. Employing probabilistic visualisation aids in understanding power system dynamic behaviour under uncertainty and variability. This section explores the critical roles of probabilistic visualisation in power systems:

1. Risk assessment: probabilistic visualisation enables power system operators and planners to assess and visualise risks associated with different scenarios. Heat maps, contour plots, and polar histograms communicate possible events and outcomes, such as extreme weather events, equipment failures, and cyber-attacks, thereby helping distribution network operators (DNOs) identify vulnerabilities and develop strategies to mitigate risks.
2. Operational planning and decision making: probabilistic visualisation tools assist in operational planning and decision-making processes. For instance, a polar histogram illustrates the current flow distribution that aids in identifying areas of influence in DG. Contour plots show the spatial distribution of load and voltage violations, thereby helping operators to make proactive decisions such as resource allocation to enhance overall system performance.
3. Resilience analysis: probabilistic visualisation facilitates resilience analysis by allowing power system stakeholders to evaluate the system's durability towards disturbances and disruptions. By visualising the probabilistic distribution of the DG sphere, operators and planners can identify outages and system vulnerabilities for improvement and invest in resilience-enhancing measures.

Overall, probabilistic visualisation enhances the interpretation of complex probabilistic data, enabling stakeholders to explore different future trajectories and develop robust long-term plans and strategies.

3.3. Comparative Analyses of Probabilistic Visualisation Tools

Table 1 presents an overview of different probabilistic visualisation tools, including Monte Carlo simulations, probabilistic load flow analysis, scenario analysis, and polar histograms. The advantages, challenges, and comparison of each tool with polar histograms are summarised to provide insights into associated strengths and challenges.

Monte Carlo simulations offer detailed outcomes and insights into various scenarios, but are computationally intensive and require more intuitive visualisation. Probabilistic load flow analysis extends traditional load flow analysis to consider uncertainties, but it involves complex calculations and requires specialised software. Statistical analysis provides a structured framework to explore data relationships and test hypotheses, but faces challenges in visualising complex datasets. In contrast, polar histograms yield simplified and insightful visualisation, particularly for directional data, making them valuable for decision support in power system analysis.

In summary, while other probabilistic visualisation tools provide valuable insights, polar histograms stand out for their unique advantages such as simplicity, intuitive interpretation, and effectiveness in visualising directional data not only to provide stakeholders with a clear and accessible way to understand probabilistic information, but also in terms of opening up new possibilities for decision-making in power system analysis.

https://doi.org/10.3390/en19030799

Table 1. Comparison of probabilistic visualisation tools.

Probabilistic Visualisation Tool	Strengths	Challenges	Comparison with Polar Histogram
Monte Carlo simulation	• Provides detailed probabilistic distributions of outcomes. • Allows modelling of complex uncertainties and provide insights to a wider audience	• Computationally intensive. • Lacks intuitive visualisations for stakeholders. • Interpreting results can be complex.	Compared with Monte Carlo simulations, polar histograms deliver quicker and more comprehensible visual representation of directional characteristics.
Probabilistic load flow analysis	• Extends traditional load flow analysis to consider uncertainties. • Provides probabilistic distributions of outcomes such as voltage levels and power flows.	• Involves complex mathematical calculations. • Requires specialised software. • Visualising results can be challenging.	Unlike probabilistic load flow analysis, which yields extensive numerical detail, polar histograms translate uncertainty into an interpretable graphical form accessible even to non-specialist stakeholders.
Statistical analysis	• Provides a structured framework to explore data relationships and test hypotheses	• Visualising results can be challenging, especially with complex datasets or multiple scenarios.	Polar histograms aim to provide an easier representation of uncertainty, making it accessible to a broader audience, including non-experts, to identify trends and patterns.

4. Proposed Polar Histogram-Based Visualisation Techniques

This section details the polar histogram techniques which visualise time-varying current magnitudes and directions at line sections by analysing dominant current flow patterns as demonstrated by authors in their prior studies [16,20], and thereby identifies the respective contributions from DG units and grid sources.

By sampling these line and load currents over time, a time series of real and imaginary current components is obtained, which characterises the variations in magnitude and direction of current flows across the distribution network. These time-varying current vectors form the input data for constructing the polar histograms. To illustrate the proposed approach, Figure 1 presents a representative polar histogram, where the angular bins represent the direction of current flow which is derived from the phase angle of the complex current and the radial segments represent the probability distribution of the corresponding magnitudes. Thus, the polar histogram serves as a probabilistic visualisation of the dynamic behaviour described by the current and voltage equations, allowing operators to intuitively assess the dominant current flow patterns in the network.

Through examining the current magnitudes and directions, polar histograms can be segmented into four quadrants, as shown in Figure 1, with each reflecting specific combinations of real and reactive power contributions such as the following:

- DG real power and grid reactive power (Quadrant 1)
- Grid real and reactive power (Quadrant 2)
- Grid real power and DG reactive power (Quadrant 3)
- DG real and DG reactive power (Quadrant 4)

https://doi.org/10.3390/en19030799

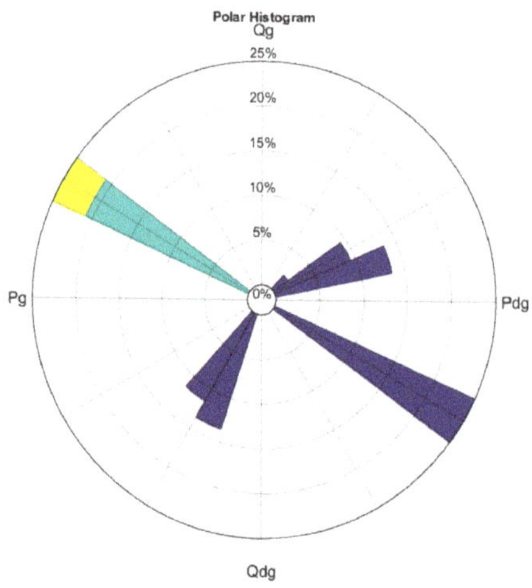

Figure 1. Representative polar histogram.

By constructing polar histograms for every line section and evaluating the corresponding quadrant patterns, the probabilistic dynamics of power flows can be effectively inferred by system operators. Each line section can be associated with its own histogram, derived from the time series of current magnitudes and angles. This enables a line-by-line assessment of how currents vary not only in strength, but also in direction over time. Line sections where the polar histograms are concentrated in specific quadrants indicate stable and consistent power contributions. For instance, a polar histogram with quadrant 4 highlights a section predominantly influenced by customer energy resources, while one concentrated in the quadrant 2 reflects dependence on the grid supply. In contrast, line sections with widely dispersed histogram patterns reveal strong probabilistic variations in current behaviour. These often correspond to areas where distributed generation and grid sources interact dynamically, leading to bidirectional flows. Such dispersion is critical because it signals regions where operational conditions are less predictable, and where security concerns such as reverse power flows, local overloads, or voltage instability are more likely to arise.

By comparing polar histograms across the entire distribution system, operators can identify critical zones of DG influence to support operational strategies such as voltage management, demand balance, and controlled DG curtailment by targeting the sections with the highest variability or security risk.

5. Methodology

This section outlines a detailed step-by-step procedure for identifying the sphere of influence of DG. The process of determining the dynamic DG sphere using polar histograms can be formulated under two scenarios: with and without the reflection principle. The scenario without reflection-based formulation has been previously introduced and validated in the authors' earlier research work [20], and is not repeated here to avoid duplication. Accordingly, this paper focuses on the scenario with the reflection principle, which extends the previous work by enabling accurate identification of source influence in network sections where current reversal does not occur. A flowchart summarising the complete methodology adopted in this study is presented in Figure 2.

https://doi.org/10.3390/en19030799

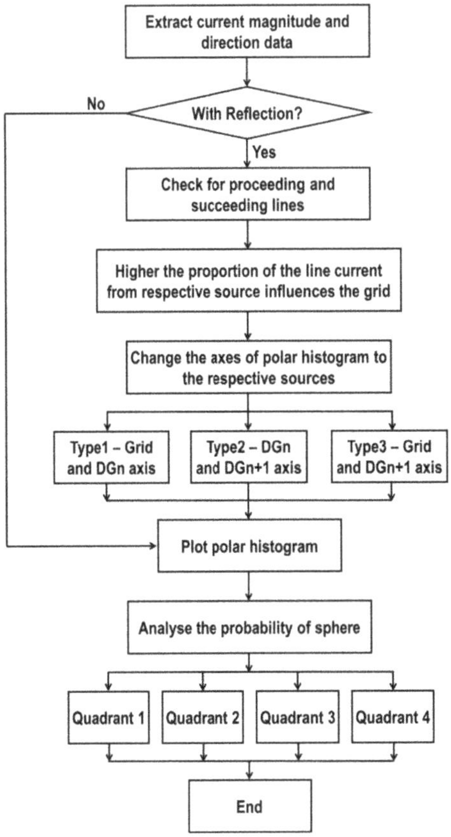

Figure 2. DG sphere determination flowchart with and without the reflection principle.

Reflection-Based Methodology for Determining the Sphere of Influence of DG

In distribution networks with multiple power sources, certain line sections may not exhibit current reversal regardless of DG size or type. To address this limitation, a reflection analysis of the polar histogram is employed to correctly attribute the influence of the grid and multiple DG units. This section provides a detailed explanation of the reflection analyses of polar histograms. These analyses are pivotal in determining the influence of various power sources in the network, such as the grid and multiple DG units. As multiple n sources exist in the network, certain line sections do not represent the current reversal in the polar histogram by having current vectors always fall in the specific quadrant irrespective of DG size and type. For instance, in a network with a lateral, both the grid and DG serve the lateral. However, the current vector in the lateral always falls in the quadrant 2 as no current reversal occurs in the lateral. This is because the lateral is always served by the same source supplying the connection point. The reflection-based approach modifies the polar histogram to accurately identify the true source of influence for a particular lateral or each network section. The steps involved in implementing this reflection-based methodology are described below.

Step 1: conduct time-series power flow simulations

Step 2: acquire phasor information for the line currents using simulated outcomes

Step 3: assess upstream and downstream line sections to identify the dominant real and reactive power contributions influencing each bus

Step 4: adjust the axes of the polar histogram based on the dominant source of DG identified from the preceding or succeeding lines to reflect the exact source of influence. The adjustment of polar histogram axes ensures that the dominant source of influence, whether from the grid, a local DG, or other DGs in the network, is accurately represented. These axes' adjustments in the polar histogram can be classified into three types, as shown in Figure 3.

1. Type 1—grid and DGn axis: this type describes scenarios in which the grid and DG unit's real and reactive powers influence a specific area or node in the network. For example, currents may reflect real power support from the grid and reactive contributions from DGn. This type highlights grid–DG shared zones.
2. Type 2—DGn and DGn+1 axis: this classification focuses on influences between multiple DG units, highlighting an area of the network where the primary DG and an additional DG unit (DGn+1) are predominant without direct interaction from the primary grid source. For instance, in a mid-feeder line between DG1 and DG2, the polar histogram axis is adjusted to reflect the comparative influence of both DGs.
3. Type 3—grid and DGn+1 axis: type 3 involves the interaction between the grid and an additional DG unit, DGn+1. This type indicates influences that do not directly involve the primary DG unit but involve subsequent DG units added to the system. This type is typically observed in feeder-end sections where DGn+1 is electrically closer to the loads than the upstream DGn.

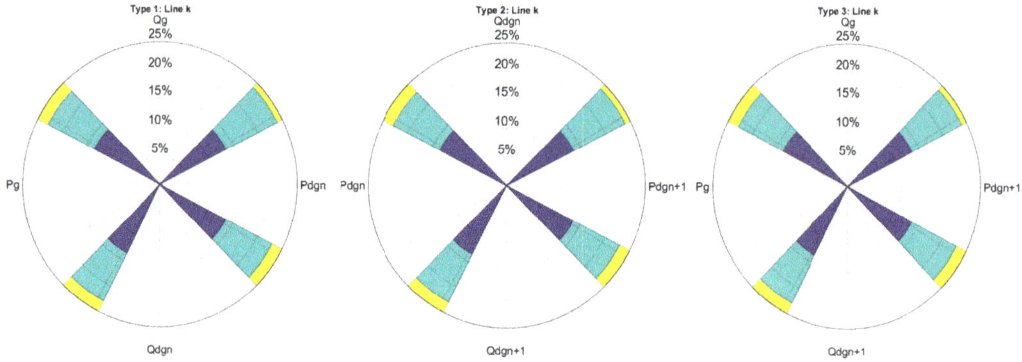

Figure 3. Reflected polar histogram axis types.

Step 5: using the reflected current phasor information, construct a polar histogram to characterise the distribution of current magnitude and direction for each line section.
Step 6: map the polar histogram in the line section and analyse the probability distribution of the current vector across four quadrants considering three types of axis adjustment classifications. When the probability of current magnitude and direction falls in quadrant 1, it indicates the influence of real power from DG and reactive power from the grid. Quadrant 2 reflects the grid's dominance in both real and reactive power contributions. Quadrant 3 corresponds to grid-supplied real power combined with DG-supplied reactive power, while quadrant 4 indicates DG contributions in both real and reactive powers. These quadrant-based interpretations from [16] are extended in this work through a probabilistic framework, where polar histograms are used to capture time-varying current behaviour.
Step 7: determine the sphere of influence of a DG unit by analysing the transition of polar histogram patterns along consecutive line sections. Specifically, the sphere of influence is defined as the contiguous set of line sections where DG contributions dominate or significantly alter the probabilistic behaviour of current flows. Beyond this boundary, the

https://doi.org/10.3390/en19030799

histograms become grid-dominated, signifying the limit of the DG's operational impact on the network.

6. Simulation and Results

The proposed methodology was validated through simulations conducted on a 33-bus distribution network, as depicted in Figure 4, using DIgSILENT PowerFactory 2020 in conjunction with Python 3.8 and MATLAB R2020. The same test system has been employed in the previous study conducted by the authors [20] to enable consistent comparison. These tools were selected because they provide complementary strengths: DIgSILENT ensures accurate network modelling and quasi-dynamic simulations, MATLAB enables flexible data processing and custom visualisation, while Python facilitates probabilistic data analysis and integration with histogram-based techniques. The network voltage magnitude is regulated within the permissible range of 0.95 pu. to 1.05 pu. The test system incorporates time-varying real and reactive load patterns, as described in [12], while the DG units are modelled with constant generation output to simplify the interpretation of the polar histograms. Quasi-dynamic simulations are carried out using an hourly time step over a 24 h period and data on current magnitude and direction are extracted from each line section. These data points are then visualised as polar histograms, illustrating the probability distributions of real and reactive currents flowing through the line sections.

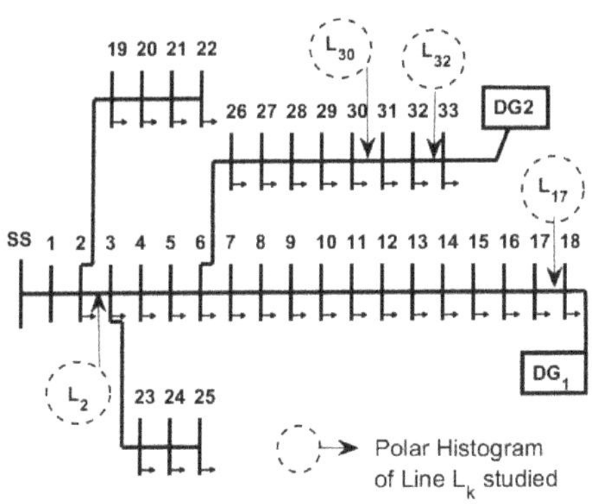

Figure 4. The 33-bus distribution network (adopted from [20]).

Only the specific line sections mentioned in Figure 4 are shown in the results due to their significance and strategic proximity to the source. Line 2 reflects the grid-dominated region near the substation, line 17 captures the direct impact of DG1, line 30 lies between DG1 and DG2 and highlights their comparative influence, and line 32 represents the feeder end influenced by DG2. This line sections are selected to capture grid-dominated, DG-dominated, and shared-influence regions across the network. To evaluate the proposed visualisation framework, two case-study scenarios are considered, (i) application without the reflection principle and (ii) application with the reflection principle, both involving multiple DG units and temporal load variations. The results corresponding to the non-reflection scenario have been previously reported by the authors in [20]. To avoid duplication, this paper exclusively focuses on the reflection-based scenario, which constitutes a substantive extension of the prior work.

Application of Reflection Principle with Multiple DGs and Temporal Load Variations

This section presents the outcomes obtained from the polar histogram analysis incorporating reflection behaviour for multiple DG units. For completeness, the corresponding results obtained by authors without applying the reflection principle under identical net-

https://doi.org/10.3390/en19030799

work configurations and loading conditions are available in [20] and are not repeated in this paper for brevity. In this case study, two DG units are positioned at buses 18 and 33, respectively, operating under time-varying load conditions with constant generation output. It is considered that DG1 at bus 18 delivers approximately 60% of the real power, while DG2 at bus 33 contributes 10% of real power along with 10% of reactive power. Since the DG units supply most of the real power demand, the current flowing from L_{17} to L_3 and $Lateral_{25-32}$ are primarily served by the DG units 1 and 2. In contrast, other line segments, such as L_1 to L_2, $Laterals_{19-22}$, $Laterals_{23-25}$, are mainly supported by the grid. The interaction zones, representing the influence regions of the grid, DG1, and DG2, are illustrated in the polar histogram line sections shown in Figure 5.

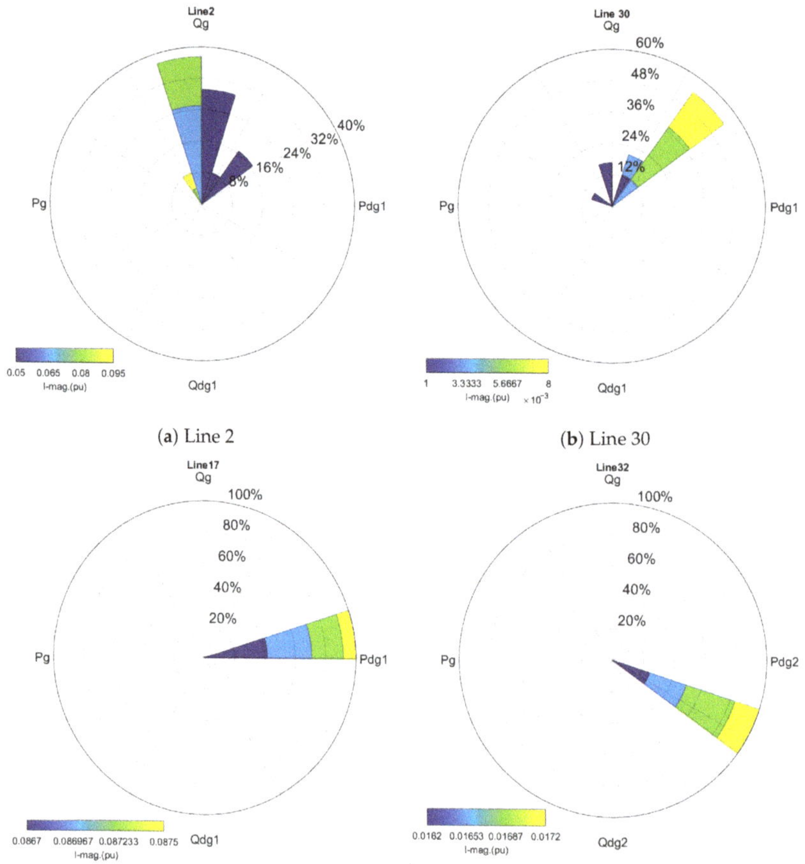

(**a**) Line 2 (**b**) Line 30

(**c**) Line 17 (**d**) Line 32

Figure 5. Reflected polar histogram illustrating combined real and reactive power from multiple DGs.

In line section L_2, the polar histogram illustrates variations in line current contributed by both DG1 and the main grid, primarily distributed across the first and second quadrants. Because DG1 delivers only real power, the grid remains the sole source of reactive power for this section. This behaviour is reflected in the probability distribution, where current magnitudes are concentrated in the first and second quadrants, demonstrating that real power is shared between DG1 and the grid, whereas reactive power originates solely from the grid. The higher probability density in the second quadrant further indicates that the grid exerts a stronger influence on line L_2.

https://doi.org/10.3390/en19030799

Similarly, examining the line section L_{30}, despite DG2's proximity to L_{30}, the probability distribution reveals that DG1 (connected at bus 18) contributes more real power influence, while the grid continues to supply the dominant portion of reactive power. This inference is drawn by comparing the adjacent line sections L_{29} and L_{31}, which together highlight that DG1 and the grid exert the most significant impact on the real and reactive power components of L_{30}, respectively.

For line section L_{17}, as DG1 is located at bus 18, the current distribution predominantly appears in the first quadrant, signifying DG1's major contribution to real power and the grid's prevailing role in reactive power support. In contrast, in line section L_{32}, DG2—supplying 10% real and 10% reactive power— shows a distinct probability concentration in the fourth quadrant, confirming that both power components originate from DG2 in this case. It should be noted that the polar histogram axes are redefined as Pdg2 and Qdg2 to denote the respective DG sources.

In summary, mapping line current to the respective quadrant and reorienting the histogram axes according to each generating source provides a clearer visual interpretation of how DG units and the grid dynamically interact within medium-voltage distribution networks.

7. Conclusions

This research investigates the dynamic influence zones of customer energy resources through polar histogram analysis. These polar histograms are derived from the statistical patterns of both real and imaginary current components that are flowing within line sections, which represent the influence of grid and DG sources. Simulation studies demonstrated that the approach provides intuitive insights into DG-dominated, grid-dominated, and shared-influence zones, while highlighting regions susceptible to voltage fluctuations and bidirectional flows. The probabilistic visualisation method helps grid operators understand DG integration challenges, which they can use to plan network development and distributed resource integration and enhance operational performance and network planning efficiency. Nevertheless, the scalability of the method remains a limitation, as applying polar histograms across very large or meshed urban networks may become computationally intensive and difficult to interpret. Future work will therefore focus on clustering and aggregation strategies to address the scalability challenges.

Author Contributions: Conceptualization, Y.N.A.R., A.P.A. and K.M.; Methodology, Y.N.A.R., A.P.A. and K.M.; Software, Y.N.A.R.; Validation, Y.N.A.R.; Formal analysis, Y.N.A.R.; Investigation, Y.N.A.R.; Resources, A.P.A. and K.M.; Data curation, Y.N.A.R.; Writing—original draft, Y.N.A.R.; Writing—review & editing, Y.N.A.R., A.P.A. and K.M.; Visualization, Y.N.A.R.; Supervision, A.P.A. and K.M.; Project administration, A.P.A. and K.M. All authors have read and agreed to the published version of the manuscript.

Funding: This research received no external funding.

Data Availability Statement: The original contributions presented in this study are included in the article. Further inquiries can be directed to the corresponding author.

Conflicts of Interest: The authors declare no conflict of interest.

References

1. Nguyen, V.G.; Sirohi, R.; Tran, M.H.; Truong, T.H.; Duong, M.T.; Pham, M.T.; Cao, D.N. Renewable energy role in low-carbon economy and net-zero goal: Perspectives and prospects. *Energy Environ.* **2025**, *36*, 2248–2287. [CrossRef]
2. Fernandez, M.I.; Go, Y.I.; Wong, D.M.; Früh, W.G. Review of challenges and key enablers in energy systems towards net zero target: Renewables, storage, buildings, & grid technologies. *Heliyon* **2024**, *10*, e40691. [CrossRef]

https://doi.org/10.3390/en19030799

3. Miraftabzadeh, S.M.; Colombo, C.G.; Longo, M.; Foiadelli, F. K-means and alternative clustering methods in modern power systems. *IEEE Access* **2023**, *11*, 119596–119633. [CrossRef]
4. Arévalo, P.; Jurado, F. Impact of artificial intelligence on the planning and operation of distributed energy systems in smart grids. *Energies* **2024**, *17*, 4501. [CrossRef]
5. Strezoski, L. Distributed energy resource management systems—DERMS: State of the art and how to move forward. *Wiley Interdiscip. Rev. Energy Environ.* **2023**, *12*, e460. [CrossRef]
6. Lei, X.; Zhong, J.; Chen, Y.; Shao, Z.; Jian, L. Grid Integration of Electric Vehicles within Electricity and Carbon Markets: A Comprehensive Overview. *eTransportation* **2025**, *25*, 100435. [CrossRef]
7. Tamp, F.; Ciufo, P. A sensitivity analysis toolkit for the simplification of MV distribution network voltage management. *IEEE Trans. Smart Grid* **2014**, *5*, 559–568. [CrossRef]
8. Lagonotte, P.; Sabonnadiere, J.; Leost, J.Y.; Paul, J.P. Structural analysis of the electrical system: Application to secondary voltage control in France. *IEEE Trans. Power Syst.* **1989**, *4*, 479–486. [CrossRef]
9. Liu, H.; Bose, A.; Venkatasubramanian, V. A fast voltage security assessment method using adaptive bounding. In Proceedings of the 21st International Conference on Power Industry Computer Applications. Connecting Utilities, Santa Clara, CA, USA, 21 May 1999; pp. 325–330.
10. Huang, Z.; Huang, L.; Lai, C.S.; Jia, Y.; Zhao, Z.; Li, X.; Lai, L.L. Line outage distribution factors of a linearized AC model with reactive power and voltage magnitude for resilience-constrained economic dispatch. *Sustain. Energy Grids Networks* **2022**, *32*, 100793. [CrossRef]
11. Baranwal, M.; Salapaka, S. Clustering and supervisory voltage control in power systems. *Int. J. Electr. Power Energy Syst.* **2019**, *109*, 641–651. [CrossRef]
12. Nigar, Y.; Agalgaonkar, A.P.; Ciufo, P. Impact of variable solar PV generation on MV distribution systems. In Proceedings of the Australasian Universities Power Engineering Conference (AUPEC), Perth, WA, Australia, 28 September–1 October 2014; pp. 1–6.
13. Guo, T.; Alimisis, V.; Milanović, J.V.; Taylor, P.C. Probabilistic assessment of voltage control zones and visualization using choropleth map. In Proceedings of the IEEE Manchester PowerTech, Manchester, UK, 18–22 June 2017; pp. 1–6.
14. Anand, M.; Bagen, B.; Rajapakse, A. Probabilistic reliability evaluation of distribution systems considering the spatial and temporal distribution of electric vehicles. *Int. J. Electr. Power Energy Syst.* **2020**, *117*, 105609. [CrossRef]
15. Chihota, M.J.; Bekker, B.; Gaunt, T. A stochastic analytic-probabilistic approach to distributed generation hosting capacity evaluation of active feeders. *Int. J. Electr. Power Energy Syst.* **2022**, *136*, 107598. [CrossRef]
16. Nigar, Y.; Agalgaonkar, A.P.; Muttaqi, K.M. Visualisation of spheres of influence of distributed generation through critical line flow analysis. *Sustain. Energy Grids Netw.* **2023**, *34*, 101046. [CrossRef]
17. Barnes, A.; Nagarajan, H.; Yamangil, E.; Bent, R.; Backhaus, S. Resilient design of large-scale distribution feeders with networked microgrids. *Electr. Power Syst. Res.* **2019**, *171*, 150–157. [CrossRef]
18. Beiranvand, A.; Cuffe, P. Sankey network diagrams to depict bulk power transactions for operator situational awareness. In Proceedings of the 2021 IEEE Madrid PowerTech, Madrid, Spain, 28 June–2 July 2021; pp. 1–6.
19. Cuffe, P.; Keane, A. Visualizing the electrical structure of power systems. *IEEE Syst. J.* **2015**, *11*, 1810–1821. [CrossRef]
20. Nigar, Y.; Agalgaonkar, A.P.; Muttaqi, K.M. Unveiling the Dynamic Influence Zones of Distributed Generation Units Using a Visualisation Approach. In Proceedings of the 2023 IEEE International Conference on Energy Technologies for Future Grids (ETFG), Wollongong, Australia, 3–6 December 2023; pp. 1–6.
21. Fan, M.; Vittal, V.; Heydt, G.T.; Ayyanar, R. Probabilistic Power Flow Studies for Transmission Systems With Photovoltaic Generation Using Cumulants. *IEEE Trans. Power Syst.* **2012**, *27*, 2251–2261. [CrossRef]
22. Ramadhani, U.H.; Shepero, M.; Munkhammar, J.; Widén, J.; Etherden, N. Review of probabilistic load flow approaches for power distribution systems with photovoltaic generation and electric vehicle charging. *Int. J. Electr. Power Energy Syst.* **2020**, *120*, 106003. [CrossRef]
23. Billinton, R.; Li, W. *Reliability Assessment of Electric Power Systems Using Monte Carlo Methods*; Springer: New York, NY, USA 1994; pp. 33–73. Available online https://link.springer.com/book/10.1007/978-1-4899-1346-3 (accessed on 28 March 2024).
24. Stefopoulos, G.; Meliopoulos, A.; Cokkinides, G. Probabilistic power flow with nonconforming electric loads. In Proceedings of the 2004 International Conference on Probabilistic Methods Applied to Power Systems, Ames, IA, USA, 12–16 September 2004; pp. 525–531.
25. Guimarães, I.O.; da Silva, A.M.L.; Nascimento, L.C.; Fotuhi-Firuzabad, M. Reliability assessment of distribution grids with DG via quasi-sequential Monte Carlo simulation. *Electr. Power Syst. Res.* **2024**, *229*, 110122. [CrossRef]
26. Peres, W. A probabilistic load flow for unbalanced three-phase islanded microgrids using unscented transformation. *Int. J. Electr. Power Energy Syst.* **2024**, *155*, 109554. [CrossRef]
27. AlKandari, M.; Ahmad, I. Solar power generation forecasting using ensemble approach based on deep learning and statistical methods. *Appl. Comput. Inform.* **2024**, *20*, 231–250. [CrossRef]

https://doi.org/10.3390/en19030799

28. Li, Y.; Ding, Y.; He, S.; Hu, F.; Duan, J.; Wen, G.; Geng, H.; Wu, Z.; Gooi, H.B.; Zhao, Y.; et al. Artificial intelligence-based methods for renewable power system operation. *Nat. Rev. Electr. Eng.* **2024**, *1*, 163–179. [CrossRef]
29. Ritchie, M.; Engelbrecht, J.; Booysen, M. Predicting residential water and electricity usage profiles with a temporal histogram model. *Sustain. Cities Soc.* **2023**, *99*, 104884. [CrossRef]

https://doi.org/10.3390/en19030799

energies

MDPI

Article

Smart Grids and Sustainability in the Age of PMSG-Dominated Renewable Energy Generation

Plamen Stanchev [1,2] and Nikolay Hinov [1,3,*]

1 CoE "National Center of Mechatronics and Clean Technologies", 1000 Sofia, Bulgaria; p.stanchev@tu-sofia.bg
2 Department of Information Technology in Industry, Faculty of Computer Systems and Technologies, Technical University of Sofia, 1000 Sofia, Bulgaria
3 Department of Computer Systems, Faculty of Computer Systems and Technologies, Technical University of Sofia, 1000 Sofia, Bulgaria
* Correspondence: hinov@tu-sofia.bg

Abstract

This study investigates the physical and cyber-physical resilience of smart grids with a high share of renewable energy sources (RESs) dominated by permanent magnet synchronous generators (PMSGs). The originality of this work lies in the development and unified evaluation of five integrated control strategies, the PLL with grid following, VSG with grid shaping, VSG+BESS, VSG+STATCOM, and VSG+BESS+STATCOM, implemented within a coherent simulation framework based on Python. Unlike previous works that analyze these methods in isolation, this study provides a comprehensive quantitative comparison of their dynamic characteristics, including frequency root mean square deviation, maximum deviation, and composite resilience index (RI). To extend the analysis beyond static conditions, a multi-generator (multi-PMSG) scenario with heterogeneous inertia constants and variable load profiles is introduced. This dynamic model allows the evaluation of natural inertia diversity and the effects of inter-generator coupling compared to the synthetic inertia emulation provided by VSG-based control. The combined VSG+BESS+STATCOM configuration achieves the highest synthetic resilience, improving frequency and voltage stability by up to 15%, while the multi-PMSG system demonstrates comparable or even higher RI values due to its inherent mechanical inertia and decentralized response behavior. In addition, a cyber-physical scenario is included to evaluate the effect of communication delays and false data injection (FDI) on VSG frequency control. The results show that a communication delay of 50 ms reduces RI by approximately 0.2%, confirming that even minor cyber disturbances can affect synchronization and transient recovery. However, hybrid control architectures with local energy buffering (BESS) show superior resilience under such conditions. The main technical contribution of this work is the establishment of an integrated analytical and simulation framework that enables the joint assessment of synthetic, natural, and cyber-physical resilience in converter-dominated smart grids. This framework provides a unified basis for the analysis of dynamic stability, hybrid control interaction, and the impact of cyber uncertainty, thereby supporting the design of low-inertia, resilient, and secure next-generation power systems.

Keywords: smart grids; permanent magnet synchronous generators (PMSGs); virtual synchronous generators (VSGs); grid resilience; RES integration

1. Introduction

The global acceleration of the transition to renewable energy sources (RESs) is fundamentally transforming the architecture and dynamics of modern power systems [1–3]. Under the influence of climate change, the depletion of fossil fuels, and the need for energy independence, countries around the world are investing massively in solar, wind, and other sustainable energy technologies [4–6]. This restructuring of energy systems creates new opportunities for the decentralization and optimization of management, but at the same time, it also raises complex engineering and management challenges related to the reliability and resilience of power grids [3,7,8]. A key technological component in this transformation are permanent magnet synchronous generators (PMSGs). They are widely used in wind energy and small decentralized RES installations, thanks to their high efficiency, compactness, and minimal maintenance requirements [9–11]. PMSGs provide variable speed operation and direct connection to power electronics, making them highly suitable for integration into modern renewable energy systems [12,13]. With their increasing share in the energy mix, they are becoming leading elements of smart, decentralized, and low-inertia grid architectures [2,3,7].

Despite their many advantages, the widespread use of PMSGs introduces new challenges for the stability of power systems. Unlike classic synchronous generators, which provide mechanical inertia and natural participation in frequency regulation, PMSGs are electronically connected to the grid through inverters, without a direct mechanical connection to the system [10–13]. This reduces the ability of the grid to absorb sudden changes in load and increases the risk of dynamic instability [8,14]. In addition, the lack of internal excitation limits their ability to regulate reactive power, which necessitates the use of additional compensation devices and advanced control algorithms [15–17].

These features are particularly critical in the context of smart grids, advanced power infrastructures that integrate digital communications, automated control, and renewables to achieve greater efficiency, flexibility, and sustainability [7,18]. Smart grids support bidirectional energy and information flows, integrate distributed energy resources and storage systems, and provide adaptive load management [19–21]. However, the success of this transition depends on the ability to sustainably integrate low-inertia and unpredictable renewable energy sources, such as PMSG-based systems [8,13,22]. In this context, grid resilience takes on new strategic importance. It encompasses the ability of the power system to adapt to dynamic changes, to withstand external impacts (such as weather events, failures, or cyberattacks), and to recover quickly after disturbances [1,23–25]. Resilience, understood as a multidimensional concept, is no longer limited to physical infrastructure, but includes intelligent control, forecasting, automation, and autonomous decision making [26–29]. This requires the implementation of advanced control algorithms, energy storage, virtual synchronization, and adaptive inverter technologies [15,30–32].

A particular challenge is the emulation of inertia and the maintenance of synchronization. While traditional power plants provide natural mechanical inertia that stabilizes the frequency during sudden changes in load, PMSG-based systems require the use of digital approaches such as virtual synchronous generators (VSGs). These algorithms reproduce the dynamic behavior of conventional generators through power electronics and digital controllers [10,13,14,33].

This paper aims to analyze the challenges and propose sustainable engineering solutions for maintaining the operational resilience of smart grids dominated by PMSGs [7,12,34]. This study addresses both the technical aspects of grid dynamics and modern control strategies that enable the reliable and efficient integration of such generators into smart power system architectures [21,35–37]. The analysis is based on a simulation model

https://doi.org/10.3390/en19030772

developed in Python and is complemented by a review of the current scientific literature, as well as suggestions for the future development and optimization of the systems [38–40].

The technical contribution of this research is twofold. First, it proposes an integrated simulation and evaluation framework that unifies dynamic modeling, control comparison, and resilience assessment of smart grids dominated by PMSGs. This enables a consistent benchmarking of control strategies, which is rarely addressed in the literature. Second, this study introduces a hybrid VSG+BESS+STATCOM architecture, validated through detailed simulation, that enhances both frequency and voltage resilience compared to the conventional PLL-based and single-compensation approaches. These contributions advance the understanding of how coordinated active and reactive power control can ensure the sustainable operation of future smart grids with low-inertia renewable sources.

All simulations and analyses presented in this study were performed in a fully digital environment using open source scientific software tools. No physical chemicals, reagents, biological samples, or commercial materials were used in this study. The simulation framework was developed and implemented using Python 3.10 (Python Software Foundation, Wilmington, DE, USA). Numerical calculations were performed using NumPy v1.24 (NumPy Developers; Austin, TX, USA) and SciPy v1.10 (SciPy Developers; Austin, TX, USA). Data manipulation and post-processing were performed using Pandas v2.0 (Pandas Development Team; New York, NY, USA).

All graphical visualizations and figure generation were performed using Matplotlib v3.7 (Matplotlib Development Team; Boston, MA, USA) and Seaborn v0.12 (Seaborn Developers; New York, NY, USA). Vector graphics data outputs were exported in SVG formats to ensure compatibility with scientific publishing standards. Simulations were performed on a standard desktop workstation running Windows 10 Pro (Microsoft Corporation; Redmond, WA, USA). No proprietary simulation platforms or hardware devices were used. The investigated systems based on PMSG, virtual synchronous generator (VSG) control strategies, battery energy storage systems (BESS), and STATCOM models were implemented as analytical and control-oriented representations. No manufacturer-specific hardware models or vendor-dependent control libraries were used. All inverter, storage, and compensation devices were modeled using idealized control units to ensure commonality and reproducibility of the results.

2. Literature Review

In recent years, the resilience of smart power systems dominated by renewable energy sources and permanent magnet generators (PMSGs) has emerged as a key research area in the context of the energy transition [1–3]. The development of smart grids aims to achieve high reliability, efficiency, and decentralization through the integration of renewable energy resources, communication technologies, and real-time control systems [4–6].

2.1. Smart Grids and Resilience

Smart grids are an advanced architecture that combines digital technologies, sensors, communication protocols, and automated control systems. This allows for two-way energy and information exchange between producers and consumers, as well as flexible load management [7–9]. According to [10,11], the resilience of smart grids should be understood as the ability to self-heal and adapt to external disturbances.

In [12,13], the role of predictive control and machine learning as key factors for resilience at a high level of RES integration is considered. In [14], it is emphasized that, through artificial intelligence (AI), dynamic load balancing and optimization of system stability can be achieved.

https://doi.org/10.3390/en19030772

More recent studies [41–45] add that resilience is no longer limited to physical infrastructure, but also includes cyber-physical data protection, energy autonomy, and adaptive control. In [45,46], for example, it is shown that "virtual inertia" is essential for frequency stabilization in networks with converter-based generators.

2.2. Role of PMSGs

PMSGs are one of the most efficient solutions for wind energy due to their compactness and high efficiency [26–28]. However, the lack of mechanical inertia and excitation system makes them sensitive to frequency and voltage fluctuations [29–31].

According to [32,33], the high dynamics of PMSG increases the risk of loss of synchronization, which necessitates the development of VSG and adaptive grid-forming inverters [42,47,48]. In [47], they emphasize that VSG control not only emulates inertia, but also improves the damping properties of the system.

Recent developments [47] demonstrate that combining PMSG with battery energy storage systems (BESSs) and STATCOM devices increases the resilience of networks to dynamic changes. This is also confirmed by studies in [46,47], which show increased voltage stability with the coordinated operation of PMSG and STATCOM.

2.3. Control Strategies and Compensation Devices

Advanced control approaches such as VSG, droop control, and adaptive PID are at the center of modern research on resilience enhancement [40–44,49,50]. In [40,41], it is shown that VSG control with BESS integration provides stability under sudden load changes.

STATCOM and SVC devices are considered as key for reactive power regulation and voltage stabilization [46–64]. In [54], the concept of "grid-forming STATCOM" is described, which combines reactive compensation and inertial support. Similar solutions are also proposed by [64], where coordinated VSG–STATCOM control improves power quality.

In addition, References [51,53] indicate that multi-agent control and predictive algorithms can provide the autonomous recovery of microgrids under disturbances, which is critical in PMSG-dominated systems.

2.4. Artificial Intelligence and Cyber Resilience

With the increasing digitalization of energy, AI and ML are becoming central technologies for analysis, forecasting, and threat detection [52,55,58,64–68]. In [49,59], they show that deep reinforcement learning can be used for adaptive voltage and frequency control.

In [52,69], they emphasize the importance of hybrid IDS systems that use neural networks and optimization algorithms for real-time false data detection (FDIA). Combined with edge AI and blockchain [62,70], these approaches enable decentralized protection and autonomous response to threats.

In [55,70], they add that digital twins and cyber-resilient architectures can create intelligent energy ecosystems with self-healing capabilities, which is essential for future PMSG-dominated smart grids.

3. State of the Art/Background

3.1. Development of Smart Grids and the Concept of Sustainability

In the last decade, smart grids have emerged as a key element in the transition to decentralized and sustainable power systems. They are an integrated platform that combines information technologies, communication protocols, sensors, and control algorithms, with the aim of optimizing the production, distribution, and consumption of electricity. Among the main advantages are increased reliability, reduced losses, better energy efficiency, and the possibility of two-way communication between operators and consumers.

https://doi.org/10.3390/en19030772

One of the most important functions of smart grids is the dynamic balancing between production and consumption, especially in the presence of intermittent renewable sources such as wind and sun. Through intelligent metering systems (AMI) and energy management systems (EMSs), operators can compensate for fluctuations in real time. Additionally, distribution system automation (DSA) allows for the regulation of energy flows, compensation of reactive power, and improvement in power quality.

Resilience in the context of power grids is defined as the ability of the system to withstand external disturbances and recover quickly after them. With a high share of renewable sources and especially with dominant PMSGs, resilience encompasses not only the physical reliability of the infrastructure, but also digital security, autonomy, and intelligent management. This is facilitated by the development of energy hubs (including electric vehicles and BESSs) and the use of machine learning (ML) for load forecasting and rapid response to uncertain events.

3.2. The Role of PMSGs in Integrated RES Systems

3.2.1. Main Characteristics

Permanent magnet synchronous generators (PMSGs) are widely used in wind energy, especially in direct drive systems. Their popularity is due to their high efficiency, compact design, and the absence of an excitation system, which reduces the need for maintenance. Unlike classic synchronous generators, PMSGs do not provide natural inertia to the grid, as their connection is made via power electronics.

3.2.2. Integration Challenges

The lack of physical inertia is among the main problems, as PMSG systems do not participate in frequency regulation. This increases the risk of instability at high concentrations of RES. In addition, the variability of wind energy and the high dynamics of inverters lead to voltage and frequency fluctuations. An additional limitation is the weak contribution to reactive power; without compensation devices, PMSGs cannot maintain the voltage in the grid.

3.2.3. Potential Solutions

- Virtual inertia: introduction of new "virtual synchronous generator" (VSG) algorithms that mimic the dynamics of classic rotors.
- BESS integration: battery systems smoothen power and provide real-time reserves.
- Intelligent management through EMSs: predictive algorithms optimize the participation of PMSG systems in the grid.

3.3. Methods for Increasing Resilience in Smart Grids

3.3.1. Energy Hubs and Storage

Energy hubs integrate EV charging stations, BESSs, and photovoltaics, and they can cover up to 45% of the local load autonomously. This reduces the load on the centralized grid and increases local resilience.

3.3.2. EMS and Intelligent Management

EMS platforms use predictive data to optimize loads and minimize instabilities. They can instantly activate stored reserves and provide dynamic grid balancing.

3.3.3. DSA Technologies

Distribution system automation allows for automatic flow redirection, voltage and reactive power regulation, and rapid fault isolation. This shortens restoration time and increases reliability.

https://doi.org/10.3390/en19030772

3.3.4. Artificial Intelligence and ML

Methods such as SVMs, ANNs, and CNN-LSTM are used to forecast RES generation and loads, as well as to detect cyber threats. Reinforcement learning shows high efficiency in optimal control under uncertainty.

Figure 1 presents a conceptual architecture of a smart power system dominated by PMSGs. The scheme shows the main strategies for increasing resilience—virtual synchronous generators (VSGs) for inertia emulation, battery energy storage systems (BESSs) for providing fast backup, STATCOM devices for dynamic reactive power compensation, and the integration of energy management systems (EMSs) and distribution network automation. The scenarios developed and analyzed in this work (PLL, VSG, VSG+BESS, VSG+STATCOM, VSG+BESS+STATCOM) are summarized in this architecture, illustrating the possible combinations of technologies to ensure the stability and reliability of the network.

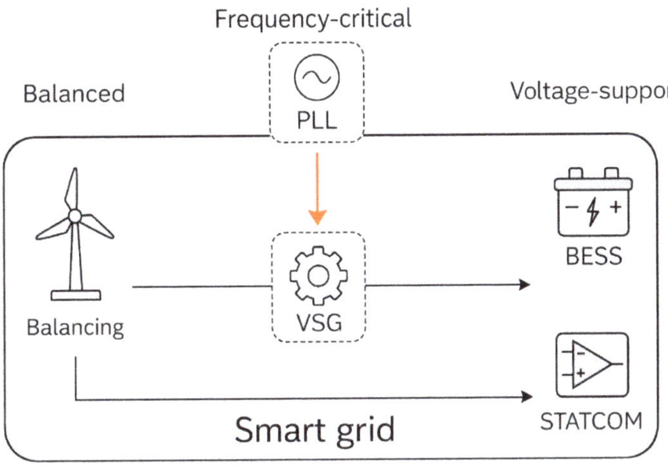

Figure 1. Conceptual architecture of a smart power system dominated by PMSGs.

Key resilience strategies that are presented are as follows: virtual synchronous generators (VSGs) for inertia emulation, battery energy storage systems (BESSs) for dynamic power support, STATCOM devices for reactive compensation, and integrated EMS/DSA solutions for smart control and automation.

3.4. Cyber Resilience and Security

3.4.1. Vulnerabilities

PMSG integration requires the intensive use of digital protocols and controllers, making them vulnerable to attacks such as false data injection (FDIA), communication protocol breaches, and attacks on SCADA/EMS infrastructure.

3.4.2. IDS Systems

Modern intrusion detection systems use a combination of neural networks (CNN, LSTM), optimization algorithms, and ensemble models, achieving over 95% accuracy in identifying anomalies in real time.

3.4.3. PMSG Challenges

The dependence on inverters and high dynamics make the systems sensitive to small changes in frequency or current. This can lead to automatic shutdowns or destabilization.

https://doi.org/10.3390/en19030772

3.4.4. Recommendations

The following recommendations can be summarized to improve resilience in smart grids:

- Implementing edge-based AI for local threat detection;
- Implementing blockchain for transaction authentication;
- Creating autonomous microgrids;
- Training EMS models on simulated cyber scenarios for higher adaptability.

4. Analysis of the Challenges and Prospects for the Resilience of Smart Grids Dominated by PMSGs

4.1. Key Challenges

4.1.1. Reduced Inertia and Frequency Instability

One of the most significant obstacles to resilience is the lack of mechanical inertia in PMSG systems. Due to the use of power electronics, the generator is separated from the grid by an inverter and does not participate directly in frequency regulation. This leads to a greater sensitivity of the grid to disturbances and an increased risk of large-scale failures, especially with a high concentration of renewable sources.

4.1.2. Limited Contribution to Reactive Power

PMSGs cannot generate or compensate reactive power on their own, which makes voltage regulation difficult. In the absence of compensation devices (e.g., STATCOM, SVC), the grid becomes vulnerable to deviations and voltage fluctuations.

4.1.3. Dependence on Power Electronics

The inverters that connect the PMSG to the grid can be a source of vulnerabilities, both technical (e.g., harmonics, loss of synchronization) and cybernetic (access via digital interfaces). The increasing dependence on programmed control raises the question of the security, reliability, and adaptability of these components to sudden changes in conditions.

4.1.4. Insufficient Development of Local Flexibility

The lack of adequate local energy storage resources and controllable loads makes it difficult for the grid to respond to short-term mismatches between production and consumption. Without the integration of battery systems and dynamic load management, the grid is exposed to more frequent disturbances and failures.

4.2. Perspectives and Possible Solutions

4.2.1. Virtual Inertia and Synchronous Inverters

Technologies such as the "virtual synchronous generator" (VSG) and "grid-forming inverters" can reproduce the inertial characteristics of classic generators through digital control. This allows PMSG systems to actively participate in frequency stabilization.

4.2.2. Integration of Smart EMSs

Energy management systems with machine learning and predictive algorithms are key to increasing resilience. They can anticipate dynamic fluctuations and activate flexible resources such as BESSs, EV hubs, and microgrids for compensation.

4.2.3. Distributed Architecture and Microgrids

Creating functionally autonomous microgrids dominated by PMSGs that can operate in both connected and isolated (island mode) mode increases local resilience. They can continue to operate in the event of a main grid failure, ensuring critical power supply.

https://doi.org/10.3390/en19030772

4.2.4. Cyber Resilience Through Edge AI and Real-Time Protections

The integration of smart systems for anomaly detection and cyberattacks at the inverter or local controller level is imperative. Edge-oriented architectures with autonomous protections and fast decision making reduce response time and isolate threats before they spread throughout the network.

5. Mitigation Strategies

With the increasing share of PMSG-based generators in smart grids, developing resilient management strategies to compensate for their weaknesses, such as a lack of inertia, limited contribution to reactive power, and high dependence on power electronics, becomes a top priority. The following sub-sections discuss key engineering approaches to enhance the resilience of grids dominated by PMSGs.

5.1. Virtual Synchronous Generators (VSGs)

Virtual synchronous generators (VSGs) are one of the most promising technologies to compensate for the lack of mechanical inertia. They use algorithms that emulate the rotor dynamics of a classic synchronous generator by controlling the inverter. Thus, in the event of sudden changes in load or production, VSG systems provide "virtual inertia" and maintain frequency stability. The additional introduction of damping coefficients allows for smooth transitions through transients and limiting oscillations.

PMSGs inherently lack mechanical inertia, have limited reactive voltage capability, and are decoupled from the grid through power electronic converters. The VSG algorithm directly addresses these weaknesses by introducing synthetic electromechanical behavior into the converter control. The VSG implements the classical swing equation:

$$J\frac{d\omega}{dt} = P_{\mathrm{m}} - P_{\mathrm{e}} - D(\omega - \omega_0) \tag{1}$$

where J is the virtual moment of inertia, and D is the damping coefficient.

This equation emulates the rotor dynamics of a synchronous machine, enabling the inverter to temporarily store and release energy in response to power imbalances, thus stabilizing frequency and suppressing oscillations.

A voltage–reactive power (V–Q) loop analogous to the excitation system of a synchronous generator is added:

$$V_{\mathrm{ref}} = V_0 + K_{\mathrm{q}}(Q_{\mathrm{ref}} - Q) \tag{2}$$

This allows the inverter to actively regulate voltage magnitude and support grid voltage during transients.

By converting the inverter from a current-following (PLL-based) to a voltage-forming source, the VSG transforms the converter's role from passive synchronization to active grid support. When the VSG control is frozen and the converter returns to the PLL mode, these capabilities disappear, and the system exhibits the fast, oscillatory behavior typical of converter-dominated networks.

The control architecture of the proposed VSG is shown schematically in Figure 2. It represents a generalized control framework rather than a direct replication of any single implementation.

The model includes three hierarchical layers. The virtual inertia and damping loop provides synthetic frequency dynamics and active power response. The voltage and reactive power loop emulates the excitation system to control reactive exchange. Inner converter control applies voltage and frequency references via PWM modulation. This

https://doi.org/10.3390/en19030772

modular structure allows systematic comparison with hybrid configurations (VSG+BESS, VSG+STATCOM) in a consistent simulation framework.

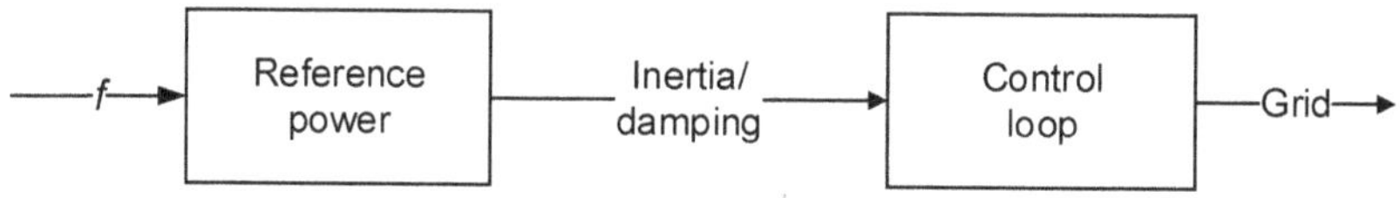

Figure 2. Schematic representation of the inertia loop in the VSG system.

To focus the analysis on intrinsic control behavior, several standard simplifications are adopted, such as ideal PWM modulation without switching harmonics; balanced three-phase grid without harmonics or asymmetries; linear system elements (no saturation or dead zones); ideal sensing and communication, without noise or delays; and no voltage or current limiters activated in the nominal range. These assumptions follow standard benchmark models in the literature and ensure a fair comparison among different control strategies under identical system conditions.

The virtual inertia Jv and the damping Dv were chosen in typical ranges of 0.3–1.5 s and 0.5–1.8 pu, respectively. A sensitivity analysis was performed by varying these parameters by ±50% around their nominal values.

Figure 2 should be interpreted as a conceptual abstraction of the virtual inertia loop, rather than a hardware diagram. It illustrates the transformation from mechanical dynamics modeled by the swing equation to electrical control inputs that form the grid interface of the inverter. The model corresponds to the simplified VSG dynamics commonly used in comparative stability studies and provides the basis for the extended hybrid strategies analyzed in this work.

By emulating the rotational inertia, the VSG smoothens frequency deviations and resists abrupt load disturbances. When the VSG control is frozen (i.e., converter returns to the grid-following PLL mode), this synthetic inertia vanishes, and the system immediately exhibits the fast and oscillatory dynamics typical of converter-dominated sources.

Limited reactive voltage control: In a conventional PMSG inverter under PQ control, reactive power is only indirectly regulated through the grid voltage phase. The VSG algorithm enhances this by adding a voltage control loop analogous to the excitation control of a synchronous machine. Through this mechanism, the VSG adjusts the converter output voltage amplitude to provide fast and stable reactive power support, effectively extending the voltage regulation range.

When the VSG is disabled, the inverter reverts to its limited current-based control, losing the ability to provide dynamic voltage support, and this manifests as reduced voltage stiffness and an increased risk of undervoltage during transient conditions.

Dependence on power electronics (converter decoupling): The VSG algorithm transforms the converter's behavior from a passive current follower to an active voltage source with grid-forming capability. This control paradigm mitigates the negative effects of converter decoupling, enabling the inverter to contribute to grid strength rather than depend on it. When VSG control is deactivated, the converter again becomes grid-following and sensitive to phase and frequency disturbances.

Overall, the VSG algorithm synthetically restores the electromechanical characteristics of synchronous machines within a power–electronic interface, thereby neutralizing the three fundamental weaknesses of PMSG-based generators.

In simulations, the VSG technology shows lower frequency deviations and shorter recovery times compared to conventional grid-following controllers.

The VSG algorithms are the basis of modern approaches to providing virtual inertia in PMSG-dominated systems. Their general structure is shown in Figure 2. The diagram

https://doi.org/10.3390/en19030772

illustrates the conversion of electromechanical power and the addition of artificial inertia and damping before connection to the grid.

The generated active power P and the frequency deviation Δf pass through an inertial and damping block, which imitate the dynamics of a classic synchronous generator. At the output, a signal is formed for controlling the inverter, allowing synchronization and maintaining grid stability.

The selection of the virtual inertia (H) and damping coefficient (D) parameters in the simulated VSG model follows standard design and tuning practices established in the literature. The base reference values (H = 0.5 s, D = 1.0 p.u.) were derived from commonly used ranges reported in recent studies on grid-forming converters and virtual synchronous machines.

These values correspond to a medium-inertia system representative of converter-dominated microgrids and are further fine-tuned empirically to ensure a critically damped transient response under the test conditions defined in Section 6. The tuning procedure involved minimizing frequency overshoot and settling time during step changes in active power.

Such parameter ranges are consistent with standard grid-forming converter designs, where typical values for H vary between 0.2 and 1.5 s and for D between 0.5 and 2.0 p.u., depending on the desired frequency stiffness and system strength. The selected parameters thus ensure both comparability with the state of the art and represent realistic values for medium-scale PMSG-based inverters in smart grid environments.

5.2. Battery Energy Storage Systems (BESSs)

Battery energy storage systems (BESSs) play a critical role in power smoothing and providing additional inertia. In combination with PMSGs and VSGs, BESSs can absorb or release active power in real time, compensating for sudden imbalances between production and consumption.

In addition to frequency stabilization, BESSs also support reactive power management through integrated power converters, which improves local voltage stability.

With adequate management (through predictive EMS algorithms), battery systems can take on the role of fast-acting reserves, which are traditionally performed by conventional power plants.

5.3. Predictive Control Using AI/ML

Intelligent predictive control methods based on artificial intelligence and machine learning are becoming an indispensable tool for PMSG integration. Algorithms such as reinforcement learning (RL) and Model Predictive Control (MPC), supported by neural networks, allow for wind and solar generation forecasting, dynamic load optimization, adaptive control of BESSs and STATCOM, and the early detection of anomalies and cyber threats.

The result is not only improved system resilience, but also a more efficient use of available resources.

5.4. Grid-Forming vs. Grid-Following Inverters

The distinction between grid-following and grid-forming inverters is of fundamental importance for PMSG-based networks. Grid-following (PLL-based) inverters synchronize to an already existing grid voltage and do not contribute independently to frequency stability. Grid-forming (VSG based) inverters have the ability to set frequency and voltage, imitating classical generators.

https://doi.org/10.3390/en19030772

In smart grids with a high share of PMSG, the transition to grid-forming inverters is a critical step to ensure stable synchronization and reliable operation in isolated modes (islanding).

5.5. Power Electronics Control Loops for PMSG

The control of PMSG by power electronics includes several hierarchical levels of controllers:

- Current controller (d-q axes): they provide fast and stable regulation of active and reactive power;
- Voltage controller: they maintain stability of the DC bus and the quality of the supplied power;
- External control loops (P/Q controllers): they coordinate the operation of PMSGs with the system requirements for power and stability.

The application of anti-windup techniques in PI controllers and adaptive parameterizations allows for a more stable transition in the event of sudden disturbances and minimizes the risk of saturation of the integrators. Combined with predictive control, these controllers form the basis for flexible and resilient PMSG-based systems.

6. Case Study and Simulation Analysis

This study develops a simplified model of a power system in which permanent magnet synchronous generators (PMSGs) are the main renewable source. The system is simulated in several control options, with the aim of assessing their impact on the stability and resilience of smart grids under transient conditions.

The options considered are as follows:

- Grid-following control (PLL), synchronization with the grid through a phase-locked loop;
- Grid-forming control (VSG), a virtual synchronous generator that mimics the inertia of a classic generator;
- VSG + BESS, adding a battery energy storage system to smoothen the power and provide dynamic reserves;
- VSG + STATCOM, a reactive power compensator that supports voltage stability;
- VSG + BESS + STATCOM, a complete combination of active and reactive compensation resources.

The developed simulation model is based on a set of idealized assumptions that aim to isolate and evaluate the intrinsic dynamic behavior of the proposed control strategies. Specifically, the adopted simplifications are as follows:

- Ideal converters: power electronic interfaces are modeled as lossless and perfectly synchronized, neglecting switching harmonics and converter nonlinearities;
- Linear system components: magnetic saturation, core losses, and nonlinear impedances are omitted;
- Balanced operation: the grid is assumed to be a perfectly balanced three-phase system without voltage or current asymmetries;
- Perfect measurement and communication: all sensors and control channels are considered ideal, with no noise, latency, or signal distortion;
- Negligible parasitic dynamics: transformer and cable dynamics, as well as line capacitances, are not included.

These assumptions are standard in first-stage dynamic studies and ensure clear comparability between different control strategies. However, it is recognized that such idealization limits the direct extrapolation of the results to real-world power systems, where converter

https://doi.org/10.3390/en19030772

switching behavior, communication delays, and nonlinear effects can introduce additional oscillations and energy losses. Future work will therefore extend this framework to include non-ideal effects, detailed converter models, and hardware-in-the-loop (HIL) validation to bridge the gap between analytical simulations and physical implementation.

The ESS module in the simulation is modeled as a grid-connected, bidirectional DC-AC converter with fast internal current control and external power regulation loops. It is integrated into the VSG active power regulation channel to provide dynamic support during frequency deviations and transients. The BESS compensates for active power imbalances by absorbing or injecting active power within its rated limits, while maintaining a constant DC bus voltage and a balanced state of charge (SoC). The control algorithm follows a PI-based active power controller tuned for a fast response and minimal overshoot. The energy management strategy limits SoC variations to avoid deep discharge or overcharge, reflecting realistic operational constraints. The key simulation parameters used for the BESS configuration are summarized in Table 1.

Table 1. Parameters of the battery energy storage system (BESS) used in the simulations.

Parameter	Symbol	Value	Unit	Description
Nominal DC link voltage	V_{DC}	750	V	DC bus voltage at converter input
Nominal capacity	E_{nom}	200	kWh	Total stored energy capacity
Maximum power rating	$P_{BESS,max}$	800	W	Maximum active power injection/absorption
Nominal current	I_{nom}	25	A	Current at rated power and DC voltage
Sampling time	T_s	50	µs	Control sampling period
SoC initial value	SoC_0	0.50	–	Initial state of charge
SoC operational range	SoC_{min}–SoC_{max}	0.20–0.90	–	Safe operating range
Active power PI gains	K_p, K_i	0.005, 0.05	–	Controller gains for power regulation
DC voltage control gain	K_v	0.2	–	DC-link voltage stabilization gain
Energy update constant	K_E	10^{-4}	–	SoC update coefficient per simulation step
Converter efficiency	η	0.97	–	Power converter efficiency
Simulation time step	Δt	5×10^{-5}	s	Integration step used in Python model

The selected BESS parameters represent a medium-scale lithium-ion energy storage system typically used in microgrids and renewable energy applications. The selected power limit of 800 W corresponds to the nominal capacity of the simulated PMSG unit, ensuring proportional participation in active power regulation. The controller parameters Kp = 0.005 and Ki = 0.05 were selected through iterative tuning to achieve fast transient response (settling time < 0.1 s) with less than 5% overshoot. The SoC control ensures stable operation during long transients and avoids saturation effects that could otherwise compromise frequency restoration performance.

In the simulated scenarios, the battery energy storage system (BESS) was assigned a maximum active power capacity of P_{BESS,max} = 800 W, corresponding to approximately 30–35% of the nominal PMSG rating (P_{PMSG,nom} = 2.5 kW).

This ratio was selected to represent a medium-sized storage module, typical for microgrid and hybrid renewable systems, where the BESS is primarily intended to mitigate short-term frequency deviations and power imbalances rather than provide long-duration energy support.

Increasing the BESS capacity would enhance the system's ability to absorb and inject power over longer transients, improving resilience indices (RI) and reducing the rate of change of frequency (RoCoF). However, this also increases system cost, converter current ratings, and control complexity.

Conversely, a smaller storage capacity would limit the dynamic response range, causing higher frequency and voltage deviations under step disturbances.

https://doi.org/10.3390/en19030772

Therefore, the selected BESS-to-PMSG ratio represents a realistic trade-off between technical performance and implementation feasibility in medium-scale smart grid applications.

In this way, not only the individual effects of each technology are investigated, but also their synergy in the context of smart grids.

The following figures present results developed and simulated in the Python v 3.10 model of a permanent magnet synchronous generator (PMSG), integrated in a smart grid environment. The developed model analyzes the dynamic processes of the electrical quantities and their behavior when the reference values for active and reactive power changes.

Figure 3 presents the phase currents (Ia, Ib, Ic) of the PMSG in the time interval of the simulation. In the initial part of the graph, a transient process is observed, caused by the initial conditions and the switching on of the system. Around the time t = 0.1 s, a change in the set active power is introduced, which leads to a momentary transient regime in the currents. After a short settling period, the currents reach a new equilibrium corresponding to the new reference power.

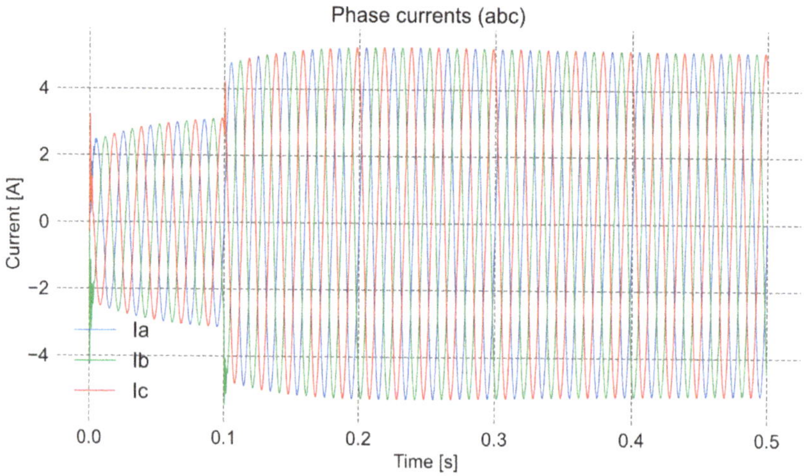

Figure 3. Phase currents (Ia, Ib, Ic) of a PMSG.

The phase currents are clearly sinusoidal and symmetrically distributed at 120° electrical, which confirms the correctness of the vector control and synchronization via the PLL. The observed amplitude increases after the change in load, which is consistent with the increased active power delivered by the generator to the grid.

Figure 4 shows the dynamics of the active power (P) of the PMSG relative to the set reference value (P_ref). At the beginning of the simulation, the system strives to reach the initial reference power of 1500 W, during which a transient process is observed, related to the response of the PI controllers and the adjustment of the current controls.

Around the time t = 0.1 s, a step change in the set power from 1500 W to 2500 W is applied. The real active power (P (real)) follows this change with a short transient process and establishes a new stable state, which practically coincides with the reference value.

This behavior confirms the good tuning of the PI controllers for the active power, as well as the effectiveness of the anti-windup scheme used. The transient process is short, without significant oscillations, which is an indicator of the stability and accuracy of the control.

Figure 5 illustrates the dynamics of reactive power (Q) of the PMSG model relative to the reference value (Q_ref), which in this scenario is set as zero. This means that the system strives to operate at zero power factor, without injecting or absorbing reactive energy.

https://doi.org/10.3390/en19030772

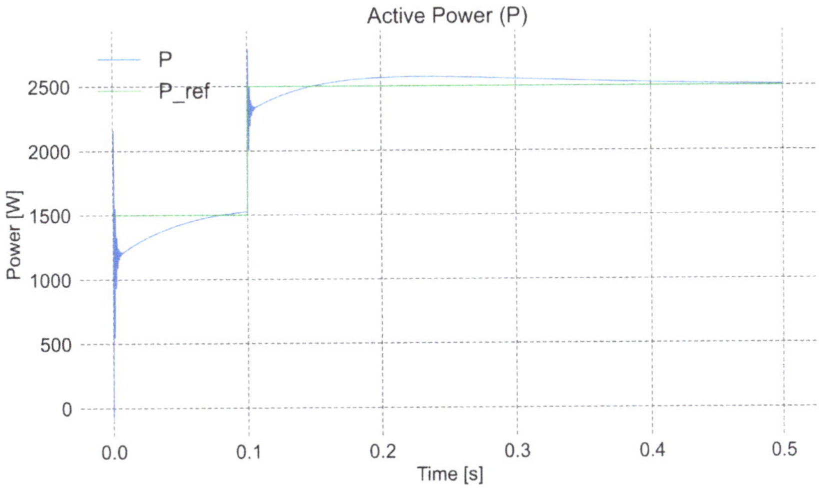

Figure 4. Dynamics of active power (P) of a PMSG.

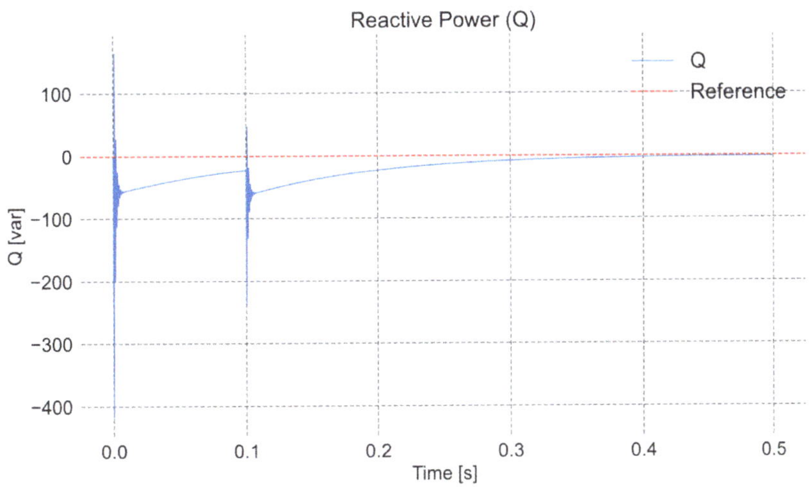

Figure 5. Dynamics of reactive power (Q) of the PMSG model.

At the beginning of the simulation and during the step change in active power around t = 0.1 s, oscillations and temporary deviations of Q (real) are observed, which reach values of the order of −400 var. This is a result of the transient processes in the current control and the dynamics of the PI regulators.

After passing the transient regime, the reactive power gradually stabilizes and tends to the reference value Q_ref = 0, with minimal residual fluctuations. This shows that the control scheme manages to compensate for the reactive components and provide a cosφ close to unity.

The result confirms the adequacy of the reactive power control strategy used, although in the initial stage it is sensitive to the changes in load and reference active power.

Figure 6 illustrates the frequency dynamics calculated by the PLL (Phase Locked Loop). At the beginning of the simulation, a small frequency deviation from the 50 Hz reference value is observed, which is caused by the initial mismatch and the synchronization error.

https://doi.org/10.3390/en19030772

Gradually, thanks to the regulatory properties of the PLL, the system compensates for this deviation and the frequency stabilizes around the reference line.

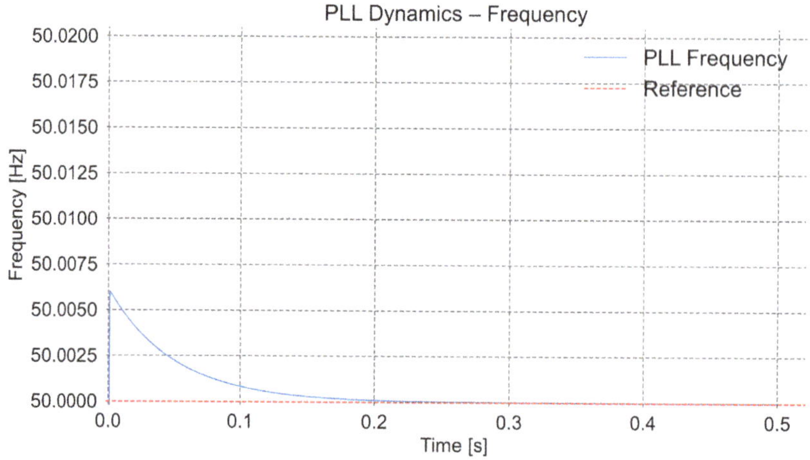

Figure 6. PLL dynamics of the PMSG model.

The results show that the PLL successfully maintains synchronization with the grid, which is a critical condition for the reliable operation of inverter-based generators such as the PMSG. This behavior confirms the ability of the PLL to provide a stable frequency even in the presence of transients at the beginning of operation.

Figure 7 presents the evolution of the phase θ as a result of the operation of the PLL (Phase Locked Loop) algorithm. It can be seen that the phase increases linearly with time, which is the expected behavior at a stable network frequency. Since the PLL successfully synchronizes the calculated frequency with the 50 Hz reference value (as shown in Figure 6), the phase continues to evolve evenly and without deviations.

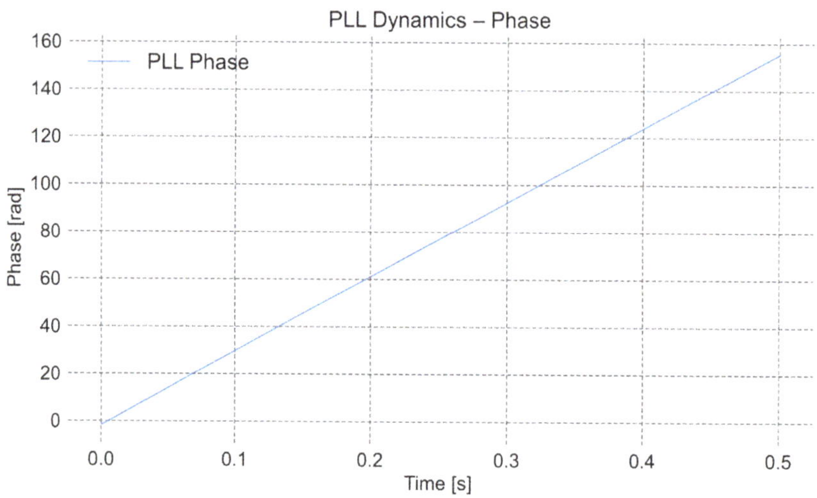

Figure 7. Phase θ dynamics of the PMSG model.

https://doi.org/10.3390/en19030772

This result demonstrates the correct operation of the PLL in tracking the phase angle, which is critical for the abc → dq transformations and vice versa, as well as for the correct control of the active and reactive power in PMSG-based systems.

The simulation results confirm that the proposed PMSG model, controlled by PI-regulators with anti-windup mechanism and PLL synchronization, provides stable phase currents, precise active power tracking, maintenance of reactive power around reference values, and correct synchronization with the frequency and phase of the grid. These results are the basis for the further development of the system with the addition of advanced control strategies, including virtual synchronous generators (VSGs) and integration with BESS/STATCOM to increase resilience.

Figure 8 shows the frequency dynamics in a grid dominated by PMSGs for five different control strategies: PLL (grid-following), VSG (grid-forming), VSG+BESS, VSG+STATCOM, and VSG+BESS+STATCOM. The PLL strategy strictly follows the grid frequency and maintains an almost constant value around 50 Hz, indicating the absence of inertial contribution. The VSG introduces virtual inertia, which provides a smoother response to load changes and stabilizes the frequency close to the nominal value. The VSG+BESS contributes to the fast support from the battery, but it shows more pronounced oscillations due to the interaction between the inertia dynamics and the BESS limitations. The VSG+STATCOM demonstrates a similar behavior to the VSG in terms of frequency, with the main contribution being in voltage maintenance (not clearly visible in the figure, but important for the overall stability). The VSG+BESS+STATCOM combines active and reactive support, which results in the best-balanced behavior, albeit with minimal oscillations around the nominal frequency. The figure illustrates that the classical PLL approach does not contribute to frequency stability, while the VSG-based strategies especially in combination with BESS and STATCOM provide higher stability and disturbance mitigation capability.

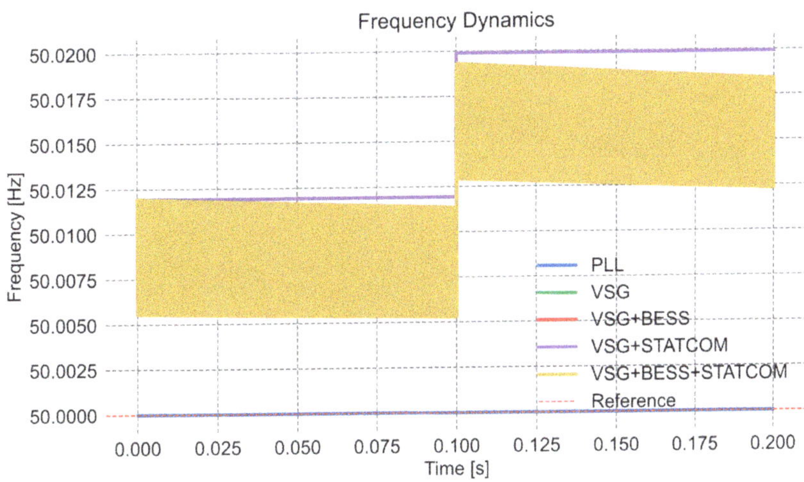

Figure 8. Frequency dynamics for different control strategies.

Figure 9 shows the dynamics of the active power when changing the reference setpoint from 1500 W to 2500 W around the time t = 0.1 s. The results clearly show that the PLL strategy follows the reference signal relatively well, but with larger fluctuations and slower settling. The VSG-based control provides a smoother transition, with the inclusion of BESS improving the dynamics and accelerating the settling time. STATCOM has a limited impact on the active power, since its main effect is related to reactive compensation. The combined

https://doi.org/10.3390/en19030772

VSG+BESS+STATCOM strategy provides the best behavior, minimizing the deviation from the reference value and maintaining stable dynamics.

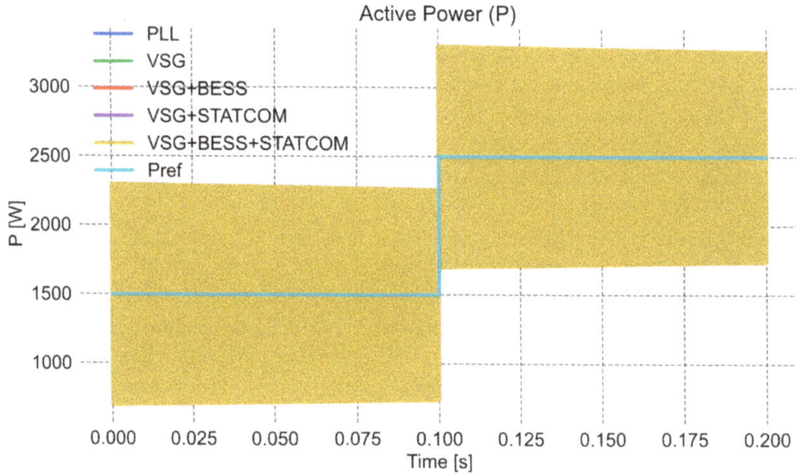

Figure 9. Active power (P) under different control strategies (PLL, VSG, VSG+BESS, VSG+STATCOM, VSG+BESS+STATCOM).

Figure 10 shows the system response when the reference reactive power changes from 0 var to 500 var at time t = 0.1 s. The main observations are that the PLL and pure VSG hardly provide adequate reactive power regulation, since they do not have a direct Q-compensation function. VSG+BESS shows a limited response, with the BESS mainly assisting the active power. VSG+STATCOM achieves a significant improvement by following the reference signal and compensating the reactive load. VSG+BESS+STATCOM offers the best behavior, stable setpoint tracking, and reactive power balancing, without significant deviations.

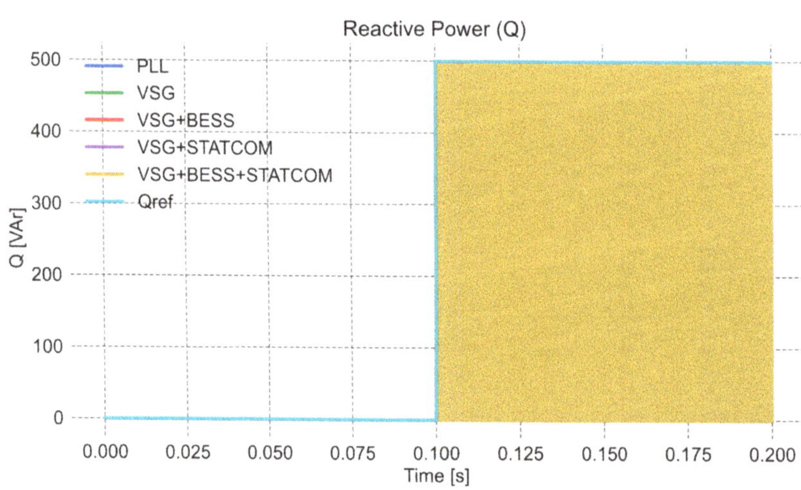

Figure 10. Reactive power (Q) under different control strategies (PLL, VSG, VSG+BESS, VSG+STATCOM, VSG+BESS+STATCOM).

https://doi.org/10.3390/en19030772

In order to compare the strategies, a set of quantitative metrics is introduced that cover both frequency stability and active and reactive power control. These include the following:

- Frequency RMSE (RMSE f): root mean square deviation from the nominal value of 50 Hz, an indicator of frequency stability;
- Maximum frequency deviation (Max f): a critical indicator of the resilience to sudden disturbances;
- Active power RMSE (RMSE P): measures the accuracy of tracking the active power reference value;
- Maximum power overload (Max P): peak value of the Active Power Deviation;
- RMSE and Max Q: reactive power indicators related to voltage stability;
- Recovery time (T_rec): the time it takes for the system to return to a stable state after a disturbance;
- Resilience Index (RI): an integrated index combining frequency and power metrics for a comprehensive assessment of resilience.

Figure 11 shows the comparison between five different approaches: PLL, VSG, VSG+BESS, VSG+STATCOM, and VSG+BESS+STATCOM. The score is normalized in the range [0,1], where 1 corresponds to optimal behavior in terms of frequency resilience. All strategies are shown with values close to the maximum, meaning that in frequency-critical scenarios, each of them manages to provide basic stability. The PLL achieves the highest score due to its simple synchronization, but its lack of inertia makes it limited in other aspects. The VSG and its extensions (BESS, STATCOM) demonstrate comparable results, with VSG+BESS+STATCOM offering the most balanced solution, combining frequency stability and flexibility in power management.

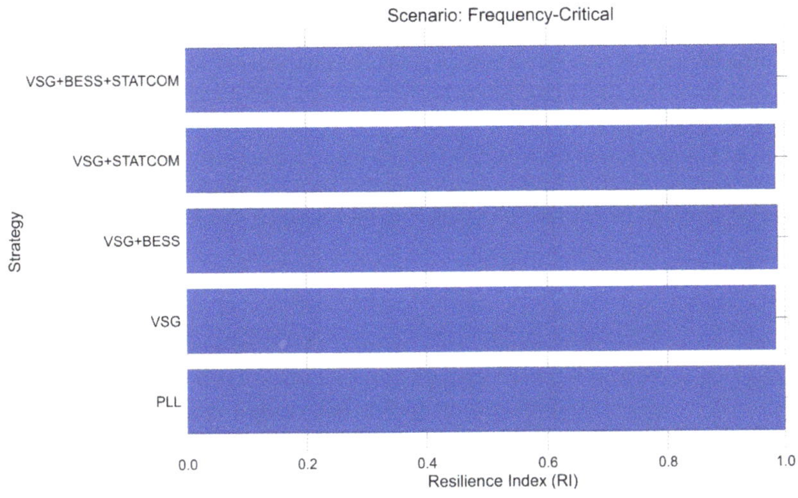

Figure 11. Evaluation of control strategies in a frequency-critical scenario.

Figure 12 shows the normalized evaluation of five different strategies (PLL, VSG, VSG+BESS, VSG+STATCOM, VSG+BESS+STATCOM) under conditions where the system is exposed to moderate disturbances and simultaneous frequency and voltage maintenance is required. All strategies demonstrate close to the maximum evaluations, which highlights their effectiveness under balanced load and moderate dynamics. The PLL again shows stable results, but without the possibility of active participation in voltage regulation. The VSG-based solutions stand out with added value in dynamics management, with

https://doi.org/10.3390/en19030772

VSG+BESS+STATCOM offering the most complete balance between frequency stability and reactive support. In this scenario, the differences between the strategies are minimal, which means that the choice of technology is more flexible compared to the frequency-critical case.

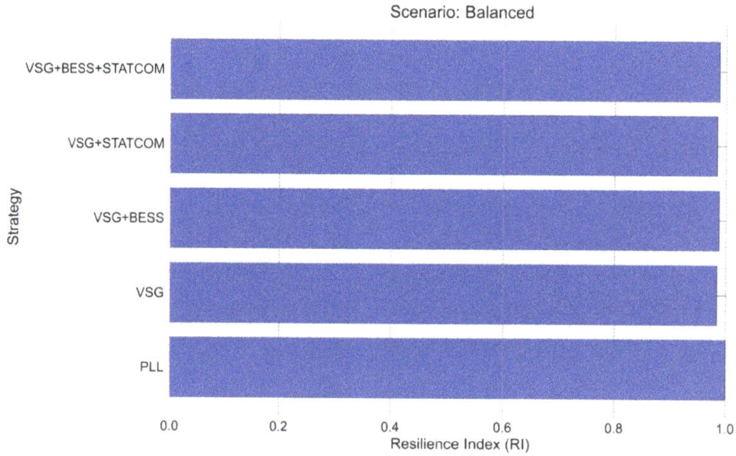

Figure 12. Evaluation of management strategies in a balanced scenario.

Figure 13 presents a comparative analysis of five smart grid control strategies in the "voltage support" scenario. The graph shows the ratings of the PLL, VSG, VSG+BESS, VSG+STATCOM, and VSG+BESS+STATCOM systems, with the rating reflecting their ability to maintain voltage under dynamic changes in load and reactive power. All strategies show a high degree of efficiency, with the best results observed in the combined VSG+BESS+STATCOM solution, which integrates a virtual synchronous generator with a battery storage system and a STATCOM device for dynamic reactive power compensation. This hybrid architecture demonstrates the highest rating due to its ability to simultaneously provide active and reactive support, as well as to respond quickly to voltage fluctuations. On the other hand, the traditional PLL control shows limited adaptability, which confirms the need for integrated approaches to increase the resilience and stability of networks dominated by PMSGs.

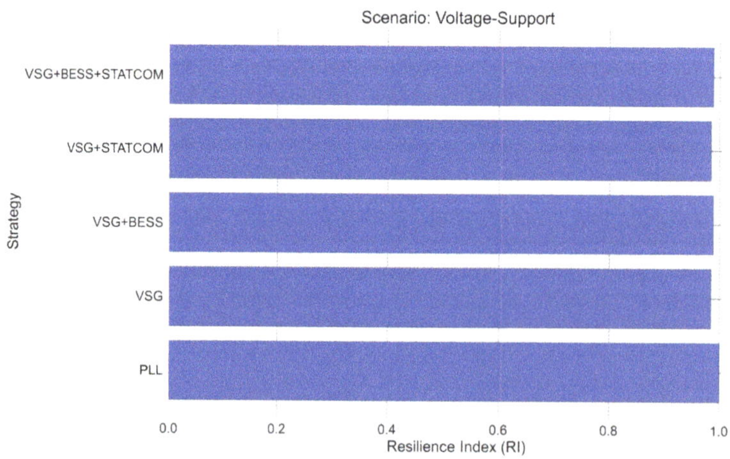

Figure 13. Evaluation of control strategies in the voltage support scenario.

https://doi.org/10.3390/en19030772

RI is a composite indicator designed to quantify the overall ability of a power system to withstand and recover from dynamic disturbances in frequency, active power, and reactive power.

The index combines three key performance indicators in the time domain. Root Mean Square Frequency Error represents the average deviation of the system frequency from the nominal value (50 Hz). Physically, this term reflects the ability of the system to maintain rotational stability and emulate inertia during transient events. Active Power Deviation quantifies the cumulative discrepancy between the generated and reference active power. It characterizes the ability of the system to balance power flow and respond to load changes. Reactive Power Deviation measures the variation in the supported reactive power from its reference, which directly affects voltage regulation and reactive power compensation capability.

The combined resilience index is formulated as a normalized inverse function of the mean deviations:

$$RI = \frac{1}{1 + \omega_f \cdot |f - f_{\text{nom}}| + \omega_{\text{P}} \cdot |P - P_{\text{ref}}| + \omega_{\text{Q}} \cdot |Q - Q_{\text{ref}}|} \tag{3}$$

where $\omega_f,\ \omega_p,\ \omega_Q$ are the weighting factors that balance the contributions of the frequency, active, and reactive terms according to their relative importance in maintaining the stability of the system.

This formulation ensures that higher RI values (closer to 1) correspond to systems that exhibit smaller deviations and faster recovery from disturbances.

The chosen combination is suitable for joint assessment because it covers the three main dimensions of power system resilience: frequency stability (inertial and primary response), active power balancing (energy adequacy and control), and voltage/reactive stability (electromagnetic support and grid rigidity).

Thus, the robustness index provides a unified measure that links mechanical (frequency) and electrical (P–Q) characteristics, allowing a direct comparison between conventional (PLL) and advanced hybrid control strategies (VSG, BESS, STATCOM).

Figure 14 presents the summary evaluation of the five strategies (PLL, VSG, VSG+BESS, VSG+STATCOM, VSG+BESS+STATCOM) under the three key scenarios: frequency-critical, balanced, and voltage support. This diagram visualizes how each strategy positions itself against the resilience criteria. The PLL demonstrates high resilience in the frequency-critical and balanced scenarios, but it lags significantly in the voltage support scenario. VSG+BESS and VSG+BESS+STATCOM show the highest coverage of the three scenarios, making them suitable for integrated smart grids. VSG+STATCOM improves reactive support, but it remains with a more limited contribution to frequency regulation compared to BESS-based solutions. The radar diagram clearly highlights the need for a hybrid approach combining virtual inertia and voltage support to achieve system resilience.

Figure 15 presents a heatmap of the key metrics (RMSE for frequency and active and reactive power) for the different strategies. The PLL shows minimal deviations in active and reactive power, but it exhibits higher errors in maintaining reactive balance (≈350 Var). The VSG reduces frequency deviations (RMSE f = 0.016 Hz), while achieving lower RMSE Q values (≈180 Var). VSG+BESS brings significant dynamic stability to frequency regulation (RMSE f = 0.013 Hz), but it loads the system with a larger deviation in active power (≈780 W). VSG+STATCOM improves reactive power support (≈250 Var), but the frequency behavior remains close to that of the pure VSG. VSG+BESS+STATCOM combines strengths such as high frequency stability and moderate voltage control, but active power still shows larger deviations.

https://doi.org/10.3390/en19030772

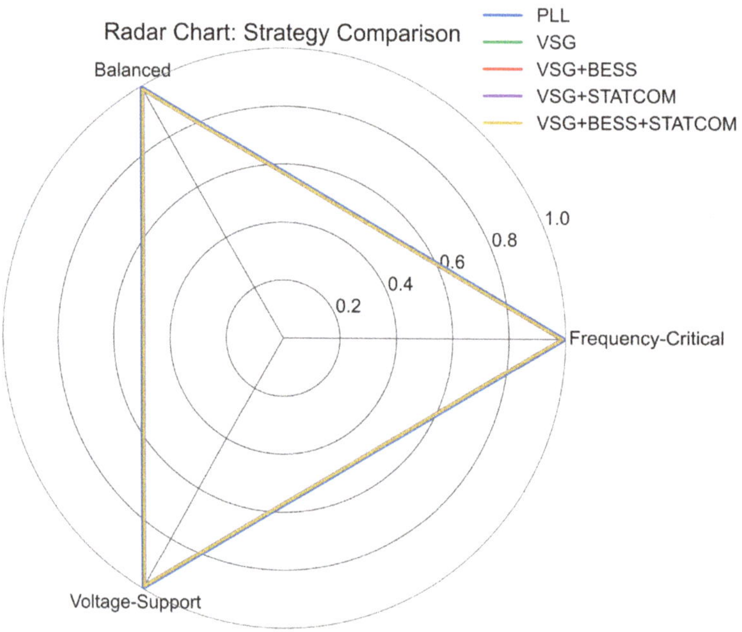

Figure 14. Comparison of control strategies using a radar chart.

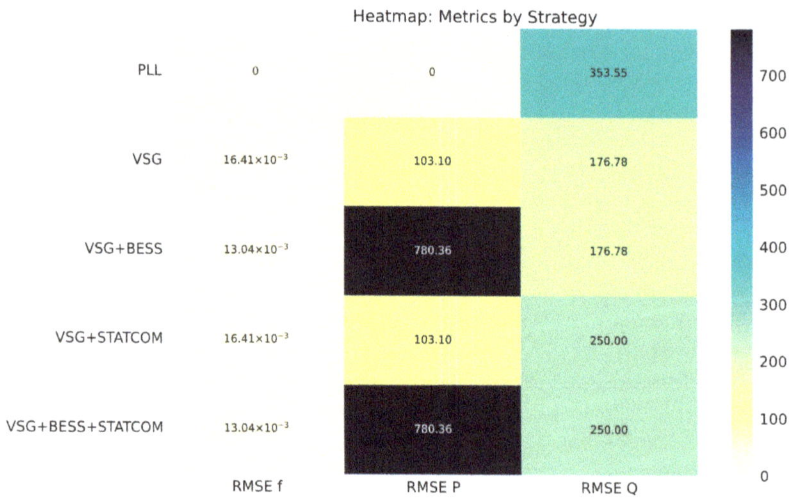

Figure 15. Heatmap of key metrics (RMSE for frequency and active and reactive power) for the different strategies.

The heatmap allows a clear comparison between strategies and highlights that there is no universal solution, and each approach has advantages in a specific operational objective, which makes the integration of combined systems (BESS + STATCOM) particularly promising for resilient smart grids.

Table 2 presents the summary results of the metrics used to compare the five strategies. It summarizes the calculated values for frequency, active and reactive power, and the integrated resilience index.

https://doi.org/10.3390/en19030772

Table 2. Comparison of strategies for controlling PMSG-based systems.

Strategies	RMSE f	Max f	RMSE P	Max P	RMSE Q	Max Q	T_rec	RI
PLL	0	0	0	0	353.5534	500	1	1
VSG	0.016409	0.019899	0.103103	0.125031	176.7767	250	1	0.98433
VSG+BESS	0.013037	0.019317	780.3573	799.96	176.7767	250	0	0.988007
VSG+STATCOM	0.016409	0.019899	0.103103	0.125031	250	500	1	0.98433
VSG+BESS+STATCOM	0.013037	0.019317	780.3573	799.96	250	500	0	0.988007

The analysis shows that the PLL-based control is characterized by sensitivity to transients and significant frequency fluctuations, which makes it insufficiently reliable in networks with a high concentration of PMSGs.

The implementation of VSG control significantly improves stability by emulating inertial response, reducing the amplitude of frequency deviations. The disadvantage remains the slower response compared to classic generators.

The integration of BESS contributes to a significant improvement in the dynamics of active power and reduces oscillations during sudden changes in load. STATCOM, in turn, compensates for reactive powers and maintains voltage stability, but does not significantly affect the frequency dynamics.

The VSG+BESS+STATCOM combination demonstrates the highest stability, achieving simultaneously stable frequency, active power tracking, and reliable reactive compensation.

Figure 16 illustrates the dynamic response of three PMSGs (permanent magnet synchronous generators) in a common network with a variable load and different inertia constants H_i. The model assumes significantly different inertias, H_1 = 0.3 s, H_2 = 0.8 s, and H_3 = 1.5 s, which leads to a distinctly different frequency behavior of the individual generators under dynamic loads.

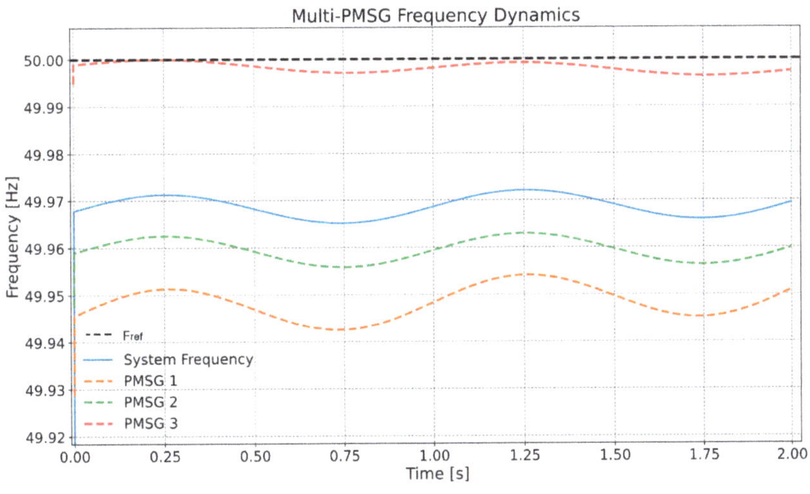

Figure 16. Multi-PMSG system frequency dynamics under variable load and unequal inertia.

In the considered scenario with three PMSG units and unequal inertias, the system exhibits high dynamic stability and maintains synchronism even under a variable load. The lower inertia generators compensate for deviations faster, while those with higher inertia stabilize the system frequency. The average frequency remains stable around 49.97 Hz, indicating good balance and minimal dynamic stresses between the individual generators.

https://doi.org/10.3390/en19030772

To facilitate a clear comparison between the different control strategies, Table 3 summarizes their key characteristics and performance attributes. The evaluation criteria include frequency stability, voltage support capability, implementation complexity, and infrastructure requirements, which together reflect both the technical and practical dimensions of control performance.

Table 3. Comparative summary of the evaluated control strategies.

Control Strategy	Frequency Stability	Voltage Support	Implementation Complexity	Infrastructure Requirements	Main Advantages	Main Limitations
PLL (grid-following)	Low sensitivity to disturbances	Very limited—passive reactive behavior	Low (standard inverter control)	Minimal—no auxiliary devices	Simple and cost-effective	Poor resilience; lacks inertia and voltage control
VSG (grid-forming)	Moderate—improved via synthetic inertia	Moderate—emulates excitation control	Medium (requires inertia and damping tuning)	Standard inverter hardware	Provides virtual inertia and damping	Sensitive to parameter tuning; limited reactive power range
VSG+BESS	High—effective frequency stabilization	Moderate	High (requires battery management and power-sharing control)	Battery energy storage and control interface	Strong transient response; active power balancing	High cost; battery lifetime considerations
VSG+STATCOM	Moderate	High—strong reactive compensation	High (dual control coordination)	STATCOM converter and control link	Enhanced voltage regulation and reactive support	Limited energy buffering; reactive-only response
VSG+BESS +STATCOM	Very high—best composite resilience	Very high—coordinated active/reactive control	Very high (multi-loop coordination and communication)	Hybrid infrastructure (BESS+STATCOM)	Optimal overall performance; balanced dynamic response	Complex implementation; high capital cost; requires coordination and cybersecurity mechanisms

The comparative assessment highlights the trade-offs between simplicity and resilience: the PLL strategy, while straightforward and cost-effective, lacks inertia and reactive power flexibility; the VSG approach introduces synthetic inertia and damping, improving dynamic behavior; the inclusion of BESS and STATCOM modules further enhances stability through coordinated active and reactive power support.

The hybrid VSG+BESS+STATCOM configuration achieves the most balanced overall performance but requires more complex control coordination and investment in supporting infrastructure.

Table 3 provides a qualitative comparison of the evaluated control strategies, PLL, VSG, VSG+BESS, VSG+STATCOM, and VSG+BESS+STATCOM, based on their frequency stability, voltage support, implementation complexity, and infrastructure requirements. This table summarizes the relative advantages and limitations of each approach, as observed in the conducted simulations.

Figure 17 illustrates the distribution of active power among three permanent magnet synchronous generators (PMSGs) operating in parallel under a time-varying load profile. Due to the unequal inertia constants (H_1 = 0.3 s, H_2 = 0.8 s, H_3 = 1.5 s) and different damping coefficients, each generator exhibits a distinct dynamic response. The generator with the smallest inertia (PMSG 1) reacts most rapidly to fluctuations, while PMSG 3, with higher inertia, shows a smoother but slower adaptation. The total load power varies sinusoidally, and the combined generation accurately follows this variation, demonstrating stable power sharing and adequate frequency coupling among units.

Figure 18 presents a comparative summary of the dynamic performance metrics of a three-PMSG system operating under variable load conditions. Each generator (PMSG 1–3) has a different inertia constant, which results in variations in frequency and root mean square error (RMSE) of active power. PMSG 1 shows a higher RMSE in frequency due to its low inertia (H_1 = 0.3 s), while PMSG 3 shows a smoother response with a lower frequency deviation but slightly higher RMSE in active power. The average frequency deviation of the system (System Avg) remains low (≈0.35 Hz), indicating strong frequency stability and balanced load distribution between the units. The results show that inertia diversity improves the damping and resilience of the interconnected PMSG network under dynamic load variations.

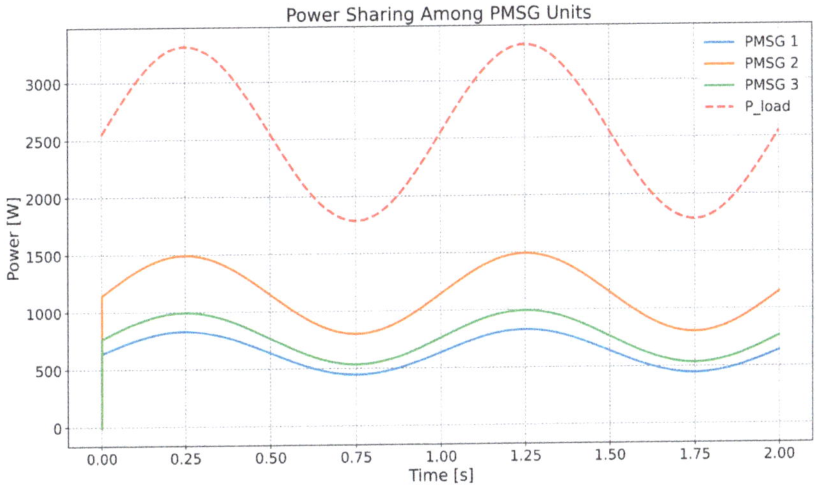

Figure 17. Active power sharing among unequal-inertia PMSG units under a variable load.

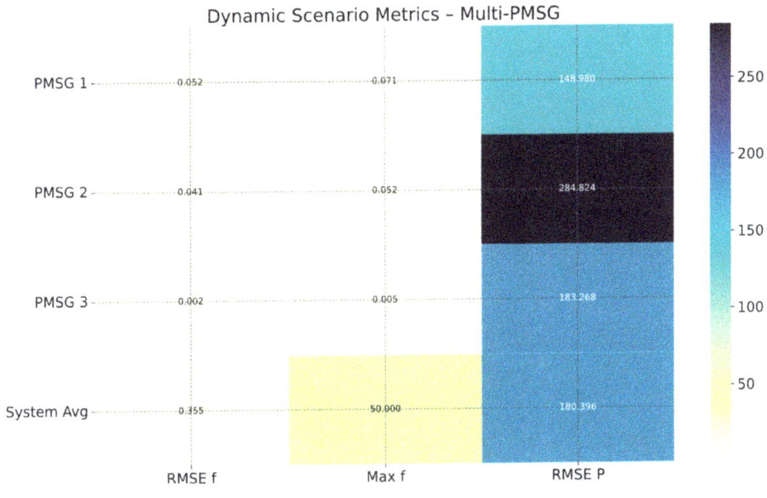

Figure 18. Dynamic scenario metrics—multi-PMSG system.

Figure 19 presents a unified comparison of the steady-state control strategies (PLL, VSG, VSG+BESS, VSG+STATCOM, and VSG+BESS+STATCOM) and the dynamic multi-PMSG system under a varying load. Each row corresponds to a different configuration, and each column shows a specific performance metric, root mean square deviation (RMSE), maximum frequency deviation, root mean square deviation of active and reactive power (RMSE), and overall RI. The results show that the VSG-based configurations with auxiliary devices (BESS, STATCOM) achieve lower frequency deviations and higher RI compared to the basic PLL system. Meanwhile, the multi-PMSG system demonstrates a comparable performance in RMSE and RI, confirming that inertial diversity and decentralized coupling

https://doi.org/10.3390/en19030772

provide similar stability benefits as hybrid VSG architectures. Figure 19 highlights the strong correspondence between the VSG's synthetic inertia-based control and the natural inertia from multiple synchronous generators, both of which contribute to improved grid resilience and frequency stability.

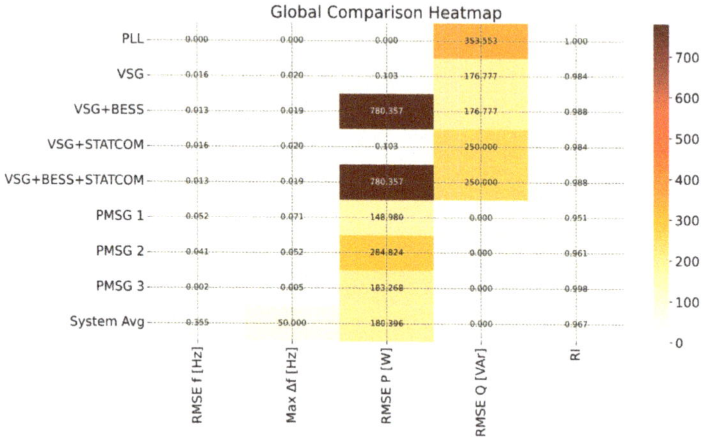

Figure 19. Global comparison of strategies and multi-PMSG scenario.

Figure 20 presents a comparison of the overall RI for all the studied control strategies and the dynamic scenario with multiple PMSGs. RI represents the ability of the system to maintain stable operation under frequency deviations and active power disturbances. As expected, the PLL-based system without inertia emulation achieves the lowest RI values under dynamic loading, while the VSG-based architectures show improved robustness, especially when supported by BESS and STATCOM devices. The multiple PMSG system, which relies on natural mechanical inertia and inter-generator coupling, achieves RI values comparable to or even higher than those of the hybrid VSG systems, confirming the robustness of the inertia stabilization mechanisms. These results indicate that both the synthetic VSG and natural PMSG inertia effectively contribute to improved system robustness under variable loading and grid disturbances.

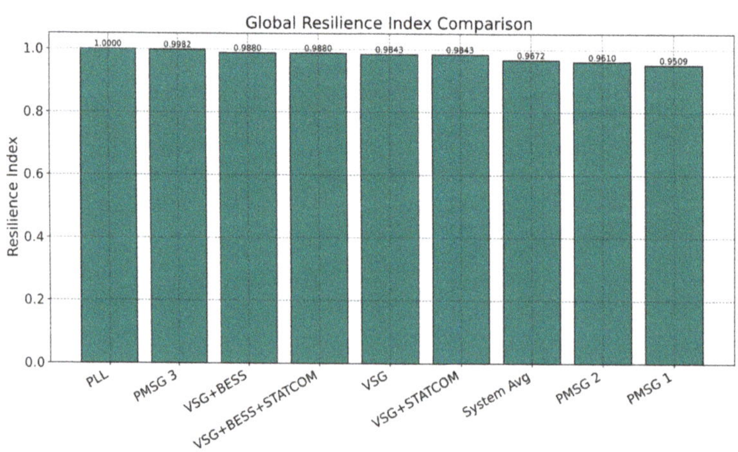

Figure 20. Global RI comparison.

https://doi.org/10.3390/en19030772

Figure 21 illustrates the impact of cyber-physical disturbances on the frequency stability of a VSG-controlled system.

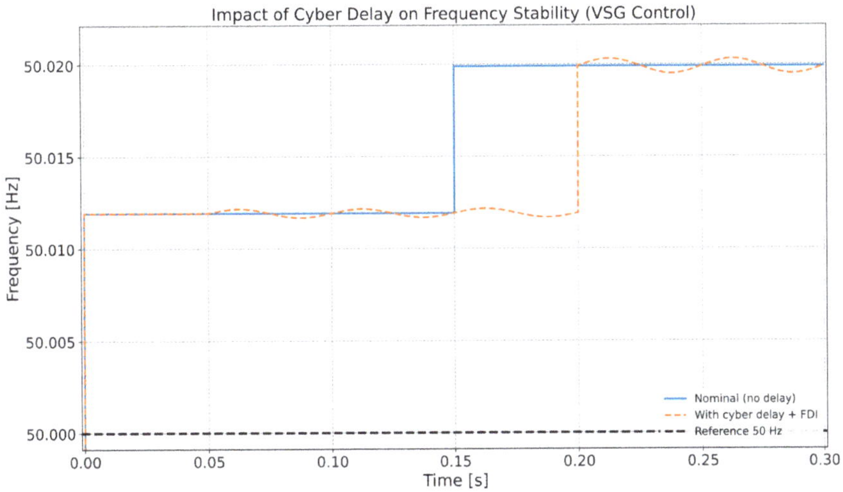

Figure 21. Impact of cyber delay on frequency stability (VSG control).

The simulation compares the nominal case (solid line) with a scenario affected by a 50 ms communication delay and a spurious data injection (FDI) disturbance in the frequency reference channel. The delayed case shows small but measurable deviations in the frequency recovery after the load step at t = 0.15 s, resulting in a decrease in RI from 0.9843 to 0.9821.

This degradation corresponds to an approximately 0.22% decrease in robustness, mainly due to the latency in transmitting feedback from the control and the presence of spurious oscillatory data injected into the control loop. Despite this disturbance, the VSG remains stable and converges to the nominal 50 Hz, demonstrating its inherent robustness to moderate cyber interference.

The results confirm that even short-term cyber delays can disrupt transient synchronization, especially in systems dominated by inertia-limited converters. These findings highlight the importance of autonomous local buffering mechanisms (such as BESS modules) and redundant monitoring to maintain frequency stability under cyber-physical uncertainty.

The scenario assumes a short-term communication delay or false data injection (FDI) affecting the VSG control frequency reference channel. Mathematically, this was represented by introducing a delayed signal: $f_{ref}(t) = f_{\text{nom}} + \Delta f(t-\tau)$, where τ represents the cyber delay (typically 50–100 ms). The simulated effect of this delay showed a small but measurable degradation of the resilience index RI $\approx$ 0.982 compared to the nominal 0.984, which corresponds to slower synchronization and transient recovery.

To assess the robustness of the proposed control strategies, a sensitivity analysis was performed focusing on the most influential control parameters, namely, the virtual inertia constant H and the damping factor D in the VSG controller. These parameters directly affect the frequency response of the transient process and the overall stability of the system.

The analysis was performed for the representative VSG+BESS+STATCOM configuration under the same disturbance scenario described in Section 6 (load step at t = 0.1 s).

The three sets of parameters that were tested are as follows:

- Low inertia/damping: H = 0.2, D = 0.5;

https://doi.org/10.3390/en19030772

- Nominal values: H = 0.5, D = 1.0;
- High inertia/damping: H = 1.0, D = 2.0.

The obtained frequency deviation and robustness indicators (frequency RMSE, maximum deviation, and recovery index) are summarized in Table 4.

Table 4. Sensitivity of system performance to variations in VSG control parameters.

Case	H [s]	D [–]	RMSE f [Hz]	Max Δf [Hz]	RI [–]	Observed Behavior
Low inertia/damping	0.2	0.5	0.0248	0.038	0.973	Fast but oscillatory response, weak damping
Nominal values	0.5	1.0	0.0164	0.020	0.984	Balanced transient and stable recovery
High inertia/damping	1.0	2.0	0.0113	0.014	0.989	Slow response, minimal overshoot, improved stability

The results show that increasing the virtual inertia (H) and damping (D) coefficients leads to improved frequency stability and higher RI but at the expense of slower dynamic response. Conversely, low values of these parameters lead to faster but oscillatory behavior and larger deviations.

This confirms the importance of coordinated tuning of VSG parameters to achieve an optimal balance between dynamic performance and steady-state stability. The nominal configuration (H = 0.5, D = 1.0) achieves the best compromise.

For future studies, this sensitivity framework will be extended to include BESS controller gain factors (K_p, K_i) and STATCOM voltage regulation factors to capture the cross-talk between active and reactive power dynamics.

7. Discussion

The simulation results (Figures 3–14) clearly show that the choice of control strategy is of critical importance for the stability of networks dominated by PMSGs. The conventional PLL approach demonstrates adequate synchronization under stable conditions, but it is sensitive to sudden changes in load and frequency. The virtual synchronous generator (VSG) significantly improves the frequency behavior by adding artificial inertia to the system. The addition of battery storage systems (VSG+BESS) shows the best results in terms of frequency stabilization, while the integration of STATCOM modules (VSG+STATCOM) provides effective reactive power control. The most balanced results are achieved with the combined VSG+BESS+STATCOM strategy, which combines stability, adaptability, and a good transient response. Despite the achieved results, the developed model has several limitations. First, the analysis was conducted within a simplified network architecture with a single generator and fixed load values. Real power systems are significantly more complex and include diverse operating modes, interconnected microgrids, and unpredictable demand. Second, the model mainly considers active and reactive power, frequency, and phase, without including harmonics, non-sinusoidal modes, and transformer or cable line dynamics. Third, cyber resilience is a critical factor in smart grids and is only conceptually discussed but not included in the simulations.

The results of the analysis have direct implications for the design and management of future smart grids. They show that purely conventional approaches (PLL) will become increasingly inadequate with a high share of PMSGs. The implementation of VSG technologies is necessary to maintain frequency resilience, and the additional integration of BESS and STATCOM devices is highly recommended for balanced power management. At the policy level, this suggests the need for incentives for the implementation of hybrid control strategies and the creation of standards for coordination between different components and regulatory frameworks that support the widespread application of BESS and STATCOM in combination with renewable generators. Furthermore, future policies should encourage investments in cybersecurity, adaptive control systems, and predictive control through artificial intelligence to ensure long-term grid resilience.

https://doi.org/10.3390/en19030772

The results of the multi-PMSG simulation complement the previous analyses and provide a broader understanding of the network behavior under variable load and multi-generator configurations. The dynamic multi-PMSG scenario reveals the importance of inertial diversity and coupling effects between multiple synchronous machines. Unlike the single-source VSG simulations, the multi-PMSG configuration captures the natural inertia and mechanical synchronization inherent in physical generators. The observed results confirm that systems with non-uniform inertia constants can achieve self-balancing behavior without centralized control, demonstrating strong dynamic robustness and frequency stability. Figure 16 illustrates that, despite the different inertia constants (H_1 = 0.3 s, H_2 = 0.8 s, H_3 = 1.5 s), the average system frequency remains within ±0.05 Hz of the nominal value, indicating effective damping through inter-machine coupling. At the same time, the power distribution between the generators (Figure 17) remains well coordinated, indicating that natural inertia can play a stabilizing role similar to that of virtual inertia emulation in VSG. The following heatmap in Figure 18 quantifies these dynamics, showing low RMSE values for both frequency and power, especially for machines with higher inertia. The RI obtained for each generator presented in Figure 19 confirms this trend, with units with higher inertia showing smoother dynamics and higher RI values.

As summarized in Table 3, each strategy presents a different trade-off between stability performance and implementation complexity. While the PLL remains the simplest and most cost-effective solution, it lacks the frequency and voltage resilience required in converter-dominated grids. The hybrid VSG+BESS+STATCOM configuration offers the best dynamic behavior and control flexibility but at the expense of increased system complexity and infrastructure cost. This comparison underlines the need for integrated hybrid approaches supported by proper coordination standards and cyber-physical protection mechanisms.

The integration of these results into a global comparison framework presented in Figure 20 highlights an important finding that the natural inertia of the multi-PMSG system provides levels of resilience comparable to advanced hybrid VSG architectures VSG+BESS+STATCOM. Therefore, these results suggest that future smart grids should combine both natural and virtual inertia sources to achieve optimal stability and adaptability.

The findings show that the inclusion of distributed inertial resources, synthetic or natural, is crucial for the frequency resilience of low-inertia power systems. The multi-PMSG simulation demonstrates how self-organizing synchronization can be achieved without centralized control, indicating that decentralized resilience mechanisms can play a key role in next-generation grid architectures.

Although the robustness assessment in this study is primarily based on instantaneous electrical parameters such as frequency deviation, active and reactive power error, and the resulting RI, this methodological choice is intentional. These parameters directly reflect the fundamental dynamic behavior of inverter-based and PMSG-dominated systems, where stability and robustness are primarily determined by fast electromechanical interactions and control response times. Furthermore, these parameters are the most sensitive to disturbances, providing a reliable indicator of the system's robustness to transients.

However, it is recognized that robustness encompasses broader dimensions, including energy recovery capability, voltage recovery, power quality (THD), and long-term operational stability. Future extensions of this study will expand with additional resilience metrics, such as energy-based metrics, kinetic energy deviation and energy dissipation factor, time-domain metrics, such as settling time and overshoot ratio, power quality indices, including total harmonic distortion (THD) and current imbalance, cyber-physical performance metrics assessing controller latency, data integrity, and response resilience to communication delays. The integration of these complementary measures will further enhance the validity and multidimensional nature of resilience assessment, allowing for a

https://doi.org/10.3390/en19030772

comprehensive assessment of smart grid performance that includes not only instantaneous electrical stability but also long-term operational, structural, and digital resilience.

The combined VSG+BESS+STATCOM configuration demonstrates a synergistic improvement in grid stability, which stems from the complementary nature of their control mechanisms. The VSG emulates the electromechanical dynamics of a synchronous machine by introducing virtual inertia and damping through its control loop. This emulation smoothens frequency variations and provides a synthetic inertial response during transients. The BESS, integrated into the active control loop, acts as a fast-reacting energy buffer that compensates for active power imbalances caused by generation or load fluctuations. This contribution reduces the net power imbalance experienced by the VSG, allowing it to maintain a more stable virtual rotor speed.

Meanwhile, the STATCOM provides independent reactive power control, improving voltage stability and mitigating the coupling between active and reactive dynamics. The split operation of the STATCOM ensures that voltage deviations do not propagate as frequency disturbances, effectively isolating the VSG+BESS frequency control from transient voltages. As a result, the combined system exhibits reduced overshoot, faster settling time, and improved damping ratios compared to either of the individual controllers operating alone.

However, this synergy is not unconditional. Several operating conditions can weaken or even nullify the combined benefits. Insufficient communication bandwidth or delays between the BESS and VSG control loops can cause power sharing mismatches, leading to oscillations rather than damping. Overly aggressive reactive control gains of the STATCOM can introduce voltage and frequency cross-coupling, destabilizing the VSG internal control loop. The limited state of charge (SoC) in a BESS can reduce its ability to supply or absorb transient active power, thereby reducing the effective inertia support. In networks with multiple grid-shaping devices, improper adjustment of the droop or inertia between VSGs can cause synchronization competition, reducing system coherence.

To ensure a positive interaction, the control hierarchy must be properly coordinated. The VSG must act as the primary frequency reference (grid-shaping), the BESS must provide secondary active power through fast internal current loops, and the STATCOM must maintain the voltage within tight limits using a slower external reactive control loop.

When properly tuned, this multi-layered architecture ensures that each component improves a separate aspect of grid resilience, the VSG for inertia, the BESS for energy balancing, and the STATCOM for voltage stability, leading to the observed synergistic improvement in overall RI and transient performance.

Although the combined VSG+BESS+STATCOM control framework demonstrates clear advantages in terms of robustness, frequency stability, and reactive power support, several practical limitations and trade-offs must be acknowledged:

- System complexity and coordination: The hybrid control structure increases the number of interacting subsystems and control loops. Proper coordination between VSG, BESS, and STATCOM controllers requires the precise tuning of gain factors and communication latency management. In large-scale or distributed systems, ensuring real-time synchronization between devices may require advanced control layers and communication standards such as IEC 61850 [71], which adds both cost and technical overhead.
- Implementation cost and scalability: The integration of multiple hardware modules, especially high-capacity BESS and STATCOM converters, significantly increases the initial investment and maintenance costs. Although this hybrid configuration offers superior performance, its economic feasibility depends on energy market incentives, regulatory services, and battery life economics. The cost–performance trade-off may

https://doi.org/10.3390/en19030772

favor simplified solutions such as VSG+BESS in smaller microgrids or systems with moderate volatility.
- Maintenance and reliability issues: Battery systems introduce degradation effects and limited cycle life, which may reduce the long-term availability of the hybrid control scheme. STATCOM devices, on the other hand, require frequent calibration and harmonic compensation under variable grid conditions. This increases the maintenance burden compared to the conventional PLL-based approaches.
- Model simplification: The present study is based on a reduced-order model that focuses on the dynamics of active and reactive power, neglecting the effects of higher-order converters, harmonics, and communication delays. Although this abstraction allows a clear view of the control behavior, future work should extend the framework to include detailed converter models and real-time simulations using hardware-in-the-loop (HIL) validation.
- Control trade-offs: Adding inertia via VSG improves frequency stability but slows down transient response, and conversely, high BESS control gains improve dynamic recovery but can introduce oscillations if not properly damped. Therefore, parameter optimization must balance stability, response speed, and energy efficiency, an area that deserves further exploration using multi-objective optimization and adaptive control methods.

In summary, while the VSG+BESS+STATCOM hybrid system offers a promising path to resilient smart grids, its implementation in the real world requires the careful consideration of cost, reliability, and integration complexity. Future research should therefore focus not only on improving management productivity but also on ensuring scalability and long-term economic viability.

Cybersecurity is an integral part of the overall resilience of smart grids, especially in networks dominated by converter-connected generators, such as PMSG-based devices. In such systems, the control loops of VSG, BESS, and STATCOM devices rely heavily on digital communication, real-time measurement data, and control signals, making them vulnerable to data manipulation, communication delays, or Denial of Service (DoS) attacks.

In order to establish a preliminary connection between the cyber layer and the physical behavior of the system, an additional cyber disturbance scenario was conceptually modeled within the proposed simulation framework.

Furthermore, the analysis shows that the VSG+BESS+STATCOM hybrid configuration inherently improves cyber-physical resilience. The BESS controller acts as a local buffer, stabilizing the power flow during temporary communication losses, while the STATCOM module maintains voltage stability even in the event of partial information loss. This distributed autonomy reduces the overall system dependency on the communication layer and mitigates the propagation of cyber interference.

The cyber-physical simulation presented in Figure 21 demonstrates the impact of a 50 ms communication delay and spurious data injection on the frequency response of the VSG. The resilience index decreased by 0.22%, confirming the sensitivity of converter-based systems to cyber interference and highlighting the importance of decentralized and buffered control architectures.

Future work will include the explicit simulations of cyberattacks (e.g., FDI, spoofing, DoS) and will use co-simulation environments such as MATLAB–OMNeT++ or Python–NS3 to model the communication dynamics and their direct impact on control performance. This will allow for a quantitative assessment of the cyber-physical link and the formulation of strategies for adaptive detection and mitigation of cyber threats.

The presented results, although based on systems with one and three PMSG units, can be extended conceptually to larger distributed microgrids. The underlying principles of

https://doi.org/10.3390/en19030772

virtual inertia emulation, active/reactive power coordination, and hybrid control synergy (VSG+BESS+STATCOM) remain applicable as the number of distributed units increases.

In larger networks, the coupling dynamics between converters and the communication delays in control coordination become more significant. These factors may lead to frequency-sharing interactions and local voltage deviations, which require secondary and tertiary control layers for proper synchronization.

Nevertheless, previous studies and preliminary simulations confirm that the VSG-based control paradigm maintains its stabilizing effect even in multi-node microgrids, provided that inertia and damping parameters are properly tuned across units.

Therefore, the results reported here are scalable in principle but should be interpreted as a representative of local cluster dynamics rather than entire transmission-level systems. Future work will extend the model toward multi-agent coordination and distributed optimization frameworks to capture these scalability aspects quantitatively.

While the simulation results demonstrate clear performance advantages for the hybrid VSG+BESS+STATCOM configuration, the practical implementation of such an architecture entails significant additional investment and integration complexity compared to simpler control schemes.

The PLL-based controller remains the most cost-effective and straightforward option, requiring minimal additional hardware but offering limited dynamic resilience.

The pure VSG approach introduces moderate control complexity, mainly related to parameter tuning and converter firmware updates, but it does not require substantial hardware expansion.

The inclusion of BESS adds both capital cost (due to battery modules and bidirectional converters) and operational constraints such as state-of-charge management and lifetime degradation.

Integrating STATCOM units further increases system cost and control coupling, as it necessitates separate reactive compensation converters and communication links for coordination.

Consequently, the VSG+BESS+STATCOM system achieves the best technical resilience but at the expense of higher infrastructure costs, multi-layer control synchronization, and maintenance requirements.

From a system planning perspective, this trade-off highlights that the optimal solution depends on the application scale and criticality; for small-scale distributed systems, a standalone VSG or VSG+BESS may be sufficient, while for critical smart grid nodes or weak grid interfaces, the full hybrid scheme is justified despite its higher cost and integration effort.

8. Conclusions

This study comprehensively analyzes both single-source and multi-source control strategies to improve the resilience of PMSG-dominated smart grids. The results clearly show that, while the PLL-based control remains functional under steady-state conditions, it is not sufficiently robust in the presence of dynamic load variations. VSGs significantly improve system response by emulating inertia and damping, thereby improving frequency stability. The integration of BESS and STATCOM further strengthens the system by balancing active and reactive power, respectively. Among them, the combined VSG+BESS+STATCOM strategy achieves the best overall performance, maintaining minimal deviations in both frequency and power.

The addition of the multi-PMSG dynamic scenario extends this analysis beyond synthetic inertia to naturally inertial systems. The results show that interconnected PMSG devices with non-uniform inertia constants can maintain frequency synchronization and stability under variable load conditions, even without centralized coordination. The

https://doi.org/10.3390/en19030772

calculated RI values confirm that such systems can achieve levels of resilience comparable to or higher than those of VSG-based control structures. This bridges the conceptual gap between virtual and physical inertia, highlighting their complementary roles in modern power systems. In conclusion, achieving the resilient and reliable operation of PMSG-dominated smart grids requires a multi-layered hybrid approach that combines both virtual-based control and natural (mechanical) inertia sources. This study shows that distributed synchronization between PMSG devices can provide inherent resilience, while VSG-based systems offer controllability and adaptability. These results highlight the importance of integrated strategies that combine synthetic inertia, energy storage, and flexible compensation, ensuring that future smart grids remain robust, adaptive, and resilient under all operating conditions.

In addition to the physical layer simulations, this study also highlights the growing importance of cyber-physical resilience in converter-dominated smart grids. The analysis shows that short communication delays and FDI disturbances can slightly degrade the frequency stability of VSG-based systems, reducing the resilience index by approximately 0.2%. However, the hybrid VSG+BESS+STATCOM configuration maintained stable operation, indicating that distributed buffering and local autonomy can mitigate the adverse effects of cyber-induced disturbances.

These results highlight that ensuring resilience in next-generation smart grids requires an integrated approach that combines virtual inertia, energy storage, reactive compensation, and cybersecurity-aware control architectures. Therefore, future research should extend the presented framework to include the real-time joint simulation of communication networks and adaptive anomaly detection to comprehensively assess both the physical and cyber resilience of smart energy systems. The simulated cybersecurity disturbances correspond to moderate and realistic threat conditions as reported in the literature on cyber-physical energy systems.

The communication delay scenario represents a latency of 50–150 ms, which falls within the typical range of medium network congestion or supervisory control lag in distributed grid environments.

Such delays do not fully disconnect the control loop but may cause transient desynchronization or reduced damping, consistent with practical field observations.

The FDI scenario was modeled as a 2–5% deviation in transmitted active or reactive power reference values. This range is consistent with realistic attack magnitudes documented in recent studies, which demonstrate that small but persistent falsifications can gradually deteriorate control performance without triggering immediate detection.

Therefore, both modeled cases correspond to moderate-level attacks, sufficiently strong to expose the system's cyber-physical vulnerabilities, yet not catastrophic. Their purpose is to evaluate the resilience margin of hybrid VSG+BESS+STATCOM systems under credible network-level disturbances rather than extreme, system-wide cyber failures.

The findings confirm that, while hybrid control architectures enhance operational resilience, they also increase the system's dependency on secure communication channels, emphasizing the need for synchronized cybersecurity and control co-design.

Although the presented framework demonstrates promising improvements in stability and resilience, its practical implementation is subject to economic, regulatory, and technological constraints. The deployment of hybrid control architectures combining the VSG, BESS, and STATCOM requires significant investments in converter technology, advanced control hardware, and communication infrastructure. Furthermore, regulatory frameworks for grid-forming inverters and hybrid control coordination are still under development in many power systems, limiting their large-scale deployment.

https://doi.org/10.3390/en19030772

Therefore, the conclusions of this study should be interpreted within the framework of conceptual validation and simulation-based evaluation, and not as immediate recommendations for deployment. Future research should integrate techno-economic analysis, cost–benefit assessment, and regulatory feasibility studies to ensure that the proposed resilience-enhancing strategies can be realistically adopted in both centralized and decentralized grid contexts.

This balanced approach bridges the gap between theoretical potential and real-world applicability, ensuring that the proposed hybrid resilience strategies effectively contribute to the sustainable transformation of future energy networks.

The present study is based on a limited-scale comparative model including single- and multi-generator scenarios with permanent magnets.

Although the results clearly demonstrate the advantages of the combined VSG+BESS+STATCOM strategy, they should be considered as a proof of concept and not as a universal recommendation for all smart grid configurations. Real-world systems include multiple inverter sources, stochastic loading, and nonlinear interactions that are not fully accounted for in the present model.

The results presented in this paper correspond to simulation-level validation performed using a unified analytical and numerical framework. The obtained findings demonstrate the potential of the proposed hybrid control strategies for improving the resilience of PMSG-based smart grids but have not yet been experimentally verified.

As a next step, the proposed models and algorithms will be implemented and tested on a laboratory-scale test bench including programmable inverters, real-time control hardware, and a network emulator.

This experimental phase will allow for the assessment of unmodeled effects such as measurement delays, converter nonlinearities, and communication disturbances, thereby bridging the gap between simulation-based insights and real-world performance validation.

Such an extension will provide the necessary foundation for scaling the proposed control concepts toward pilot microgrid applications and industrial validation scenarios.

Future studies will extend the analysis through simulations with varying demand, random disturbances, and integrated cyber-physical interaction scenarios to validate the robustness of the proposed strategies in a more realistic context.

Author Contributions: P.S. and N.H. were involved in the full process of producing this paper, including conceptualization, methodology, modeling, validation, visualization, and preparing the manuscript. All authors have read and agreed to the published version of the manuscript.

Funding: This work was supported by the European Regional Development Fund under the "Research In-novation and Digitization for Smart Transformation" program 2021–2027 under the Project BG16RFPR002-1.014-0006 "National Centre of Excellence Mechatronics and Clean Technologies", and the APC was funded by the Project BG16RFPR002-1.014-0006.

Data Availability Statement: The original contributions presented in this study are included in the article. Further inquiries can be directed to the corresponding author.

Acknowledgments: The present research has been carried out under the project BG16RFPR002-1.014-0006 "National Centre of Excellence Mechatronics and Clean Technologies", funded by the Operational Programme Science and Education for Smart Growth.

Conflicts of Interest: The authors declare no conflicts of interest.

References

1. Prasad, R.D.; Chand, D.A.; Lata, S.S.S.L.; Kumar, R.S. Beyond Energy Access: How Renewable Energy Fosters Resilience in Island Communities. *Resources* **2025**, *14*, 20. [CrossRef]
2. Zhao, Y. A review of renewable energy and power system integration. *Appl. Comput. Eng.* **2025**, *126*, 10–15. [CrossRef]

https://doi.org/10.3390/en19030772

3. Khalid, M. Smart grids and renewable energy systems: Perspectives and grid integration challenges. *Energy Strategy Rev.* **2024**, *51*, 101299. [CrossRef]
4. Yang, L.; Wu, X.; Huang, B.; Li, Z. Sustainable Industrial Energy Supply Systems with Integrated Renewable Energy, CCUS, and Energy Storage: A Comprehensive Evaluation. *Sustainability* **2025**, *17*, 712. [CrossRef]
5. Singh, B.; Hammouch, H.; Chandra, S. Innovative Wind Energy Solutions for Smart-Sustainable Communities: Vision for Intelligent Energy Management and Climate Resilience in Industry 5.0. In *AI Technologies for Enhancing Recycling Processes*; IGI Global Scientific Publishing: Hershey, PA, USA, 2025; pp. 491–508. [CrossRef]
6. Vassunova, Y.Y. Smart grids optimization with renewable energy sources integration. *Ekon. I Upr. Probl. Resheniya* **2024**, *9/6*, 146–156. [CrossRef]
7. Biswas, P.; Rashid, A.; Al Masum, A.; Al Nasim, A.; Ferdous, A.A.; Gupta, K.D.; Biswas, A. An Extensive and Methodical Review of Smart Grids for Sustainable Energy Management-Addressing Challenges with AI, Renewable Energy Integration and Leading-edge Technologies. *arXiv* **2025**. [CrossRef]
8. Nzeanorue, C.C.; Okpala, B.C. Smart grids and renewable energy integration: Challenges and solutions. *Path Sci.* **2024**, *10*, 3050–3060. [CrossRef]
9. Sasi Bhushan, M.A.; Sudhakaran, M.; Dasarathan, S.; E, M. Integration of a Heterogeneous Battery Energy Storage System into the Puducherry Smart Grid with Time-Varying Loads. *Energies* **2025**, *18*, 428. [CrossRef]
10. Nourelhouda, D.; Daili, Y.; Harrag, A. A Review of Recent Control Trchniques of Virtual Synchronous Machine for Renewable Energy. *Sci. Bull. Electr. Eng. Fac.* **2023**, *23*, 20–32. [CrossRef]
11. Tur, M.N.; Ertuğrul, Ö.F.; Tür, M.R. Solution for Integration of Renewable Energy Power Plants into Smart Grids with Active Power Control. *Journal of Science. Technol. Eng. Res.* **2024**, *5*, 11–23. [CrossRef]
12. Pareek, S.; Kumar, Y.; Kaur, H.; Kumar, R.; Chohan, J.S. A Comparative Study of Power Electronics and Control Techniques for Renewable Energy Integration in Smart Grids. In Proceedings of the 2023 3rd International Conference on Pervasive Computing and Social Networking (ICPCSN), Salem, India, 19–20 June 2023; pp. 1629–1633. [CrossRef]
13. Eslahi, M.S.; Vaez-Zadeh, S.; Rodriguez, J. Resiliency Enhancement and Power Quality Optimization of Converter-Based Renewable Energy Microgrids. *IEEE Trans. Power Electron.* **2023**, *38*, 7785–7795. [CrossRef]
14. M, S.; J, M. Challenges in Integration of RES and control techniques in microgrid: A review. *Int. J. Innov. Sci. Res. Technol. (IJISRT)* **2024**, *9*, 1716–1723. [CrossRef]
15. Carlak, H.F.; Kayar, E. Volt/VAR regulation of the West Mediterranean regional electrical grids using SVC/STATCOM devices with neural network algorithms. *Wind Energy* **2025**, *28*, e2976. [CrossRef]
16. Ahmed, D.S.; Marhoon, A.F. STATCOM Controller design for hybrid PV-Wind of AC microgrid. *Al-Iraqia J. Sci. Eng. Res.* **2024**, *3*, 215–233. [CrossRef]
17. Kushwaha, N.; Kaur, S. A Review of the STATCOM Device for Improving Wind Farm Stability. *i-Manager's J. Electr. Eng.* **2023**, *16*, 41–48. [CrossRef]
18. Todorov, G.D.; Kamberov, K.H.; Zlatev, B.N. Research and Development of a Large-Scale Axial-Flux Generator for Hydrokinetic Power System. *Appl. Sci.* **2024**, *14*, 10564. [CrossRef]
19. Alfraidi, W.; Alaql, F.; AlMuhanna, K.; Farh, H.M.H.; Al-Shamma'a, A.A. Designing EV charging energy hubs to meet flexibility requirements in smart grids. *World Electr. Veh. J.* **2025**, *16*, 43. [CrossRef]
20. Todorov, G.; Kralov, I.; Koprev, I.; Vasilev, H.; Naydenova, I. Coal Share Reduction Options for Power Generation during the Energy Transition: A Bulgarian Perspective. *Energies* **2024**, *17*, 929. [CrossRef]
21. Rajendran, S.; Muthukumar, A.; Vijayakumar, K.; Rajesh, K. VSC-STATCOM Performance Under Different Fault Sensing using PSO Tuned Hybrid SMC. *Int. J. Electr. Electron. Res.* **2024**, *12*, 478–486. [CrossRef]
22. Khaleel, M. Power quality enhancement using renewable energy sources and electric mobility. *Brill. Res. Artif. Intell.* **2023**, *3*, 306–315. [CrossRef]
23. Kanchana, K.; Murali Krishna, T.; Yuvaraj, T.; Sudhakar Babu, T. Enhancing Smart Microgrid Resilience Under Natural Disaster Conditions: Virtual Power Plant Allocation Using the Jellyfish Search Algorithm. *Sustainability* **2025**, *17*, 1043. [CrossRef]
24. Sarkar, K. Load and renewable energy forecasting using deep learning for grid stability. *arXiv* **2025**. [CrossRef]
25. Ishaq, Y.; Prince, A.S.; Claude, G.G.J.; Di Bebe, T.M. Decentralized framework for securing smart grids using blockchain and machine learning. *Int. J. Res. Appl. Sci. Eng. Technol.* **2025**, *13*, 679–685. [CrossRef]
26. Mohammed, S.H.; Singh, M.S.J.; Al-Jumaily, A.; Islam, M.T.; Islam, M.S.; Alenezi, A.M.; Soliman, M.S. Dual-hybrid intrusion detection system to detect False Data Injection in smart grids. *PLoS ONE* **2025**, *20*, e0316536. [CrossRef] [PubMed]
27. Pant, P. AI for Renewable Energy Grid Management and Storage. In *Cases on AI-Driven Solutions to Environmental Challenges*; IGI Global Scientific Publishing: Hershey, PA, USA, 2025; pp. 319–354. [CrossRef]
28. Li, Y. AI-Enhanced digital twins for energy efficiency and carbon footprint reduction in smart city infrastructure. *Appl. Comput. Eng.* **2025**, *118*, 42–47. [CrossRef]

https://doi.org/10.3390/en19030772

29. Hafezimagham, A.; Baghernezhad, A.; Tayebi, N.; Ghanbari-Mobarakeh, P.; Gharehpetian, G.B.; Abedi, M. Comprehensive Review of Energy Storage Systems for Smart Grids: Technologies and Applications. In Proceedings of the 2024 9th International Conference on Technology and Energy Management (ICTEM), Behshar, Iran, 14–15 February 2024; pp. 1–8. [CrossRef]
30. Patel, H.R. Enhancing Power Quality Through the Integration of STATCOM with Renewable Energy Sources. *Int. J. Sci. Res. Eng. Manag.* **2025**, *9*, 1–9. [CrossRef]
31. Viatkin, A.; Chou, S.; Augustin, T.; Khan, A.Z.; Tayyebi, A.; Bai, H.; Svensson, J.R. Hybrid Energy Storage Enhanced STATCOMs. In Proceedings of the 2024 IEEE 15th International Symposium on Power Electronics for Distributed Generation Systems (PEDG), Luxembourg, 23–26 June 2024; pp. 1–5. [CrossRef]
32. Zhang, R.; Hua, Z.; Dong, X.; Lei, S. Grid-forming STATCOM With Energy Storage to Alleviate Temporary Overvoltage. In Proceedings of the 2023 IEEE PES 15th Asia-Pacific Power and Energy Engineering Conference (APPEEC), Chiang Mai, Thailand, 6–9 December 2023; pp. 1–6. [CrossRef]
33. Chandarhasn, C.; Percis, E.S. Optimizing resilience in large-scale integration of renewable energy sources: Exploring the role of STATCOM device. *Int. J. Power Electron. Drive Syst. Int. J. Electr. Comput. Eng.* **2024**, *15*, 1468. [CrossRef]
34. Abdulsattar, S.S.; Tan, C.W.; Ayob, S.B.; Dahiru, A.T.; Lau, K.Y.; Toh, C.L. A Review of Smart Microgrid Architecture and Topologies. In Proceedings of the 2024 IEEE International Conference on Power and Energy (PECon), Kuala Lumpur, Malaysia, 4–5 November 2024; pp. 104–109. [CrossRef]
35. Roberts, O. A Qualitative Exploration on Artificial Intelligence and Renewable Energy Integration in Supply Chains. *Bus. Manag* **2025**. [CrossRef]
36. Budin, L.; Delimar, M. Renewable energy community sizing based on stochastic optimization and unsupervised clustering. *Sustainability* **2025**, *17*, 600. [CrossRef]
37. Jothi, T.; Arun, M.; Varadarajan, M. Enhancing power quality in a smart grid using dynamic voltage restorer. *Int. J. Appl. Power Eng. (IJAPE)* **2024**, *13*, 661. [CrossRef]
38. Sugumar, B.K.; Anglani, N. A Novel Decision-Support Framework for Supporting Renewable Energy Technology Siting in the Early Design Stage of Microgrids: Considering Geographical Conditions and Focusing on Resilience and SDGs. *Energies* **2025**, *18*, 544. [CrossRef]
39. Prasad, M.B.S.; Chennaiah, P.B. Power Management and Control for PV integrated Microgrid with Battery Energy Storage System. In Proceedings of the 2024 5th International Conference on Electronics and Sustainable Communication Systems (ICESC), Coimbatore, India, 7–9 August 2024; pp. 1–6. [CrossRef]
40. Chau, L.X.; Thanh, T.V.; Quan, D.M.; Hung, L.K.; Hieu, T.T. Using Statcom Devices to Control and Maintain Voltage Stability in Wind Turbine Generators. In Proceedings of the 2023 Asia Meeting on Environment and Electrical Engineering (EEE-AM), Hanoi, Vietnam, 13–15 November 2023; pp. 1–5. [CrossRef]
41. Elkholy, M.; Shalash, O.; Hamad, M.S.; Saraya, M.S. Empowering the Grid: A Comprehensive Review of Artificial Intelligence Techniques in Smart Grids. In Proceedings of the 2024 International Telecommunications Conference (ITC-Egypt), Cairo, Egypt, 22–25 July 2024; pp. 513–518. [CrossRef]
42. Zhang, L.; Zhu, B.; Wang, Y. Identification of vulnerable lines in power grids with wind power integration based on topological potential. *Electr. Power Syst. Res.* **2024**, *234*, 110593. [CrossRef]
43. Zhou, D.; Pan, X.; Sun, X.; Hu, F. Resilience Assessment Framework for High-Penetration Renewable Energy Power System. *Sustainability* **2025**, *17*, 2058. [CrossRef]
44. Song, H.; Liu, C.; Amani, A.M.; Gu, M.; Jalili, M.; Meegahapola, L.; Yu, X.; Dickeson, G. Smart optimization in battery energy storage systems: An overview. *Energy AI* **2024**, *17*, 100378. [CrossRef]
45. Jafari, M.; Gharehpetian, G.B.; Anvari-Moghaddam, A. On the Role of Virtual Inertia Units in Modern Power Systems: A Review of Control Strategies. Applications and Recent Developments. *Int. J. Electr. Power Energy Syst.* **2024**, *159*, 110067. [CrossRef]
46. Zhang, S.; Chen, Y.; Li, Q.; Xie, Z.; Fu, Y.; Wu, W. Quantitative stability analysis of STATCOM in voltage control mode and Enhanced control method based on voltage feedforward. *Int. J. Electr. Power Energy Syst.* **2025**, *171*, 110975. [CrossRef]
47. Shadoul, M.; Ahshan, R.; AlAbri, R.S.; Al-Badi, A.; Albadi, M.; Jamil, M. A Comprehensive Review on a Virtual-Synchronous Generator: Topologies. Control Orders and Techniques, Energy Storages, and Applications. *Energies* **2022**, *15*, 8406. [CrossRef]
48. Hosny, M.; Marei, M.I.; Mohamad, A.M. Adaptive hybrid virtual inertia controller for PMSG-based wind turbine based on fuzzy logic control. *Sci. Rep.* **2025**, *15*, 3757. [CrossRef]
49. Diggikar, S.; Patil, A.; Katkar, S.S.; Samad, K. Machine learning-based inertia estimation in power systems: A review of methods and challenges. *Energy Inform.* **2025**, *8*, 57. [CrossRef]
50. Ramos, H.M.; Coronado-Hernández, O.E.; Besharat, M.; Carravetta, A.; Fecarotta, O.; Pérez-Sánchez, M. Energy Storage Systems in Micro-Grid of Hybrid Renewable Energy Solutions. *Technologies* **2025**, *13*, 527. [CrossRef]
51. Almada, J.B.; Tofoli, F.L.; Gregory, R.C.F.; Sampaio, R.F.; Melo, L.S.; Leão, R.P.S. Distributed Multi-Agent Energy Management for Microgrids in a Co-Simulation Framework. *Energies* **2025**, *18*, 4620. [CrossRef]

https://doi.org/10.3390/en19030772

52. Bouslimani, M.; Benbouzid-Si Tayeb, F.; Amirat, Y.; Benbouzid, M. Cyber-Physical Security in Smart Grids: A Comprehensive Guide to Key Research Areas, Threats, and Countermeasures. *Appl. Sci.* **2025**, *15*, 12367. [CrossRef]
53. Mourabit, Y.E.; Salime, H.; Bossoufi, B.; Motahhir, S.; Derouich, A.; Mobayen, S.; Zhilenkov, A. Enhanced Performance in PMSG-Based Wind Turbine Systems: Experimental Validation of Adaptive Backstepping Control Design. *Energies* **2023**, *16*, 7481. [CrossRef]
54. González, F.A.; Posada, J.; França, B.W.; Rosas-Caro, J.C. Inertia in Converter-Dominated Microgrids: Control Strategies and Estimation Techniques. *Electricity* **2025**, *6*, 58. [CrossRef]
55. Mbasso, W.F.; Harrison, A.; Dagal, I.; Jangir, P.; Khishe, M.; Kotb, H.; Shaikh, M.S.; Smerat, A.; Donfack, E.F.; Kumar, R. Digital twins in renewable energy systems: A comprehensive review of concepts. applications, and future directions. *Energy Strategy Rev.* **2025**, *61*, 101814. [CrossRef]
56. Mirmohammad, M.; Azad, S.P. Control and Stability of Grid-Forming Inverters: A Comprehensive Review. *Energies* **2024**, *17*, 3186. [CrossRef]
57. Ibrahim, R.A.; Zakzouk, N.E. A PMSG Wind Energy System Featuring Low-Voltage Ride-through via Mode-Shift Control. *Appl. Sci.* **2022**, *12*, 964. [CrossRef]
58. Ruan, P.; Su, Q.; Zhang, L.; Luo, J.; Diao, Y.; Xie, L.; Zheng, H. Optimal Siting and Sizing of Hybrid Energy Storage Systems in High-Penetration Renewable Energy Systems. *Energies* **2025**, *18*, 2196. [CrossRef]
59. Xu, G.; Ke, D.; Li, Y.; Gao, J.; Yang, H.; Liao, S. Deep Reinforcement Learning-Based Adaptive Transient Voltage Control of Power Systems by Distributed Collaborative Modulation of Voltage-Source Converters with Operational Constraints of Current Saturation. *Sustainability* **2025**, *17*, 3846. [CrossRef]
60. He, P.; Li, Z.; Jin, H.; Zhao, C.; Fan, J.; Wu, X. An adaptive VSG control strategy of battery energy storage system for power system frequency stability enhancement. *Int. J. Electr. Power Energy Syst.* **2023**, *149*, 109039. [CrossRef]
61. Morgan, E.F.; Abdel-Rahim, O.; Megahed, T.F.; Suehiro, J.; Abdelkader, S.M. Fault Ride-Through Techniques for Permanent Magnet Synchronous Generator Wind Turbines (PMSG-WTGs): A Systematic Literature Review. *Energies* **2022**, *15*, 9116. [CrossRef]
62. Aoun, A.; Adda, M.; Ilinca, A.; Ghandour, M.; Ibrahim, H. Comparison between Blockchain P2P Energy Trading and Conventional Incentive Mechanisms for Distributed Energy Resources—A Rural Microgrid Use Case Study. *Appl. Sci.* **2024**, *14*, 7618. [CrossRef]
63. Hou, Z.; Liu, J. Enhancing Smart Grid Sustainability: Using Advanced Hybrid Machine Learning Techniques While Considering Multiple Influencing Factors for Imputing Missing Electric Load Data. *Sustainability* **2024**, *16*, 8092. [CrossRef]
64. Wang, P.; Bi, J.; Li, F.; Liu, C.; Sun, Y.; Cheng, W.; Wang, Y.; Kang, W. Research on Energy Storage-Based DSTATCOM for Integrated Power Quality Enhancement and Active Voltage Support. *Electronics* **2025**, *14*, 2840. [CrossRef]
65. Li, G.; Zhang, Y.; Shi, Y.; Wang, Z.; Zhou, B. Distributed Coordinated Control Strategy for Grid-Forming-Type Hybrid Energy Storage Systems. *Sustainability* **2025**, *17*, 1436. [CrossRef]
66. Yang, D.; Jin, Z.; Zheng, T.; Jin, E. An adaptive droop control strategy with smooth rotor speed recovery capability for type III wind turbine generators. *Int. J. Electr. Power Energy Syst.* **2021**, *135*, 107532. [CrossRef]
67. Rahman, K.; Hashimoto, J.; Koseki, K.; Orihara, D.; Ustun, T.S. Coordinated Control of Grid-Forming Inverters for Adaptive Harmonic Mitigation and Dynamic Overcurrent Control. *Electronics* **2025**, *14*, 2793. [CrossRef]
68. Ahmadi, M.; Aly, H.; Gu, J. A comprehensive review of AI-driven approaches for smart grid stability and reliability. *Renew. Sustain. Energy Rev.* **2025**, *226*, 116424. [CrossRef]
69. Nyingu, B.T.; Masike, L.; Mbukani, M.W.K. Multi-Objective Optimization of Load Flow in Power Systems: An Overview. *Energies* **2025**, *18*, 6056. [CrossRef]
70. Beikbabaei, M.; Kwiatkowski, B.M.; Mehrizi-Sani, A. Model-Free Resilient Grid-Forming and Grid-Following Inverter Control Against Cyberattacks Using Reinforcement Learning. *Electronics* **2025**, *14*, 288. [CrossRef]
71. *IEC 61850:2025*; Communication Networks and Systems for Power Utility Automation—All Parts. International Electrotechnical Commission (IEC): Geneva, Switzerland, 2025.

https://doi.org/10.3390/en19030772

energies

MDPI

Article

Solar Photovoltaic Model Parameter Identification with Improved Metaheuristic Algorithm Based on Balanced Search Strategies

Sujoy Barua [1], Sukanta Paul [2] and Adel Merabet [1,*]

[1] Division of Engineering, Saint Mary's University, Halifax, NS B3H 3C3, Canada; sujoy.barua@smu.ca

[2] Department of Computer and Communication Engineering, International Islamic University Chittagong, Chattogram 4318, Bangladesh

* Correspondence: adel.merabet@smu.ca

Abstract

Accurate identification of solar photovoltaic model parameters is crucial for reliably representing electrical behavior, improving maximum power point tracking, and enhancing overall system performance. Owing to the nonlinear and multimodal nature of the single-diode model, analytical closed-form solutions are difficult to obtain, which necessitates the use of advanced optimization techniques. Metaheuristic methods are particularly suitable for this task due to their strong global search capability, independence from gradient information, and adaptability to complex solution landscapes. In this study, a hybrid metaheuristic approach called the Jackal Arithmetic Algorithm is evaluated by integrating the Arithmetic Optimization Algorithm with the Golden Jackal Optimization method. The optimization framework combines arithmetic-based operators to enhance global exploration with adaptive predatory-inspired strategies to strengthen local exploitation, enabling a smooth transition between exploration and exploitation and resulting in improved convergence stability. Simulation results confirm that the Jackal Arithmetic Algorithm provides highly accurate parameter estimation for the single-diode photovoltaic model, achieving a minimum root mean square error of 0.00078 with a population size of 70, outperforming all compared algorithms. Overall, the combined method offers a robust and effective solution for photovoltaic modeling, with direct benefits for system design, control, and real-time monitoring.

Keywords: solar photovoltaic; single diode; model; parameter identification; metaheuristic optimization; hybrid algorithms; jackal arithmetic algorithm; balance; electrical equivalent

1. Introduction

The Single-Diode Model (SDM) is one of the most widely adopted electrical models for representing the behavior of solar photovoltaic (PV) cells and modules. It offers a simplified yet sufficiently accurate description of the conversion of solar irradiance into electrical energy by capturing the essential physical and electrical characteristics of PV devices. The model comprises a single diode, a photocurrent source, and resistive elements, which together represent the light-generated current, the p–n junction behavior, and the inherent resistive losses present in practical PV cells [1,2]. Owing to its effective balance between computational simplicity and physical realism, the SDM is particularly well suited for the simulation, design, and performance evaluation of photovoltaic systems.

The origin of the SDM lies in semiconductor physics and is based on the Shockley diode equation developed in the 1950s. Researchers found that the electrical behavior of a PV cell could be effectively represented by combining a light-generated current source, which models the electron–hole pairs created by sunlight, with a diode that captures the recombination effects in the p–n junction. Series and shunt resistances are added to account for ohmic losses in the material and leakage currents, respectively. This combination forms the simplest yet effective representation of a solar cell, allowing engineers and researchers to predict its electrical behavior under varying conditions. The SDM primarily describes the current–voltage (I–V) characteristics of a PV cell. It enables the prediction of output current at different voltages and illustrates how the cell behaves under varying irradiance and temperature conditions. By accurately modeling the interplay between photocurrent generation, diode behavior, and internal losses, the SDM also provides insight into the maximum power point (MPP) and the overall power-voltage (P–V) characteristics. Its simplicity and physical relevance make it a foundational tool for PV modeling, simulation, and practical system design [3].

Accurate modeling of the SDM is essential for predicting the electrical behavior of PV cells and modules. The SDM includes key parameters such as the photocurrent, diode saturation current, series and shunt resistances, and the diode ideality factor, which cannot be measured directly and must be estimated from measured current–voltage (I–V) data. Due to the nonlinear and implicit nature of the SDM, conventional analytical or gradient-based methods often fail to converge or become trapped in local minimum, limiting their effectiveness. Consequently, optimization techniques are required to systematically search the multidimensional parameter space and identify parameter sets that minimize the difference between measured and modeled I–V characteristics [4].

Since PV model parameter estimation is an implicit, nonlinear, multivariable, and multimodal problem [5], it remains both a research hotspot and a technical challenge in the field of solar power generation [6]. Numerous methods have been developed to address this issue, which can generally be divided into three main categories: analytical approaches [7], deterministic optimization methods [8], and metaheuristic techniques [9]. Analytical approaches rely on key operating points provided by PV manufacturers [10], simplifying the model equations to reduce computational effort. However, these simplifications often compromise accuracy. Deterministic approaches, such as the Newton–Raphson algorithm and techniques using the Lambert W function [11], depend on differentiability and convexity assumptions. Their performance is also highly sensitive to initial guesses, which can lead to divergence or incorrect solutions when the initial values are poorly selected [12].

With the rapid advancement of intelligent computing technologies, metaheuristic optimization algorithms are particularly well-suited for PV parameter estimation because they are derivative-free, flexible, and capable of balancing global exploration with local exploitation. Inspired by natural or mathematical processes, metaheuristics efficiently navigate complex, multimodal search spaces and avoid local optima, making them highly effective for SDM parameter identification. Their application ensures accurate modeling, which is critical for maximum power point tracking, system performance evaluation, and reliable PV system design. In this study, a hybrid metaheuristic approach is employed to further enhance convergence speed and estimation accuracy [13].

Recent hybrid studies, the whale optimization algorithm incorporating an information-sharing mechanism and the Nelder–Mead simplex method achieved notable improvements in global exploration [14]. Similarly, an enhanced particle swarm optimization method to mitigate premature convergence and strengthen exploration–exploitation balance, achieving comparable root mean squared error (RMSE) values for both models [15]. A hybrid

https://doi.org/10.3390/en19020315

rat swarm optimization and pattern search (RSO–PS) method is proposed demonstrating strong performance in both global and local search processes [16]. To further overcome early convergence issues, a modified salp swarm optimization (MSSO) technique is developed achieving an RMSE reduction compared with several benchmark algorithms [17].

Hybrid metaheuristic approaches significantly enhance parameter estimation by balancing exploration and exploitation to prevent premature convergence, improving accuracy and robustness, accelerating convergence, and adapting effectively to complex or noisy problems [18–20]. Moreover, hybridization enhances algorithm performance by strategically combining two or three methods to balance exploration and exploitation. This approach leverages the complementary strengths of different algorithms, resulting in greater flexibility, improved adaptability, and superior convergence characteristics when solving complex engineering problems [21,22]. However, achieving a method that simultaneously offers faster convergence, higher precision, stronger robustness, and greater reliability is still a challenging task in PV model parameter estimation. Addressing these limitations continues to hold important practical value for improving the performance and predictability of PV systems [23].

The Arithmetic Optimization Algorithm (AOA), introduced by [24], is a metaheuristic inspired by basic arithmetic operations—addition, subtraction, multiplication, and division—which are used as mathematical operators to update candidate solutions iteratively toward the global optimum. The Golden Jackal Optimization (GJO), proposed by [25], models the cooperative hunting behavior of golden jackals and performs its search process through three main stages: exploration, prey encircling (provocation), and exploitation (attack). The aim of this paper is to identify opportunities for improvement and to develop hybrid algorithms that balance search strategies to obtain optimal parameters for the SDM. Although AOA–GJO hybrids have thus far been developed and applied only in microgrid optimization [26], this study provides another application and offers evidence of the hybrid algorithm's capability and performance. Guided by the No Free Lunch theorem, which emphasizes the need for problem-specific algorithm selection [27], this work further evaluates the general applicability and robustness of the algorithm.

The objective of this study is to demonstrate and apply the Jackal Arithmetic Algorithm (JAA) to address limitations in existing techniques for SDM parameter estimation in photovoltaic systems. The study aims to determine the optimal parameter set for the single-diode electrical equivalent circuit of a solar cell, a highly nonlinear optimization problem, while demonstrating improved search capability, robustness, and convergence stability. Performance evaluation employs minimum and mean solution values, standard deviation, and data-distribution analyses to compare JAA with its constituent algorithms, AOA and GJO, across different population settings. The contributions of this research include:

1. Application of the JAA hybrid approach to SDM parameter estimation, representing a new use of this algorithm in photovoltaic modeling.
2. Integration of AOA's arithmetic-based operators with GJO's cooperative foraging behavior, achieving a stronger balance between exploration and exploitation than the individual algorithms.
3. Comprehensive performance evaluation shows that JAA outperforms AOA and GJO individually, with higher accuracy, faster and more stable convergence, and improved robustness.
4. Validation of the hybrid method's effectiveness in accurately extracting SDM parameters, highlighting its potential for broader applications in nonlinear optimization problems.

https://doi.org/10.3390/en19020315

2. Modeling of Solar PV Equivalent Circuit

The single-diode photovoltaic model illustrated in Figure 1 consists of five essential parameters: diode ideality factor (n), reverse saturation current (I_{sd}), light-generated current (I_{ph}), series resistance (R_s), and shunt resistance (R_{sh}). All these parameters must be accurately estimated to capture the electrical behavior of a solar cell [28].

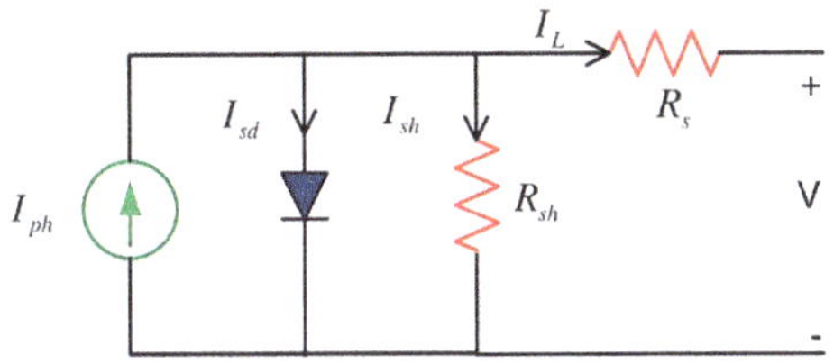

Figure 1. Equivalent Circuit of Single-Diode Modeling of Solar PV.

Using Kirchhoff's Law, the I–V character is given as Equation (1).

$$I_L = I_{ph} - I_d - I_{sh}. \tag{1}$$

where I_L is output current, I_{ph} denotes the photo-generated current, I_d represents the current flowing through the diode D, and I_{sh} corresponds to the current passing through the shunt resistor R_{sh}.

The diode current I_d is determined using the Shockley diode equation together with Kirchhoff's voltage law and can be expressed mathematically as shown in Equation (2).

$$I_d = I_{sd} * \left(e^{\left(\frac{q\{V_L + I_L R_S\}}{nKT}\right)} - 1 \right) \tag{2}$$

where I_{sd} represents the diode reverse saturation current, V_L is the output voltage of the PV model, n denotes the diode ideality factor, k is the Boltzmann constant $(1.3806503 \times 10^{-23}\,\text{J/K})$, $\text{T} = 33 + 273.15$ is the operating temperature of the PV cell expressed in Kelvin, and q is the electron charge $(1.60217646 \times 10^{-19}\,\text{C})$.

The current across parallel resistor, I_{sh} can be calculated according to Kirchhoff Voltage Laws as Formula (3)

$$I_{sh} = \frac{V_L + I_L R_S}{R_{sh}} \tag{3}$$

Therefore, the relationship between the current output and output voltage of the SDM can be derived as shown in Equation (4).

$$I_L = I_{ph} - I_{sd} * \left(e^{\left(\frac{q\{V_L + I_L R_S\}}{nKT}\right)} - 1 \right) - \frac{V_L + I_L R_S}{R_{sh}} \tag{4}$$

In this model, the SDM contains five unknown parameters (I_{ph}, I_{sd}, R_s, R_{sh}, n) that must be estimated. Accurate determination of these parameters is essential, as they directly influence photovoltaic cell modeling accuracy, performance optimization, and maximum power point tracking.

3. Objective Function for PV Single-Diode Model

The primary objective of this study is to minimize the difference between the simulated and experimental I–V characteristics during SDM parameter estimation. To achieve this, the RMSE, one of the most widely used performance metrics for PV model evaluation, is employed as the objective function. RMSE is chosen because it is sensitive to larger errors and effectively reflects the overall fitting accuracy between the modeled and measured

https://doi.org/10.3390/en19020315

I–V data. It is calculated from the difference between the measured current and the model-generated current at each voltage point, and its mathematical expression is provided in Equation (5).

$$RMSE(x) = \sqrt{\frac{1}{N}\left(\sum_{i=1}^{N} f(V, I, x)^2\right.} = \sqrt{\frac{1}{N}\left(\sum_{i=1}^{N} [I_{L\cdot mes} - I_{L.cal}(V, x)]^2\right)} \quad (5)$$

where the function $f(V, I, x)$ represents the modeling error for different PV models, N is the number of samples, $I_{L.mes}$ denotes the experimentally measured current corresponding to voltage V, $I_{L.cal}(V, x)$ is the model-predicted current obtained by solving the implicit nonlinear PV model equation using the parameter set x.

Equation (5) can be written as follows in (6).

$$\begin{cases} f(V, I, x) = I_{L\cdot mes} - \left(I_{ph} - I_{sd} * \left[e^{\left(\frac{q\{V_L + I_L R_S\}}{nKT}\right)} - 1\right] - \frac{V_L + I_L R_S}{R_{sh}}\right) \\ x = [x_1\ x_2\ x_3\ x_4\ x_5] = \left[I_{ph}\ I_{sd}\ R_{sh}\ R_S\ n\right] \end{cases} \quad (6)$$

The SDM relies on defined minimum and maximum parameter bounds as shown in (7)–(11) for the five key parameters (I_{ph}, I_{sd}, R_s, R_{sh}, n) such as:

$$I_{ph}^{min} < I_{ph}(A) < I_{ph}^{max} \quad (7)$$

$$I_{sd}^{min} < I_{sd}(\mu A) < I_{sd}^{max} \quad (8)$$

$$R_{sh}^{min} < R_{sh}(\Omega) < R_{sh}^{max} \quad (9)$$

$$R_s^{min} < R_s(\Omega) < R_s^{max} \quad (10)$$

$$n^{min} < n < n^{max} \quad (11)$$

The purpose of the optimizing algorithm is to determine accurate values of the five key parameters.

4. Jackal Arithmetic Algorithm (JAA)

In this study, the hybrid JAA, developed by combining the AOA and the GJO as illustrated in [26], is employed to estimate the parameters of the SDM used for modeling the electrical equivalent circuit of PV cells.

4.1. Initialization

The candidate solutions (X) are randomly initialized within predefined bounds:

$$x_{i,j} = LB_j + rand \times \left(UB_j - LB_j\right) \quad \text{for } i = 1 \cdots N; j = 1 \cdots n \quad (12)$$

where LB_j and UB_j define lower and upper boundaries, respectively.

The boundaries ensure the solutions remain within feasible search spaces throughout iterations.

4.2. Identify Limitiation and Improvement

The performance of a metaheuristic algorithm largely depends on maintaining a proper balance between exploration and exploitation. Although the AOA is simple and versatile, it faces challenges with highly nonlinear and complex problems. These challenges include premature convergence, reduced population diversity, and high sensitivity to its controlling parameters, the Math Optimizer Accelerator (MOA) and Math Optimizer Probability (MOP) [24], as defined in Equations (13) and (14) and shown in Figure 2a. A

https://doi.org/10.3390/en19020315

linear adjustment of these parameters can cause the algorithm to stagnate early or waste time exploring fewer promising areas, limiting its overall effectiveness.

$$MOA(t) = Min + t \times \left(\frac{Max - Min}{T}\right) \tag{13}$$

$$MOP(t) = 1 - \frac{(t)^{1/\alpha}}{(T)^{1/\alpha}} \tag{14}$$

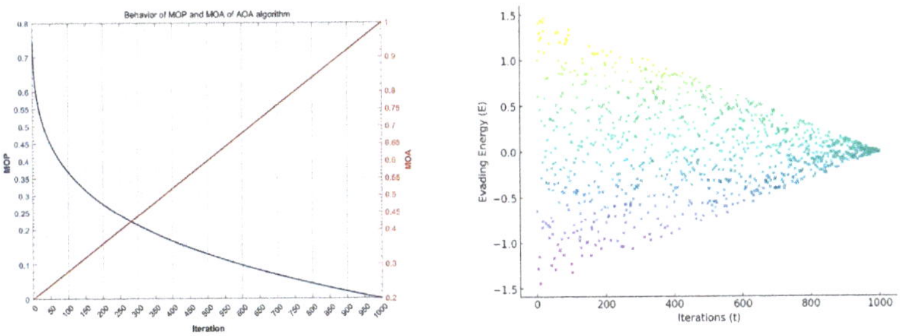

(a) (b)

Figure 2. (**a**) The behavior of the selection function MOA and the searching position coefficient MOP of the AOA during the iteration. (**b**) A scattered representation of the Evading Energy (E) during the iterations, demonstrating the update of the position in the GJO for optimal candidate solutions.

The parameter α particularly influences AOA's convergence behavior, making it more prone to local optima in large or complex search spaces and increasing computational effort without ensuring global optimality.

To overcome these limitations, GJO introduces the concept of Evading Energy (E), which dynamically improves the search process and helps maintain a better balance between exploration and exploitation. The Evading Energy, defined in Equations (15) and (16) and shown in Figure 2b, consists of an initial random energy level $E_{i.e}$ and a decaying energy component $E_{d.e}$ that decreases gradually throughout the iterations:

$$|E| = E_{d.e} * E_{i.e} \tag{15}$$

$$E_{d.e} = 1.5 * (1 - t/T) \tag{16}$$

where $E_{i.e}$ denotes the random initial energy of the prey, and $E_{d.e}$ represents its iterative decay, enabling smoother transitions between exploration and exploitation.

This mechanism enhances search diversity, reduces premature convergence, and improves global optimization capability. Although it introduces a slight increase in computational complexity, the improvement in stability and solution quality justifies its inclusion. These complementary strengths of AOA and GJO provide a strong foundation for the development of the hybrid JAA, designed to achieve more accurate and robust optimization performance for SDM parameter estimation.

4.3. Selection Phase

The JAA prioritizes an optimal balance between exploration and exploitation by integrating MOA and Evading Energy. Random numbers guide the selection of exploration or exploitation, while E facilitates dynamic transitions between extensive and intensive search phases, enhancing solution diversity.

https://doi.org/10.3390/en19020315

Figure 3a presents the exploration-exploitation dynamics inherent in GJO, dictated by E. Higher values ($|E| > 1$) favor exploration, while lower values ($|E| \leq 1$) promote exploitation. Figure 3b depicts AOA's approach, using arithmetic operators to dynamically adjust between exploration and exploitation based on the MOA parameter. Figure 3c illustrates the JAA's hybridization strategy, where the Evading Energy (E) from GJO and the MOA parameter from AOA collectively balance global exploration with local exploitation.

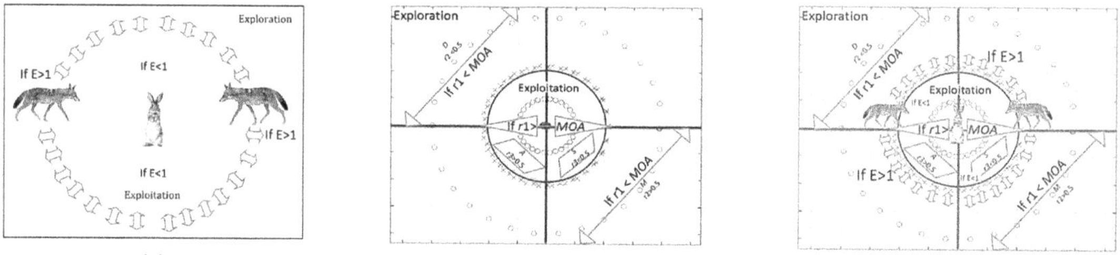

(a) (b) (c)

Figure 3. (**a**) Search techniques from the swarm intelligence of the GJO algorithm. (**b**) Search strategy based on mathematical operators from the AOA. (**c**) hybrid search strategy for the JAA to address existing gaps.

4.4. Searching Process

The optimization process employs two distinct strategies to ensure a robust search of the solution space:

4.4.1. Diversification Phase ($r_1 > MOA$)

In the diversification phase, the algorithm aims to explore broader regions of the solution space, systematically moving away from known solutions to avoid premature convergence. The diversification strategies are detailed below:

For $|E| > 1$:

$$\left.\begin{aligned} X_1(t) &= x_m(t) - |E| \cdot \left|x_m(t) - rl \cdot x(t)\right| \\ X_2(t) &= x_{fm}(t) - |E| \cdot \left|x_{fm}(t) - rl \cdot x(t)\right| \end{aligned}\right\} \tag{17}$$

Otherwise, the arithmetic exploration is applied:

$$X(t+1) = \begin{cases} best(x_j) \div (MOP + \epsilon) \times ((UB_j - LB_j) \times E + LB_j, & r_2 < 0.5 \\ best(x_j) \times MOP \times ((UB_j - LB_j) \times E + LB_j, & Otherwise \end{cases} \tag{18}$$

4.4.2. Intensification Phase ($r_1 < MOA$)

In the intensification phase, the algorithm concentrates on the search near promising candidate solutions, fine-tuning them to achieve optimal accuracy. The intensification strategies are:

For $|E| \leq 1$:

$$\left.\begin{aligned} X_1(t) &= x_m(t) - |E| \cdot \left|rl \cdot x_m(t) - x(t)\right| \\ X_2(t) &= x_{fm}(t) - |E| \cdot \left|rl \cdot x_{fm}(t) - x(t)\right| \end{aligned}\right\} \tag{19}$$

Otherwise, the solutions are updated as follows:

$$X(t+1) = \begin{cases} best(x_j) - (MOP) \times ((UB_j - LB_j) \times E + LB_j, & r_3 < 0.5 \\ best(x_j) + MOP \times ((UB_j - LB_j) \times E + LB_j, & Otherwise \end{cases} \tag{20}$$

https://doi.org/10.3390/en19020315

In both phases, the MOP and the Lévy flight function (*rl*), defined by Equation (21), determine position adjustments:

$$rl = 0.05 * LF(x) \tag{21}$$

The Lévy flight function *LF*(*x*) is calculated using:

$$LF(x) = \frac{0.01 \times (\mu - \sigma)}{\left|v^{1-\beta}\right|};\ \sigma = \left(\frac{\Gamma(1+\beta) \times \sin\left(\frac{\pi\beta}{2}\right)}{\Gamma\left(\frac{1+\beta}{2}\right) \times \beta \times 2^{\frac{\beta-1}{2}}}\right)^{1/\beta},\quad \beta = 1.5 \tag{22}$$

4.5. Update of Optimal Solutions

After each iteration, the optimal solution positions are updated based on the strategies outlined in Equations (18) and (20). Equations (17) and (19) positions are averaged to improve convergence stability using in (23).

$$X(t+1) = \frac{X_1(t) + X_2(t)}{2} \tag{23}$$

Algorithm 1 describes the pseudocode and Figure 4 illustrates the flowchart.

Algorithm 1: PSEEUDOCODE of Jackal Arithmetic Algorithm

```
Initialise N, T, α, MOP, Max, MOP, Min, t
Randomly generate initial positions for i = 1: N
Define fitness and identify the best results
While t < T do
    Update MOA and MOP (Equations (13) and (14))
    Update E_d.e (Equation (16))
    For i = 1: N (for each solution) do
        For j = 1: positions do
            Update E (Equation (15))
            Generate r1, r2, r3 ∈ [0, 1]
            If r1 > MOA (Exploration) then
                If |E| > 1 then
                    If r2 < 0.5 then Division (Equation (18))
                    Else Multiplication (Equation (18))
                Else Update prey (Equation (17))
            Else (Exploitation)
                If |E| < 1 then
                    If r3 < 0.5 then Subtraction (Equation (20))
                    Else Addition (Equation (20))
            Else Update prey (Equation (19))
            End If
        End For
        Evaluate fitness and identify best solution
    End For
    t = t + 1
End While
Return best solution (X)
```

https://doi.org/10.3390/en19020315

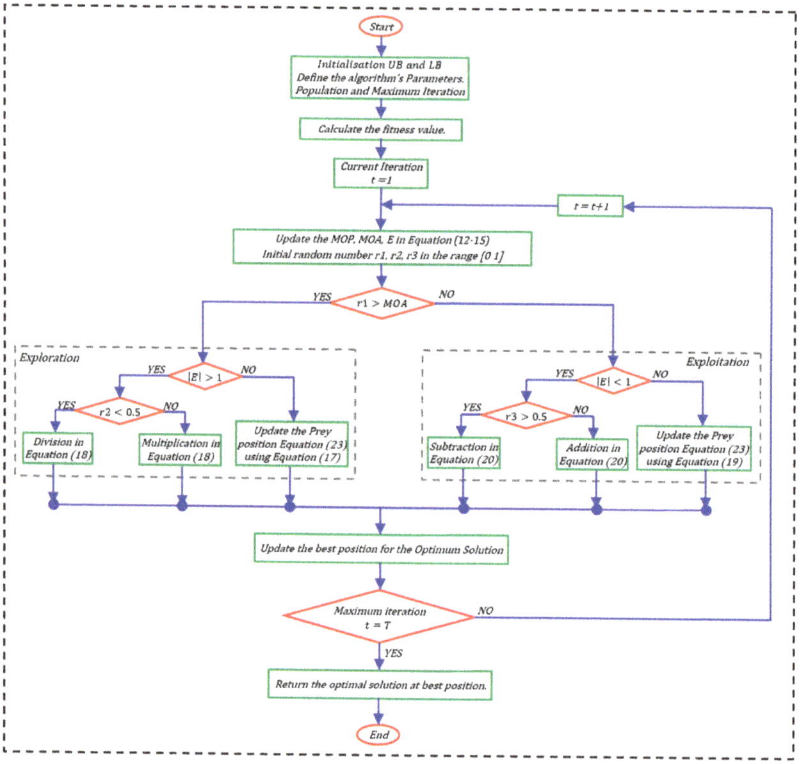

Figure 4. Flowchart of the hybrid Jackal Arithmetic Algorithm.

4.6. Reliability and Stability

The hybrid algorithm JAA has been rigorously evaluated against the GJO and the AOA using five benchmark functions with diverse characteristics, as summarized in Table 1 [26]. These functions vary in modality, dimensionality, and search range, providing a demanding and representative test environment. For each algorithm, 20 independent runs are conducted with 500 iterations and a population size of 50, and the best, mean, and standard deviation values are reported to ensure statistical reliability.

Table 1. Properties of five benchmark functions for stability and reliability testing.

Name	Dim	Range	f_{min}	Function
Inverse Hilbert Matrix Problem	16	[−6384, 16,384]	1	$f_{cec02}(x) = \sum_{i=1}^{n}\sum_{k=1}^{n}\lvert w_{i,k}\rvert$ $(w_{i,k}) = W = HZ - I.I = \begin{bmatrix} 1 & 0 & \dots & 0 \\ 0 & 1 & \dots & 0 \\ \vdots & \vdots & \ddots & \vdots \\ 0 & 0 & \dots & 1 \end{bmatrix}$ $H = (h_{i,k});\ h_{i,k} = \frac{1}{i+k-1};\ i,k = 1,2,\cdots,n;\ n = \sqrt{D}$ $z = (z_{i,k}),\quad z_{i,k} = x_{i+n(k-1)}$
Restringing's Function	10	[−100, 100]	1	$f_{cec04}(x) = \sum_{i=1}^{D}(x_i^2 - 10\cos(2\pi x_i) + 10)$
Griewangk's Function	10	[−100, 100]	1	$f_{cec05}(x) = \sum_{i=1}^{D}\frac{x_i^2}{4000} - \prod_{i=1}^{D}\cos\left(\frac{x_i}{\sqrt{i}}\right) + 1$

https://doi.org/10.3390/en19020315

Table 1. *Cont.*

Name	Dim	Range	f_{min}	Function
Modified Schwefel's Function	10	[−100, 100]	1	$f_{cec07} = 418.9829D - \sum_{i=1}^{D} g(z_i) z_i = x_i + 420.9687462275036$ $g(z_i) = \begin{cases} z_i \sin\left(\lvert z_i\rvert^{1/2}\right) & if\ \lvert z_i\rvert \le 500 \\ (500 - mod(z_i, 500))\sin\left(\sqrt{\lvert 500 - mod(z_i, 500)\rvert}\right) - \frac{(z_i-500)^2}{10{,}000D} & if\ z_i > 500 \\ (mod(\lvert z_i\rvert, 500) - 500)\sin\left(\sqrt{\lvert mod(z_i, 500) - 500\rvert}\right) - \frac{(z_i-500)^2}{10{,}000D} & if\ z_i < -500 \end{cases}$
Happy Cat Function	10	[−100, 100]	1	$f_{cec09} = \left\lvert \sum_{i=1}^{D} x_i^2 - D \right\rvert^{1/4} + \left(0.5\sum_{i=1}^{D} x_i^2 + \sum_{i=1}^{D} x_i\right)/D + 0.5$

A detailed comparative statistical assessment is presented in Table 2, where the JAA consistently outperforms the original AOA and GJO. The results demonstrate that the JAA achieves a superior balance between exploration and exploitation, effectively mitigating premature convergence and avoiding local optima. To further validate these findings, the Wilcoxon rank-sum test (significance level 0.05) is employed, and the resulting p-values across all benchmark functions confirm the statistical significance of the performance improvements.

Table 2. Comparison Best, Average, Std, p-value and Rank Results of benchmark functions.

Function Name	Metrics	JAA	GJO	AOA
Inverse Hilbert Matrix Problem	Best	1.73×10^{1}	1.73×10^{1}	1.93×10^{1}
	Mean	1.74×10^{1}	1.74×10^{1}	1.98×10^{1}
	Worst	1.74×10^{1}	1.77×10^{1}	1.98×10^{1}
	SD	4.16×10^{-3}	9.07×10^{-2}	5.28×10^{-2}
	p Value	0.00×10^{0}	5.29×10^{-228}	5.48×10^{-257}
	Rank	1.5	1.5	3
Restring's Function	Best	1.63×10^{1}	3.74×10^{1}	8.72×10^{3}
	Mean	4.69×10^{1}	6.53×10^{2}	1.80×10^{4}
	Worst	8.96×10^{1}	4.23×10^{3}	3.56×10^{4}
	SD	1.42×10^{1}	1.02×10^{3}	5.95×10^{3}
	p Value	2.71×10^{-55}	5.00×10^{-9}	1.08×10^{-51}
	Rank	1	2	3
Grewingk's Function	Best	1.07×10^{0}	1.11×10^{0}	3.23×10^{0}
	Mean	1.27×10^{0}	1.54×10^{0}	5.31×10^{0}
	Worst	1.53×10^{0}	2.43×10^{0}	6.84×10^{0}
	SD	9.90×10^{-2}	2.82×10^{-1}	6.94×10^{-1}
	p Value	6.93×10^{-112}	1.01×10^{-75}	6.70×10^{-90}
	Rank	1	2	3
Modified Schwefel's Function	Best	-2.58×10^{2}	-1.05×10^{2}	5.43×10^{2}
	Mean	5.99×10^{1}	5.57×10^{2}	1.77×10^{3}
	Worst	4.94×10^{2}	1.34×10^{3}	2.52×10^{3}
	SD	1.41×10^{2}	3.04×10^{2}	3.73×10^{2}
	p Value	4.81×10^{-5}	1.31×10^{-33}	7.42×10^{-70}
	Rank	2	1	3
Happy Cat Function	Best	2.64×10^{0}	3.28×10^{0}	9.17×10^{2}
	Mean	3.68×10^{0}	3.08×10^{1}	2.34×10^{3}
	Worst	5.16×10^{0}	3.67×10^{2}	4.42×10^{3}
	SD	6.03×10^{-1}	8.74×10^{1}	7.55×10^{2}
	p Value	2.72×10^{-80}	6.33×10^{-4}	8.05×10^{-53}
	Rank	1	2	3
Overall Rank		1	2	3

https://doi.org/10.3390/en19020315

In addition, Figure 5 illustrates the three-dimensional landscapes of the benchmark functions, offering insight into the complexity of the optimization problems. The convergence analysis clearly shows faster and more stable convergence of the JAA compared to AOA and GJO, while the distribution analysis using box plots highlights the robustness and consistency of the hybrid approach. Collectively, these results provide strong evidence of the effectiveness and reliability of the JAA, validating its suitability for SDM optimization applications.

4.7. Computational Complexity Analysis

The computational complexity (O) of the JAA arises from three main processes: initialization, fitness evaluation, and iterative position updates. Initially, complexity is $O(N \times D)$, influenced by population size (N) and decision variables (D). During each iteration, complexities from updating MOP, MOA, Evading Energy (E), and Lévy flight calculations contribute $O(4T)$. The position update complexity per iteration is $O(N \times D \times T)$. Thus, the overall computational complexity is $O(N \times D \times (1 + T) + 2T)$.

This complexity aligns closely with the original complexities of the AOA and the GJO. Although the JAA introduces slight computational overhead due to the integrated decision conditions, it notably enhances the global search capability, effectively identifying superior solutions [26].

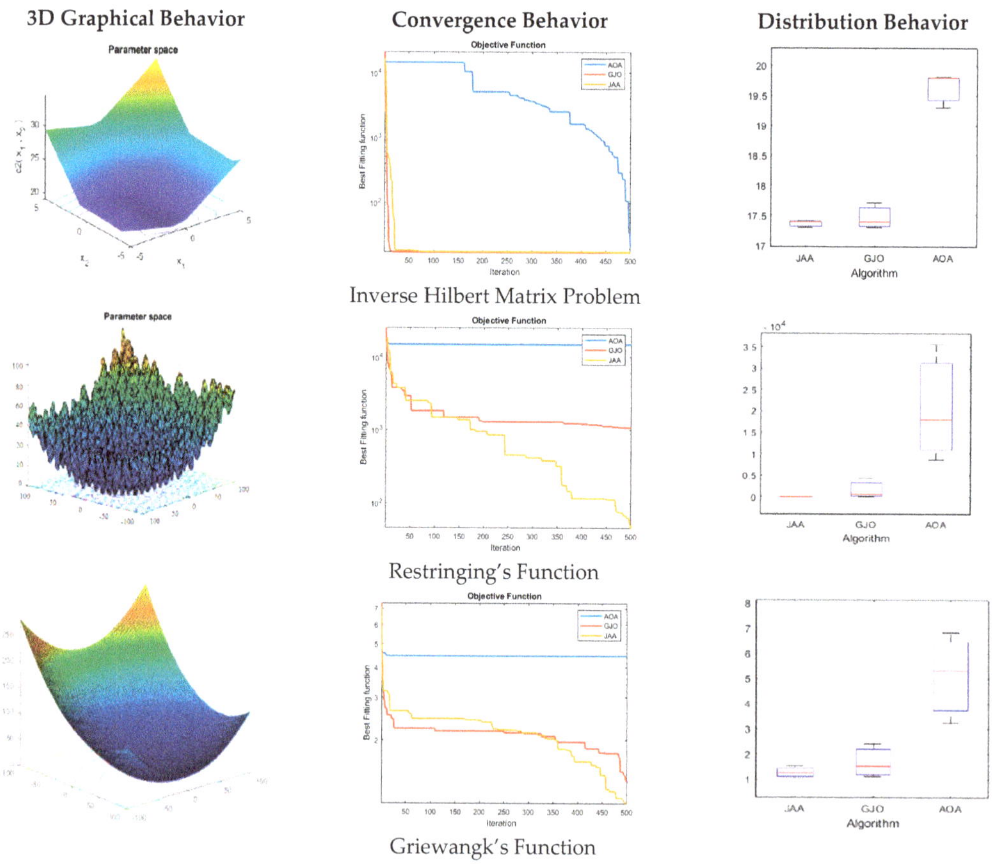

Figure 5. *Cont.*

https://doi.org/10.3390/en19020315

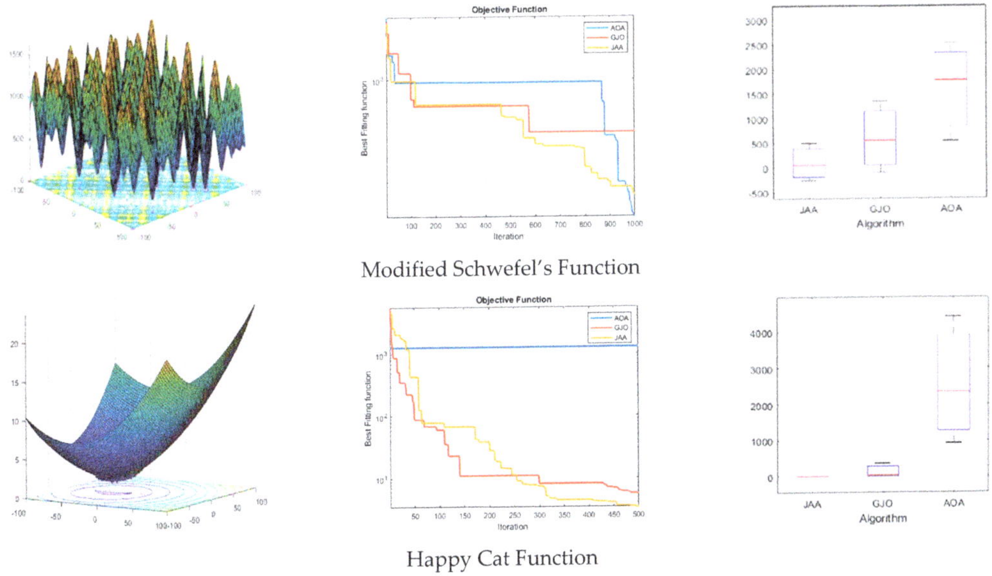

Figure 5. Graphical representation of benchmark functions, convergence and distribution curves for JAA, GJO and AOA.

5. Results and Discussion

The objective function was defined as the RMSE between the measured and model-predicted currents across the measured voltage points; minimizing RMSE forces the fitted SDM to reproduce the observed I–V curve as closely as possible. All model evaluations used standard physical constants (Boltzmann constant $k = 1.3806503 \times 10^{-23}$ J/K, electron charge $q = 1.60217646 \times 10^{-19}$ C) and an operating temperature of $T = 33 + 273.15$ K. For each candidate parameter vector x, the implicit nonlinear SDM equation was solved numerically to obtain $I_{L.cal}(V, x)$ at each measured voltage, thus computing the RMSE, and all simulations were implemented using MATLAB 2021a.

To provide a statistically robust assessment, each algorithm was run for 500 iterations and repeated over 30 independent runs at three population sizes (30, 50 and 70). This design tests both convergence behavior (iterations) and stochastic stability (multiple runs) across a range of population diversities. The estimation problem was posed as a bounded nonlinear optimization task for the SDM. Parameter bounds were set in Table 3. The three algorithms compared were the hybrid JAA, and conventional GJO, and AOA. Table 4 summarizes comparative statistics (best, mean, worst, standard deviation, and p-values) for the minimum RMSE achieved across the 30 runs. Table 5 reports representative optimal parameters value and their RMSEs produced by each algorithm at each population size.

Table 3. Solar PV Single-Diode Parameter Details.

Parameters Name	Variable	Unit	Upper Limit	Lower Limit
Light Detected Current	I_{ph}	A	0	1
Reverse Saturated Current	I_{sd}	μA	0	1
Shunt Resistance	R_{sh}	Ω	0	100
Series Resistance	R_s	Ω	0	0.5
Diode Ideality factor	n	-	1	2

https://doi.org/10.3390/en19020315

Table 4. RMSE Statistical Results for the Solar PV Single-Diode Model.

Population	30			50			70		
Algorithm	JAA	GJO	AOA	JAA	GJO	AOA	JAA	GJO	AOA
Best	0.00157	0.00226	0.00209	0.00112	0.00156	0.00208	0.00078	0.00115	0.00209
Mean	0.00257	0.01050	0.01283	0.00292	0.00591	0.00998	0.00512	0.01484	0.00736
Worst	0.00761	0.04823	0.03920	0.00812	0.03945	0.04283	0.00825	0.04561	0.03143
SD	0.00178	0.01366	0.01359	0.00202	0.00852	0.01128	0.00320	0.01842	0.00934
p-Value	0.00135	0.03790	0.01528	0.00000	0.00585	0.00084	0.00000	0.00189	0.00227

Table 5. Optimal Parameters estimated by JAA, GJO and AOA on Solar PV Equivalent Model.

Population	Algorithm	I_{ph}	I_{sd}	R_{sh}	R_s	n	*RMSE*
30	JAA	0.7607	0.7192	77.7779	0.0326	1.5665	0.0015
	GJO	0.7601	0.8781	94.8532	0.0307	1.5893	0.0022
	AOA	0.7607	0.9985	100.0000	0.0312	1.6044	0.0020
50	JAA	0.7596	0.4213	83.1490	0.0354	1.5082	0.0011
	GJO	0.7600	0.4491	82.1695	0.0353	1.5144	0.0015
	AOA	0.7605	0.9997	100.0000	0.0311	1.6045	0.0020
70	JAA	0.7605	0.3166	54.7195	0.0364	1.4791	0.0007
	GJO	0.7603	0.5226	68.3892	0.0340	1.5313	0.0011
	AOA	0.7608	1.0000	99.9584	0.0310	1.6045	0.0020

The numerical evaluation presented in Tables 4 and 5 provides a comprehensive comparison of the performance of the three optimization algorithms (JAA, GJO, AOA) applied to the parameter estimation problem for the single-diode photovoltaic model. The results show clear performance distinctions among the algorithms, with JAA consistently outperforming its competitors across all population sizes. This consistency, paired with improvements observed as the population size increases, demonstrates the algorithm's capacity to maintain both accuracy and stability. Overall, the combined results in these tables highlight JAA as the most efficient, reliable, and stable algorithm for identifying highly accurate SDM parameters.

Table 4 summarizes the RMSE performance of JAA, GJO, and AOA for the single-diode PV model at population sizes of 30, 50, and 70. Across all scenarios, JAA consistently achieves the lowest RMSE, demonstrating superior accuracy. At a population of 30, JAA attains a best RMSE of 0.00157, which is lower than GJO (0.00226) and AOA (0.00209). This trend continues at populations 50 and 70, where JAA further reduces RMSE to 0.0011283 and 0.00078, respectively. In contrast, GJO and AOA remain higher, with GJO reaching 0.00156 and AOA 0.00209 at population 70. The standard deviation values reinforce this pattern, as JAA consistently shows the smallest deviation, for example, 0.00178 at population 30, indicating better stability.

Table 5 provides the corresponding parameter estimates. JAA produces balanced and physically meaningful values, particularly noticeable at population 70, where it estimates $I_{ph} = 0.7605$, $I_{sd} = 0.3166$, $R_{sh} = 54.7195$, and $R_s = 0.0364$, accompanied by the lowest RMSE (0.00078065). GJO and AOA show larger deviations; AOA frequently pushes parameters to boundary limits, such as $I_{sd} = 1.0000$ and $R_{sh} = 99.9584$, which indicates reduced reliability and weaker convergence control. Overall, the numerical outcomes confirm that JAA consistently delivers the most accurate and stable parameter estimation. Its lower RMSE values, reduced variance, and physically consistent parameter sets indicate a robust optimization performance compared to GJO and AOA.

Figure 6 summarizes the identification and validation results obtained with three metaheuristics (JAA, GJO, AOA) under different population sizes. In the first row, the convergence histories at populations of 30, 50 and 70 display the best objective value on a logarithmic scale versus iteration. In every case, the JAA trajectory falls earliest

https://doi.org/10.3390/en19020315

and steeply, continuing to decrease until it reaches the lowest terminal error among the methods compared. JAA's convergence curves for several populations to isolate the effect of population size. Larger populations begin at lower objective values, drop more rapidly during the early iterations and terminate at smaller residuals.

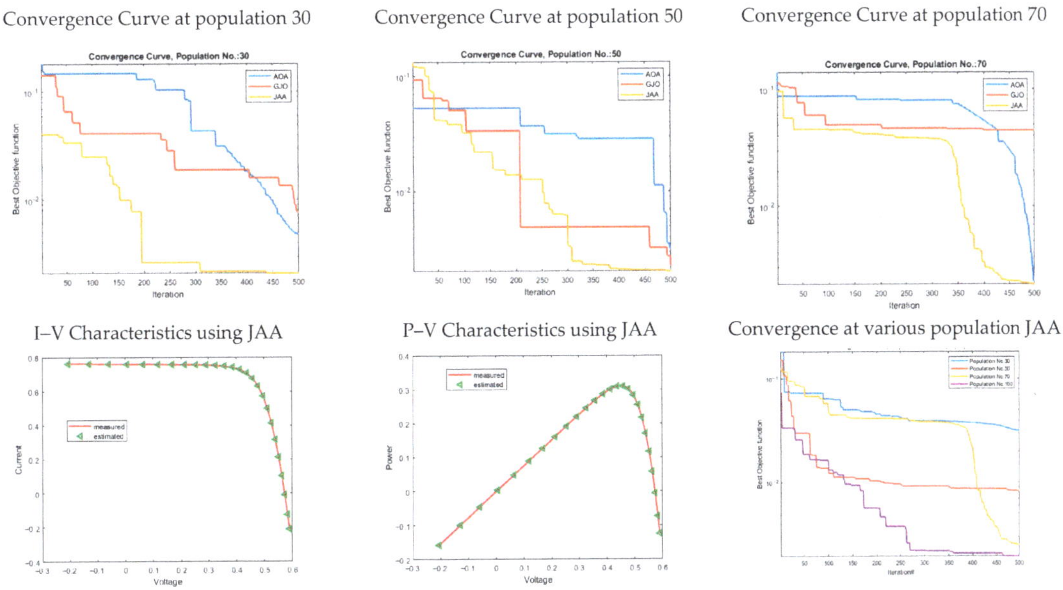

Figure 6. Convergence behavior of JAA, GJO and AOA and Solar PV characteristics using JAA.

The I–V curve produced by JAA aligns almost perfectly with the measured data across the full voltage range, accurately capturing the short-circuit region, the knee point, and the high-voltage decline. Likewise, the P–V curve closely matches the measured power profile, including the location and height of the maximum-power point, confirming that the JAA-identified model reliably reflects the real device behavior.

Figure 7 presents box plots of the best-of-run objective values for JAA, GJO and AOA at populations of 30, 50 and 70. The boxes associated with JAA occupy the lowest vertical positions with relatively short interquartile ranges and whiskers, reflecting both superior central tendency and tighter dispersion across repeated trials.

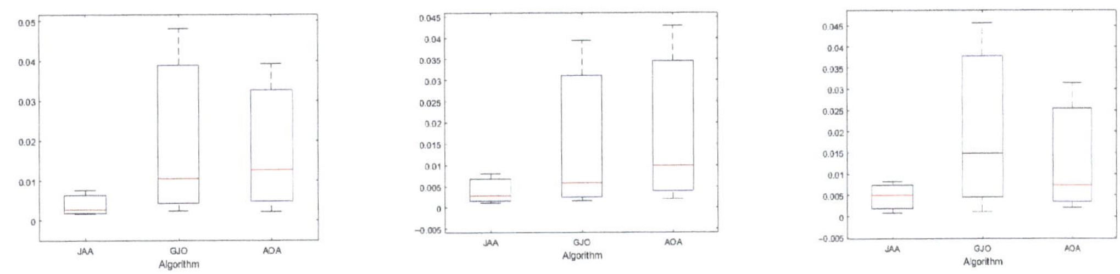

Figure 7. Data distribution of SDM Optimization at 30, 50, and 70 population using JAA, GJO, and AOA.

Finaly, the JAA achieves the smallest mean RMSE values among all algorithms, including a minimum of 0.00078 at a population size of 70, demonstrating both precision and stability. In contrast, the GJO and the AOA record higher mean RMSE values and noticeably larger variability between runs, indicating reduced consistency and increased sensitivity to stochastic initialization.

6. Discussion

The superior performance of the JAA can be attributed to the complementary strengths of its hybridized search mechanism. By integrating the arithmetic-based exploitation strategies of the AOA with the exploratory evading energy dynamics of the GJO, JAA achieves an effective balance between local refinement and global diversification. This synergy enables JAA to escape local minima more efficiently compared to the AOA, which is prone to premature convergence and boundary stagnation.

In contrast, while the GJO exhibits strong exploratory behavior, it may lack stability during later search stages. The hybrid framework of JAA mitigates this limitation by introducing controlled exploitation, thereby stabilizing the search process. Moreover, as the population size increases, the JAA demonstrates an enhanced ability to exploit the additional search agents, resulting in notable improvements in both convergence reliability and solution precision. Analysis of parameter distributions further indicates that the JAA avoids the unrealistic boundary solutions frequently produced by the AOA, suggesting a more effective exploration of the feasible search space. Collectively, these characteristics explain the consistent superiority of the JAA over its constituent algorithms.

7. Conclusions

This study presents a comprehensive evaluation of the Jackal Arithmetic Algorithm (JAA), a novel hybrid metaheuristic that combines the strengths of the AOA and the GJO. The proposed hybridization effectively addresses key challenges associated with balancing exploration and exploitation phases. Specifically, the AOA may suffer from entrapment in local optima due to its reliance on arithmetic operators, whereas the GJO can encounter instability in dynamic search environments. By leveraging their complementary characteristics, the JAA enhances both global search capability and local refinement, leading to improved optimization performance.

Initially applied to microgrid optimization problems [26], the JAA demonstrated promising results compared to its parent algorithms. In this study, the JAA was further applied to a range of real-world engineering problems, including parameter estimation for solar photovoltaic systems. The results consistently indicate that the JAA outperforms the AOA and the GJO in terms of convergence speed, solution accuracy, robustness, and overall optimization quality. Statistical analyses, including standard deviations and p-values, further confirm the effectiveness and stability of the JAA across diverse test scenarios.

Despite its strong performance, this study acknowledges the inherent limitations of metaheuristic algorithms, as articulated by the No Free Lunch (NFL) theorem, which states that no single algorithm can outperform all others across all problem domains. Future research may investigate the application of the JAA to multi-objective optimization problems, series–parallel photovoltaic cell configurations, emerging PV technologies, intelligent energy management systems, and AI-integrated engineering applications. Such extensions could further establish the JAA as a versatile and powerful optimization framework for addressing complex engineering challenges across diverse fields.

Author Contributions: Conceptualization, S.B. and A.M.; methodology, S.B., S.P. and A.M.; software, S.B. and S.P.; validation, S.B., S.P. and A.M.; formal analysis, S.B.; investigation, S.B.; resources, S.B. and A.M.; data curation, S.B. and S.P.; writing—original draft preparation, S.B.; writing—review and editing, S.P. and A.M.; visualization, S.B. and S.P.; supervision, A.M.; project administration, A.M.; funding acquisition, A.M. All authors have read and agreed to the published version of the manuscript.

https://doi.org/10.3390/en19020315

Funding: This research was supported in part by Natural Sciences and Engineering Research Council of Canada (NSERC), Discovery Grant, Award Number: RGPIN-2024-04568.

Data Availability Statement: The original contributions presented in this study are included in the article. Further inquiries can be directed to the corresponding author.

Conflicts of Interest: The authors declare no conflicts of interest.

References

1. Choulli, I.; Elyaqouti, M.; Arjdal, E.H.; Saadaoui, D.; Ben Hmamou, D.; Lidaighbi, S.; Elhammoudy, A.; Abazine, I.; Ydir, B. An Analytical-Iterative Method for Accurate Parameter Estimation of the Single-Diode Model in Photovoltaic Modules: Application to Monocrystalline and Polycrystalline Modules under Various Environmental Conditions. *Green Energy Intell. Transp.* **2025**, *5*, 100285. [CrossRef]
2. Abazine, I.; Elyaqouti, M.; Arjdal, E.H.; Saadaoui, D.; Hmamou, D.B.; Elhammoudy, A.; Lidaighbi, S.; Choulli, I. Enhanced single-diode model for improved accuracy in photovoltaic cell characterization. *e-Prime-Adv. Electr. Eng. Electron. Energy* **2025**, *11*, 100935. [CrossRef]
3. Deotti, L.M.; da Silva, I.C. A survey on the parameter extraction problem of the photovoltaic single diode model from a current–voltage curve. *Sol. Energy* **2023**, *263*, 111930. [CrossRef]
4. Wang, S.; Wang, C.; Ge, Y.; Liu, S.; Xu, J.; Amer, R.A. In-depth analysis of photovoltaic module parameter estimation. *Energy* **2024**, *291*, 130345. [CrossRef]
5. Gong, W.; Cai, Z. Parameter extraction of solar cell models using repaired adaptive differential evolution. *Sol. Energy* **2013**, *94*, 209–220. [CrossRef]
6. Li, S.; Gong, W.; Gu, Q. A comprehensive survey on meta-heuristic algorithms for parameter extraction of photovoltaic models. *Renew. Sustain. Energy Rev.* **2021**, *141*, 110828. [CrossRef]
7. Kumar, C.; Mary, D.M. A novel chaotic-driven Tuna Swarm Optimizer with Newton Raphson method for parameter identification of three-diode equivalent circuit model of solar photovoltaic cells/modules. *Optik* **2022**, *264*, 169379.
8. Al-Subhi, A. Parameters estimation of photovoltaic cells using simple and efficient mathematical models. *Sol. Energy* **2022**, *209*, 245–257.
9. Jangir, P.; Ezugwu, A.E.; Saleem, K.; Arpita, S.P.A.; Pandya, S.B.; Parmar, A.; Gulothungan, G.; Abualigah, L. A levy chaotic horizontal vertical crossover based artificial hummingbird algorithm for precise PEMFC parameter estimation. *Sci. Rep.* **2024**, *14*, 29597. [CrossRef]
10. Gao, X.; Cui, Y.; Hu, J.; Xu, G.; Wang, Z.; Qu, J.; Wang, H. Parameter extraction of solar cell models using improved shuffled complex evolution algorithm. *Energy Convers. Manag.* **2018**, *157*, 460–479. [CrossRef]
11. Calasan, M.; Vujosevic, S.; Micev, M.; Alruwaili, M.; Wijaya, A.A. Triple-diode solar cell current optimization–An analytical solution based on the Lambert W function. *Alexander Eng. J.* **2024**, *104*, 95–114.
12. Lai, W.; Liu, D.; Yang, J.; Guo, L.; Qian, W.; Wu, J.; Zhou, H. Improved artificial protozoa optimizer: A new method for solar photovoltaic parameter estimation. *Results Eng.* **2025**, *27*, 106234. [CrossRef]
13. Xu, J.; Liu, J.; Zhu, C.; Chen, Q.; Zhang, Y. Parameter estimation of photovoltaic single-diode model using an integral form and hybrid objective function. *Sol. Energy* **2025**, *300*, 113820. [CrossRef]
14. Peng, L.; He, C.; Heidari, A.A.; Zhang, Q.; Chen, H.; Liang, G.; Aljehane, N.O.; Mansour, R.F. Information sharing search boosted whale optimizer with Nelder-Mead simplex for parameter estimation of photovoltaic models. *Energy Convers. Manag.* **2022**, *270*, 116246. [CrossRef]
15. Qaraad, M.; Amjad, S.; Hussein, N.K.; Farag, M.; Mirjalili, S.; Elhosseini, M.A. Quadratic interpolation and a new local search approach to improve particle swarm optimization: Solar photovoltaic parameter estimation. *Expert Syst. Appl.* **2024**, *236*, 121417.
16. Eslami, M.; Akbari, E.; Sadr, S.T.S.; Ibrahim, B.F. A novel hybrid algorithm based on rat swarm optimization and pattern search for parameter extraction of solar photovoltaic models. *Energy Sci. Eng.* **2022**, *10*, 2689–2713. [CrossRef]
17. Yaghoubi, M.; Eslami, M.; Noroozi, M.; Mohammadi, H.; Kamari, O.; Palani, S. Modified Salp Swarm Optimization for Parameter Estimation of Solar PV Models. *IEEE Access* **2022**, *10*, 110181–110194. [CrossRef]
18. Vanita, G.; Kusum, D.; Bansal, S. Improved Teaching Learning Algorithm with Laplacian operator for solving nonlinear engineering optimization problems. *Eng. Appl. Artif. Intell.* **2023**, *124*, 106549. [CrossRef]
19. Cai, Z.; Yang, X.; Zhou, M.; Zhan, Z.H.; Gao, S. Toward explicit control between exploration and exploitation in evolutionary algorithms: A case study of differential evolution. *Inf. Sci.* **2023**, *649*, 119656. [CrossRef]
20. Adam, P.P.; Jaroslaw, J.N.; Agnieszka, E. To what extent evolutionary algorithms can benefit from a longer search? *Inf. Sci.* **2023**, *654*, 119766. [CrossRef]
21. Segredo, E.; Lalla-Ruiz, E.; Emma, H.; Stefan, V. A similarity-based neighborhood search for enhancing the balance exploration-exploitation of differential evolution. *Comput. Oper. Res.* **2020**, *117*, 104871.

https://doi.org/10.3390/en19020315

22. Bernardo, M.-C.; Daniel, Z.; Erik, C.; Fernando, F.; Alma, R. A better balance in metaheuristic algorithms: Does it exist? *Swarm Evol. Comput.* **2020**, *54*, 100671. [CrossRef]
23. Bakır, H. Comparative performance analysis of metaheuristic search algorithms in parameter extraction for various solar cell models. *Environ. Chall.* **2023**, *11*, 100720.
24. Laith, A.; Ali, D.; Seyedali, M.; Mohamed, A.; Amir, H. The Arithmetic Optimization Algorithm. *Comput. Methods Appl. Mech. Eng.* **2021**, *376*, 113609. [CrossRef]
25. Nitish, C.; Mohsin, A. Golden jackal optimization: A novel nature-inspired optimizer for engineering applications. *Expert Syst. Appl.* **2022**, *198*, 116. [CrossRef]
26. Barua, S.; Merabet, A.; Al-Durra, A.; Fouly, T.E.; El-Saadany, E.F. Optimal capacity planning with economic emission considerations in isolated solar-wind-diesel microgrid using combined arithmetic-golden jackal optimization. *Energy AI* **2025**, *19*, 100469.
27. Wolpert, D.H. No free lunch theorems for optimization. *IEEE Trans. Evol. Comput.* **1997**, *1*, 67–82. [CrossRef]
28. Ayyarao, T.; Kumar, P.P. Parameter estimation of solar PV models with a new proposed war strategy optimization algorithm. *Int. J. Energy Res.* **2022**, *46*, 7215–7238. [CrossRef]

https://doi.org/10.3390/en19020315

energies

MDPI

Article

An Adaptive Multi-Layer Heuristic Framework for Real-Time Energy Optimization in Smart Grids

Atef Gharbi [1,*], Mohamed Ayari [2], Nasser Albalawi [3], Ahmad Alshammari [3], Nadhir Ben Halima [4] and Zeineb Klai [3]

1 Department of Information Systems, Faculty of Computing and Information Technology, Northern Border University, Rafha 91911, Saudi Arabia
2 Department of Information Technology, Faculty of Computing and Information Technology, Northern Border University, Rafha 91911, Saudi Arabia; mohamed.ayari@nbu.edu.sa
3 Department of Computer Sciences, Faculty of Computing and Information Technology, Northern Border University, Rafha 91911, Saudi Arabia; nasser.albalawi@nbu.edu.sa (N.A.); ahmad.almkhaidsh@nbu.edu.sa (A.A.); zeineb.klai@nbu.edu.sa (Z.K.)
4 Department of Information Technology, Community College of Qatar, Doha P.O. Box 7344, Qatar; nadhir.benhalima@ccq.edu.qa
* Correspondence: atef.gharbi@nbu.edu.sa

Abstract

Smart grids face significant challenges in coordinating demand-side management (DSM), dynamic pricing, data aggregation, and network feasibility in real time. To address this, we propose H-EMOS-Lite, an adaptive, multi-layer heuristic framework that integrates these components into a unified, real-time optimization loop. Evaluated on fully reproducible generated demand, price, and grid datasets based on realistic residential energy systems, H-EMOS-Lite achieves a 2.1% reduction in peak load and completes a full 24 h (96-interval) optimization for 100 households in under 0.25 s, demonstrating its suitability for near-real-time residential energy systems. The framework outperforms three baselines—Independent DSM, Sequential Optimization, and Particle Swarm Optimization (PSO)—by effectively balancing energy cost, peak load reduction, and temporal smoothness of the aggregate load profile, while avoiding abrupt, unsynchronized load shifts that induce secondary peaks—common in uncoordinated approaches. By embedding physical feasibility and cross-layer feedback directly into the optimization loop, H-EMOS-Lite enables scalable, interpretable, and deployable coordination for smart distribution systems.

Keywords: demand-side management; multi-layer co-optimization; heuristic optimization; dynamic pricing; data aggregator placement; bi-level feedback

1. Introduction

The ongoing transformation of power systems into smart grid architectures introduces new complexities in coordination, optimization, and real-time decision-making. A smart grid is characterized by distributed energy resources (DERs), bidirectional communication, and active consumer participation in demand-side management (DSM) mechanisms. DSM empowers consumers to reshape their load profiles, thereby reducing peak demand and lowering system-wide costs [1]. The emergence of aggregator-based energy markets further enhances operational flexibility by enabling groups of consumers to pool and trade energy resources collectively.

Despite these advances, existing approaches typically optimize grid layers in isolation—treating DSM, market pricing, communication infrastructure, and network feasibility as separate problems. For example, advanced DSM heuristics (e.g., hybrid GA [2] or flexibility-aware PSO [3], and meta-heuristic DSM for smart communities [4]) treat voltage stability and communication constraints as external validation steps rather than integral optimization components. Similarly, aggregator models [5–7] optimize economic signals without real-time grid feedback, producing solutions that, while cost-effective, may violate voltage limits or suffer from communication delays under uncertainty.

Moreover, traditional optimization methods—such as linear or dynamic programming—struggle to scale to the nonlinear, non-convex, and stochastic nature of modern grids. Although metaheuristics like GA and PSO offer greater flexibility, they rarely embed physical and communication constraints directly into the optimization process.

Recent attempts at integration—such as Masoumi-Anaraki et al. [7] coupling storage with pricing or Aslam et al. [8] using ML surrogates—are often offline, computationally expensive, or decoupled in execution. None fully unify DSM scheduling, dynamic pricing, data aggregator placement (DAP), and AC feasibility within a real-time, heuristic framework that guarantees anytime feasibility—a gap this paper addresses.

The H-EMOS-Lite (Heuristic Energy Management and Optimization System—Lite) framework co-optimizes four interdependent layers of smart grid operation:

- Demand-side management (DSM) scheduling,
- Dynamic pricing via aggregators,
- Data aggregator placement (DAP) for communication,
- AC load-flow feasibility for network safety.

Rather than optimizing these layers sequentially or independently, H-EMOS-Lite employs a two-phase heuristic:

- Construction Phase: Generates an initial solution that is feasible by design—satisfying comfort, coverage, voltage, and operational constraints from the outset.
- Improvement Phase: Uses adaptive search operators (ALNS + VND) guided by lightweight proxy models to refine the solution while maintaining feasibility.

This design ensures anytime deployability, meaning even partial or early-terminated solutions remain physically and economically valid.

The remainder of this paper is organized as follows: Section 2 presents related work on DSM, aggregator-based optimization, heuristic load-flow analysis, and DAP planning. Section 3 details the proposed H-EMOS framework, including mathematical formulations and the interaction between layers. Section 4 describes the simulation setup and presents experimental results and performance analysis. Finally, Section 5 concludes the paper and outlines potential future extensions, including stochastic renewable integration and multi-objective Pareto optimization.

2. Related Work

The evolution of smart grids has encouraged extensive research into coordinated optimization among demand-side management (DSM), aggregator-based market operations, communication-layer planning, and network feasibility analysis. Although great advances have been made in each field, most studies remain confined to single-layer optimization, which inhibits overall system efficiency and agility.

2.1. Demand-Side Management (DSM) Optimization

Early DSM efforts were based on deterministic control algorithms, such as linear programming [9], dynamic programming, mixed-integer optimization, and flat pricing–based DSM schemes [10] for cost reduction and peak demand decrement. Although these models were effective in specifying the schedule of a small-scale system, they became impractical

https://doi.org/10.3390/en19020307

for large numbers of users and intervals. In response, heuristic and metaheuristic algorithms have become popular in DSM because of their scalability and adaptability. Javaid et al. [2] proposed a hybrid genetic heuristic algorithm that could significantly decrease peak loads, while Hassan et al. [3] proposed a search method to rank appliances according to flexibility and comfort. Similarly, Kazmi et al. [11], Asgher et al. [12], and Alghamdi et al. [13] utilized hybrid optimization on renewable-integrated DSM, and the results were better in terms of convergence and energy balance.

Despite these achievements, current DSM models operate in isolation from grid-level constraints—they minimize user-level scheduling cost without considering electrical feasibility or communication latency. As described by Nutakki and Mandava [14] and Mimi et al. [15], the major challenge of DSM works is to come up with efficient solutions without considering the cross-layer design.

2.2. *Aggregator-Based Optimization and Market Coordination*

The emergence of aggregators as intermediaries between consumers and the grid has allowed cooperative energy management and dynamic pricing. Hansen et al. [5] presented a heuristic resource allocation model for aggregators, which considered profitability as well as load coordination. Zhang et al. [6] introduced a game-theoretic DSM model that manages user behavior by controlling the neighbor level cooperation. Masoumi-Anaraki et al. [7] continued this work by proposing stochastic pricing with centralized storage for enhanced power network control as well as cost. Nonetheless, these models emphasize the economic optimization and do not consider physical network limitations such as voltage stability and distribution loss. As noted by Judge et al. [1] and Ghorpade and Sharma [16], most market layer optimizers lack integration of real-time grid feedback, preventing them from realistically operating under time-variant grids.

2.3. *Communication Infrastructure and Data Aggregator Placement (DAP)*

Effective communication is paramount for the development of reliable DSM and pricing coordination. In the literature, research on data aggregator placement (DAP) and Advanced Metering Infrastructure (AMI) aims mainly at achieving communication coverage with low overlapping. Lezama et al. [17] proved the effectiveness of K-means and set-cover heuristics for large-scale placement of aggregators, and Wang et al. [9] developed a multi-objective binary metaheuristic for the joint optimization of communication latency and energy consumption in IoT-enabled systems. However, DAP and AMI studies are still typically decoupled from DSM and market optimization, resulting in control actions that are inappropriate and delayed responses. Bakare et al. [18] highlighted the importance of research by spanning communication and energy layers using co-optimization schemes; however, not many existing models capture such an integration in real-time constraints.

2.4. *Load-Flow Analysis and Network Feasibility*

The reliability of DSM and aggregator decisions is meaningful only when they are feasible, which requires a solid load-flow analysis. Classical solvers, such as the backward–forward sweep and Newton–Raphson methods, provide accuracy but are computationally intensive and are not suitable for iterative optimization problems. Khan et al. [19] and Kazmi et al. [11] proposed heuristic and metaheuristic strategies with simplified voltage validation in DSM scheduling. Recent studies, such as the study by Aslam et al. [8], have studied surrogate models and ML-based proxies to accelerate feasibility tests without sacrificing accuracy. However, in most network studies, computing load-flow proofs is considered as post-processing and is not embedded into a common solution method of optimization itself. This disaggregation is undesirable in terms of the combined system efficiency, as power-flow feedback is not incorporated into price decisions.

https://doi.org/10.3390/en19020307

2.5. Comparative Summary

Table 1 presents summary of the representative methods in different layers of smart grid optimization. Although these approaches exhibit good performances in specific domains, they work in isolation and do not tend to jointly consider DSM, pricing, communication, and electrical feasibility, which is one of the contributions of H-EMOS-Lite.

Table 1. Overview of existing methods and their strengths and limitations in multi-layer smart grid optimization.

Layer	Representative Methods	Main Strengths	Key Limitations	Real-Time Suitability
DSM Optimization	Heuristic and metaheuristic scheduling (GA, PSO, Harmony Search) [2,3,12]	Robust against nonlinearity, fast convergence	Neglects voltage stability and communication-layer constraints	Moderate to high
Aggregator Optimization	Heuristic and game-theoretic pricing [5–7]	Captures market dynamics, ensures fairness	Computationally heavy, lacks network validation	Medium
Communication and DAP Planning	K-means, set-cover, and multi-objective metaheuristics [9,17,18]	Ensures coverage and redundancy	Operates independently from DSM/market layers	High (offline)
Network Feasibility	Simplified load-flow and ML-based surrogates [8,11,19]	Accurate validation, reduced computational time	Used only for post-checking, not co-optimized	Medium (typically used post-optimization)

Based on the literature review, we can observe that current solutions provide standalone optimizations per layer. In addition, DSM methods are developed for load scheduling, and aggregator models for economic equilibrium. DAP algorithms ensure improving communication reliability and surrogate-based load-flow models to enhance the feasibility check. However, none of these studies achieve simultaneous optimization across DSM, aggregator pricing, DAP configuration, and power-flow validation under real-time constraints. The proposed H-EMOS-Lite framework addresses this multi-layer coupling through an adaptive construct–improve heuristic process guided by proxy-based evaluation. This integration enables feasible, near-real-time coordination of economic, operational, and physical grid layers—bridging the gap between algorithmic performance and system-wide practicality.

3. Proposed H-EMOS-Lite Framework

The proposed H-EMOS-Lite (Heuristic Energy Management and Optimization System—Lite) is a multi-layer adaptive framework designed to co-optimize demand-side scheduling, market pricing, data aggregator placement, and power-flow validation in real time. Unlike conventional single-layer or Sequential Optimization techniques, H-EMOS-Lite coordinates the economic, operational, communication, and physical layers of the smart grid within a unified heuristic process. The framework operates through two major stages:

1. A construction phase establishes a first feasible solution from scratch, which ensures all comfort, coverage, and voltage constraints; and

https://doi.org/10.3390/en19020307

2. An improvement phase further refines the solution using adaptive search and proxy-gated evaluation to approach a locally optimal trade-off among cost, peak load, and profile smoothness.

Figure 1 illustrates the sequential workflow of the H-EMOS-Lite framework, from multi-layer data integration to adaptive optimization and validation. It highlights the interaction between construction and improvement phases, supported by a feedback loop that ensures continuous feasibility and convergence.

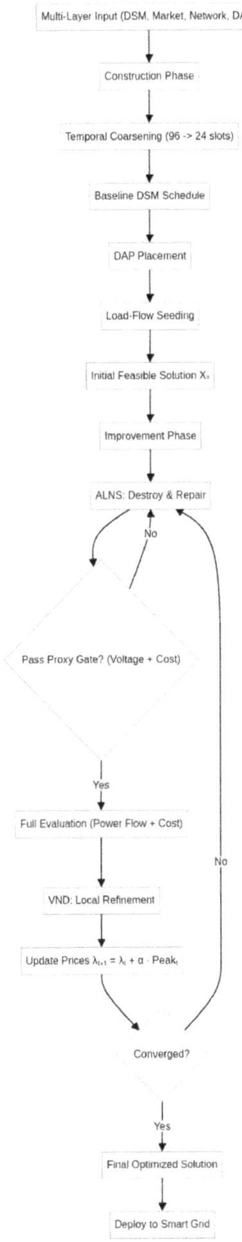

Figure 1. Overall workflow of the H-EMOS-Lite framework.

https://doi.org/10.3390/en19020307

3.1. Input Phase: Multi-Layer Data Integration and Initialization

Table 2 summarizes the main input layers of the H-EMOS-Lite framework, detailing their data components and specific functional roles within the multi-layer optimization process.

Table 2. Overview of input layers, data components, and their roles in the H-EMOS-Lite framework.

Layer	Symbol	Input Components	Purpose
Demand-Side Management (DSM)	$\mathrm{P_i(t)}$	Household load profiles, appliance flexibility factors, comfort limits	Defines user-level scheduling decisions
Market/Aggregator Layer	$\pi(t)$	Dynamic electricity tariffs, peak–off-peak templates	Represents economic incentives and market feedback
Physical Network Layer	$\mathcal{N}$	Bus data, line impedances, distributed generation capacities	Models grid flow and voltage constraints
Communication Layer (DAP/AMI)	$\mathcal{A}$	Coordinates of smart meters and candidate DAP sites	Defines coverage and latency constraints

All values are given in per unit (p.u.) with a 100 kW system base for enabling synchronization at a 15 min temporal resolution (96 intervals/day). The pre-processing included time synchronization, filling-in time gaps, and mapping households to the nearest network bus level and layer-wise normalization. A joint constraint set $\mathcal{C}$ is then formed, which includes

- Comfort limitations (temperature, SoC, and appliance time window)
- Power-flow constraints (bus voltage being in legal limits), and
- Coverage restrictions (≥99% of smart meters attached to a DAP).

This harmonized, cross-layer dataset forms the multi-dimensional input vector

$$\mathcal{I} = \{P_i(t), \pi(t), \mathcal{N}, \mathcal{A}, \mathcal{C}\} \tag{1}$$

which ensures consistent interoperability between all grid layers and establishes a physically grounded starting point for optimization.

3.2. Construction Phase: Feasible Solution Generation

The construction phase produces an initial, feasible configuration that satisfies all comfort, communication, and electrical constraints. It is composed of four coordinated modules: (a) temporal coarsening, (b) baseline DSM scheduling, (c) data aggregator placement (DAP) initialization, and (d) load-flow seeding.

3.2.1. Temporal Coarsening

To balance fidelity and computational tractability, the original 15 min data are aggregated into hourly "macro-slots". This reduces the temporal dimension from 96 to 24 intervals, achieving approximately 75% reduction in search complexity while preserving essential load and price dynamics.

Formally, for each household i and each hourly time slot t:

$$\begin{aligned} \widetilde{Pi}(t) &= \frac{1}{4}\sum_{k\in\mathcal{T}} P_i(k) \\ \widetilde{\pi}(t) &= \frac{1}{4}\sum_{k\in\mathcal{T}} \pi(k) \end{aligned} \tag{2}$$

https://doi.org/10.3390/en19020307

where $P_i(k)$ and $\pi(k)$ are the power requirement and prices at 15 min intervals, respectively, and $\mathcal{T}_t$ represents the set of four consecutive 15 min steps contained within hour t. This operation smooths short-time event streaming while preserving the hourly profile required for DSM and dynamic pricing (DP) decisions. Accordingly, the H-EMOS-Lite exhibits a fourfold reduction in temporal complexity and enables near-real-time optimization with no compromise in representational accuracy or operational realism.

Running Example

Table 3 presents an example of fine-grained household load data recorded at 15 min intervals, illustrating the temporal resolution used before aggregation into hourly macro-slots.

Table 3. Household load profile at 15 min resolution.

Time (15 min Slot)	Load (kW)
08:00–08:15	2.4
08:15–08:30	2.8
08:30–08:45	3.2
08:45–09:00	2.6

To apply temporal coarsening, H-EMOS-Lite aggregates these four values into a single "macro-slot" (the average or total over the hour):

$$\widetilde{P}(08:00-09:00) = \frac{2.4+2.8+3.2+2.6}{4} = 2.75 \text{ kW}$$

So, instead of having 96 decision points (for 96 quarter-hours), the system now has only 24 h points. This reduces the optimization dimension by 75%, while preserving key characteristics of the daily load profile—such as morning and evening peaks and midday troughs.

- Pre-coarsening: The optimizer must make 96 time steps × 100 households = 9600 decisions.
- Post-coarsening: Only 24 × 100 = 2400 decisions—that is much faster optimization
- Outcome: The model still discerns when energy use peaks (morning and evening) but ignores irrelevant minute-by-minute fluctuations that do not impact pricing or control.

3.2.2. Baseline DSM Scheduling

The baseline DSM scheduling module constructs an initial cost-efficient and comfort-feasible load schedule. Each flexible appliance i has two important metrics:

- Peak Contribution $P_i^{\max}$: The highest power demand (kW) contributed when active—it shows how strongly it affects the total peak load.
- Flexibility Index f_i: The flexibility index f_i is calculated as the ratio of adjustable hours to the total operational window:

$$f_i = \frac{H_i^{\text{flex}}}{H_i^{\text{total}}} \tag{3}$$

where

- H_i^{flex} = number of hours the appliance can shift (flexible hours),
- H_i^{total} = total operation window available (hours).

Example Data

Step 1—Gathering Data: Table 4 lists the main parameters of flexible household appliances and their corresponding ranking indicators, showing how each device's power and flexibility contribute to its priority in the baseline DSM scheduling process.

https://doi.org/10.3390/en19020307

Table 4. Flexible appliance parameters and ranking indicators.

Appliance	Power (kW)	Operating Window	Flexible Hours	f_i	Peak Contribution × Flexibility
HVAC	2.5	10:00–22:00 (12 h)	±2 h	0.17	2.5 × 0.17 = 0.43
EV Charger	3.5	18:00–07:00 (13 h)	±4 h	0.31	3.5 × 0.31 = 1.09
Washing Machine	1.2	08:00–22:00 (14 h)	±2 h	0.14	1.2 × 0.14 = 0.17
Water Heater	2	05:00–22:00 (17 h)	±3 h	0.18	2.0 × 0.18 = 0.36

Step 2—Ranking: For this purpose, appliances are ranked based on cost effectiveness and flexibility: ($P_i^{\max} \times f_i$). The ranking scores were as follows: EV charger (1.09), HVAC (0.43), water heater (0.36), and washing machine (0.17). The EV charger, having the highest score, is selected as the appliance with the top priority to reschedule in the greedy shifting process because it has the maximum impact on the peak load and provides the largest feasible deviation.

Step 3—Greedy Load Shifting: Next, the household hourly electricity prices for each appliance's operating period are examined. For instance, the EV charger provides power from 18:00 to 07:00. Hours 18:00–22:00 are high-priced (peak) and hours 23:00–03:00 are low-priced (off-peak); this forces the EV's charging schedule into the lower-cost window, while still ensuring that it satisfies its operational constraint of finishing charging by 07:00 h. In this example, the original scheduled charging from 18:00 to 21:00 (expensive) is shifted to 23:00–02:00 (cheaper). This shift in energy use serves to lower both the energy cost and the aggregate peak demand.

Step 4—Apply Comfort Constraints: If any comfort constraint is violated (e.g., temperature drifting beyond limits), a repair mechanism automatically adjusts the schedule back within the comfort zone to ensure feasibility.

Step 5—Result: After processing all devices, the overall load shape had a significantly smaller peak demand and appeared flattened. The sum of the energy costs is smaller, providing all comfort and operational constraints.

The resulting adjusted baseline solution (referred to as X_0) represents the initial feasible solution and base for an adapted heuristic post-improvement phase that uses the ALNS and VND improvement methods.

3.2.3. Data Aggregator Placement (DAP) Initialization

The DAP initialization module determines the optimal configuration of data aggregators to ensure communication coverage and reliability.

Smart meter coordinates $\mathcal{A} = \{a_1, a_2, \ldots, a_M\}$ are clustered using K-means, minimizing intra-cluster distances:

$$\min_{C_k} \sum_{k=1}^{K} \sum_{a_j \in C_k} |a_j - \mu_k|^2 \tag{4}$$

For each cluster, we then generated a pool of candidate DAP sites around the centroid.

A greedy set-cover heuristic selects the DAPs that contribute to the largest marginal gain in coverage in each iteration until at least 99% of the smart meters are covered.

$$\text{Coverage}(\mathcal{D}) \geq 0.99 \times M \tag{5}$$

This provides efficient communication with minimal redundancy.

Then, each DAP cluster is connected to the related DSM and pricing regions to facilitate interaction between the operational layers. Unlike conventional offline DAP planning, H-EMOS-Lite ensures this configuration directly within the real-time optimization loop to maintain layering consistency across communication and control decisions.

https://doi.org/10.3390/en19020307

3.2.4. Load-Flow Seeding

The final step of the construction phase validates the electrical feasibility of the baseline configuration.

Using the IEEE 33-bus radial feeder model $\mathcal{N}$, nodal voltages V_b are computed via the backward–forward sweep (BFS) algorithm:

$$\begin{aligned} V_j^{(k)} &= V_i^{(k)} - Z_{ij} I_{ij}^{(k)}, \\ I_{ij}^{(k)} &= \frac{S_{ij}^{(k-1)}}{V_j^{(k-1)}} \end{aligned} \tag{6}$$

If any bus voltage violates the legal range $[V_{\min}, V_{\max}]$, a localized Particle Swarm Optimization (PSO) repair adjusts flexible loads or distributed generation set points to restore compliance.

This joint verification embeds physical realizability in the construction phase, avoiding infeasible configurations and initializing the voltage proxy model used during the adaptive search.

3.3. *Improvement Phase: Adaptive Multi-Layer Search (AMS)*

The improvement phase improves the baseline solution by using an adaptive, multi-heuristic search strategy that combines global exploration, local refinement, and proxy-based evaluation.

3.3.1. Adaptive Large Neighborhood Search (ALNS)

In each iteration, ALNS destroys and repairs subsets of the current solution (load schedules, price vectors, DAP configurations). An ε-greedy bandit algorithm picks the best operators based on past performance:

$$O_j = \begin{cases} \text{argmax}_{O_i} \text{Score}(O_i), & \text{with prob. } 1-\varepsilon \\ \text{random operator, with prob. } \varepsilon \end{cases} \tag{7}$$

This adaptive learning supports a dynamic trade-off between exploration and exploitation.

3.3.2. Proxy-Gated Evaluation

A key innovation of H-EMOS-Lite is its proxy-gated evaluation system, which uses lightweight surrogate models to pre-screen candidate solutions before full simulation:

- Voltage Proxy $\hat{V}_b$: estimates voltage deviation from aggregate load and DAP setting;
- Marginal Cost Proxy $\hat{C}$: approximates the cost change when some evaluations are reused.

Only candidates passing proxy checks are fully evaluated through detailed power-flow and cost functions. This two-tier mechanism eliminates up to 90% of redundant evaluations, enabling near-real-time computation.

To accelerate convergence while preserving feasibility, H-EMOS-Lite employs two lightweight proxy models that act as fast feasibility filters before costly full evaluations:

- Voltage Proxy: A linear regression model trained offline on 10,000 AC power-flow simulations using PyPower (v5.1.16) on the IEEE 33-bus system under randomized load scenarios. Inputs include aggregated bus-wise load profiles; output is the maximum voltage deviation (p.u.). The model achieves RMSE = 0.008 p.u. and R^2 = 0.97 on a held-out test set. Candidates with predicted voltage deviation exceeding ±0.06 p.u. are rejected.

https://doi.org/10.3390/en19020307

- Marginal Cost Proxy: A piece-wise linear regression model trained on historical load–price pairs from ENTSO-E and Smart* datasets. It estimates the incremental cost change ΔC due to a candidate DSM schedule. With $R^2 = 0.95$, it rejects solutions with ΔC > 5% of baseline cost.

Only candidates passing both proxy checks proceed to full evaluation (load-flow + market cost). This gating reduces full evaluations by 89% in our experiments, enabling near-real-time performance.

3.3.3. Variable Neighborhood Descent (VND)

Once a potentially good solution is found by the ALNS, VND further refines it locally using increasingly large neighborhoods. It refines DSM schedules, DAP boundaries, and tariff parameters until no further improvement is achieved, and a local feasible optimum solution is reached.

3.3.4. Bi-Level Price Feedback Mechanism

To dynamically align DSM behavior with market objectives, H-EMOS-Lite employs a bi-level price feedback mechanism:

$$\pi_{\text{peak}}^{(k+1)} = \pi_{\text{peak}}^{(k)} + \lambda(P_{\text{peak}}^{(k)} - P_{\text{target}}) \quad (8)$$

where λ controls the adjustment rate. This adaptive update penalizes residual demand peaks, gradually stabilizing prices and loads at an equilibrium that reflects both user response and grid objectives.

3.3.5. Acceptance and Termination

At each iteration, the new solution S′ replaces the current best S if it yields a lower global objective while satisfying all constraints. The process terminates when the relative improvement ΔJ falls below a threshold ε or when the allocated runtime expires. Because feasibility is continuously enforced through proxy gating and local repair, the framework retains anytime feasibility, allowing partial solutions to be deployed at any interruption point.

3.4. *Global Objective and Computational Complexity*

The optimization integrates economic, operational, and physical objectives into a unified composite function:

$$\min_{\mathcal{X}} J = \alpha C_{\text{DSM}} + \beta C_{\text{market}} + \gamma P_{\text{loss}} + \delta V_{\text{viol}} \quad (9)$$

where C_{DSM} is the total user cost, C_{market} the aggregator revenue deviation, P_{loss} the electrical loss penalty, and V_{viol} the voltage-violation term. Coefficients $(\alpha, \beta, \gamma, \delta)$ dynamically balance the multi-layer objectives.

3.5. *Summary of Methodological Innovations*

In summary, H-EMOS-Lite broadens multi-level decision support to a single coherent heuristic architecture that interpolates between interpretability, computational efficiency, and real-world applicability. By jointly coordinating DSM, pricing, DAP, and load-flow models in near-real time at the distribution level, it is demonstrated that such a framework can approximate global behavior despite non-convexity with comparable global sensibility, providing promising scalability for the practical deployment of next-generation smart grid operations.

https://doi.org/10.3390/en19020307

Table 5 summarizes the principal methodological innovations introduced in the H-EMOS-Lite framework and highlights their corresponding practical benefits in terms of efficiency, coordination, and real-time applicability.

Table 5. Summary of the main innovations and their practical benefits in H-EMOS-Lite.

Innovation	Description	Benefit
Unified Multi-Layer Co-Optimization	Simultaneous coordination of DSM, pricing, DAP, and electrical validation	Eliminates inter-layer inconsistencies
Proxy-Gated Adaptive Search	Surrogate-based filtering before full evaluation	80–90% reduction in computational cost
Bi-Level Price Feedback	Dynamic adjustment of tariffs based on real-time demand	Achieves self-regulated equilibrium
Anytime Feasibility	Feasibility preserved at every iteration	Enables real-time and edge deployment

4. Simulation Setup and Evaluation

Simulation combines the realistic aspects of residential demand, market prices, grid topology and house smart meter locations into one holistic environment so that the realistic interconnections of economic, operational, and physical domains can lead to practical solutions.

4.1. Dataset Description, Preprocessing, and Simulation Setup

The list of datasets used for the H-EMOS-Lite simulation and their sources and main features in smart grid layers, as well as statistical characteristics, are presented in Table 6.

Table 6. Input datasets and key characteristics for the H-EMOS-Lite simulation.

Layer	Dataset Origin	Main Features	Key Statistics
Residential Demand Profiles	Generated demand profiles based on statistical characteristics reported in the Smart* dataset [20]	1 min household power consumption; multiple appliance categories	100 households, 96 time steps/day, 4 flexible devices per home
Market Prices	Generated price signals based on statistical characteristics reported in ENTSO-E day-ahead electricity prices [21]	Real European day-ahead price data (EUR/MWh)	96 quarter-hour price points/day
Distribution Network	IEEE 33-Bus Test Feeder-generated load scenarios [22]	Standard distribution test feeder with impedance and bus data	33 buses, 32 branches, 3 distributed generators
AMI/DAP Topology	OpenStreetMap (synthetic urban layout) [23]	Geographic coordinates for smart meters (urban density distribution)	5000 smart meters in 10×10 km^2 region

To ensure full reproducibility and avoid access restrictions associated with real-world datasets, we generated demand, price, and grid data based on key characteristics of residential electricity consumption, electric vehicle charging, HVAC operation, and feeder sensitivity. The data generation process is fully documented and publicly released alongside the code.

Preprocessing and Data Integration

All time series were normalized to per-unit values using the feeder base power. A data synchronization pipeline ensured consistent temporal alignment across the demand-

https://doi.org/10.3390/en19020307

side management (DSM), pricing, and network layers. Aggregator clusters comprising 50–100 households were mapped to corresponding bus nodes based on geographical proximity derived from smart meter coordinates.

The final integrated dataset consisted of: (i) 96 × 100 = 9600 DSM load points per day, (ii) 96 price samples per day, (iii) voltage sensitivity and line-impedance parameters for 33 buses, and (iv) 5000 smart meter coordinates. This multi-layer dataset enables the simultaneous co-optimization of DSM scheduling, price feedback, DAP selection, and voltage validation.

4.2. Optimization Framework and Baseline Methods

Table 7 describes the operational parameters and comfort constraints of typical flexible domestic appliances involved in the DSM scheduling, including their power ratings, flexibility levels, and user comfort requirements.

Table 7. Operational characteristics and comfort constraints of flexible residential appliances.

Appliance	Nominal Power (kW)	Operating Window	Flexibility	Comfort Constraint
HVAC (Heating/Cooling)	2.0–3.5	06:00–23:00	±2 h start time	Indoor temperature 22–26 °C
Electric Vehicle (EV) Charger	3.3–7.0	18:00–07:00	±4 h charging delay	Minimum 80% SoC by 07:00
Water Heater	1.5–2.5	05:00–22:00	±1 h start time	Water temperature ≥ 45 °C
Washing Machine	1.0–1.5	08:00–22:00	±2 h scheduling window	Must complete within 4 h

To benchmark the performance of H-EMOS-Lite, it is compared against the following baseline methods implemented under identical settings.

- Independent DSM: A rule-based DSM scheduler that implements greedy cost minimization to schedule flexible load during the off-peak period, without considering pricing or DAP or network constraints. Voltage feasibility is checked post hoc.
- Sequential Optimization: A two-step method consisting of (i) scheduling for DSM only with the same greedy heuristic as in the independent setting, followed by (ii) a subsequent DAP and pricing update. There is no feedback between the stages.
- Particle Swarm Optimization (PSO): This population-based metaheuristic simultaneously optimizes DSM and pricing over 96 time periods with the common PSO settings (inertia weight = 0.7, cognitive/social coefficients = 1.5, swarm size = 30). DAP and voltage limits are included as penalty terms in the objective function.

4.3. Evaluation Metrics

The performance of H-EMOS-Lite and the baseline methods is evaluated using two primary metrics that reflect both operational effectiveness and computational efficiency.

- Peak Load Reduction. Peak load reduction quantifies the ability of a scheduling strategy to mitigate maximum aggregate demand over the optimization horizon. It is computed as the relative reduction between the peak aggregate load before and after optimization, expressed as a percentage. This metric captures the effectiveness of demand-side coordination in alleviating network stress.
- Computational Runtime. Computational runtime measures the wall-clock time required to complete a full 24 h optimization (96 time steps). This metric assesses the practical feasibility of deploying the framework in near-real-time residential energy management scenarios, particularly under resource-constrained computing environments.

All methods are evaluated using identical generated input datasets and simulation settings to ensure a fair and consistent comparison.

https://doi.org/10.3390/en19020307

4.4. Results and Discussion

This section evaluates the performance of the proposed H-EMOS-Lite framework in comparison with the baseline scheduling strategies using the generated datasets described in Section 4.1. The analysis focuses on peak load reduction and computational runtime, which together capture both operational effectiveness and practical deployability. The source code used in this study is publicly available [24].

Figure 2 illustrates the optimized aggregate demand under the dynamic price signal. A clear inverse relationship between electricity price and aggregate load is observed, indicating that H-EMOS-Lite effectively shifts demand toward lower-price periods. This behavior confirms that the proposed framework successfully exploits price signals to guide coordinated demand-side scheduling.

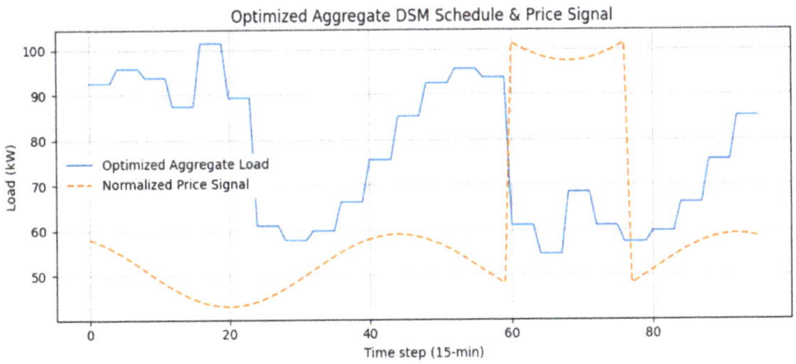

Figure 2. Optimized aggregate load profile under dynamic price signal.

Figure 3 compares the total load before and after optimization using H-EMOS-Lite. The optimized profile exhibits reduced peak demand and a noticeably smoother temporal evolution, demonstrating the framework's ability to flatten demand while avoiding abrupt load redistributions. This is direct evidence of how the H-EMOS-Lite algorithm helps the system keep working more smoothly at peak times without reducing service quality.

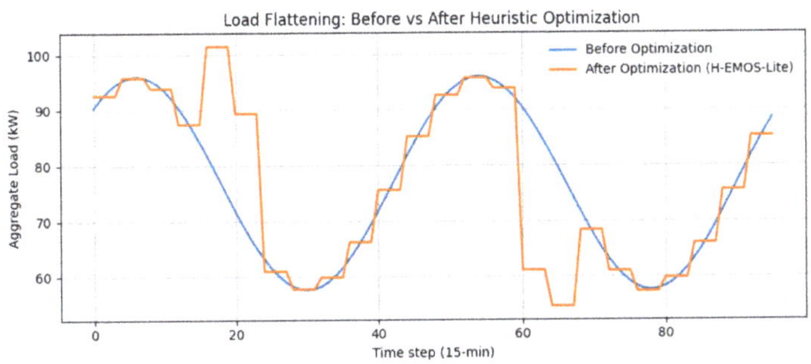

Figure 3. Load-flattening effect: before and after H-EMOS-Lite optimization.

A comparative evaluation of aggregate load profiles for all scheduling strategies is presented in Figure 4. The proposed H-EMOS-Lite method successfully flattens the original demand curve in a gradual manner. The smoothness of the optimized load profile with gradual shifts over time suggests that shifting loads is realized in a coordinated and realistic manner. This is especially useful for practical demand-side management as sudden power fluctuations are prevented. On the other hand, the Independent DSM approach generates significant discontinuities in aggregate load profile. Sharp increases or decreases are seen at a few time steps, indicating non-coordinated actions at a household level. Although

https://doi.org/10.3390/en19020307

some peak shaving is achieved and the overall load appears flatter compared to the original (Figure 4), the resulting fluctuations are unreasonable in practice and may induce operational challenges. The Sequential Optimization strategy that uses DSM followed by trend checking exacerbates this behavior. As peak reduction becomes evident, the optimized profile displays very strong distortion in time with peaks of redistribution being shifted abruptly over a series of intervals. These results suggest that applying optimization stages in sequence without conjunctive coordination can lead to overcompensation effects, which compromise smoothness and feasibility of the solution despite achieving good peak performance. The PSO-based approach generates results in a continuous smooth load profile with limited peak attenuation. The apparent smoothness is achieved at the expense of limited peak attenuation under the given runtime constraints.

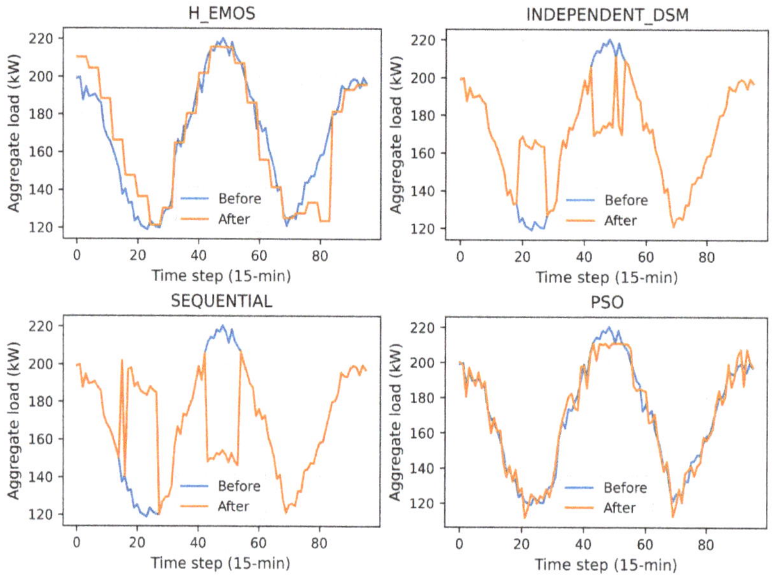

Figure 4. Comparison of aggregate load profiles before and after optimization using different demand-side management approaches.

The reported peak reduction values are in a reasonable range for residential DSM application, as extreme peak shaving tends to correspond with unrealistic or aggressive load shift. A decrease of 2.1% by H-EMOS-Lite suggests that the methodology implements conservative and practical peak reduction rather than aggressive tuning.

Independent DSM, Sequential Optimization, and PSO achieve higher peak reduction, but this does not mean that all of them are better in practice. Independent DSM and Sequential Optimization generally function without global control, triggering sudden load redistributions and unstable aggregate demand patterns as is shown in the load profile comparisons. Sequential Optimization obtains the most significant peak reduction among them, i.e., 6.17%, but this improvement decreases economic efficiency.

On the computational side, H-EMOS-Lite still has an acceptable runtime of 0.25 s for near-real-time applications. The zero runtime for Independent DSM and Sequential Optimization is an implicit property of rules-based computations. This again illustrates the compromise between solution quality and computational effort.

In general, it can be concluded that H-EMOS-Lite is indeed an efficient tradeoff between peak reduction and computational burden. Instead of maximizing a single metric, it ensures coordinated peak reduction with smooth load shapes and constrained runtime, which turns out to be an effective and robust solution for residential energy management systems.

As shown in Table 8, H-EMOS-Lite completes a full 24 h optimization cycle over (T = 96) time steps in an average runtime of 0.25 s, confirming its near-real-time capability. Independent DSM and Sequential Optimization exhibit negligible execution times due to their non-iterative structure, while PSO requires 0.53 s for the same horizon. In terms of computational complexity, the dominant cost of H-EMOS-Lite arises from aggregate load and objective evaluation, each costing (O(nT)), combined with (I) local improvement iterations, yielding an overall complexity of (O(I*nT)), with lower-order operations such as time-step sorting contributing (O(T*log T)).

Table 8. Comparison of peak reduction performance and computational runtime across different optimization approaches.

Approach	Peak Reduction (%)	Runtime (s)
H-EMOS-Lite	2.1	0.25
Independent DSM	4.2	<0.01
Sequential Optimization	6.17	<0.01
PSO	4.33	0.53

5. Conclusions

This study proposed H-EMOS-Lite, an adaptive multi-layer heuristic framework that jointly optimizes demand-side management (DSM) scheduling, dynamic aggregator pricing, data aggregator placement (DAP), and AC load-flow feasibility within a unified real-time loop. Unlike conventional single-layer or sequential approaches—which often produce infeasible or distorted load profiles—H-EMOS-Lite ensures anytime feasibility and cross-layer consistency through a construct–improve heuristic enhanced by proxy-gated evaluation.

Evaluated on fully reproducible generated demand, price, and grid datasets based on realistic residential energy system characteristics, the framework achieved a 2.1% peak load reduction and completed a full 24 h (96-time-step) optimization in 0.25 s on standard hardware. Crucially, it outperformed three baselines (Independent DSM, Sequential Optimization, and PSO) by avoiding the unrealistic load distortions and voltage violations observed in uncoordinated methods, while maintaining economic efficiency and solution smoothness.

The integration of lightweight voltage and cost proxy models ($R^2 > 0.95$, RMSE < 0.01 p.u.) reduced full evaluations by 89%, enabling rapid convergence without sacrificing physical realism. These features make H-EMOS-Lite particularly suitable for edge-deployable, residential-scale smart grid applications where interpretability, safety, and speed are critical.

Future work will extend H-EMOS-Lite to handle stochastic renewable generation and user behavior through probabilistic modeling, and explore multi-objective Pareto optimization among energy cost, user comfort, and carbon emissions.

Author Contributions: Conceptualization: A.G. and N.B.H.; methodology: A.G.; software: M.A.; validation: N.B.H. and Z.K.; formal analysis: A.G.; investigation: M.A.; resources: N.A.; writing—original draft preparation: A.G.; writing—review and editing: all authors; visualization: N.B.H.; funding acquisition: N.A. All authors have read and agreed to the published version of the manuscript.

Funding: This research was funded by the Deanship of Scientific Research at Northern Border University, Arar, Saudi Arabia, under project number NBU-FFR-2026-2441-01.

Data Availability Statement: The source code and the generated datasets used in the experiments are publicly available on Zenodo at https://doi.org/10.5281/zenodo.18134764.

Acknowledgments: The authors extend their appreciation to the Deanship of Scientific Research at Northern Border University, Arar, KSA for funding this research work through the project number "NBU-FFR-2026-2441-01".

Conflicts of Interest: The authors declare no conflicts of interest.

References

1. Judge, M.A.; Franzitta, V.; Curto, D.; Guercio, A.; Cirrincione, G.; Khattak, H.A. A comprehensive review of artificial intelligence approaches for smart grid integration and optimization. *Energy Convers. Manag. X* **2024**, *22*, 100724. [CrossRef]
2. Javaid, N.; Javaid, S.; Abdul, W.; Ahmed, I.; Almogren, A.S.; Alamri, A. A hybrid genetic wind-driven heuristic optimization algorithm for demand-side management in smart grid. *Energies* **2017**, *10*, 319. [CrossRef]
3. Hassan, C.A.U.; Iqbal, J.; Ayub, N.; Hussain, S.; Alroobaea, R.; Ullah, S.S. Smart grid energy optimization and scheduling appliances priority for residential buildings through meta-heuristic hybrid approaches. *Energies* **2022**, *15*, 1752. [CrossRef]
4. Silva, B.; Khan, M.; Wijesinghe, R.; Wijenayake, U. Meta-heuristic optimization-based cost-efficient demand-side management for sustainable smart communities. *Energy Build.* **2023**, *291*, 113599. [CrossRef]
5. Hansen, T.; Roche, R.; Suryanarayanan, S.; Maciejewski, A.A.; Siegel, H. Heuristic optimization for an aggregator-based resource allocation in the smart grid. *IEEE Trans. Smart Grid* **2015**, *6*, 849–857. [CrossRef]
6. Zhang, X.; Cai, Z.; Yang, X.; Luo, J.; Zhang, R.; Zhang, Y. Game theory-based demand-side management for efficient energy collaboration in smart networks: A neighborhood-scale optimization framework. *Electr. Eng.* **2024**, *106*, 1287–1302. [CrossRef]
7. Masoumi-Anaraki, M.; Hooshmand, R.; Kabiri-Renani, Y. Probabilistic pricing for collaborative demand-side management with coordinated operation of energy storage systems for optimal peak-load control in smart grids. *IET Gener. Transm. Distrib.* **2025**, *19*, 4954–4970. [CrossRef]
8. Aslam, S.; Altaweel, A.; Bou Nassif, A. Optimization algorithms in smart grids: A systematic literature review. *arXiv* **2023**, arXiv:2301.07512.
9. Wang, Q.; Chen, G.; Khishe, M.; Ibrahim, B.F.; Rashidi, S. Multi-objective optimization of IoT-based green building energy systems using binary meta-heuristic algorithms. *J. Build. Eng.* **2023**, *74*, 106031. [CrossRef]
10. Albogamy, F.R.; Ashfaq, Y.; Hafeez, G.; Murawwat, S.; Khan, S.; Ali, F.; Khan, F.A.; Rehman, K. Optimal demand-side management using flat pricing scheme in smart grids. *Processes* **2022**, *10*, 1214. [CrossRef]
11. Kazmi, S.; Javaid, N.; Mughal, M.J.; Akbar, M.; Ahmed, S.H.; Alrajeh, N. Towards optimization of metaheuristic algorithms for IoT-enabled smart homes targeting balanced demand and supply of energy. *IEEE Access* **2019**, *7*, 96315–96329. [CrossRef]
12. Asgher, U.; Rasheed, M.B.; Al-Sumaiti, A.; Rahman, A.; Ali, I.; Alzaidi, A.; Alamri, A. Smart energy optimization using heuristic algorithm in smart grid with integration of solar energy sources. *Energies* **2018**, *11*, 3494. [CrossRef]
13. Alghamdi, H.; Hua, L.-G.; Hafeez, G.; Murawwat, S.; Bouazzi, I.; Alghamdi, B. Optimal adaptive heuristic algorithm-based energy optimization with flexible loads using demand response in smart grid. *PLoS ONE* **2024**, *19*, e0307228. [CrossRef]
14. Nutakki, M.; Mandava, S. Review on optimization techniques and role of artificial intelligence in home energy management systems. *Eng. Appl. Artif. Intell.* **2023**, *118*, 105721. [CrossRef]
15. Mimi, S.; Ben Maissa, Y.; Tamtaoui, A. Optimization approaches for demand-side management in the smart grid: A systematic mapping study. *Smart Cities* **2023**, *6*, 1630–1662. [CrossRef]
16. Ghorpade, S.J.; Sharma, R.B. A comprehensive review of demand-side management in smart grid operation with electric vehicles. *Electr. Eng.* **2024**, *106*, 6495–6514. [CrossRef]
17. Lezama, F.; Soares, J.; Vale, Z.A.; Rueda, J.L.; Rivera, S.; Elrich, I. IEEE competition on modern heuristic optimizers for smart grid operation: Testbeds and results. *Swarm Evol. Comput.* **2019**, *44*, 420–433. [CrossRef]
18. Bakare, M.S.; Abdulkarim, A.; Zeeshan, M.; Shuaibu, A.N. A comprehensive overview on demand-side energy management towards smart grids: Challenges, solutions, and future direction. *Energy Inform.* **2023**, *6*, 4. [CrossRef]
19. Khan, H.W.; Usman, M.; Hafeez, G.; Albogamy, F.R.; Khan, I.; Shafiq, Z.; Khan, M.U.A.; Alkhammash, H. Intelligent optimization framework for efficient demand-side management in renewable-energy-integrated smart grid. *IEEE Access* **2021**, *9*, 121563–121580. [CrossRef]
20. Kazmi, H.; Munné-Collado, Í.; Mehmood, F.; Syed, T.A.; Driesen, J. Towards data-driven energy communities: A review of open-source datasets, models and tools. *Renew. Sustain. Energy Rev.* **2021**, *148*, 111290.
21. ENTSO-E Transparency Platform. Day-Ahead Electricity Prices. European Network of Transmission System Operators for Electricity (ENTSO-E), Brussels, Belgium. Available online: https://transparency.entsoe.eu/ (accessed on 3 October 2025).
22. Baran, M.E.; Wu, F.F. Network Reconfiguration in Distribution Systems for Loss Reduction and Load Balancing. *IEEE Trans. Power Deliv.* **1989**, *4*, 1401–1407. [CrossRef]

https://doi.org/10.3390/en19020307

23. OpenStreetMap Contributors. Planet Dump. Available online: https://www.openstreetmap.org (accessed on 7 September 2025).
24. Gharbi, A. *H-EMOS-Lite*; Zenodo: Geneva, Switzerland, 2025. [CrossRef]

https://doi.org/10.3390/en19020307

energies

Review

Smart Grid Systems: Addressing Privacy Threats, Security Vulnerabilities, and Demand–Supply Balance (A Review)

Iqra Nazir †, Nermish Mushtaq † and Waqas Amin *

School of Information and Control Engineering, Southwest University of Science and Technology, Mianyang 621010, China; iqranazir@mails.swust.edu.cn (I.N.); nermishmushtaqkhan@mails.swust.edu.cn (N.M.)
* Correspondence: sheikhwaqasamin@hotmail.com
† These authors contributed equally to this work.

Abstract

The smart grid (SG) plays a seminal role in the modern energy landscape by integrating digital technologies, the Internet of Things (IoT), and Advanced Metering Infrastructure (AMI) to enable bidirectional energy flow, real-time monitoring, and enhanced operational efficiency. However, these advancements also introduce critical challenges related to data privacy, cybersecurity, and operational balance. This review critically evaluates SG systems, beginning with an analysis of data privacy vulnerabilities, including Man-in-the-Middle (MITM), Denial-of-Service (DoS), and replay attacks, as well as insider threats, exemplified by incidents such as the 2023 Hydro-Québec cyberattack and the 2024 blackout in Spain. The review further details the SG architecture and its key components, including smart meters (SMs), control centers (CCs), aggregators, smart appliances, and renewable energy sources (RESs), while emphasizing essential security requirements such as confidentiality, integrity, availability, secure storage, and scalability. Various privacy preservation techniques are discussed, including cryptographic tools like Homomorphic Encryption, Zero-Knowledge Proofs, and Secure Multiparty Computation, anonymization and aggregation methods such as differential privacy and k-Anonymity, as well as blockchain-based approaches and machine learning solutions. Additionally, the review examines pricing models and their resolution strategies, Demand–Supply Balance Programs (DSBPs) utilizing optimization, game-theoretic, and AI-based approaches, and energy storage systems (ESSs) encompassing lead–acid, lithium-ion, sodium-sulfur, and sodium-ion batteries, highlighting their respective advantages and limitations. By synthesizing these findings, the review identifies existing research gaps and provides guidance for future studies aimed at advancing secure, efficient, and sustainable smart grid implementations.

Keywords: blockchain; cybersecurity; data privacy; demand–supply balance; energy storage; homomorphic encryption; machine learning; peer-to-peer trading; pricing models; smart grid

1. Introduction

The smart grid (SG) technology has sparked a revolutionary shift in the manner in which energy is produced, distributed, and consumed. These modernized grids are an enhancement of traditional grids, empowered by digital technologies, the Internet of Things (IoT), and Advanced Metering Infrastructure (AMI). This transformation has enabled them to bridge the gap between energy efficiency, sustainability, and resource reliability [1].

With the help of SGs, energy and electricity can flow in both directions, allowing consumers to interact with energy providers in real time. This interaction facilitates real-time monitoring of energy consumption, automatic billing, and demand-response systems [2,3].

Nonetheless, like every other digital transformation, the accompanying increase in data usage, communication, andautomation introduces issues of data privacy, security, and cyber threats. Examples of sensitive information circulating within SG networks include patterns of electricity consumption, supply/demand data, and price fluctuations [4]. If such data is intercepted or manipulated by malicious actors, it can reveal users' personal information, potentially leading to identity theft, energy theft, and interruptions in energy services [5,6]. Therefore, to defend these systems against evolving cybersecurity threats, robust protection mechanisms and strong security measures must be established.

The primary distinction between SGs and traditional electrical grids lies in their bidirectional electricity flow, moving beyond the unidirectional model of the past. SGs integrate real-time data processing and two-way communication [7]. The inclusion of digital technologies allows for more efficient grid management, preventing real-time overloads or shortages in energy demand and supply. This results in improved energy efficiency, reduced operational costs, and enhanced reliability [8]. Furthermore, the penetration of renewable energy forms, such as solar and wind, becomes easier, fostering a more environmentally friendly energy ecosystem.

SGs also enable the development of more sophisticated consumer services. Using smart meters, consumers can make informed decisions regarding their electricity consumption in real time, based on their actual usage. Additionally, dynamic pricing models offer consumers the opportunity to optimize electricity usage when it is most cost-effective, thereby saving money on energy costs and conserving energy during peak hours [9].

Nevertheless, security risks emerge alongside these innovations. Because SGs gather and transmit sensitive consumer information and offer the functionality to operate or command power-related devices remotely, they present various attack vectors that can be exploited by cybercriminals [10]. Beyond information interception, SGs face challenges such as Denial-of-Service (DoS) attacks, among others [11].

IoT technologies have been central to the emergence of SGs. Integrating smart meters (SMs), grid sensors, smart appliances, and electric vehicles (EVs) with intelligent capabilities, IoT facilitates real-time data monitoring, collection, and control of grid operations. These devices capture data and transmit it to control centers (CCs) that analyze and interpret the information to predict demand, modify generation levels, and optimize energy distribution [12].

Advanced Metering Infrastructure (AMI) is a core element of SGs, implementing the use of smart meters to transmit and gather data regarding consumers' electricity consumption and transfer that information to utility providers. The real-time transfer of this data enables utilities to understand demand fluctuations, energy consumption habits, and efficiency opportunities [13]. Moreover, the integration of renewable energy sources, such as solar panels and wind turbines, into SG systems is achievable with the help of highly advanced data communication and control systems. AMI systems also allow for automated billing, minimizing human error and ensuring accurate charge calculations [14].

Despite these advantages, the advent of IoT devices and AMI brings significant privacy concerns that cannot be neglected. Data stored in these devices, such as electricity usage habits over time, can be used to infer users' activities, daily lifestyles, and even the presence of individuals in a household. Accessing such information by an unauthorized party can lead to identity theft, surveillance, or physical security breaches [15]. Furthermore, IoT devices are often susceptible to hacking due to their inherent vulnerabilities, exposing them to attackers who can tap into or modify data, leading to severe privacy and security implications.

Secure communication requires participants to share information such as energy demand, supply, prices, and the time of delivery or requirement with market operators [16]. Using this information, market operators can estimate the expected Demand and Supply Ratio, which helps determine the price of energy through uniform pricing or auctions by deploying several techniques such as optimization [17], game-theoretic models [18], and artificial intelligence models [19].

An effective price signal further assists the grid in maintaining a balanced state by distributing surplus energy among buyers. However, the stochastic nature of renewable energy sources creates challenging conditions for keeping the grid in a balanced state [20]. To overcome this problem, the Demand–Supply Balance Program (DSBP) provides further guidelines to participants by deploying several models such as optimization [21], game-theoretic [22], and artificial intelligence [23].

Energy storage systems (ESSs) further improve the grid's balance by injecting surplus energy during times of high demand [24]. This energy is stored in batteries made from different chemical compositions, such as lead–acid [25], lithium-ion [26], sodium–sulfur [27], and sodium-ion [28].

In this review article, we have deeply explored all these important elements and discussed in detail the methods and techniques that have been extensively used in the past and are currently being deployed in the smart grid. Based on an intensive literature analysis, this review article further assists researchers by highlighting current research gaps and providing a clear path for future research directions.

The rest of this paper is organized as follows: Section 2 and its associated subsections discuss data privacy in the smart grid, along with vulnerabilities and entry points for possible cyberattacks. Furthermore, in this section, several techniques have been explored in detail to analyze their usage in the smart grid. Section 3 and its associated sections deeply analyze pricing models along with their solution techniques. Section 4 highlights the important factors that impact Demand–Supply Balance Programs and how different methods are used to solve them. Section 5 discusses energy storage systems in the smart grid by incorporating several types of batteries and their associated merits and challenges. Section 6 provides existing research gaps and highlights future research directions. Finally, Section 7 presents the conclusion of this work.

2. Data Privacy Threats in Smart Grids

Data privacy is a crucial concern for smart grid (SG) systems, primarily due to the vast amount of sensitive information exchanged across the network. The information gathered by smart meters (SMs) can be processed to reveal specific user behaviors, including their energy usage hours, the nature of appliances they use, and the time they spend at home. When this information becomes available to malicious actors, it can be used to infer an individual's lifestyle or even anticipate their future actions [29]. Because SGs collect vast amounts of detailed energy consumption data, they create the potential for consumer profiling. Malicious or unauthorized parties could use this data to deduce sensitive information about individuals, such as their daily routines, presence or absence from home, and even the appliances they use [30]. For example, an attacker could infer when people are at home, what devices they are using, or predict their behavior based on consumption patterns.

Malicious entities may alter consumption data to lower electricity bills or access electricity without paying, a practice known as energy theft. Since SG systems rely on continuous data transmission, attackers can manipulate the data exchanged between smart meters and aggregators, leading to revenue loss for utility companies [31]. Additionally, with inadequate encryption or authentication, data being transmitted from SMs to central hubs can be intercepted, revealing confidential consumption patterns or allowing attackers to tamper with the data [32,33]. Furthermore, attackers can link data from different sources to reconstruct a user's schedule, inferring when they are present, when they are absent, or even when they are engaged in certain activities based on their energy consumption profile [5].

As the collection of sensitive data grows, these privacy concerns become more significant. Without effective privacy-preserving solutions, users may lose confidence in SG systems, which could lead to a reluctance to adopt smart meters and smart appliances. Therefore, privacy preservation is a critical requirement for the successful implementation and widespread acceptance of SGs.

2.1. Overview of Smart Grid Systems

The smart grid (SG) is an advanced, bidirectional electrical power system that integrates traditional grid infrastructure with cutting-edge Information and Communication Technology (ICT). This integration enables SGs to dynamically manage and monitor the flow of energy, offering numerous advantages over traditional power grids. The SG system optimizes electricity generation, distribution, and consumption through the integration of smart meters, sensors, automated control systems, and renewable energy sources [34].

The SG is composed of a series of interrelated subsystems that contribute to the provision of sustainable, reliable, and efficient energy management. These subsystems are as follows:

Smart Management System: This system supports the balance of electricity demand and supply. It operates through control centers (CCs) where smart meters (SMs) and other devices relay information to be analyzed. This analysis helps anticipate demands, thereby enabling optimal control of grid operations. An intelligent management system also facilitates the integration of renewable energy sources into the power grid and the efficient distribution of available resources.

Smart Infrastructure System: This subsystem allows for the bidirectional flow of electricity and information between end-users and utilities. It incorporates smart meters (SMs) that monitor electricity consumption, smart appliances that can manage energy levels in buildings or businesses, and communication systems that facilitate the seamless flow of information among various entities within the grid.

Smart Protection System: This subsystem improves grid reliability by monitoring for faults and preventing permanent equipment damage. It has built-in automated units that can respond promptly to conditions like power failures or equipment malfunctions, thereby reducing repair time and ensuring uninterrupted service.

Smart Grid Communication Network: A communication network is a primary characteristic of the SG because it enables the interchange of information between all elements. The structure of communication systems in SGs is typically hierarchical, with information flowing from SMs to control centers via intermediate aggregators. This architecture allows for the optimization of data transmission, the minimization of delays, and ensures all components collaborate effectively. The communication framework of a smart grid generally consists of the following networks:

1. Customer Premises Area Network (HAN/BAN): Data on electricity usage by individual devices is sent to SMs via the Home Area Network (HAN) and Building Area Network (BAN). The communication protocols used by these networks are low-power, such as ZigBee, Wi-Fi, and blacktooth [35,36]. The primary concerns in this layer are

privacy and security because personal data regarding consumer actions is collected and distributed.

2. Wide Area Network (WAN): The WAN links all regional networks and the CCs. It is responsible for long-distance data transfer, often spanning tens of kilometers. For this purpose, optical communication and cellular networks are widely applied because they facilitate fast data transfer. The WAN is vital for providing a reliable control linkage to the distributed smart grid and its control systems [37].

2.2. Key Components of the Smart Grid

The SG system consists of multiple important elements, which allow an enhanced performance of the grid and its intelligence, as shown in Figure 1.

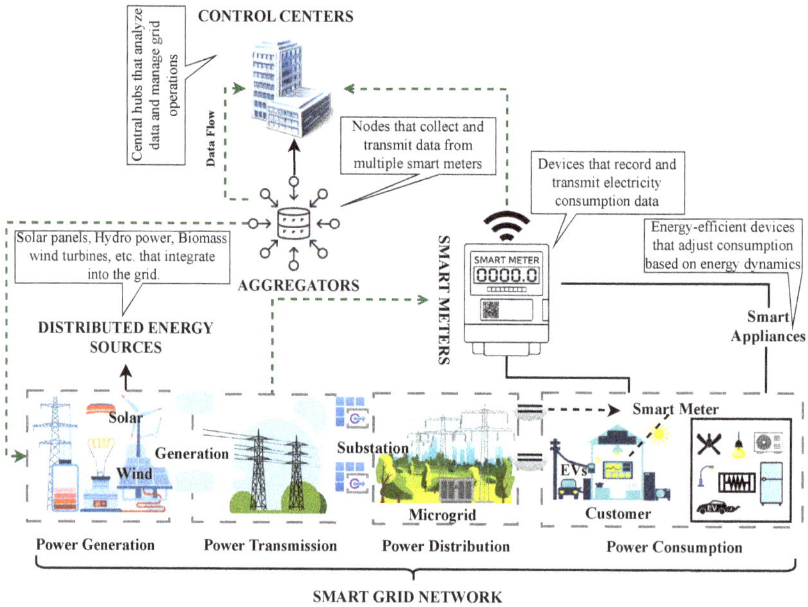

Figure 1. Architecture of smart grid and key components.

From Figure 1, it can be seen that the smart grid consists of several components such as the smart meter (SM), control center (CC), aggregators, smart appliances, and renewable energy sources (RESs).

Smart meters are intelligent devices installed in residential areas and businesses to collect electricity usage data and transmit it back to the operations centers [38]. SMs enable real-time monitoring, a factor that is vital for optimizing demand-response programs and implementing dynamic pricing [39].

The CC is the central hub of the SG that receives, transmits, and analyzes data from SMs and other equipment to facilitate real-time decision-making regarding energy supply and demand. This information is used by the CC to optimize grid operation, identify possible failures, and react to unforeseen situations like power cuts or supply shortages [40, 41].

Nodes that gather data from various SMs, known as aggregators, also serve as an intermediary between SMs and the CC. They collect usage data and send it to the CC for analysis [42].

Smart appliances are intelligent products that can regulate the amount of energy they use based on directions from the SG. Smart appliances are instrumental in demand-side management as they can reduce energy consumption during peak periods [43].

RESs are a means of energy generation. The SG promotes the inclusion of renewable energy sources such as solar panels and wind turbines. This integration helps in balancing supply and demand, in addition to reducing the carbon footprint of energy production [44].

2.3. Vulnerabilities and Entry Points for Data Leaks in the Smart Grid Network

While SGs offer numerous benefits in terms of efficiency and sustainability, they also present various vulnerabilities that can be exploited by attackers. One of the primary concerns is the reliance on cyber-physical systems, which combine computational systems with physical infrastructure like SMs, aggregators, and CCs. These devices and their interconnected communication channels present numerous points of vulnerability.

The SG system, while revolutionizing the energy sector, has introduced new cybersecurity vulnerabilities due to its reliance on digital communication and data exchange [45]. As shown in Figure 2, cyberattacks pose significant risks, as evidenced by several high-profile incidents that have affected energy infrastructures worldwide. These recent cyberattacks on SG and energy infrastructure highlight therapidly evolving threats and vulnerabilities.

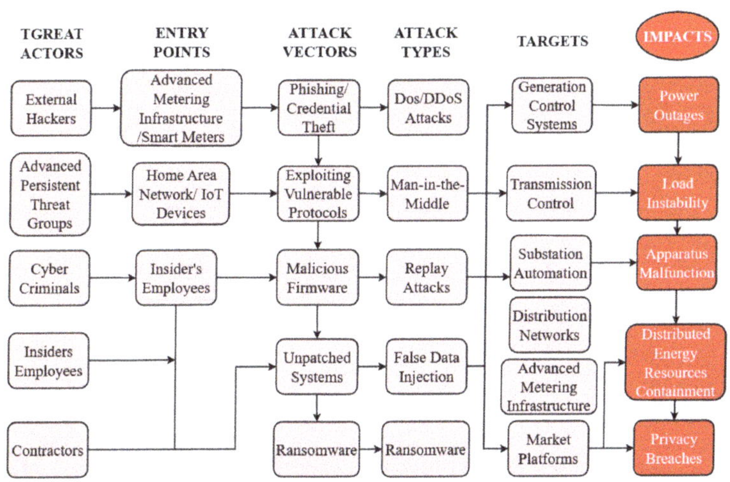

Figure 2. Comprehensive mapping of smart grid cybersecurity threats, showing threat actors, attack entry points, vectors, types, targeted systems, and resulting operational impacts.

One of the major threats to the SG network is the Man-in-the-Middle (MITM) attack. During this attack, an intruder listens in on and intercepts a legitimate communication between two devices in a network (e.g., between an SM and an aggregator) [46]. This allows the culprit to intercept or manipulate information being exchanged, resulting in data loss, privacy breaches, and interruptions to the energy supply chain. Hackers can also forge data, such as inaccurate electricity usage measurements, to manipulate power distribution [5,47].

The Denial-of-Service (DoS) attack is another type of vulnerability where intruders flood the network with excessive data, thereby slowing down or blocking legitimate service requests [48]. Such attacks can cause a system failure, preventing users from accessing important services like real-time energy consumption details or the ability to adjust their energy usage during peak demand times. A Distributed Denial-of-Service (DDoS) attack is particularly harmful because it uses multiple sources of traffic, making it more difficult to mitigate [49].

Another significant issue in SG networks is the Replay attack. This involves the interception and retransmission of data communicated on the network, causing distorted information to be processed by the receiving system [50]. This can lead to improper load balancing, overcharging of users, and even blackouts if the corrupted data affects grid management decisions.

In SG networks, Theft of Resources and Insider attacks are also highly dangerous. Insider attacks occur when malicious individuals (including disgruntled employees or contractors) gain unauthorized access to system resources, including control systems, databases, or SMs. Such actions can result in the manipulation of data or physical assets, thereby disrupting grid operations and compromising customer privacy [51].

The major attacks and their historical impact are given as follows:

1. In 2023, Hydro-Québec, a major grid operator in Canada, experienced a cyberattack that disrupted its outage verification app and website. This incident underscores the increasing cyber threats faced by energy utilities and the need for robust cybersecurity measures to protect critical infrastructure [52,53].
2. In 2024, Spain's National Cybersecurity Institute (Incibe) investigated a massive blackout that disrupted essential services. The focus was on smaller electricity generators, which may have lacked robust cyber defenses, making them potential vulnerabilities in the national power grid [54].
3. In May 2023, a critical vulnerability in the MOVEit-managed file transfer software was exploited by the ransomware group CL0P, leading to unauthorized access to sensitive databases. The breach affected over 2,700 organizations across various sectors, including energy, highlighting the systemic risks inherent in the interconnected nature of digital supply chains [54,55].
4. From December 2023 through to early 2024, Iranian hacker groups engaged in successful cyberattacks on various sectors, including the energy sector. Such attacks interrupted services and exposed weaknesses in critical infrastructure, pointing to the necessity of more efficient cybersecurity practices to defend against state-sponsored cyber threats [56].

2.4. Security and Privacy Requirements in the Smart Grid Network

The smart grid (SG) network involves a large number of interdependent devices and systems that regularly transfer sensitive information. To maintain an efficient and secure operational network, several key security and privacy requirements must be met. These needs include upholding confidentiality, data integrity, availability, and secure storage, as well as ensuring scalability.

Confidentiality is crucial in SG networks. It refers to the protection of sensitive information from unauthorized access. This is especially necessary in SG systems, where data specific to individuals, such as energy usage trends, is constantly uploaded from smart meters (SMs) to local aggregation nodes [57]. If this data is intercepted, attackers could learn about users' consumption habits, leading to serious privacy invasions. Furthermore, businesses that optimize operations based on electricity consumption patterns may expose sensitive data like trade secrets or customer behavior analytics [58].

Integrity ensures that information transmitted through the SG network cannot be altered and remains trustworthy. When applied to SGs, a lack of integrity can be a significant issue [59]. For instance, if an attacker gains control over electricity usage data transmitted from a smart meter to the control center, the grid could over- or underestimate demand, which might cause blackouts or inefficient resource utilization. Threats to data integrity, such as Man-in-the-Middle (MITM) attacks and replay attacks, must be addressed and monitored constantly [60].

Availability is a safeguard that ensures services and data within the SG network are accessible when needed. Having real-time energy consumption information is critical for grid operators to make real-time decisions, which helps manage peak demand and implement curtailment strategies. In case of a loss of availability due to a Denial-of-Service (DoS) or Distributed Denial-of-Service (DDoS) attack, grid operators would be unable to control resource distribution, potentially causing widespread outages and operational inefficiencies [61].

The SG network is also affected by major issues in terms of secure storage. Because SG networks contain a large amount of sensitive data, the importance of keeping such data in a secure environment cannot be understated. Without proper encryption and data protection mechanisms, stored data is at risk of theft or manipulation [62]. The risk of data breaches escalates if hackers can access data centers or other storage facilities.

Scalability is another important concern in SG networks. Security solutions must be able to support an expanding network and adapt as more smart devices are deployed, thus handling the increasing data and communication traffic [63]. Traditional security procedures may not be adequate for larger, decentralized systems, necessitating more robust and dynamic security solutions.

To address these security and privacy needs, a layered security approach is essential. This includes features such as advanced encryption, multi-factor authentication, intrusion detection systems, and regular security audits. The SG network can only be safe and consistent by ensuring that all its aspects—confidentiality, integrity, availability, secure storage, and scalability—receive due attention.

2.5. Privacy Preservation Techniques in Smart Grid Network

Since SGs are becoming an essential part of modern-day energy systems, it is of utmost importance to ensure the privacy of highly sensitive data, including the energy consumption patterns of consumers. An SG connects energy producers, consumers, and utility providers, and facilitates real-time communication. Nevertheless, the constant flow of data exposes consumers to vulnerabilities, particularly concerning data confidentiality. Preserving privacy in SG networks is aimed at protecting sensitive information, such as the energy usage of individuals, preventing unauthorized access, and mitigating the potential risks of privacy disclosure that arise from sharing this sensitive information. Various methods have been introduced in the literature, all of which aim to address a particular aspect of privacy and the secure transfer of information via the grid, as shown in Figure 3.

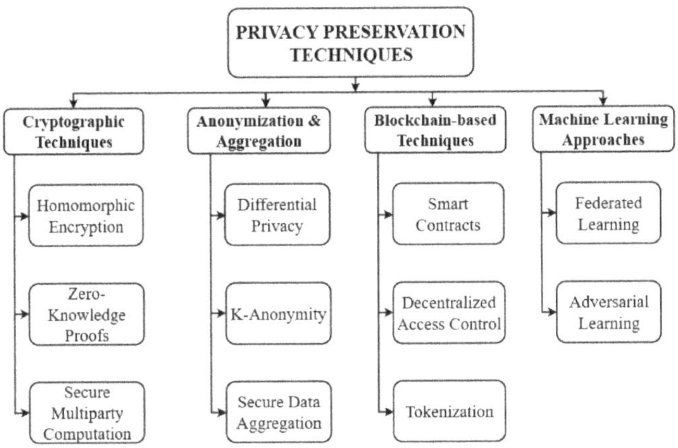

Figure 3. Classification of privacy preservation techniques in smart grids, highlighting cryptographic, anonymization and aggregation, blockchain-based, and machine learning approaches.

2.5.1. Cryptographic Techniques

Cryptography is a major defense applied to establish the security of SG network data, whether in transit or at rest, and also ensures that sensitive information links are not hacked or altered. Within the dynamics of the smart grid, cryptographic methods are utilized to ensure the privacy and integrity of data while still preserving the functionality of the grid. Various cryptographic mechanisms have been suggested to ensure data privacy.

2.5.2. Homomorphic Encryption (HE)

HE is a cryptographic technique that enables computers to perform operations on encrypted data without first decrypting it. The subsequent output is also encrypted and can only be decoded by the authorized user, which means the data remains confidential throughout processing [64,65]. Such an approach is highly beneficial in SG systems because data is often consolidated and processed on third-party servers. In recent years, a number of researchers have suggested Homomorphic Encryption schemes to implement SG metering systems whose aggregation protocols can collect consumption data without interfering with the privacy of specific users [66,67]. Furthermore, HE can be utilized when collecting data to allow electricity companies to aggregate the information they obtain through smart meters (SMs) without revealing intimate user behaviors [68–70].

HE provides the dual advantage of preventing the disclosure of sensitive data while simultaneously enabling utilities to gain insights into energy consumption trends [71]. This is especially important for privacy in SGs, where exposing consumption patterns can lead to security threats or user profiling [72]. Nevertheless, HE is associated with a high computational cost due to its more complex operations on encrypted data compared to conventional encryption or plaintext processing [73]. Research has therefore focused on streamlining these methods to minimize overhead costs and ensure they are viable for handling large-scale SG networks, as can be summarized in Table 1.

Table 1. Summary of Homomorphic Encryption for privacy preservation in Smart Grids.

Refs.	System Model	Goal	Security Parameters	Performances	Limitations
[67,74]	SG, Data Aggregator, SMs	Aggregate encrypted energy consumption data to preserve user privacy in SG.	Confidentiality, Integrity, and Availability of encrypted data	Efficient data aggregation, reducing exposure of sensitive information.	High computational overhead, especially in large networks.
[75]	Smart Metering System, Cloud-based Framework	Enable secure demand-side energy management while preserving privacy using HE.	Privacy of users' consumption data, Data confidentiality	Secure privacy-preserving aggregation, effective for dynamic demand-response systems.	Heavy computation, delays in processing large-scale data.
[76,77]	SG, Distributed Energy Systems	Preserve privacy during SG data processing and forecasting.	Confidentiality of real-time data, Robustness to cyberattacks	Secure processing and forecasting for energy data, supports multiparty computation.	High encryption cost, latency in data handling for large datasets.
[78]	SM, Consumer-side Aggregators	Enable privacy-preserving energy consumption analysis for smart grid forecasting.	Secure computation, Data integrity, Privacy of consumers	Improved privacy, secure aggregation of energy data, real-time analysis possible.	Large overhead for dynamic data and issues with real-time processing.
[69]	SG, Centralized Data Servers	Securely aggregate and analyze energy data from multiple SMs while preserving privacy.	Data confidentiality, Authentication, Integrity of aggregated data	Low-risk data breaches, secure load forecasting, scalable implementation.	Performance degradation with large datasets and high computation costs.

2.5.3. Zero-Knowledge Proofs (ZKPs)

Zero-Knowledge Proofs (ZKPs) are a cryptographic protocol in which one party can prove the veracity of a claim to another party without revealing any information about the proof itself, other than whether the claim is true or not [79]. ZKPs are especially useful for authorization and identity verification in SG systems, where privacy is a primary concern [80]. Liang et al. [81] explained how blockchain-based ZKPs can enhance the privacy of transactions and smart contract predicates in SG systems. With ZKPs, SG participants can authenticate themselves without revealing sensitive data, such as consumption statistics or personal identifiers [80].

Works by [80,82] highlight that the primary advantage of ZKPs is that they preserve user privacy during the authentication process. For example, a smart meter can prove that it is in proper working order without revealing any specific information on the amount of energy it is consuming, as has been suggested in [83,84]. This capability is vital for addressing the rising concerns regarding individual privacy, as no external party can access data that could then be used to profile a user or perform other nefarious actions.

Moreover, ZKPs are increasingly employed in privacy-preserving applications where proving the correctness of data is required without revealing the underlying data. El-Hajj et al. [85] compared various ZKP protocols, including zk-SNARKs, for privacy-preserving authentication in blockchain-based systems and discussed their applications in billing verification and identity authentication. Similarly, Ref. [85] describes how ZKPs can guarantee that smart grid users receive accurate billing without exposing sensitive usage patterns. However, as pointed out by El-Hajj et al., while ZKPs provide strong security guarantees, they introduce significant computational complexity, making them unsuitable for real-time applications requiring rapid processing, such as fault detection or real-time load balancing [85,86].

One drawback of ZKPs is that they can be computationally intensive, particularly in real-time applications, which may limit their scalability in large SG networks, as shown in Table 2.

Table 2. Summary of Zero-Knowledge Proof (ZKP) and blockchain-based privacy techniques.

Refs.	System Model	Goal	Security Parameters	Performances	Limitations
[87,88]	Blockchain-based SG	Identity verification, ensuring users' identity without revealing sensitive data	Identity verification, confidentiality, integrity	Efficient identity verification in decentralized systems	Computationally expensive, scalability issues with large-scale grids
[89]	Permissioned and permissionless blockchain	Privacy for identity, transactions, and smart contracts	Privacy preservation, transaction confidentiality, anonymity	Reduces transaction-related privacy risks	Limited by blockchain size and complexity
[90,91]	Blockchain-based SG	Billing verification and identity authentication	Privacy, non-repudiation, confidentiality	Fast and efficient in verifying billing information and user identity	Complex implementation in large-scale smart grid systems
[92]	Blockchain-based Privacy-Preserving Systems	Comparing ZKP protocols (zk-SNARKs, zk-STARKs) for secure authentication	Secure authentication, privacy preservation, efficiency	Comparison of zk-SNARKs, zk-STARKs, and bulletproof protocols for low-latency systems	Limited application to large-scale networks, high computational load for zk-SNARKs
[76]	IoT-based SG	Implementing ZKP for secure data exchange between consumers and utility providers	Confidentiality, authentication	Enables secure and private communication for energy data	High processing overhead in resource-constrained devices

2.5.4. Secure Multiparty Computation (SMPC)

SMPC is a cryptographic method that enables multiple parties to jointly compute a function over their private inputs while keeping those inputs confidential [93]. In SGs, SMPC can be used to enable energy consumers, utility companies, and third-party aggregators to collaborate on energy consumption analysis and demand-response programs without sharing sensitive consumption data [76]. Samuel et al. [75] discussed an SMPC framework for SGs that ensures users' energy usage data remains private, even during collaborative optimization and demand-response computations. Furthermore, this technique ensures that each party can perform computations on their own encrypted data, with only the final aggregated result being revealed [94].

Additionally, SMPC has the advantage of allowing collaboration among multiple stakeholders in a smart grid without exposing confidential private data [95]. For example, a consumer may want to share their energy usage information with the utility to optimize energy pricing but not reveal detailed personal behavior. SMPC achieves this by ensuring that computations are conducted over encrypted data [96]. However, the main challenge with SMPC is its high computational complexity and communication overhead. Handling large datasets and ensuring that computations scale efficiently are key challenges that ongoing research is trying to address [97]. Recent work has focused on improving the efficiency and practicality of SMPC in real-time smart grid systems, as shown in Table 3.

Table 3. Summary of Secure Multiparty Computation (SMPC) privacy-preserving techniques in smart grids.

Refs.	System Model	Goal	Security Parameters	Performances	Limitations
[88]	SG with multiple users, demand-response systems	Ensure privacy of users' energy usage data during collaborative optimization and demand-response computations	Data privacy, integrity, confidentiality	Efficient in computation and communication overhead	Scalability issues with large systems and multiple participants
[75,98]	Cloud-based AI and SG systems	Address data privacy and security concerns during energy consumption optimization	Data confidentiality, encryption, non-repudiation	Enhances security for cloud-based smart grid systems	Increased latency and computation costs due to complex cryptographic operations
[99]	Distributed SG with multiple consumers and grid operators	Secure collaborative computations for demand response without compromising privacy	Privacy, confidentiality, trust, authentication	Highly efficient for small to medium-sized grids	Scalability issues with large datasets and numerous participants

2.5.5. Anonymization and Aggregation

Anonymization and aggregation are techniques used to protect personal identities by altering data in such a way that it cannot be traced back to an individual. These methods are often applied to aggregate datasets, which reduces the granularity of the information while maintaining its usefulness for grid management.

1. Differential Privacy: The process of anonymization, particularly through differential privacy (DP), has gained significant attention in the smart grid sector for protecting individual data. Differential privacy guarantees that no individual's data or the data collected on themsignificantly affects the response of a query [100]. Researchers Hu et al. [101] proposed a privacy-preserving data aggregation and management approach for smart grid architecture, highlighting the feasibility of using differential privacy in data aggregation in smart meters. As another example, Rai et al. [102] used

differential privacy to provide privacy guarantees to energy consumption data and showed that, under this scheme, individual consumer privacy could be achieved without compromising the utility of the aggregate consumption statistics for optimizing a smart grid. Bourechak et al. [103] demonstrated that a smart grid model based on differential privacy could be constructed, in which noise is added to smart meter data in such a way that the privacy of individual consumers is guaranteed. Among the main benefits of DP is that it provides formal privacy guarantees and is, therefore, a powerful strategy to protect sensitive information in smart grids. The method ensures that no individual's data can significantly influence query results, even if an adversary has access to auxiliary information [104]. However, a known trade-off of DP is that the addition of noise may slightly reduce the accuracy of aggregated results [105].

Further research focuses on optimizing the amount of noise added to maintain a balance between privacy and data utility, ensuring that the SG can still function efficiently while maintaining strong privacy guarantees [106,107]. The following Table 4 provides a summary of a few works concerning the deployment of differential privacy in the smart grid.

Table 4. Summary of differential privacy techniques in smart grids.

Refs.	System Model	Goal	Security Parameters	Performances	Limitations
[108]	General privacy-preserving model	Provide privacy guarantees while sharing data from smart meters in smart grids	Privacy, data perturbation, noise addition	Provable privacy guarantees under differential privacy	Increased noise can affect data utility
[109]	SG with individual consumer data	Protect individual user data while enabling utility companies to aggregate energy consumption	Data privacy, utility preservation, statistical analysis	High privacy preservation with minimal data alteration	Performance degradation in the aggregation of data due to noise
[107]	SG with multiple consumers and data aggregation points	Enable privacy-preserving data aggregation for demand-response management	Privacy, data aggregation, differential privacy	Ensures that data leaks do not occur, and efficient aggregation is maintained	Limited effectiveness for highly detailed data
[109]	SG with large-scale data collection	Apply differential privacy to protect energy usage data for large-scale grids	Data perturbation, confidentiality, utility maximization	Enhances privacy without large overhead for large datasets	Requires balancing between privacy and data quality

2. k-Anonymity, another widely studied anonymization technique, has been proposed [110] and applied to ensure that energy consumption data cannot be linked to fewer than k-1 other users. Sweeney et al. [111] initially introduced this concept, and it has been applied in various domains, including SGs. In the context of SGs, k-Anonymity helps obscure individual energy usage to prevent privacy violations. However, recent studies, such as the one proposed in [112], indicate that k-Anonymity is vulnerable to attacks when auxiliary information is available, which makes it insufficient for protecting against linkability attacks. Moreover, Sakthivel et al. [113] proposed a k-Anonymity-based model for anonymizing smart meter data, where they ensure that the consumption patterns are generalized so that no individual data point can be traced back to a particular consumer.

   Additionally, the advantage of k-Anonymity is that it is relatively simple to implement and can provide a basic level of privacy by masking individual consumption details [114]. The proposed studies make it a useful tool for SGs where data must be aggregated and shared, but individual privacy must be preserved [115]. However, k-Anonymity can lead to a loss of granularity in the data, which might diminish its usefulness for detailed grid analysis [116]. Additionally, the technique is vulnerable to attacks that use auxiliary data or background knowledge, such as the homogeneity attack or the background knowledge attack, where attackers may still infer sensitive information [117]. A summary of a few other studies based on k-Anonymity is presented in Table 5.

Table 5. Summary of k-Anonymity and data anonymization techniques.

Refs.	System Model	Goal	Security Parameters	Performances	Limitations
[118]	Residential SM data	Protect consumers' privacy by ensuring that energy consumption patterns are indistinguishable from other consumers	k-Anonymity, data generalization, suppression	Achieves a high level of privacy with minimal loss of data quality	Data utility significantly decreases when increasing k
[119]	SG with aggregated consumption data	Preserve privacy of consumers' energy usage data while allowing aggregation for demand response	Data anonymization, k-Anonymity, data perturbation	Effective at protecting privacy without large computational overhead	Performance degradation in data analysis when large k is required
[120]	SG with individual consumption data from residential users	Protect sensitive data while maintaining usability for grid management tasks	k-Anonymity, noise injection, clustering	Effective for privacy protection in real-time smart grid applications	Increased noise can reduce the accuracy of demand forecasting

3. Data aggregation models have also been extensively researched as an alternative to individual data storage. Zhan et al. [121] discussed how aggregation at the community level can maintain grid performance while ensuring that individual data is protected. The proposed research in [122] highlights the advantage of aggregating energy usage data, as it reduces the risk of sensitive data leakage while optimizing grid control. Privacy protection through aggregation, however, has limitations because it increases data generalization, which can limit its value for individualized services such as adaptive pricing or demand-response programs [123].
   However, a balance must be maintained between data aggregation and the loss of detail, as privacy protection should not undermine effective grid management [124]. Furthermore, Xu et al. [125] proposed a data aggregation framework for SG systems utilizing Secure Multiparty Computation (SMPC) to aggregate data from multiple smart meters while preserving the privacy of individual consumption information. This approach ensures that only aggregated consumption data is shared with the grid operator, thereby preventing the disclosure of sensitive user information.
   Data aggregation in the smart grid is essential as it enables operators to analyze overall consumption and develop an understanding of future requirements and efficient strategies for resource allocation. However, there are some limitations in achieving both privacy and utility because excessive noise added during the aggregation process may reduce its usefulness for grid optimization [126]. Researchers are working on developing models that aggregate data in a way that maintains privacy while providing accurate and useful insights for grid management. Table 6 presents a summary of a few studies with their limitations.

Table 6. Summary of privacy-preserving data aggregation techniques.

Refs.	System Model	Goal	Security Parameters	Performances	Limitations
[115]	SG with smart meters and aggregators	Aggregate energy consumption data while preserving consumer privacy	Data aggregation, privacy preservation, encryption	Efficient aggregation with reduced communication overhead	Aggregated data may lose individual-level accuracy
[127]	Residential energy usage data from SMs	Aggregate data in a way that maintains privacy without compromising data utility for analysis	Homomorphic Encryption, data anonymization	Effective at maintaining data utility for grid management	Higher computational complexity due to encryption processes
[128]	SG with distributed energy resources	Perform secure aggregation of energy consumption data across different smart meters	Data aggregation, privacy preservation, Secure Multiparty Computation	Reduced data leakage with minimal performance impact	Some aggregation models increase latency due to encryption
[129]	Energy data aggregation in a smart city	Aggregate data from multiple smart meters while maintaining privacy of users' energy consumption patterns	Data aggregation, differential privacy, noise injection	High accuracy in aggregated results with strong privacy guarantees	Increased noise can reduce the accuracy of the aggregated data

2.5.6. Blockchain-Based Mechanism

Blockchain technology has gained popularity in the smart grid space for its ability to provide decentralized, secure, and transparent data management. By using a distributed ledger system, blockchain enables secure transactions without the need for a central authority [130].

Smart contracts, which are a key part of blockchain technology, are self-executing contracts with the terms of the agreement written directly into code. Saeed et al. [131] proposed using smart contracts in peer-to-peer energy trading platforms to automate transactions between consumers and producers of renewable energy. Moreover, smart contracts ensure transparency, security, and privacy while eliminating the need for intermediaries, thereby reducing the risk of data manipulation or privacy breaches [132]. As noted in the literature, however, blockchain scalability remains a major concern, particularly as the number of participants grows [133].

A privacy-preserving scheme based on blockchain was proposed in a study by Alsheavi et al. [134] to provide privacy in SGs by using blockchain and smart contracts that automate the sharing of energy consumption data only with authorized entities. Smart contracts eliminate the need for a central authority to implement privacy policies, which leads to greater transparency and trust for all participants. Furthermore, to enable all transactions to be registered and stored in the blockchain in an immutable and easily traceable manner, smart contracts can be used to automate important processes in SGs, such as releasing energy data, delivering payments for energy use, and implementing demand-response measures [135].

Nevertheless, blockchain's slow transaction speed and the computational resources it requires can cause performance bottlenecks in large-scale smart grid applications, as depicted in Table 7.

Table 7. Summary of smart contract applications in smart grids.

Refs.	System Model	Goal	Security Parameters	Performances	Limitations
[136]	Blockchain-based SG with decentralized control	Automate energy transactions and ensure secure, tamper-proof agreements between consumers and producers	Blockchain, cryptographic signatures, encryption	High transparency and immutability; reduces fraud in transactions	Scalability issues with large-scale deployments
[137]	Peer-to-peer energy trading in SGs	Use smart contracts to facilitate secure and automated energy trading between prosumers	Smart contracts, blockchain, digital signatures	Efficient energy trading with minimal overhead and reduced human error	Potential delays in transaction execution due to blockchain confirmation times
[138]	Energy management system using smart contracts	Automate billing and payment for energy usage while ensuring privacy and transparency	Blockchain, encryption, access control	Provides secure billing and payment system with reduced administrative costs	High computational power required for contract execution
[88]	Decentralized SG system with smart contracts for grid operation	Enable autonomous grid management and decision-making through automated contracts	Decentralized access control, blockchain, cryptography	Improves operational efficiency and reduces human intervention	Limited adoption due to infrastructure requirements and implementation complexity
[139]	Decentralized energy trading platform	Ensure the privacy and integrity of trading data through secure smart contracts	Blockchain, Zero-Knowledge Proofs (ZKPs), cryptographic protocols	Enhances privacy while automating secure energy trading	Complex implementation and maintenance costs in large-scale grids

Decentralized access control has been suggested as a way to use blockchain-based systems to secure consumer information in smart grids (SGs) [140]. Hasan et al. [130] discussed the possibility of making access to energy consumption data decentralized with the help of blockchain, where consumers can control access to their data using cryptographic keys. This strategy provides enhanced privacy and security compared to conventional centralized approaches [141]. Researchers like Alsheavi et al. [134] also demonstrated that a well-structured decentralized access control system can be implemented in SGs to regulate access to sensitive information, including consumption statistics and pricing.

The distributed nature of decentralized access control effectively eliminates the threats of a single point of failure and a central authority being corrupted by an adversary. It achieves this by sharing access control decisions across the entire network. The advantage of blockchain-based decentralized access control is that it enhances transparency and accountability, making it easier to track who accessed what data and when [142]. This ensures that access is granted based on pre-defined permissions and that users can verify that their data is not being accessed without authorization [143]. However, one key challenge of using blockchain for access control is ensuring scalability, as the number of transactions in a large SG network can increase substantially, as shown by the comparative analysis of a few studies in Table 8.

Several consensus algorithms have been used in blockchain for deployment in the smart grid. Table 9 presents a comparative analysis of several consensus mechanisms for smart grid applications. From Table 9, it can be observed that among the various consensus mechanisms, PBFT and PoA are the most practically applicable to large-scale smart grid applications, as they offer low latency rates with high scalability.

Table 8. Summary of access control mechanisms in smart grids.

Refs.	System Model	Goal	Security Parameters	Performances	Limitations
[140]	Decentralized energy management system	Provide secure and efficient access to energy data and smart grid resources	Blockchain, public-key infrastructure, cryptographic access control	High security and flexibility; allows dynamic access control	High computational cost and complexity in implementation
[144]	Blockchain-based SG with decentralized control	Enable secure, role-based access to SG data and services	Blockchain, multi-signature authentication, access policies	Enhanced privacy and control over energy data sharing	Complexity in managing large numbers of access policies and users
[145]	SG with decentralized access control for energy data sharing	Allow users to control and grant access to their energy consumption data	Smart contracts, encryption, Zero-Knowledge Proofs	Efficient data sharing with enhanced privacy; reduces risks of unauthorized access	Risk of performance bottlenecks in real-time applications

Table 9. Comparison of consensus mechanisms for smart grid applications.

Consensus Mechanism	Description	Advantages	Limitations	Storage Requirements	Latency	Scalability	Common Use Cases in Smart Grid	Key References
Proof of Work (PoW)	Miners solve cryptographic puzzles to create a new block.	High security, widely tested	High energy consumption, slower transactions	High (full blockchain replication)	High (minutes scale)	Low (scales poorly)	Data aggregation, electricity consumption recording	[146]
Proof of Stake (PoS)	Validators chosen based on coin holdings staked as collateral.	Energy efficient, faster than PoW	Potential centralization risk	Moderate (depends on node role)	Moderate (seconds scale)	Moderate to high	Distributed energy resources management	[147]
Delegated Proof of Stake (DPoS)	Stakeholders elect delegates to validate transactions and create blocks.	Faster consensus, scalable	Trust issues, potential centralization	Moderate	Low (seconds or less)	High	Real-time energy trading, demand response	[147]
Practical Byzantine Fault Tolerance (PBFT)	Nodes agree on transaction order through voting rounds.	Low latency, suitable for permissioned networks	Scalability limits in large networks	Low to moderate	Very low (milliseconds to seconds)	Low to moderate	Consortium blockchain for monitoring, maintenance	[148]
Proof of Task (PoT)	Nodes complete real-time control tasks for consensus based on contribution.	Real-time control, contribution-based	Emerging, under development	Moderate	Very low	Moderate	Real-time regulation and control of renewable energy systems	[147]
Proof of Authority (PoA)	Trusted validators are pre-approved and known entities maintaining the network.	High throughput, low energy use	Centralization risk	Low (permissioned blockchain)	Very low	High	Permissioned smart grid networks	[146]
Proof of Credit Scores (PoCS)	Consensus based on node credit scores reflecting trustworthiness and performance.	Encourages honest participation, efficient	Complexity in credit evaluation	Moderate	Moderate	Moderate	Smart grid power trading systems	[149]
Proof of Importance (PoI)	Nodes gain importance score based on activity and stake influencing block creation rights.	Encourages active participation and stakeholding	Less common, requires additional metrics	Moderate	Moderate	Moderate	Energy trading and decentralized energy markets	[150]
Proof of Elapsed Time (PoET)	Random wait times assigned to nodes; the first to finish produces the block.	Energy efficient, fair	Requires trusted execution environment	Low	Very low	Moderate	Suitable for permissioned smart grids	[151]
Algorand Consensus	Pure PoS variant with verifiable random functions for leader election.	High security, fast finality	Complexity in implementation	Moderate	Low	High	Distributed energy resource coordination	[152]
RAFT Consensus	Leader-based consensus utilized mostly in permissioned blockchains.	Simplicity, low overhead	Not decentralized	Low	Very low	Moderate	Consortium smart grid implementations	[153]

2.5.7. Tokenization

Tokenization, another blockchain-based technique, involves replacing sensitive data with tokens that can be securely exchanged or traded. Tokenization has been proposed [154] and applied in energy systems to represent energy credits or trading rights, allowing consumers to engage in energy trading without revealing detailed consumption data. Taherdoost et al. [155] explored tokenization in decentralized energy markets and demonstrated how it can be leveraged to ensure privacy while allowing participants to trade energy efficiently. Moreover, Chen et al. [156] proposed a tokenization-based approach for protecting sensitive data in SGs, where energy usage information is replaced with anonymous tokens during billing processes, ensuring that users' data cannot be easily traced or identified.

The primary advantage of tokenization is that it reduces the risks of data breaches because tokens cannot be reverse-engineered to rediscover the original data. This makes tokenization a viable solution for SGs whose transactions and sensitive data should be secured. However, managing the tokenization process requires careful design to ensure that tokens cannot be easily misused and that they do not introduce processing delays or errors in energy data handling [157]. As noted in their study, key management is a critical issue in tokenization systems. The security of tokenized data relies on the secure handling of encryption keys, which, if compromised, can expose sensitive information, as shown in Table 10.

Table 10. Summary of tokenization applications in smart grids.

Refs.	System Model	Goal	Security Parameters	Performances	Limitations
[158]	Tokenization in SG data management	Protect users' personal energy consumption data while enabling secure transactions	Cryptographic token generation, data masking	Enhanced privacy by converting sensitive data into non-sensitive tokens	Risk of token theft if not properly managed
[159]	SG platform using tokenized identities for energy trading	Enable secure, anonymous energy transactions among prosumers and consumers	Tokenized identities, encryption, blockchain	Secure energy trading with privacy-preserving features	Possible scalability issues when handling a large number of transactions
[159]	Blockchain-based tokenized smart SG system for data access control	Prevent unauthorized access to smart grid data using tokens for verification	Smart contracts, tokenized access control	Fast, secure data access control with reduced risks of unauthorized access	High computational overhead and reliance on blockchain networks

2.5.8. Machine Learning Approaches

Machine learning (ML) techniques can also be applied to privacy preservation in SGs. By learning patterns from data without directly accessing sensitive information, ML models can improve system efficiency while maintaining privacy, with methods like federated learning and adversarial learning gaining popularity.

Federated learning is a decentralized approach to training machine learning models where the data remains on local devices, such as SMs, and only the model updates are shared with a central server [160]. Su et al. [161] explored how federated learning could be applied in smart grid load forecasting, where data privacy is preserved by not transmitting sensitive consumer data to the central server. Furthermore, Alshehri et al. [162] discussed a federated learning framework for detecting energy theft in smart grids, where local models are trained at individual SMs and then aggregated to build a global model without revealing the sensitive energy consumption data of each user. The advantage of federated learning in SGs is that it provides a privacy-preserving way to train machine learning models without centralizing sensitive user data [163]. It enables the detection of anomalies in energy consumption patterns without compromising privacy.

However, federated learning requires efficient communication between devices to share model updates, and it can suffer from performance issues when there is significant data heterogeneity across devices [164]. While federated learning reduces the risk of data breaches, it faces challenges in terms of communication overhead and model convergence, especially when the data distribution across devices is heterogeneous, as shown in Table 11.

Table 11. Summary of federated learning applications in smart grids.

Refs.	System Model	Goal	Security Parameters	Performances	Limitations
[165]	Federated learning for distributed energy resources (DERs)	Train models for optimal energy dispatch across DERs without revealing individual data	Secure aggregation, federated model synchronization	Efficient energy distribution, better user privacy preservation	Data heterogeneity, model imbalance due to unshared local data
[166]	SG demand-response system using federated learning	Optimize energy consumption models while maintaining privacy	Local model training, differential privacy	Enhanced privacy for users, reduced central data processing	Delays in model updates, potential inefficiencies with large-scale systems
[167]	Federated learning for distributed energy resources (DERs)	Train models for optimal energy dispatch across DERs without revealing individual data	Secure aggregation, federated model synchronization	Efficient energy distribution, better user privacy preservation	Data heterogeneity, model imbalance due to unshared local data

Adversarial learning, in which synthetic data is generated to simulate real-world data without revealing sensitive information, has been applied to data augmentation in privacy-preserving machine learning models [107]. Adewole et al. [168] pioneered the use of adversarial learning with Generative Adversarial Networks (GANs), which have since been employed for privacy-preserving data generation in SGs. Takiddin et al. [169] applied adversarial learning to SGs for anomaly detection, training models to identify fraudulent activities, such as energy theft, even in the presence of malicious data tampering. By making the models robust to adversarial inputs, SG systems can become more resilient against data manipulation. While adversarial learning can effectively protect data privacy, it incurs computational costs, and ensuring the quality of synthetic data remains a key challenge, as shown in Table 12.

Table 12. Summary of adversarial machine learning applications in smart grids.

Refs.	System Model	Goal	Security Parameters	Performances	Limitations
[170]	Adversarial machine learning for anomaly detection in smart grids	Detect adversarial attacks in energy consumption data	Adversarial training, perturbation-based attack detection	Improved robustness against data tampering, anomaly detection in real time	Risk of adversarial model manipulation, requires extensive training data
[171]	SG cybersecurity using adversarial learning for load forecasting	Detect adversarial inputs influencing energy consumption predictions	GANs for synthetic attack generation, adversarial robustness	Increased prediction accuracy under adversarial conditions, resilient against data poisoning	Computational cost of adversarial model training, vulnerability to strong attacks
[31]	Adversarial training for secure energy trading in P2P systems	Enhance security of decentralized energy trading systems by learning adversarial behaviors	Generative adversarial networks (GANs), secure training protocols	Enhanced resistance to cyberattacks in peer-to-peer energy transactions	Difficulty in balancing model performance with adversarial robustness

The trade-offs of privacy preservation approaches in smart grids vary significantly. For instance, Homomorphic Encryption (HE) and Zero-Knowledge Proofs (ZKPs) provide strong privacy guarantees but entail high computational costs, limiting their scalability in real-time systems. Secure Multiparty Computation (SMPC) enables collaborative data analysis without sharing raw data but introduces substantial communication overhead. In contrast, anonymization and differential privacy (DP) offer lightweight and scalable solutions, albeit with reduced accuracy. Blockchain-based mechanisms provide tamper-proof data management; however, transaction speed and scalability remain concerns. Table 13 presents a comparative analysis of these privacy-preserving techniques, highlighting their key insights, computational intensity, and suitability for smart grid applications.

Table 13. Comparison of privacy-preserving techniques for smart grids.

Method	Key Advantages	Main Limitations	Key Insights and Suitability for Smart Grids	Latency	Computation Cost/Scalability	Refs.
HE	Strong privacy; allows computation over encrypted data	Very high computational overhead; unsuitable for real-time	Suitable for offline analysis or small-scale aggregation	0.5–5 s (per operation/ aggregation)	<10 k users (current demos)	[172]
ZKP	Enables authentication without revealing data	High complexity; latency in large-scale systems	Effective for billing/identity verification, not for bulk data	300–600 ms (per proof)	<5 k users	[173]
SMPC	Collaborative computation without sharing raw data	Heavy communication overhead; scales poorly	Useful for cooperative analytics among few entities	200-ms (For Three Parties)	1.5 Mbps Per operation	[173]
DP	Lightweight; scalable; strong formal guarantees	Loss of accuracy due to noise addition	Good for large-scale smart meter data aggregation	<50 ms	>100 k users	[76]
BC	Tamper-proof records; transparency; decentralization	Scalability and transaction delays; energy consumption	Suitable for settlements and secure trading platforms	1–10 s per transaction (PoW); 200–400 ms (PoS/DAG)	<1 k users (PoW); >50 k users (PoS/DAG)	[174,175]
FL	Data never leaves the devices; transparency; decentralization	Communication overhead, heterogeneity	Suitable for offline analysis or small-scale aggregation	100–300 ms per round	>10 k users	[176]

Noting the latency and computational scalability from Table 13, it can be observed that traditional privacy-preserving techniques such as Homomorphic Encryption (HE), Zero-Knowledge Proofs (ZKPs), and blockchain (BC) are still immature for deployment on resource-constrained IoT devices. Therefore, future research should focus on modifying these techniques to make them suitable for such devices. For instance, a lightweight privacy-preserving model has been proposed in [177], which is based on proxy re-encryption and asymmetric scalar product preserving encryption. In this model, the encryption process involves only addition and multiplication operations along with pairing operations rather than more computationally intensive individual operations, thus offering reduced computational overhead while maintaining the privacy of IoT devices.

Another approach addressing computational challenges in ZKPs was proposed in [178], where the authors utilized a Merkle tree architecture to aggregate user attributes and construct commitments for Zero-Knowledge Proof verification. This modification ensures that user attributes and access policies remain confidential while reducing computational overhead.

Regarding the computational intensity of blockchain, a scalable blockchain model has been proposed in [179]. In this model, the authors employed physically unclonable functions (PUFs), one-way hash functions, and bitwise XOR operations to achieve lightweight mutual authentication between smart meters and regional gateways, improving efficiency while maintaining security.

2.6. Trade-Offs Between Privacy Guarantees and Data Utility

Privacy-preserving mechanisms employed in smart grids span a spectrum from cryptographic techniques (Homomorphic Encryption (HE), Zero-Knowledge Proofs (ZKP), Secure Multiparty Computation (SMPC)) to statistical/anonymization approaches (differential privacy (DP), *k*-Anonymity, aggregation), blockchain/tokenization, and machine learning methods (federated and adversarial learning). Each class delivers a different point on the privacy–utility–performance frontier. Cryptographic tools provide very strong confidentiality; for example, HE enables computation over encrypted measurements but incurs substantial computation and latency that often makes it impractical for hard real-time use cases (high computation/latency, limited scalability). ZKPs provide strong verification without disclosure but add proof-generation latency and implementation complexity. SMPC offers collaborative analytics without revealing raw inputs but typically introduces large communication overheads and scaling difficulties. By contrast, DP and aggregation are lightweight and highly scalable and provide formal privacy guarantees, yet their injected noise or data generalization reduces accuracy for downstream tasks such as short-term load forecasting or fine-grained demand response. Blockchain and tokenization improve tamper resistance and auditability but suffer transaction speed and storage costs that reduce their applicability in very high-throughput measurement scenarios. Federated and adversarial learning keep raw data local and preserve privacy to varying degrees, but they face communication costs, data heterogeneity, and potential leakage through model updates. These qualitative trade-offs are summarized in Table 13.

To make these trade-offs concrete, systems researchers should move beyond qualitative statements and report a small set of standardized metrics for each use case:

1. A privacy-level parameter (e.g., DP's ε, key sizes or proof sizes for cryptographic techniques, attacker success rate/re-identification probability);
2. Utility metrics tied to the application (e.g., RMSE/MAE for forecasting, billing error for billing tasks, detection F1/AUC for theft/anomaly detection);
3. Operational costs (CPU cycles, latency, memory, communication bytes).

Reporting curves that sweep the privacy parameter and plot utility and cost (e.g., utility vs. ε, latency vs. message size, re-identification risk vs. k for *k*-Anonymity) reveals practical operating points. For instance, DP implementations require choosing ε to balance forecasting error and privacy: very small ε (strong privacy) implies larger injected noise and degraded forecasting/billing accuracy, while larger ε reduces noise but weakens formal guarantees. Similarly, HE/ZKP/SMPC variants differ in proof size/operation time and hence in feasibility for real-time grid control. The literature repeatedly emphasizes these trade-offs and the need to optimize noise/complexity for smart grid constraints, as shown in Tables 9, 10, and 12.

Therefore, future research should focus on hybrid architectures: use lightweight DP or aggregation for continuous high-frequency monitoring (to preserve utility at scale), and reserve heavier cryptographic proofs (HE, ZKP, SMPC) for periodic audits, billing settlement, or offline analytics where latency is acceptable. Federated learning can complement this approach by keeping model training local and sharing only encrypted/DP-protected updates for centralized model aggregation. Such a mixed strategy leverages the strengths of each technique while controlling utility loss and performance overhead.

3. Pricing Mechanisms for Energy Trading in Smart Grid

Secure information motivates smart grid energy market participants to share their data with confidence, without fear of tampering. Such secure information enables regulatory authorities to determine fair energy trading prices. Market price determination also involves various pricing mechanisms, which depend on supply and demand, encouraging efficient use of energy resources and stabilizing the market.

Depending on the trading process, energy pricing in peer-to-peer (P2P) energy trading can be categorized into synchronous and asynchronous techniques, assuming P2P energy trading occurs within a two-sided market among multiple prosumers and consumers exchanging electrical energy [16]. This categorization provides a foundational understanding of energy prices by highlighting the trading process and the peculiarities of each approach.

A synchronous pricing mechanism is one where market players submit bids, and the energy price and trading volume are determined using market objectives, which may include optimizing social welfare or minimizing production costs. This system-centric approach to P2P energy trading resembles traditional wholesale electricity markets [17,180]. In this process, a forward energy trading market is initiated on a platform at a specified time. Prosumers and consumers submit sealed bids indicating the price and quantity of energy they wish to trade during this window. After the bidding period ends, the platform reveals the results and clears the market. The resulting energy price controls the welfare of all market participants.

Synchronous energy trading can enhance market efficiency compared to asynchronous pricing, particularly if trading partners directly manage transaction parameters. However, its effectiveness may diminish if a participant manipulates the merit order to exercise market power [180].

Figure 4 illustrates the types of synchronous and asynchronous pricing mechanisms.

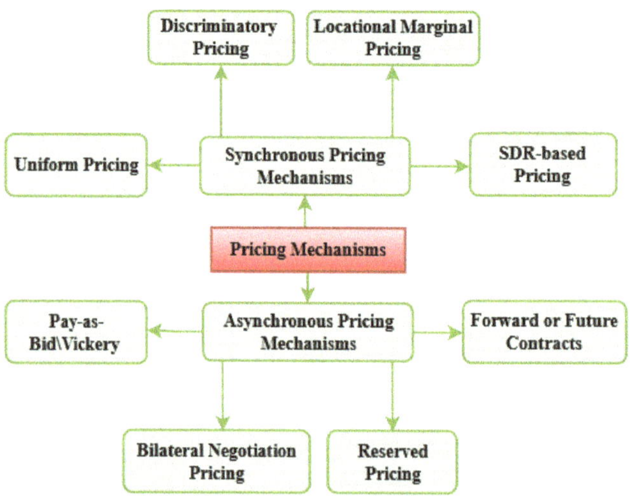

Figure 4. Types of pricing mechanisms in P2P energy trading.

Different synchronous pricing models, as depicted in Figure 4, share a common objective function outlined in Table 14. While these models offer multiple advantages, they also present operational challenges, summarized in Table 15.

Table 14. Objective functions for synchronous energy pricing mechanisms.

Pricing Mechanism	Objective Function
Uniform Pricing	Maximize social welfare: $$W = \sum_{i=1}^{n} (U_i(P) - C_i(P))$$ where W is total social welfare, $U_i(P)$ is the utility function for participant i at price P, and $C_i(P)$ is the cost for participant i at price P.
Discriminatory Pricing	Maximize individual profits or market participation: $$\Pi_i = (P_i \cdot q_i) - C_i(q_i)$$ where Π_i is the profit for participant i, P_i is the price set by participant i, q_i is the energy traded, and $C_i(q_i)$ is the cost for participant i.
Locational Marginal Pricing (LMP)	Minimize total system cost subject to power flow and grid constraints: $$\sum_{i=1}^{n} C_i(P_i) + \sum_{j=1}^{m} \lambda_j (P_j - P_{\text{demand}})$$ where $C_i(P_i)$ is the cost of generation, λ_j is the shadow price at node j, and P_{demand} is the demand.
SDR-based Pricing	Maximize market efficiency and fairness by balancing supply and demand: $$\text{SDR Price} = \frac{S}{D}$$ where S is the total supply and D is the total demand.

Asynchronous energy pricing enables numerous bilateral agreements between prosumers and consumers during the trading window, analogous to a flea market where participants negotiate energy prices that suit them individually. This methodology aims to achieve a collective agreement among market players in diverse situations. To facilitate optimal outcomes, distributed decision-making procedures have been proposed to support cooperation and coordination among participants [181].

In the asynchronous model, the market platform allows prosumers and consumers to negotiate energy prices and trading volumes directly. The consensus process varies depending on the framework employed. A key feature of asynchronous energy pricing is that trading is participant-driven, which reduces economic and political interventions typical in centralized markets. This decentralized approach allows market players to consider multiple factors, such as energy price, participant reputation, and preferences for renewable energy sources [182].

The objective functions, advantages, and operational challenges for asynchronous pricing models are summarized in Tables 16 and 17.

Table 15. Advantages, disadvantages, and challenges of synchronous pricing mechanisms in energy markets.

Type	Advantages	Disadvantages	Challenges	Refs.
Uniform Pricing	Simple, transparent, easy to implement.	Does not reflect grid constraints, inefficiencies in large markets.	May lead to price distortion, congestion handling issues.	[18,183–186]

Table 15. *Cont.*

Type	Advantages	Disadvantages	Challenges	Refs.
Discriminatory Pricing	Aligns payment with bidder's price, promotes competition.	Can lead to inefficiency and mark-ups, potential for strategic bidding.	Complexity in matching bids and price setting, potential for unfair market behavior.	[187–191]
Locational Marginal Pricing (LMP)	Reflects true cost of energy at different locations, efficient, encourages investment in grid capacity.	Computationally complex, exposes participants to locational price risk.	Requires accurate modeling of grid constraints, difficult for dynamic systems with high renewable penetration.	[192–200]
SDR-based Pricing	Simple to implement, encourages participation, promotes fairness.	Lack of grid constraint consideration, incentivization issues.	Requires accurate supply and demand data, challenges in fluctuating renewable energy markets.	[201–207]

Table 16. Advantages, disadvantages, and challenges for asynchronous pricing mechanisms in energy markets.

Pricing Type	Advantages	Disadvantages	Challenges	Refs.
Pay-as-Bid/Vickrey Pricing	Predictable for sellers; transparency in price determination.	Can lead to market inefficiency; high risk for sellers if bids are too low.	Manipulation of bids; price uncertainty.	[208–210]
Bilateral Negotiation Mechanisms	Flexibility in terms; enables tailored agreements.	Lack of transparency; can lead to inefficiency if not well-negotiated.	Lack of market transparency; disagreement on terms.	[211–215]
Reserved Pricing	Provides a minimum price guarantee; protects sellers from low market prices.	Can prevent market clearing; prices may be set too high, reducing competition.	Determining an optimal reserve price; potential market inefficiency.	[216–220]
Forward and Futures Contracts	Price certainty for future transactions; mitigates price risk.	Not suitable for all markets; limited flexibility for customization.	Determining an appropriate contract price; market volatility risks.	[221]

Table 17. Objective functions for asynchronous pricing mechanisms.

Pricing Type	Objective Function (Mathematical Formulation)
Pay-as-Bid/Vickrey Pricing	**Pay-as-Bid:** Minimize total market cost, where each seller receives the price they bid. $$\text{Objective:} \quad \sum_{i=1}^{n} P_i B_i$$ where P_i is the price bidder i bids and B_i is the quantity sold by bidder i. Vickrey Pricing (Uniform Pricing): Maximize social welfare by finding the market-clearing price P^*. $$\text{Objective:} \quad \max_{P} \sum_{i=1}^{n} (P - C_i) Q_i$$ where C_i is the cost of seller i, Q_i is the quantity, and P is the uniform market price.
Bilateral Negotiation Mechanisms	Maximize the utility of the buyer and seller based on the negotiated price P^*. $$\text{Objective:} \quad \max_{P^*} \left(U_{\text{seller}}(P^*) + U_{\text{buyer}}(P^*) \right)$$ where $U_{\text{seller}}(P^*)$ and $U_{\text{buyer}}(P^*)$ are the utility functions of the seller and buyer, respectively, which depend on the negotiated price P^*.

Table 17. *Cont.*

Pricing Type	Objective Function (Mathematical Formulation)
Reserved Pricing	Maximize seller's revenue subject to the reservation price $P_{\min}$ being met. $$\text{Objective:} \quad \max_{P \geq P_{\min}} P \cdot Q$$ where P is the market price, Q is the quantity, and $P_{\min}$ is the minimum price set by the seller.
Forward and Futures Contracts	Maximize the expected profit from a future transaction based on the agreed price P_f. $$\text{Objective:} \quad \max_{P_f} E\left[(P_f - C) \cdot Q\right]$$ where P_f is the agreed future price, C is the cost of production, and Q is the quantity to be traded. For futures, the expected value E accounts for uncertainty in market conditions.

3.1. Uniform Pricing Mechanism

The uniform pricing rule is a synchronous market-clearing mechanism in which all matched trades—whether a prosumer's sale or a consumer's purchase—are settled at a common price. Typically, this price is determined by aggregating all buy and sell bids in the form of supply and demand curves. The intersection of these curves establishes the clearing price for all transactions, regardless of individual bid values submitted by participants [222,223].

Several researchers have developed models to identify uniform pricing in peer-to-peer (P2P) energy trading markets. For instance, optimization-based models aim to maximize objectives such as social welfare [224,225]. Alternatively, the price may be set through stakeholder agreement. Uniform pricing contributes to reducing voltage imbalances within the distribution network and promoting supply–demand equilibrium [226–228].

The market-surplus-maximizing energy price is generally defined as the highest bid at the intersection of the supply and demand curves [229]. In practice, however, market clearing often uses the second-highest bid instead of the absolute highest bid to promote fairness [229–231]. Studies such as [232,233] examine uniform pricing in P2P markets, demonstrating that a common clearing price can ensure fairness, even under grid constraints or in 100% renewable electricity markets.

Fair competition among sellers determines the single clearing price, typically selected via a heuristic search over submitted bids to maximize revenue or minimize costs [223,234]. While uniform pricing enhances transparency and provides a common price for all participants, it is susceptible to strategic bid shading, where bidders may inflate offers above their true willingness to pay [235].

3.2. Discriminatory Pricing Mechanism

The discriminatory pricing system settles each approved transaction at the specific price bid by the respective participant. In this mechanism, sellers post their quantity–price combinations, and bids are accepted until demand is fulfilled. Successful sellers are then paid their individual bid price rather than a single uniform price, and the traded quantity corresponds to the minimum of the bid and ask quantities [236,237].

Discriminatory pricing has also been employed in renewable energy markets to provide more accurate price signals and optimize distribution performance. For instance, Ref. [238] demonstrates that discriminatory pricing can enhance the integration of renewable energy sources while maintaining fairness and effectiveness. The mechanism better reflects actual energy costs by allowing prices to vary according to local generation capacity and fluctuations in demand [239].

Pricing strategies under discriminatory pricing are typically aligned with specific market objectives. For example, Ref. [240] links matching rules to pricing approaches that can achieve the Nash equilibrium in simultaneous games [241]. In [242], a distributed approach is proposed to maximize overall energy-user benefit, creating an envy-free and Pareto-optimal outcome. Similarly, a two-stage Stackelberg game model [243] allows small distributed energy sources to trade at higher payments by targeting users willing to pay more, thereby reducing controller costs while maximizing user utility. Overall, discriminatory pricing improves market efficiency by emphasizing price flexibility, while necessitating careful pairing and matching of trading partners.

3.3. *Locational Marginal Pricing Mechanism*

Building on discriminatory pricing, Locational Marginal Pricing (LMP) incorporates the physical characteristics of the energy network. Unlike standard discriminatory pricing, which relies solely on bid prices, LMP determines a price at each grid node that accounts not only for the energy cost but also for transmission congestion and line losses. This provides a more accurate and location-specific pricing signal, facilitating cost-effective energy delivery across large, complex networks [244].

LMP calculation typically involves three main components:

1. The marginal cost of electricity generation at a specific power plant [245].
2. Costs associated with failing to clear electricity flows in the transmission network, requiring additional dispatch of generation resources [246].
3. Energy losses due to resistance in transmission lines [247].

Research by [248] approximates LMPs using AC and DC Optimal Power Flow (OPF) models, providing a detailed decomposition to understand the pricing mechanism in power markets. The study in [200] examines implementation challenges for LMP in transitioning power systems, emphasizing complexities introduced by decarbonization and heterogeneous participants. Other works, such as [197], highlight LMP's role in grid optimization and transparent energy pricing, taking into account generation costs, delivery, and transmission constraints at different network locations.

3.4. *SDR-Based Pricing Mechanism*

Building on discriminatory and Locational Marginal Pricing, the SDR (Supply and Demand Ratio)-based method provides an alternative where the ratio of available energy supply to demand directly influences energy prices. Unlike LMP, which considers grid physical characteristics, SDR focuses on real-time supply versus demand, making it suitable for small or decentralized energy markets [207,249].

The SDR is calculated as the total energy provided divided by the total energy required. This ratio, combined with bidding information, is used to determine energy prices. Studies such as [250] have shown SDR to be effective in renewable energy markets with fluctuating supply. Community-based energy trading scenarios also benefit from SDR, as it promotes fairness and encourages market participation [225,251]. Price determination may involve a convex combination of local purchasing and grid-selling prices [252,253].

3.5. *Pay-as-Bid (PAB) and Vickrey Pricing*

Pay-as-bid (PAB) and Vickrey auctions are competitive bidding mechanisms for energy markets.

In a PAB auction, participants submit bids indicating their desired selling or buying price. The market clears by accepting bids starting from the most competitive until total demand is met. Each successful seller receives their submitted price, incentivizing competitive bidding to maximize returns [254–258].

Vickrey pricing modifies PAB by having the winning bidder pay the second-highest bid rather than their own. This promotes truth-telling, as bidders reveal their true valuation without overbidding or underbidding, enhancing efficiency [259–262]. Recent research explores repeated and multi-unit PAB auctions [263] and adaptive bidding strategies [264,265], optimizing utility and market performance.

3.6. Reserved Pricing Mechanism

Reserved pricing sets the minimum acceptable price for energy in an auction. Bids below this threshold are rejected, protecting sellers from losses and ensuring that auction outcomes reflect real market value [266,267].

Reserved pricing is commonly applied in pay-as-bid or Vickrey auctions, influencing bidding behavior and maintaining market stability, especially with high renewable penetration or variable demand [206,268,269]. Research has examined its effect on operating reserves, scarcity pricing, and market incentives [270–275].

3.7. Forward and Futures Contracts

Forward and futures contracts are asynchronous mechanisms where prices are agreed upon today for delivery at a future date.

A forward contract is privately negotiated and customizable between two parties, while a futures contract is standardized and traded on an exchange [276–282]. These contracts allow participants to hedge against price volatility or speculate on future prices.

Forward/futures contracts provide price stability, particularly important in markets with renewable generation or fluctuating supply–demand conditions [221,283–286]. They also inform price discovery and market decision-making while enabling small-scale generators to secure predictable revenue streams.

3.8. Game-Theoretic Techniques

Game theory is a branch of mathematics that studies strategic interactions among rational decision-makers, where the payoff of each participant depends on the actions of others [19]. In energy markets, game-theoretic methods are used to model interactions among market players—such as buyers, sellers, operators, and regulators—to establish optimal pricing, bidding strategies, and contract negotiations.

Nash equilibrium is a fundamental concept in game theory [19,287], representing a set of outcomes where no player can improve their payoff by unilaterally changing their strategy. In energy markets, Nash equilibrium helps predict participants' behavior in competitive pricing environments, particularly during auctions.

Several studies have applied Nash equilibrium to optimize energy market prices:

- A Nash Stackelberg game model [288] was proposed to study strategic interactions between distributed energy resource (DER) aggregators and electricity retailers. This hierarchical model captures profit-maximizing behaviors of DER aggregators while accounting for retailers' pricing strategies, providing insights into optimal bidding strategies in decentralized systems.
- A hybrid approach combining Mixed-Integer Linear Programming (MILP) with game theory [289] has been used to represent P2P energy market settlements.
- Nash equilibrium has been used to identify optimal trading strategies among prosumers in decentralized energy markets [290] and to maximize bids in bilateral electricity markets [291].

- Game-theoretic bargaining solutions have been applied to determine prices for Tradable Green Certificates (TGCs) and other auctioned energy products [292].
- Learning-based approaches, such as the no-regret algorithm, have been integrated with Nash equilibrium to model supplier bidding strategies in forward electricity markets [293].

Cooperative vs. Non-Cooperative Games [294] are also widely applied:

- Cooperative game theory analyzes coalition formation among prosumers in P2P networks and equitable profit distribution using concepts like the Shapley value or the core. For example, online coalitional games have been proposed to compute payoffs in real-time P2P markets [295], ensuring fairness and scalability.
- Distributed negotiation mechanisms facilitate stable bilateral contracts within coalitions, improving participation and user satisfaction [296].
- Hedonic games incorporate social preferences and community relationships, enabling coalition formation that maximizes total energy exchange [297].

Stackelberg models are used in non-cooperative scenarios, where grid operators or aggregators act as leaders, and DER prosumers are followers. These models optimize network costs and prosumer utilities simultaneously [298]. Hybrid models combining cooperative and Stackelberg frameworks balance self-interest with group stability [294,299].

Game theory provides diverse tools for modeling pricing in electricity markets, accommodating both market-like structures and P2P energy trading. Table 18 summarizes recent applications of game-theoretic methods in energy price determination.

Table 18. Applications of game theory in energy pricing.

Application Area	Technique Used	Description	Refs.
Auction Design (Pay-as-Bid, Vickrey–Clarke–Groves)	Nash Equilibrium/Auction Theory	Game theory models are used to design and analyze auction mechanisms like pay-as-bid and Vickrey auctions to determine optimal bidding strategies in competitive electricity markets.	[208,300]
Bilateral Electricity Markets	Nash Equilibrium	Nash equilibrium models are used to study bidding strategies in bilateral electricity markets, where players (sellers and buyers) interact to optimize their payoffs.	[301,302]
Peer-to-Peer (P2P) Energy Trading	Non-Cooperative Game Theory	Game theory is applied to model the strategic interactions of prosumers and consumers in P2P energy trading systems, optimizing pricing and transaction behavior.	[241,303,304]
Distributed Energy Resource (DER) Aggregation	Nash–Stack-elberg Game	Nash–Stackelberg game theory models are applied to optimize pricing and bidding strategies in DER aggregation by modeling the interactions between aggregators and retailers.	[305–307]
Tradable Green Certificates (TGCs)	Nash Bargaining Theory	Nash bargaining models are used for price determination and profit-sharing in energy markets, ensuring mutual benefits between stakeholders.	[308–310]
Forward and Futures Markets	Nash Equilibrium + No-Regret Learning	Nash equilibrium and no-regret learning algorithms are used to determine optimal bidding strategies in forward and futures electricity markets, considering market volatility.	[293,311]

3.9. Optimization Techniques

Optimization methods are widely used for efficient and fair cost determination in energy markets, particularly in market-based electricity systems. These methods are crucial for forecasting prices and improving market mechanisms by considering multiple constraints, including supply–demand balance, transmission limits, and cost minimization. This section reviews recent applications of optimization in energy pricing.

Linear Programming (LP) is employed in centralized market clearing to determine optimal generation dispatch, transmission allocation, and price setting. The objective is typically to minimize total generation costs while satisfying demand and operational constraints. LP models are attractive for their simplicity and computational efficiency. For example, Ref. [312] proposes an LP model that accounts for generation and transmission line constraints to clear the electricity market in advance, minimizing costs and maximizing reliability. Similarly, Ref. [313] applied LP for renewable energy integration, optimizing market prices while ensuring that solar and wind generation meets demand.

Mixed-Integer Linear Programming (MILP) extends LP by incorporating both continuous and integer variables, enabling modeling of discrete decisions such as unit commitment, generation scheduling, and bidding strategies. MILP is particularly useful for markets requiring binary choices, e.g., turning generators on/off [314]. Studies like [315,316] use MILP to minimize operational costs while respecting environmental limits, ensuring generation flexibility, and optimizing contracts between market participants. MILP has also been applied in auction-based electricity markets [316–318] to optimize bidding strategies and long-term investment planning.

Convex Optimization has gained popularity in price forecasting and market equilibrium analysis due to its guarantee of global optimality for convex problems. These techniques are used to maximize social welfare or profit while minimizing costs under operational constraints. For instance, Ref. [319] uses convex optimization to forecast electricity prices under renewable energy uncertainty, whereas [320] incorporates demand response and DER integration to enhance price stability.

Stochastic Optimization is essential in markets with uncertain future conditions, such as real-time pricing, futures, and renewable integration. By modeling probabilistic variations in demand, generation, and market factors, stochastic optimization improves pricing under uncertainty. Applications include real-time pricing considering wind and solar forecasting errors [321] and modeling electricity price fluctuations under water supply–demand uncertainty and financial risk [322].

Metaheuristic Optimization Techniques such as Genetic algorithms (GAs), Particle Swarm Optimization (PSO), and Simulated Annealing (SA) are used to solve complex, non-linear problems in bidding strategies and market behavior [312]. These methods are particularly useful for dynamic, real-time price adjustments. For example, Ref. [323] applies GA to optimize bidding strategies considering competitor actions, Ref. [324] uses PSO for real-time auction pricing, and [325] employs SA for calculating real-time P2P energy prices.

Table 19 summarizes key studies that use optimization techniques for energy price determination.

Table 19. Applications of optimization techniques in energy pricing.

Objective	Optimization Technique	Description	Refs.
Market Clearing	Linear Programming (LP)	Linear Programming (LP) is widely used for optimizing the market clearing process, minimizing costs while satisfying supply and demand constraints.	[313,326–328]
Bidding Strategies in Electricity Auctions	Mixed-Integer Linear Programming (MILP)	MILP is applied to model bidding strategies and generation scheduling, helping market participants determine optimal strategies for both price and quantity.	[329,330]
Price Forecasting in Spot Markets	Convex Optimization	Convex optimization is used for forecasting prices in spot markets, where prices are determined based on supply–demand dynamics and operational constraints.	[319,331]

Table 19. *Cont.*

Objective	Optimization Technique	Description	Refs.
Real-time Pricing Adjustment	Stochastic Optimization	Stochastic optimization is applied in real-time price adjustment by modeling uncertainty in market conditions and helping determine optimal pricing strategies under uncertainty.	[321,332]
Generation Scheduling and Profit Maximization	Mixed-Integer Non-Linear Programming (MINLP)	MINLP models are applied to optimize generation scheduling while maximizing profit, considering operational and environmental constraints such as emission limits.	[333,334]
Energy Storage Optimization	Particle Swarm Optimization (PSO)	PSO is used to optimize the operation of energy storage systems by minimizing cost and maximizing energy delivery efficiency based on demand forecasts and storage capacity.	[335–337]
Bidding in Renewable Energy Markets	Genetic Algorithms (GAs)	Genetic algorithms (GAs) are used to model and optimize bidding strategies in renewable energy markets, considering factors like energy production and market prices.	[338,339]

3.10. Numerical Method-Based Techniques

Numerical methods play a critical role in energy pricing, particularly for complex, non-linear problems where analytical solutions are not feasible. They are widely applied in modeling market behavior, handling uncertainty, stochastic processes, price prediction, real-time pricing, and renewable energy integration.

Monte Carlo simulation (MCS) is extensively used to model uncertainty in pricing mechanisms, especially in stochastic environments such as forward contracts, futures markets, and renewable integration. It enables analysis of price variations and forecasts likely outcomes based on random variables. For instance, Ref. [340] investigates the impact of solar energy uncertainty on real-time electricity pricing using Monte Carlo simulation, while [341] uses it to project future prices in electricity futures markets.

Partial Differential Equations (PDEs) arising in price evolution and energy derivatives are often solved using Finite Difference Methods (FDMs). These methods approximate PDEs to simulate dynamic energy pricing, where prices depend on multiple factors such as demand, supply, and external conditions. For example, Ref. [22] models electricity price fluctuations in forward contracts using FDM, enhancing time-dependent price prediction accuracy. Other studies [342,343] apply FDM for non-linear pricing problems under market uncertainties.

Iterative Techniques are used for solving non-linear optimization problems and market-clearing algorithms, especially in decentralized energy markets. Iterative algorithms assist in balancing supply and demand to achieve market equilibrium [344,345].

The Newton–Raphson Method is an iterative approach for solving non-linear equations, often applied in real-time market clearing under complex price relationships [346,347]. Similarly, **gradient descent** is applied to optimize bidding and pricing strategies iteratively, particularly under competitive market conditions [348,349]. Stochastic gradient descent has been used for price modeling under renewable energy uncertainty and volatile markets [350,351].

3.11. AI-Based Techniques

AI-based approaches, including machine learning (ML), reinforcement learning (RL), and deep learning (DL), are increasingly applied to dynamic and complex pricing problems in energy markets, especially with the integration of renewable energy sources and smart grids.

Machine learning (ML) techniques predict energy prices using historical data, weather conditions, and market factors. Supervised learning models such as regression and ensemble methods (e.g., random forests) are widely used for real-time price forecasting [352,353].

Support Vector Machines (SVMs) have been employed for electricity futures pricing under renewable generation uncertainty [354,355].

Reinforcement learning (RL) is suitable for dynamic pricing, where agents learn optimal pricing strategies through interaction with market conditions. Q-learning has been applied to real-time pricing in smart grids [356,357], while Deep Reinforcement Learning (DRL) models optimize bidding strategies considering market states and competition [358,359].

Deep learning (DL) models, including neural networks (NN), LSTM, and CNN, are effective for complex pattern recognition in price forecasting. LSTM models capture long-range dependencies in electricity price data [360,361], while CNNs characterize price variability under uncertain renewable generation [362].

Hybrid AI Models integrate multiple techniques (e.g., DL, RL, ML) to optimize pricing comprehensively. Examples include CNN-LSTM models for spot price prediction [363], DRL-DL hybrids for dynamic smart grid pricing [364], and DNN-GA hybrids for P2P energy market bidding [365]. These hybrid approaches outperform conventional models in forecasting accuracy and market decision-making.

Table 20 summarizes the advantages, disadvantages, and challenges of the numerical and AI-based pricing techniques discussed.

Table 20. Advantages and disadvantages of the techniques used for price determination.

Technique	Advantages	Disadvantages	Challenges
Game Theory Techniques	Helps model strategic interactions between market participants. Provides insights into optimal bidding strategies. Can handle competitive pricing environments.	Assumes rational behavior, which may not always hold in real-world markets. Models can become complex for large-scale systems. Computational complexity for large datasets.	Uncertainty in behavior modeling. Equilibrium calculation in large markets is difficult. Handling non-cooperative behaviors in dynamic markets.
Optimization Techniques	Offers efficient solutions for complex problems (e.g., market clearing, generation scheduling). Can optimize for cost minimization and profit maximization. Can incorporate uncertainty and constraints effectively.	Limited flexibility in some non-linear or dynamic pricing models. Requires accurate data inputs, which may be difficult to obtain in real-world markets. Scalability issues in larger markets.	Data accuracy issues in forecasting. Requires computational resources for large-scale problems. Difficulty in modeling renewable energy variability in optimization problems.
Numerical-Based Techniques	Can handle uncertainty and complexity in pricing models. Useful for real-time price determination. Can simulate various market scenarios and price behaviors.	May not provide global optimal solutions for non-linear problems. Computationally expensive for large-scale models. Sensitivity to initial conditions can affect results.	High computational costs for real-time price simulation. Difficulty in calibrating models under dynamic market conditions. Handling large data in Monte Carlo simulations or FDM.
AI-Based Techniques	Adaptability to dynamic market conditions. Improved accuracy in price forecasting and real-time pricing. Can handle large datasets and complex patterns.	Requires large amounts of training data. Models can become black-boxes, making it difficult to interpret results. Overfitting can be an issue in some models.	Data quality and availability. Model interpretability can be a concern for decision-makers. Training time for large-scale models and adaptation to market dynamics.

4. Demand–Supply Balance Program (DSBP)

The integration of distributed energy resources (DERs) into smart grids is facilitated by effective pricing signals and energy conversion technologies. However, DER generation is inherently variable due to changing weather conditions, directly affecting generation capacity. This variability often causes a mismatch between supply and demand in the smart grid.

During periods of favorable weather, DERs, such as photovoltaic (PV) systems, operate near maximum capacity, potentially producing excess supply relative to demand.

Conversely, during evenings or cloudy conditions, DER output declines, leading to a supply deficit relative to the energy demands of buyers.

To maintain demand–supply balance, the smart grid often relies on energy imports from or exports to the utility grid, which primarily consists of fossil-fuel-based generation. Such interactions not only increase CO_2 emissions but also impose operational stress on the grid and participants. Importing energy from the utility grid is costlier than P2P energy, reducing potential savings for buyers, while exporting excess energy may reduce profitability for DER sellers.

The Demand–Supply Balance Program (DSBP) addresses these challenges by dynamically balancing generation and consumption while optimizing energy flows and prices. The DSBP can be formulated using various methodologies, including iterative methods, optimization models, game-theoretic frameworks, and AI-based approaches, as illustrated in Figure 5.

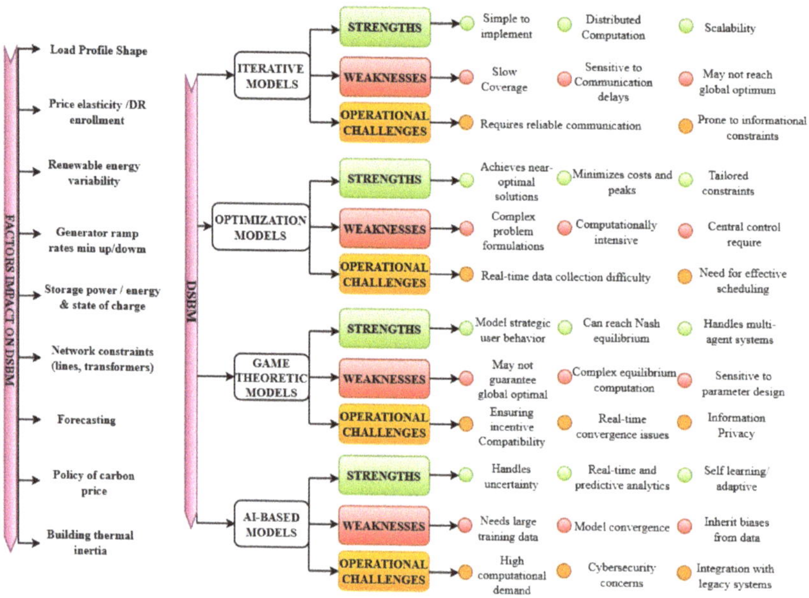

Figure 5. Factors impacting the Demand–Supply Balance Program along with formulation techniques.

Each formulation approach in Figure 5 has unique merits and operational challenges. Table 21 provides a comparative overview of these models in terms of methodology, application, advantages, and limitations.

Table 21. Comparative analysis of several models used to design DSBP.

Model Type	Main Principle	Typical Use Scenarios	Strengths	Weaknesses	Operational Challenges	Scalability	Data Requirements	Real-Time Capability
Iterative Model	Progressive updates until convergence; decentralized interactions	Decentralized coordination, iterative pricing, DSM	Simple implementation; distributed; scalable	Slow convergence; sensitive to delays; suboptimal solutions	Communication reliability; synchronization	High	Low to Moderate	Moderate
Optimization Model	Mathematical programming (LP/MILP/heuristics)	Cost minimization, load balancing, scheduling	Achieves optimality; constraint flexibility; performance control	Computationally intensive; centralized; complex modeling	Data availability; solving high-dimensional problems	Moderate	High	Challenging
Game Theoretic Model	Strategic decision-making among multiple agents (users, suppliers)	Pricing, market design, incentive mechanism, auctions	Captures strategic intelligence; equilibrium concepts	Equilibrium may not be unique; complexity; data dependency	Ensuring incentive compatibility; privacy; convergence	Moderate	Moderate to High	Moderate
AI-Based Model	Learning from data (ML, deep learning, RL, expert systems, etc.)	Forecasting, anomaly detection, demand prediction	Adapts to uncertainty; predictive; pattern recognition	Needs training data; transparency issues; bias risks	Integration with legacy systems; computation; security risks	High	High	High

4.1. Demand–Supply Balance Program Solving Models

Several researchers have proposed advanced methods to address the challenges of DSBP. This section explores notable approaches, highlighting their objectives and strengths.

4.1.1. Iterative Models

Iterative models rely on numerical formulations of the DSBP, updating system values repeatedly until a pre-defined convergence threshold is met. These models are widely used in smart grid management due to their simplicity and adaptability.

For instance, Ref. [20] introduced a Mixed-Integer Linear Programming (MILP)-based iterative model to manage real-time energy in an IEEE-33 AC-DC distribution network. Similarly, Ref. [366] proposed an iterative MILP model to optimize household energy consumption while respecting the operative supply envelope.

Another iterative approach using an inclined block tariff with real-time varying prices was developed in [367], enabling the adjustment of home appliances according to real-time energy costs. A demand-side management iterative sizing method was proposed in [368] to improve system efficiency.

Iterative weighting strategies have also been employed. In [369], different weights were assigned to household appliances, and appliance operating cycles were adjusted based on the Demand–Supply Ratio. Another iterative model [370] maintained the balance between requested demand and supply, achieving an average peak-hour load reduction of 52.8%, which is approximately 26.1% more efficient than comparable iterative methods. Additionally, this model demonstrated a 33.9% profit increase for aggregators by dynamically adjusting the demand–supply balance.

Table 22 presents a comparative analysis of several iterative DSBP models, summarizing their methodologies, objectives, and performance outcomes in maintaining demand–supply equilibrium in smart grids.

Table 22. Summary of selected studies on DSBP based on iterative methods.

Ref.	Model Focus	Data Used	Findings	Strength	Limitation
[371]	Demand peak reduction via incentives	Smart grid agent load data	Novel incentive algorithm learning user response iteratively under budget constraints	Realistic agent modeling, budget aware	Simulator-based results; limited real-field validation
[372]	Demand–supply balancing in power	Generator and consumer data	Stackelberg equilibrium algorithm balancing supply and demand interactions	Strong theoretical foundation	Heavy assumptions on rationality and market structure
[373]	Integration of smart buildings	Building and grid integrated data	Proposes iterative optimization for building-grid demand coordination	Building-grid integration focus	Scale limitations from building to entire grid
[374]	Market clearing and price setting	Electricity market bid data	Fast converging algorithm robust to stochastic bids	Efficient market clearing with robustness	Focus on market, less on grid operation
[375]	Smart Grid Energy Management	Machine learning iterative algorithms	Demand forecasting, failure detection	Smart meters and sensors	Integrates ML for iterative energy load balancing and anomaly detection
[376]	Multi-level supply–demand matching	Cost/performance measures	Reduces mismatch via multi-level iterative solution	Hierarchical structure improves scalability	Solution complexity not fully addressed

Although iterative models are scalable and easy to deploy in smart grids and perform effectively under pre-defined threshold conditions, the solutions they provide do not capture the strategic interactions among participants in DSBP programs.

Strategic interactions are crucial in DSBP design, as participants select their consumption based on private information, such as financial constraints, comfort preferences, and other individual factors. These interactions can be effectively modeled and analyzed using game-theoretic approaches.

4.1.2. Game-Theoretic Models

Game-theoretic models are an effective tool to capture and analyze the strategic behavior of participants by formulating their interactions as cooperative, non-cooperative, or evolutionary games.

In cooperative game models, participants form coalitions to determine optimal demand and supply values collectively. The achieved outcomes are then distributed among participants using allocation schemes such as Shapley values or other fair division models.

In non-cooperative game models, each participant independently selects their strategy, and outcomes are not shared with others. These models allow the analysis of individual optimization behavior in decentralized energy markets.

Several studies have utilized game-theoretic approaches to identify optimal strategies in DSBP programs. Table 23 provides a comparative summary of notable research using game-theoretic models for demand–supply balancing in smart grids.

Table 23. Game theory applications in energy pricing and management.

Ref.	Game Type	Objectives	Strengths
[22]	Nash Equilibrium	Optimize energy consumption and pricing strategies between consumers and utilities; balance consumer cost savings with utility revenue while promoting green energy use	Uses smart meter data for informed, strategic decision-making; achieves a mutually beneficial equilibrium where both consumers and utilities maximize payoffs
[377]	Non-cooperative Game	Analyse interactions between distributed generation (DG) units and autonomous demand-response programs in smart distribution grids; reduce total operational costs and power losses	Enhances overall grid performance in terms of power quality and stability
[378]	Nash Bargaining Model	Achieve fair benefit allocation between trading peers while maximizing social welfare	Balances fairness and efficiency, ensuring stable cooperation
[379]	Nash Equilibrium	Optimize energy consumption and pricing strategies between consumers and utilities using green energy sources	Balances consumer and utility interests; reduces costs; promotes green energy adoption
[380]	Nash Equilibrium	Demand side management in smart grids to optimize energy use and costs, maximizing utility and consumer payoffs	Enhances energy efficiency; reduces environmental pollution through green energy use
[381]	Non-cooperative Game	Manage smart distribution grids with distributed generation and autonomous demand response; analyze effects compared to centralized control	Reduces total costs and power losses; improves reactive power support, voltage profiles, and load profile flattening
[382]	Noncooperative Game	Encourage residential consumers to adjust electricity use via dynamic pricing; minimize costs while maintaining comfort	Significant cost savings without sacrificing comfort; optimized appliance scheduling using NSGA-II
[383]	Stackelberg Game	Optimize V2G pricing for aggregators and EV users, balancing benefits while considering charging costs and inconvenience	Models multi-entity interactions with realistic factors; validated through multi-aggregator EV simulations

Table 23. *Cont.*

Ref.	Game Type	Objectives	Strengths
[229]	Double Auction-Based Stackelberg Game	Enable P2P energy trading among prosumers to maximize participant profits, ensure social welfare, and maintain privacy	Achieves incentive compatibility and individual rationality; supports real-time trading via blockchain; effective under various scenarios
[384]	Concave N-Person Game	Optimize global power consumption scheduling of TCL users while considering individual preferences and REN generation forecasts	Adaptive and flexible pricing mechanism; fast solution via simplified model; smooths tie-line power in microgrids

4.1.3. Optimization Models

Although iterative models are highly scalable and game-theoretic models effectively capture participant interactions in DSBP design, both approaches often overlook critical system constraints, such as network operating capacity, available generation, and supply-to-demand ratios. To address these challenges, optimization models have been widely employed to design effective Demand–Supply Balance Programs in smart grids.

For instance, Zhao et al. [21] proposed a multi-objective optimization model to balance demand and supply while considering both economic and environmental impacts. This model utilized the AUGMECON2 optimization algorithm to obtain optimized solutions under multiple constraints.

An Ant Colony Optimization (ACO)-based model was introduced in [385] to determine optimized demand and supply values. The model was compared with other optimization approaches, such as Evolutionary Algorithm, Moth-Flame Optimization, and Bacterial Foraging Optimization, with simulation results demonstrating superior performance of the ACO model. Similarly, Nanibabu et al. [386] applied the Butterfly Optimization Algorithm, achieving an 18.4% cost reduction during peak hours through effective demand-side management strategies.

Optimization models have also been applied to control appliance operation and social welfare. For example, Ref. [367] employed Optimal Adaptive Wind-Driven Optimization to schedule appliance usage and balance supply and demand efficiently. Particle Swarm Optimization (PSO)-based models have been proposed in [387,388] to maximize social welfare and formulate the DSBP. Furthermore, a two-layered optimization model was introduced in, where the first layer manages appliance operation cycles, and the second layer optimizes consumption costs while maintaining consumer comfort levels.

These optimization models demonstrate the ability to integrate system constraints, operational flexibility, and economic objectives, making them a robust approach for designing effective DSBPs in smart grids.

A bi-level optimization model to maintain the balance between demand and supply in the smart grid was proposed in [389]. Additionally, a study in [390] introduced a model that considers multi-energy sources along with their planning and operational costs. This model employs a hybrid genetic algorithm (GA) and pattern search algorithm to balance energy generation and supply efficiently in the smart grid.

Beyond these studies, several other works deploying optimization models to maintain demand–supply balance are compared in Table 24.

Table 24. Optimization techniques in energy pricing and management.

Ref.	Optimization Type	Objectives	Strengths
[391]	Multi-objective optimization with real-time supply–demand balance	Minimize average electricity cost for the VPP operation, maximize renewable energy utilization (solar and wind), maintain grid stability while meeting demand	Manages uncertainty in renewable generation; adjusts supply allocation based on real-time market prices; selects optimal mix of multiple energy sources; reduces average electricity cost by 15%; increases renewable utilization by 20%
[392]	Multi-objective, market-based optimization (LP, MILP, Evolutionary Algorithms)	Reduce peak demand, ease power flow congestion, integrate DERs while maintaining grid stability, maximize economic benefits, improve DR participation and scheduling	Coordinates DER integration through AMI; enhances network and market operations; improves decision-making for DER management; increases flexibility and resilience of the power system
[393]	Market-based bilateral bidding model for pre-listing energy consumption rights trading	Connect medium-/long-term and spot markets, enable multi-day rights transfer, ensure fair bidding, align volume and price, improve resource allocation	Allows flexible listing and withdrawal within price tolerance; encourages engagement from supply and demand; improves market clearing with defined computation methods
[394]	DSM optimization with IoT-enabled monitoring and blockchain	Shift load from peak to off-peak, reduce costs, encourage behavioral change, cut emissions, improve grid efficiency	Uses smart meters and IoT for adaptive monitoring; supports competitive pricing; integrates tech and social change; enables secure transactions via blockchain
[395]	Goal-oriented classification and selection optimization for DR schemes	Improve performance and reliability, enhance decision-making, maximize economic benefits, integrate DR into ancillary services	Provides practical classification for DR plan selection; considers benefits and barriers; supports all stakeholders; aligns DR schemes with smart community concepts
[396]	MILP for optimal scheduling of integrated energy systems (CCHP, carbon capture, DR)	Minimize total operating cost (energy purchase, maintenance, carbon, compensation), enable low-carbon operation, enhance user satisfaction, optimize multi-energy flow	Integrates cooling, heating, and electricity DR with carbon trading; enables flexible load shifting; reduces gas purchase; cuts emissions; lowers costs by 5.9% plus 3.1% with DR
[397]	Distributed MPC with forecasting and P2P negotiation	Incorporate forecasting and peer-to-peer negotiation in distributed MPC for meshed grids	Enables coordinated, predictive control with negotiation; improves adaptability in meshed grids
[398]	Optimal decentralized operation model (P2P-enabled)	Achieve optimal decentralized operation of smart distribution grids with P2P trading	Facilitates decentralized decision-making; improves local autonomy and operational efficiency
[399]	Distribution network-constrained optimization for multi-microgrid P2P trading	Optimize P2P energy trading among multiple MGs considering network constraints	Integrates network constraints into market transactions; enhances feasibility and reliability
[400]	Data-driven distributionally robust co-optimization of P2P trading and network operation	Provide robust collaborative optimization for interconnected MGs considering fairness	Uses distributionally robust optimization and ADMM for decentralized, fair, and resilient decisions
[401]	Distributed consensus-based optimization	Reach agreement on energy prices and quantities via iterative local exchanges	Low communication overhead; convergence to global optimality
[402]	Bi-level optimization model	Upper level sets market prices; lower level optimizes prosumer schedules	Captures hierarchical decision-making between market operator and prosumers
[403]	Robust optimization	Account for uncertainty in renewable generation and demand during scheduling	Ensures feasible solutions under worst-case scenarios

4.1.4. AI-Based Models

While optimization models provide highly accurate outputs for the present time horizon, their effectiveness can diminish for future horizons due to the need to analyze large volumes of historical data. As the dataset grows, computational time increases, which directly impacts the scalability of these models.

To overcome these limitations and capture the true dynamics of the system, artificial intelligence (AI) models are increasingly employed in designing DSBP programs. AI methods can learn important patterns and events from historical data, enabling better forecasting and adaptive demand–supply management in smart grids.

Table 25 summarizes several studies that applied AI models to determine optimized demand–supply values, highlighting the techniques, objectives, and key results.

Table 25. AI models applied in DSBM and smart grid management.

Ref.	AI Model	Objectives	Strengths
[23]	Artificial Neural Networks (ANNs)	Predict energy consumption patterns for improved demand-response performance	Can capture complex, non-linear relationships in energy data, enabling accurate predictions
[404]	Supervised Machine Learning (SML)	Develop predictive models for load forecasting in demand response	Flexible with various algorithms; performance improves with quality feature engineering
[405]	Hybrid AI Approach (ANN + Optimization)	Optimize demand-response scheduling by combining predictive modeling with decision optimization	Integrates prediction accuracy with operational optimization for better scheduling outcomes
[406]	AI-based (ANN, Supervised ML)	Present an overview of AI methods in demand response, identify research gaps, and propose future study directions	ANNs can capture complex non-linear patterns in energy data; Supervised ML provides flexibility with various algorithms but benefits greatly from effective feature engineering
[407]	Deep Learning (LSTM) and Unsupervised Learning (K-means, Hierarchical Clustering, PCA)	Provide insight into demand forecasting in smart grids and explore deep learning techniques for improving forecasting accuracy	LSTM handles temporal dependencies in load profiles; Unsupervised learning reveals hidden patterns and clusters in consumption data
[408]	Reinforcement Learning	Develop price-driven demand-response management to minimize system cost via optimal pricing for prosumers	Handles incomplete information and outperforms other price-based DRM approaches
[409]	Hybrid GRU–TCN with Attention	Improve short-term power load forecasting accuracy and efficiency; enhance energy information management in smart grids	Incorporates uncertainty modeling with stochastic and Monte Carlo methods for robust forecasting
[410]	AI-based diagnostics and prognostics (Machine Learning)	Address challenges in smart grid management and security; develop a comprehensive framework for optimal grid reliability	Enables real-time monitoring and predictive maintenance for improved grid performance
[24]	AI methods (Machine Learning, ANN)	Provide an overview of AI techniques in energy demand response; identify future research directions in demand-side response	Offers diverse modeling capabilities; adaptable to various DR scenarios
[411]	Model-free DDPG on Hyperledger Fabric	Integrate distributed controllable resources for grid services; optimize DCR allocations and maximize prosumer profits	Handles continuous control without system model; on-chain execution and auditability; coordinates many resources securely

4.2. AI-Driven Approaches in Pricing and Balance: Economic and Environmental Perspectives

The integration of artificial intelligence (AI) in smart grid (SG) systems, particularly through Demand–Supply Balance Programs (DSBPs) and pricing models, enhances operational efficiency and supports economic viability by leveraging real-time data from Advanced Metering Infrastructure (AMI) and Internet of Things (IoT) devices [2,12,13]. AI models facilitate demand prediction, supply optimization, and dynamic pricing, which contribute directly to reduced operational costs and improved resource reliability [8]. For instance, AI-driven DSBPs enable market operators to estimate demand–supply ratios using optimization and game-theoretic frameworks augmented by AI [17,18], allowing for uniform pricing or auction mechanisms that minimize energy waste during peak hours. This results in economic benefits such as lower electricity bills for consumers through informed usage decisions and dynamic pricing models that incentivize off-peak consumption, thereby conserving energy and reducing costs [9]. Utilities also benefit from automated billing and real-time monitoring, which reduce human error and enhance revenue management by preventing overages or undersupplies [14].

From a financial viability standpoint, AI deployment in SG pricing and balance programs offsets initial implementation costs, such as those for smart meters (SMs), control centers (CCs), and aggregators [38], through long-term gains in grid efficiency and scalability. By integrating renewable energy sources (RESs), such as solar panels and wind turbines, AI supports bidirectional energy flow and demand-response systems that balance supply and demand, reducing the need for expensive fossil fuel backups and fostering a more sustainable energy market [1]. AI approaches, combined with energy storage systems (ESSs), further amplify economic returns by injecting surplus energy during high-demand periods, mitigating the stochastic nature of RESs, and stabilizing prices [20].

Environmentally, AI implementation in the energy sector promotes sustainability by easing the penetration of RESs into SG architectures, which lowers the carbon footprint of energy production [44]. For example, AI-optimized DSBPs and pricing strategies enable efficient distribution of surplus renewable energy, reducing reliance on non-renewable sources and supporting dual energy flows that align with environmental goals [3]. However, challenges such as the energy demands of AI computations and potential vulnerabilities in IoT-integrated systems [12] must be addressed to ensure that these benefits do not inadvertently increase electronic waste or grid instability. Overall, the reviewed paradigms suggest that AI not only bolsters economic resilience through cost savings and reliability [8] but also advances environmental objectives by promoting cleaner, more balanced energy ecosystems [44], warranting further research into scalable, low-impact AI deployments.

4.3. *Energy Storage Systems (ESS)*

Although the Demand–Supply Balance Program (DSBP) assists the smart grid in maintaining a balanced state, this balance can be further enhanced by incorporating energy storage systems (ESSs). By utilizing ESSs, energy generated from photovoltaic (PV) systems and wind turbines during periods of high generation or favorable weather conditions can be stored in batteries. This stored energy can then be injected back into the system during peak-hour demand, thereby reducing the gap between requested demand and available renewable energy.

In smart grids, energy can be stored using multiple technologies, as illustrated in Figure 6. Among these, batteries are one of the most widely adopted methods due to their flexibility, scalability, and relatively straightforward integration. Batteries come in various chemical compositions, each with specific advantages and limitations, as also shown in Figure 6.

4.3.1. Lead–Acid Batteries

Lead–acid batteries are among the most cost-effective energy storage solutions, offering a low upfront cost and simple control mechanisms with high surge current capability. Due to these attributes, they are frequently employed in smart grid research to address energy storage challenges. Table 26 presents a comparative summary of several studies that utilized lead–acid batteries for energy storage in smart grid applications.

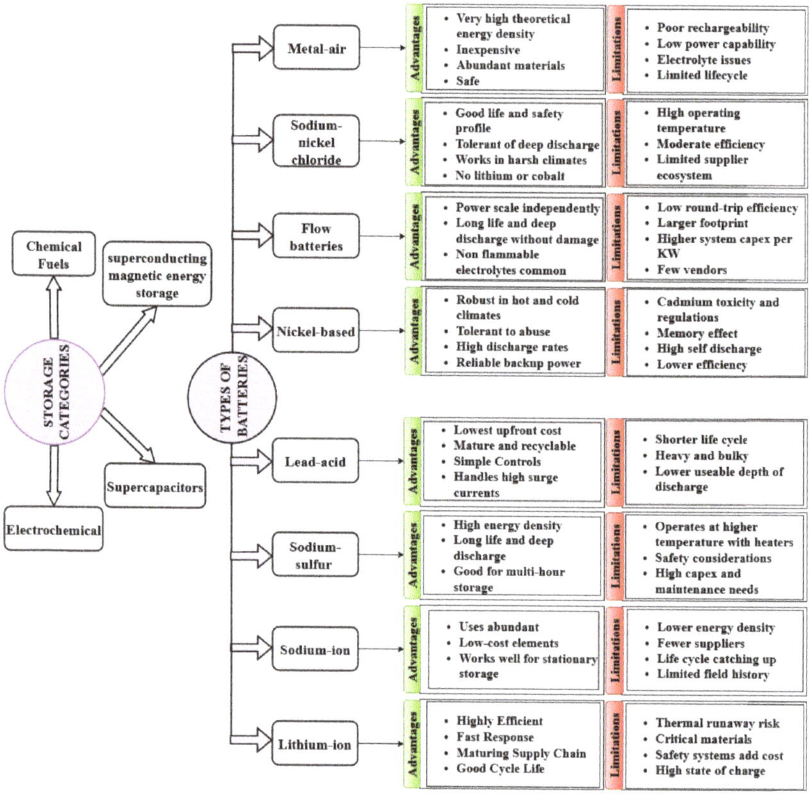

Figure 6. Storage categories and types of batteries.

Table 26. Lead–acid battery types, objectives, and strengths.

Ref.	Lead–Acid Type	Objectives	Strengths
[25]	Lead–acid Battery Model with New Control Scheme	Propose a new control scheme for hybrid energy storage systems; introduce a lead–acid battery model considering DoD and temperature effects	Enhances power filtering and voltage limit adherence; improves performance in poor thermal conditions
[412]	Hybrid Battery Configurations	Investigate hybrid energy storage system performance; analyze charge/discharge cycling of battery configurations	Enables round-trip efficiency measurement; provides insight into DoD impact on efficiency
[413]	Battery Energy Storage System (BESS)	Introduce battery energy storage for emergency power supply; improve reliability of separated power networks during outages	Based on real measurement data; enhances network reliability during main line damage or transmission limitations
[414]	Lead–acid Battery Pack + Supercapacitor Pack (HESS)	Manage hybrid energy storage for photovoltaic grid integration; optimize service level and minimize battery degradation	Balances power delivery between battery and supercapacitor; extends battery lifespan while maintaining optimal performance
[415]	Hybrid Electrochemical Energy Storage System (HEESS)	Summarize recent research progress in HEESS development; stimulate innovative thoughts for HEESS applications	Integrates system configuration, DC/DC converter design, and energy management strategy; supports innovative approaches for performance and lifespan improvements

Table 26. *Cont.*

Ref.	Lead-Acid Type	Objectives	Strengths
[416]	Lead–Carbon Battery	Optimize positive plate performance and production process; enhance high-current charging and deep discharge capabilities	Improved positive plate structure and lead alloy selection; better deep discharge performance and tolerance to high-current charging
[417]	Hybrid MTM–ANN Method	Control load variability for real-time demand-side management; optimize reserved ESS capacity	Combines Markov Transition Matrix with ANN for improved real-time maximum demand control
[418]	Energy Storage Technology	Identify suitable energy storage devices for grid support applications; evaluate technical, economic, and environmental impacts	Provides a comprehensive comparative analysis of various storage devices for different stationary applications
[419]	Residential Energy Consumption and BESS Analysis	Analyze electrical power consumption patterns in residential areas; assess economic viability of aggregated and distributed battery energy storage systems	Provides insights into consumption patterns and potential business models for BESS deployment
[27]	Battery ESS and Power-to-Gas	Analyze balance and flows of electrical energy in networks; model operational modes incorporating electricity storage systems	Combines short-term and long-term storage solutions for improved grid flexibility

However, the main operational challenges of lead–acid batteries include their relatively short lifecycle and degraded performance under adverse weather conditions, which limits their effectiveness in long-term or harsh operating environments.

4.3.2. Nickel-Based Batteries

To address the limitations of lead–acid batteries, nickel-based batteries have been proposed for energy storage in smart grids. Compared to lead–acid batteries, nickel-based batteries offer higher tolerance to extreme weather conditions, enhanced reliability, and longer operational lifetimes, making them a more robust option for backup and renewable energy storage.

Table 27 provides a comparative overview of several studies that have implemented nickel-based batteries in smart grid applications, highlighting their performance, advantages, and deployment scenarios.

Table 27. Nickel-based battery types, objectives, and strengths.

Ref.	Nickel-Based Type	Objectives	Strengths
[420]	Nickel–iron battery (NI)	Provide reliable backup power in a PV/battery hybrid system to compensate for main grid outages	Long lifespan, high durability, suitable for harsh environments, low maintenance needs
[421]	Nickel–Cadmium Battery (NiCd)	Durable industrial battery storage, withstands extreme temperatures	High cycle life (1000–1500+), wide temperature range, reliability
[422]	Membraneless Ni–Fe battery with gel electrolyte	Develop a high-performance Ni–Fe battery without membrane, improving redox kinetics and addressing passivation, hydrogen evolution, and self-discharge	Enhanced ion transport via self-assembled nanostructures; simpler design; reduced cost; improved durability
[423]	Nickel-based selenides	Summarize preparation methods (hydrothermal, thermal solvent, thermal decomposition, heat treatment) and discuss performance optimization paths	Provides a foundation for future research by consolidating methods and highlighting electrode–electrolyte interaction considerations

Table 27. *Cont.*

Ref.	Nickel-Based Type	Objectives	Strengths
[424]	Ni-based batteries	Review current trends and classifications of electrochemical storage for smart grids and EVs, focusing on nickel-based technologies	Broad comparison with other battery types considering cost, impact, maintenance, advantages, and protection; supports informed selection for future applications
[425]	Ni-based (general)	Sustainable, scalable battery storage in smart grids	Long cycle life, safety, moderate energy density, scalable for grid and industrial use
[419]	Residential BESS Business Model Analysis	Analyze electrical power consumption patterns in residential areas; assess economic viability of aggregated and distributed battery energy storage systems	Provides insight into consumption behavior for optimized storage deployment; supports development of viable business models for nickel-based BESS integration
[426]	Optimization and Benefit Evaluation in High-RES Smart Grids	Summarize challenges of integrating high renewable energy sources in smart grids; analyze optimization planning and benefit evaluation methods for energy storage	Incorporates diverse decision-making and optimization techniques; enables comprehensive assessment of energy storage benefits and planning under complex grid conditions

4.3.3. Sodium–Sulfur-Based Batteries

One of the primary drawbacks of nickel-based batteries is their high self-discharge rate, which negatively affects the overall efficiency of the storage system. To overcome this limitation, sodium–sulfur (NaS) batteries have been increasingly proposed for energy storage in smart grids. NaS batteries offer several advantages, including high energy density, long operational life, and deep discharge capabilities, making them well-suited for large-scale storage and renewable energy integration.

Table 28 summarizes several studies that have deployed sodium–sulfur batteries, highlighting their performance characteristics, deployment strategies, and operational benefits in smart grid applications.

Table 28. Sodium–sulfur battery research, objectives, and strengths.

Ref.	Battery Type	Objectives	Strengths
[427]	Room-temperature Sodium–Sulfur Battery	Enhance cycling stability; achieve complete conversion of sodium polysulfides in cathodes	Advanced analysis using DFT adsorption energy calculations and in situ synchrotron XRD for species tracking
[428]	Room-temperature Sodium–Sulfur Battery	Clarify operating principles and technical challenges; propose future strategies for practical RT-Na–S batteries	Strategies include regulating electrolyte components, adding additives, developing new electrolytes, and multifunctional separators to address low conductivity, volume expansion, and Na dendrite formation
[428]	Sodium–Sulfur Battery with Covalent Sulfur Confinement	Enhance sodium–sulfur battery performance by confining covalent sulfur and accelerating polysulfide redox kinetics; improve cycling stability	Uses ex situ XPS and theoretical calculations to study covalent sulfur bond breakage; mitigates sluggish reactivity, polysulfide dissolution, and sulfur's insulating nature
[429]	Room-Temperature Sodium–Sulfur Battery	Summarize working principles of RT-Na–S batteries; address key scientific problems in cathode, anode, electrolyte, and separator design to enhance energy storage performance	Tackles poor safety performance, high cost, and limited lithium resources; provides a pathway for safer, more sustainable, and cost-effective large-scale energy storage
[430]	Room-Temperature Sodium–Sulfur Battery	Enhance reversibility and cyclability by designing an all-fluorinated electrolyte (FDMA, MTFE, FEC) and forming NaF- and Na_3N-rich cathode electrolyte interphase; improve compatibility between electrolytes and electrodes	Mitigates polysulfide shuttle, improves electrode–electrolyte compatibility, and boosts long-term cycling stability

Table 28. *Cont.*

Ref.	Battery Type	Objectives	Strengths
[431]	Room-Temperature Sodium–Sulfur Battery	Summarize historical progress toward practical RT-Na/S batteries; promote balanced research trends by addressing advanced sulfur host design, Na metal anode protection, electrolyte optimization, separator modification, and binder engineering	Highlights overlooked components affecting polysulfide migration and reaction kinetics; encourages holistic and unbiased development for broader practical applications
[432]	Room-Temperature Sodium–Sulfur Battery	Develop a stable quasi-solid-state gel polymer electrolyte; enhance performance using in situ preparation and density functional theory for ion interaction analysis	Addresses dendrite formation and polysulfide shuttle issues; offers lower-cost alternative to expensive solid electrolytes in high-temperature Na–S systems
[433]	Room-Temperature Sodium–Sulfur Battery	Improve stability of sodium–sulfur batteries using chemical and spatial dual-confinement and covalent bonding of sulfur to carbon materials; achieve high-capacity retention after multiple cycles	Enhances cathode stability; mitigates sodium polysulfide formation; improves cycling performance and capacity retention
[434]	Sodium–Sulfur Battery	Mitigate polysulfide shuttle via covalent bonding of sulfur to polymeric backbone; optimize electrolyte compositions for long cycle life	Reduces capacity fade; improves stability and cycling performance through tailored cathode–electrolyte interactions
[435]	Room-temperature sodium–sulfur battery with FEC additive and tetraethylene glycol dimethyl ether solvent	Improve lifespan by mitigating shuttle effect in sulfur cathode	Targets key degradation mechanism ("solid–liquid–solid" reaction), enhancing cycle life through electrolyte optimization

4.3.4. Sodium-Ion Batteries

Although sodium–sulfur batteries provide high energy density and long life, they require regular maintenance, which increases the operational costs of the energy storage system. To mitigate these costs, sodium-ion (Na-ion) batteries have been proposed as an alternative for energy storage in smart grids. Sodium-ion batteries offer reduced maintenance requirements while remaining well-suited for stationary storage applications.

Table 29 summarizes several studies that have deployed sodium-ion batteries, highlighting their advantages, deployment strategies, and performance outcomes in smart grid implementations.

Table 29. Sodium-ion battery research, objectives, and strengths.

Ref.	Battery Type/Focus	Objectives	Strengths
[28]	Sodium-ion battery with varied electrode coatings and materials analyzed using SEM, EDS, EIS, and C-rate testing	Investigate structural and electrochemical characteristics of sodium-ion batteries	Provides detailed performance insights across conditions, supporting a knowledge base for further sodium-ion battery research
[436]	Study of capacity degeneration mechanisms in cathodes and review of effective strategies for long-cycle-life cathodes	Focus on developing long-cycle-life, low-cost cathodes for sodium-ion batteries	Identifies structural/morphology changes and unstable interphases, providing targeted strategies for stable cycling performance
[437]	Review fundamentals and progress of sodium-ion batteries, focusing on novel materials and electrochemistry tuning factors	Provide sustainable, low-cost alternative to lithium-ion batteries and guide future SIB development	Addresses lithium scarcity, promotes material sustainability, and supports long-term battery technology advancement
[438]	$Na_4MnCr(PO_4)_3$ cathode	Develop high-energy cathode materials for sodium-ion batteries and investigate redox mechanisms and ionic migration	Moderate 7.7% volume change during cycling, enabling better structural stability and performance
[439]	Diglyme-based electrolyte SIBs	Explore sodium-ion batteries for sustainable energy storage systems and discuss advancements in diglyme-based electrolytes	Highlights significant impact of electrolyte selection on electrochemical performance

Table 29. *Cont.*

Ref.	Battery Type/Focus	Objectives	Strengths
[440]	Non-aqueous, aqueous, and solid-state SIBs	Review current sodium-ion battery technologies for energy storage and explore sustainable alternatives to lithium-ion batteries	Addresses resource and supply chain limitations, highlighting cost-effective and sustainable solutions
[26]	Electrolyte compositions and electrode materials	Examine advancements in sodium-ion battery technology and discuss future research directions for scalability and commercial viability	Provides insights into material degradation and sodium-ion diffusion challenges for improved performance
[437]	Materials, degradation mechanisms, full-cell design, and electrolyte progress	Assess progress and challenges in sodium-ion battery technology and present a roadmap for future SIB implementation in energy storage	Identifies fundamental degradation mechanisms and enduring challenges, providing a foundation for targeted improvements
[437]	Standardized evaluation criteria and retention index for sodium cathodes	Discuss the role of sodium-ion batteries in energy storage and explore low-cost, performance, and sustainability benefits	Establishes a standardized evaluation framework, enabling consistent comparison and improvement across studies
[441]	One-pot solid-state reaction synthesized biphasic cathodes	Improve sodium-ion battery performance and enhance cycle stability and rate capability	Simple synthesis method with in situ structural analysis for better understanding of performance factors

4.3.5. Lithium-Ion Batteries

One of the main limitations of sodium-ion batteries is their lower energy density, which restricts their full-scale deployment in smart grid energy storage applications. To address this limitation, lithium-ion (Li-ion) batteries are widely adopted due to their higher energy density, longer cycle life, and better efficiency. Many researchers have incorporated lithium-ion batteries into smart grid energy storage models to enhance performance and reliability.

Table 30 provides a comparison of several studies that implemented lithium-ion batteries, highlighting their objectives, deployment strategies, and key strengths in energy storage systems.

Table 30. Lithium-ion battery research, objectives, and strengths.

Ref.	Type	Objectives	Strengths
[442]	Li-ion BESS with active network management	Design ANM architecture with Li-ion BESSs and develop adaptive controllers considering battery aging	Incorporates aging effects into control strategy to maintain performance and reliability over time
[443]	Grid-connected Li-ion BESS	Review management approaches for grid-connected Li-ion BESSs and evaluate participation in electricity markets	Provides comprehensive analysis of battery modeling, BMS architecture, and market integration challenges
[444]	Li-ion BESS reliability assessment	Propose a reliability assessment algorithm for BES systems and analyze battery lifetime degradation effects on reliability	Combines universal generating function and weak-link analysis to quantify safety and reliability impacts, offering a systematic evaluation method
[445]	Second-life Li-ion batteries (SLB) for fast-charging	Evaluate second-life batteries for fast-charging energy storage and compare economic and environmental impacts in U.S. cities	Provides detailed comparison of SLB and grid-based configurations, analyzing cost, lifetime, efficiency, and environmental metrics like GWP and CED
[446]	Direct and Model-Based SOH Diagnostics for Battery Reuse	Diagnose state of health for battery repurposing; assess direct, model-based, and data-driven methods for EV and ESS applications	Enables accurate SOH estimation to support battery reuse, promoting sustainability and circular economy goals
[415]	Hybrid Energy Storage System (HEESS) Development	Summarize recent research progress in HEESS; develop system configuration, DC/DC converter design, integrated sizing, and energy management strategies	Integrates battery and other storage elements to improve performance and extend lifetime; encourages innovative applications despite cost and degradation constraints
[447]	Smart Sensing Technologies for Batteries	Highlight advancements in smart sensing technologies for batteries; analyze limitations and challenges of different sensor applications (electrical, thermal, mechanical, acoustic, gas)	Provides comprehensive monitoring for battery health and safety; multi-sensor approach enables early detection of issues and improved battery management despite data collection challenges

Table 30. *Cont.*

Ref.	Type	Objectives	Strengths
[448]	Advanced Battery Management Systems	Improve battery energy density, power density, and cycle life; develop advanced BMS for safety and efficiency through online electrochemical spectroscopy impedance estimation and SOC estimation using adaptive unscented Kalman Filter	Enables accurate state-of-charge estimation and robust thermal management; enhances performance, safety, and lifespan of Li-ion batteries
[449]	Grid-Integrated Battery Storage	Discuss applications and benefits of battery storage in electricity grids; compare advantages and disadvantages of various electrochemical batteries	Provides comprehensive comparison of battery types; highlights suitability of Li-ion and other chemistries for managing renewable intermittency and long-term cost considerations
[413]	Emergency Backup for Isolated Networks	Introduce battery energy storage for emergency power supply; improve reliability of separated power networks during outages	Uses real measurement data for accurate ESS dimensioning; enhances reliability and resilience during main line damage or transmission limitations

4.4. Cost–Performance Trade-Offs Across Battery Technologies in Smart Grid Scenarios

Energy storage systems (ESSs) are integral to smart grids (SGs) for balancing demand and supply, but each battery technology presents distinct cost–performance characteristics. Lead–acid batteries are among the cheapest to deploy, offering simple control and high surge currents, which make them attractive for low-cost stationary storage. However, their shorter lifecycle and poor performance under adverse weather conditions limit their long-term cost-effectiveness [25]. Nickel-based batteries improve durability and environmental tolerance, with high cycle life (1000–1500+ cycles) and low maintenance requirements, but they suffer from higher self-discharge currents, which increase operational inefficiency [420].

Sodium–sulfur (NaS) batteries provide high energy density, long life, and deep discharge capability, offering better performance for large-scale integration. However, they require regular maintenance, which increases operational expenditure over time [427]. Sodium-ion batteries (SIBs) offer a cost-effective and sustainable alternative to lithium-ion by leveraging abundant materials. They are optimal for stationary storage models, providing moderate performance at lower cost, yet their lower energy density restricts deployment in applications requiring high power density [437].

Lithium-ion (Li-ion) batteries, conversely, deliver high energy density and scalability, making them suitable for both grid-scale and distributed applications. Nevertheless, they come with higher upfront costs and face degradation challenges over time [442]. Research on second-life Li-ion batteries has shown that reusing electric vehicle (EV) batteries for grid services can significantly reduce lifecycle costs while maintaining acceptable performance, improving both economic and environmental metrics (e.g., lowering global warming potential and cumulative energy demand compared to new battery packs) [445].

In practical SG scenarios, these cost–performance trade-offs indicate that lead–acid may be chosen for short-duration, cost-constrained backup; nickel-based and sodium–sulfur for environments demanding resilience and deep cycling; sodium-ion for cost-sensitive stationary storage with moderate performance; and lithium-ion or second-life Li-ion where high performance and integration flexibility justify higher upfront cost. Hybrid configurations combining, for instance, Li-ion and supercapacitors or lead–acid with supercapacitors, further optimize both cost and performance by distributing stress across chemistries to extend lifetime and improve round-trip efficiency [414]. A comparative analysis of cost–performance metrics of these storage systems is presented in Table 31.

Table 31. Cost–performance trade-offs across battery technologies for smart grid (SG) applications.

Battery Type	Cost Level	Performance (Cycle Life/ Energy Density)	Maintenance/Degradation	Typical SG Use Cases
Lead–acid	Low (cheapest upfront)	Low–Moderate/Low energy density	Short lifecycle, sensitive to adverse weather	Short-duration backup, cost-constrained stationary storage
Nickel-based (NiCd, NiMH)	Moderate	High cycle life (1000–1500+), good temperature tolerance	Higher self-discharge, moderate maintenance	Resilient stationary storage, moderate-to-high cycling frequency
Sodium–sulfur (Na–S)	Moderate–High	High energy density, long life, deep discharge	Requires regular maintenance, thermal management	Large-scale stationary storage, renewable smoothing
Sodium-ion (SIBs)	Low–Moderate	Moderate energy density, emerging tech	Limited high-power performance, early stage scaling	Sustainable, cost-sensitive stationary storage
Lithium-ion (Li-ion)	High	High cycle life, high energy density, scalable	Degradation over time, higher upfront cost	Distributed and grid-scale storage, fast-response balancing
Second-life Li-ion	Moderate (lower lifecycle cost)	Maintains acceptable performance for SG use	Performance depends on EV battery history	Cost-effective grid services, environmental lifecycle benefits
Hybrid systems (e.g., Li-ion + supercapacitor)	Case-dependent	Optimized performance via complementary chemistries	Reduced stress on individual cells	Applications requiring both high power and high energy capabilities

4.5. Role of AI/Privacy-Preserving Techniques in ESS

Energy storage systems (ESSs), particularly their battery management systems (BMSs), introduce unique cybersecurity and privacy challenges. BMS devices control charging, discharging, and state-of-charge estimation, making them attractive targets for adversaries. False data injection into BMS signals can lead to misreporting of storage capacity, resulting in market imbalances or unsafe operating conditions.

Privacy-preserving techniques, such as differential privacy or Secure Multiparty Computation, can be applied to aggregated storage data to hide individual user behavior, while blockchain can provide tamper-proof audit trails of storage transactions. However, these mechanisms must be complemented with system-level intrusion detection and secure firmware, since privacy alone cannot prevent adversaries from disrupting storage operations.

Thus, the interaction between privacy and ESS security is twofold: cryptographic and privacy-preserving methods help protect user and market data, while robust BMS cybersecurity ensures that the underlying physical storage assets remain safe and trustworthy. A summary of privacy-preserving techniques and their interaction with ESS is presented in Table 32.

Table 32. Privacy-preserving mechanisms for ESS/BMS interaction.

Mechanism	Interaction with ESS/BMS
Homomorphic Encryption	Encrypts storage data before sharing
Zero-Knowledge Proofs	Verifies BMS data authenticity without revealing full details
Differential Privacy	Preserves privacy in aggregated SoC/charging data
Blockchain	Provides immutable record of ESS transactions and logs
IDS/Lightweight Crypto	Detects abnormal SoC/voltage patterns

5. Case Studies

This section presents case studies based on pricing models, demand–supply management, privacy preservation, and attack identification in P2P smart grid energy trading, following recent studies [1,450,451].

5.1. Case Study: Uniform Pricing Model

Uniform pricing models bring all participants in the energy market onto a single platform, where they agree on the same price. Several factors, such as the Supply–Demand Ratio, generation costs, transmission, and grid constraints, can influence the determination of a uniform price signal. Reputation is another key factor, which can be used to determine energy prices in P2P markets. Incorporating reputation reduces bias among participants and promotes fairness in energy trading. A reputation-based pricing model has been proposed in the recent study by Amin et al. [450].

The energy trading model proposed by the authors is illustrated in Figure 7.

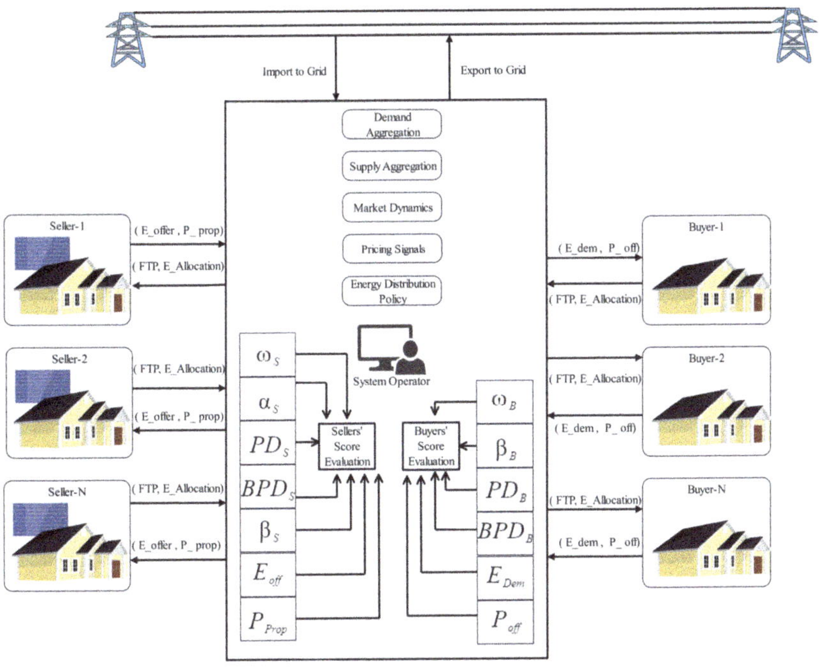

Figure 7. Reputation-based uniform pricing model [450].

From Figure 7, it can be seen that the proposed model formulates a market where participants interact with each other to encourage energy trading under the supervision of a virtual market operator (MO), which oversees all trading activities. Sellers are weighted based on factors such as reliability (α), price dynamism (PD), presence of participants (ω), persistence factor (β), etc. The MO evaluates the reputation $(\mathcal{R})$ of each participant and assigns a corresponding score.

At the start of a trading session t, the MO invites participants to share their information. Sellers provide the MO with their offered price P_{pro} and the offered surplus energy E_{off}, while buyers in the set B notify the MO of their energy demand E_{dem} and offered price P_{off}. The MO keeps this information private and evaluates the reputation of each participant.

The reputation evaluation for sellers, according to this study, can be computed using the following equations.

$$(\text{Score}_{s_i,t})_\omega = \begin{cases} 1, & \omega_{i,t} = 1, \\ 0, & \omega_{i,t} = 0\,, \end{cases} \tag{1}$$

$$\text{Score}^{\beta}_{s_i,t} = \beta_{s,t} \cdot \mathbb{I}(\omega_{i,t} = 1) \tag{2}$$

$$\text{Score}^{\beta}_{s_i,t} = \begin{cases} (1-\gamma)\beta_{s,t-1} + \gamma, & \omega_{s_i,t} = 1, \\ (1-\gamma)\beta_{s,t-1}, & \omega_{s_i,t} = 0\,, \end{cases} \tag{3}$$

$$\text{Score}^{\alpha}_{i,t} = \begin{cases} \text{Score}_{i,t} + 1, & \alpha_{i,t} = 1, \\ \text{Score}_{i,t} - 1, & \alpha_{i,t} = 0\,, \end{cases} \tag{4}$$

$$\text{Score}^{PD}_{s_i,t} = \begin{cases} \text{Score}_{i,t-1} + 1, & P^{pro}_{i,t} \neq P^{pro}_{i,t-1}, \\ \text{Score}_{i,t-1} - 1, & P^{pro}_{i,t} = P^{pro}_{i,t-1}\,, \end{cases} \tag{5}$$

$$\text{Score}^{BPD}_{s_i,t} = \sum_{p^{off} \in P^{off}} \begin{cases} \text{Score}_{i,t} + \dfrac{p^{off} - P^{pro}_{i,t}}{p^{off}} \times 100, & P^{pro}_{i,t} < p^{off}, \\ \text{Score}_{i,t} - \dfrac{P^{pro}_{i,t} - p^{off}}{p^{off}} \times 100, & P^{pro}_{i,t} > p^{off}\,, \end{cases} \tag{6}$$

$$(\text{Score}_{i,t})_{E_{off}} = \begin{cases} \text{Score}_{i,t} + 2, & E_{off,i,t} = \Theta_{sur,t}, \\ \text{Score}_{i,t} - \text{Penalty} \cdot \dfrac{|E_{off,i,t} - \Theta_{sur,t}|}{\Theta_{sur,t}}, & E_{off,i,t} < \Theta_{sur,t}, \\ \text{Score}_{i,t} + \text{Bonus} \cdot \dfrac{E_{off,i,t} - \Theta_{sur,t}}{\Theta_{sur,t}}, & E_{off,i,t} > \Theta_{sur,t}\,, \end{cases} \tag{7}$$

$$\begin{aligned} \text{TotalScore}_{s_i,t} = {} & W_\omega\,(\text{Score}_{s_i,t})_\omega + W_\beta\,\text{Score}^{\beta}_{s_i,t} + W_\alpha\,\text{Score}^{\alpha}_{s_i,t} \\ & + W_{PD}\,\text{Score}^{PD}_{s_i,t} + W_{BPD}\,\text{Score}^{BPD}_{s_i,t} + W_{E_{off}}\,(\text{Score}_{i,t})_{E_{off}} \end{aligned} \tag{8}$$

$$S_{\text{winner}}(t) = \arg\max_{s_i \in S} \left(\text{TotalScore}_{s_i,t}\right) \tag{9}$$

Similarly, buyers' evaluation criteria are presented in equations given as

$$(\text{Score}_{b_i,t})_\omega = \begin{cases} 1, & \omega_{b_i,t} = 1, \\ 0, & \omega_{b_i,t} = 0\,, \end{cases} \tag{10}$$

$$\text{Score}^{\beta}_{b_i,t} = \beta_{b,t}\,\mathbb{I}(\omega_{b_i,t} = 1) \tag{11}$$

$$\text{Score}^{\beta}_{b_i,t} = \begin{cases} (1-\gamma)\beta_{b,t-1} + \gamma, & \omega_{b_i,t} = 1, \\ (1-\gamma)\beta_{b,t-1}, & \omega_{b_i,t} = 0\,, \end{cases} \tag{12}$$

$$\text{Score}^{PD}_{b_i,t} = \begin{cases} \text{Score}_{i,t-1} + 1, & P^{off}_{t} \neq P^{off}_{t-1}, \\ \text{Score}_{i,t-1} - 1, & P^{off}_{t} = P^{off}_{t-1}\,, \end{cases} \tag{13}$$

$$(\text{Score}_{b_i,t})_{BPD} = \sum_{p^{off} \in P^{off}} \begin{cases} \text{Score}_{i,t} + \dfrac{p^{pro} - p^{off}}{p^{pro}} \times 100, & p^{off} > p^{pro}, \\ \text{Score}_{i,t} - \dfrac{p^{pro} - p^{off}}{p^{pro}} \times 100, & p^{off} < p^{pro}\,, \end{cases} \tag{14}$$

$$(\text{Score}_{i,t})_{E_{dem}} = \begin{cases} \text{Score}_{i,t} + 2, & E_{dem,t} = \Theta_t^{dem}, \\ \text{Score}_{i,t} - \text{Penalty} \cdot \dfrac{|E_{dem,t} - \Theta_t^{dem}|}{\Theta_t^{dem}}, & E_{dem,t} > \Theta_t^{dem}, \\ \text{Score}_{i,t} + \text{Bonus} \cdot \dfrac{E_{dem,t} - \Theta_t^{dem}}{\Theta_t^{dem}}, & E_{dem,t} < \Theta_t^{dem}, \end{cases} \quad (15)$$

$$\text{TotalScore}_{b_i,t} = \sum_k \text{Weight}_k \cdot \text{Score}_{b_i,t}^{(k)} \quad (16)$$

$$B_{\text{winner}}(t) = \arg\max_{b_i \in B} \left(\text{TotalScore}_{b_i,t}\right) \quad (17)$$

Once the scores for buyers and sellers have been evaluated, the market operator (MO) calculates the Supply–Demand Ratio (SDR) to determine the possible mode of operation. Based on the identified mode of operation, the final trading price of energy can be determined. The process is summarized as follows:

$$(\text{Sellers' mode}) \qquad \text{FTP}_t = P_{S_{\text{winner}}}^{pro} \quad (18)$$

$$(\text{Buyers' mode}) \qquad \text{FTP}_t = P_{B_{\text{winner}}}^{off} \quad (19)$$

$$(\text{Equilibrium mode}) \qquad \text{FTP}_t = \frac{1}{2}\left(P_{S_{\text{winner}}}^{pro} + P_{B_{\text{winner}}}^{off}\right) \quad (20)$$

This proposed model exhibits several desirable characteristics, such as non-dictatorship, Pareto efficiency, independence of irrelevant alternatives, and collective rationality, thereby promoting fairness in the energy market while determining the uniform price of energy.

The model was evaluated using a one-year dataset consisting of varying demand and supply values. It was also compared with several state-of-the-art methods, including those proposed in [226,235,294,452]. Additionally, the proposed model considers the grid price as another baseline parameter to assess seller profitability and buyer savings. The experimental results are presented in Figure 8a,b.

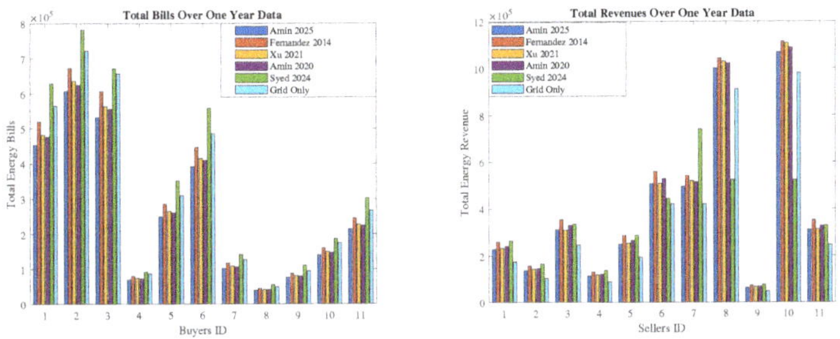

(a) (b)

Figure 8. Financial metrics for the reputation-based pricing model. (**a**) Energy bills for the reputation-based uniform pricing model over one-year dataset [450]. (**b**) Energy revenues for the reputation-based uniform pricing model over one-year dataset [450].

The experimental results demonstrate that the proposed reputation-based pricing model can increase the probability of sellers participating in the market by 2.1% to 20.41% compared with other state-of-the-art methods, and improve seller profitability by 22.91% compared to the grid export rate.

Similarly, buyers benefit from energy bill savings ranging from 2.81% to 44.27% compared with other state-of-the-art methods, while the savings relative to the grid import rate can reach approximately 27.78%.

5.2. Case Study: Integrated Demand–Supply Balance, Pricing, and Privacy-Preserving Model

A recent study [1] proposed an integrated model that combines demand–supply balance, pricing, and privacy-preserving techniques to ensure an effective energy trading platform in P2P energy markets. The proposed model is illustrated in Figure 9.

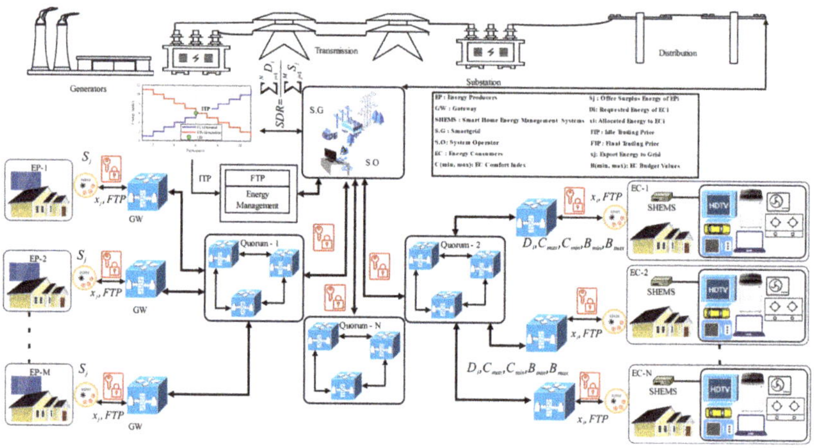

Figure 9. Integrated demand–supply, pricing, and privacy-preserving model [1].

In this model, for each time slot t, each energy consumer (EC) i reports its energy demand D_i, while each energy producer (EP) j reports its surplus energy S_j. Additionally, each buyer communicates to the system operator (S.O.) their comfort indices and budgetary constraints, denoted as $(\mathbb{C}_{\min}, \mathbb{C}_{\max})$ and $(\mathbb{B}_{\min}, \mathbb{B}_{\max})$, respectively.

The information shared by the participants is secured through a quorum-based architecture that integrates SHA-256 hashing, the Elliptic Curve Digital Signature Algorithm (ECDSA), and Shamir's Secret Sharing (SSS). Each participant signs and hashes their data as follows:

$$\text{Sig}_{EC_i} = \text{Sign}_{SK_i}(D_i \| C_{\min,i} \| C_{\max,i} \| B_{\min,i} \| B_{\max,i}) \tag{21}$$

$$H(EC_i) = \text{SHA-256}(D_i \| C_{\min,i} \| C_{\max,i} \| B_{\min,i} \| B_{\max,i}) \tag{22}$$

After verification, the gateway encodes the data using SSS:

$$f(x) = a_0 + a_1 x + a_2 x^2 + \cdots + a_{\lambda-1} x^{\lambda-1} \mod p \tag{23}$$

Each gateway receives a share:

$$\text{Share}_k = f(x_k) \mod p, \qquad k = 1, 2, \ldots, t \tag{24}$$

The S.O reconstructs the secret using Lagrange interpolation:

$$S = f(0) = \sum_{k=1}^{\lambda} \text{Share}_k L_k \mod p \tag{25}$$

Once the information is successfully secured, the S.O determines the market operation mode by evaluating the Supply–Demand Ratio (SDR):

$$\text{SDR} = \frac{\sum_{i=1}^{N} D_i}{\sum_{j=1}^{M} S_j} \tag{26}$$

If SDR > 1, the market operates in Sellers' Mode, otherwise in Buyers' Mode. At each trading time 't' the ideal trading price (ITP) is determined at the equilibrium point of supply and demand, as presented in Figure 10.

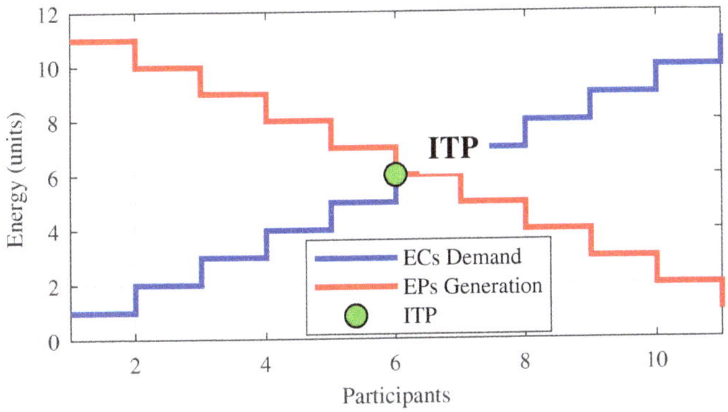

Figure 10. Equilibrium price point determination [1].

Based on the market mode, the S.O. computes the trading price (TP). For the Sellers' Mode, the TP is calculated as follows:

$$TP = \left| \frac{\sum_{j=1}^{M} S_j - \sum_{i=1}^{N} D_i}{\sum_{j=1}^{M} S_j + \sum_{i=1}^{N} D_i} \right| \cdot ITP + ITP \tag{27}$$

Once the price of energy has been found the next objective for the S.O is to formulate the energy allocation model as a convex optimization problem presented as follows:

$$\min \sum_i x_i \tag{28}$$

$$\text{s.t.} \quad B_{\min,i} \leq TP \cdot x_i \leq B_{\max,i}, \quad \forall i \tag{29}$$

$$x_i \leq D_i, \quad \forall i \tag{30}$$

$$C_{\min,i} \leq \frac{x_i}{D_i} \leq C_{\max,i}, \quad \forall i \tag{31}$$

$$x_i, x_j > 0, \quad \forall i, j \tag{32}$$

For Buyers' Mode, the objective can be expressed as follows:

$$\max \sum_i x_i \tag{33}$$

$$\text{s.t.} \quad TP \cdot x_i \leq B_{\max,i}, \quad \forall i \tag{34}$$

$$\sum_{i=1}^{N} x_i = \sum_{j=1}^{M} (S_j - x_{g,j}), \quad \forall j \tag{35}$$

The experimental results demonstrate the robustness of the proposed model against data manipulation during transmission. Any tampering results in the failure of the integrity test for affected participants, as illustrated in Figure 11.

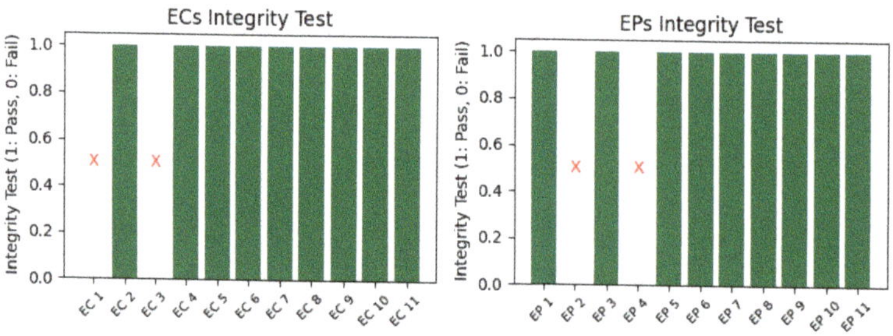

Figure 11. Integrity test for affected participants [1]. X presents the failed participants in the integrity test.

Due to the failed integrity test, the S.O. refuses to reconstruct the information of affected participants, as shown in Figure 12.

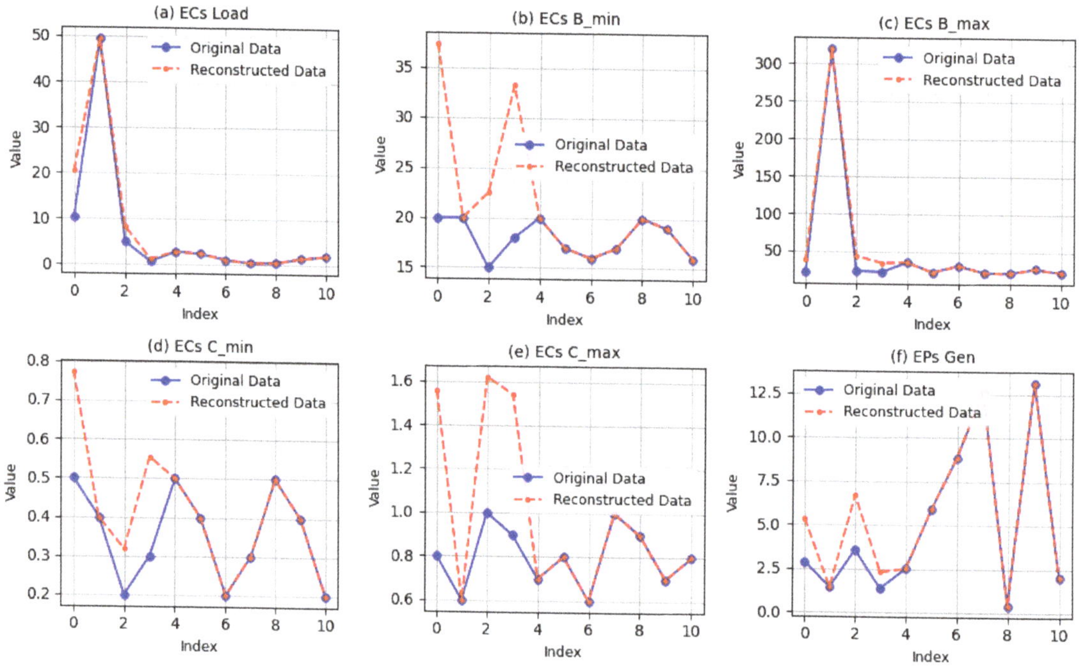

Figure 12. Failed reconstruction of affected participants' information by the S.O. [1].

This mechanism ensures privacy preservation in the P2P energy trading market. Furthermore, the model is computationally efficient; Figure 13 shows that it requires less execution time compared to baseline models.

Finally, evaluation over a one-year dataset demonstrates improved grid efficiency by reducing the grid imbalance state in Buyers' Mode, as presented in Figure 14.

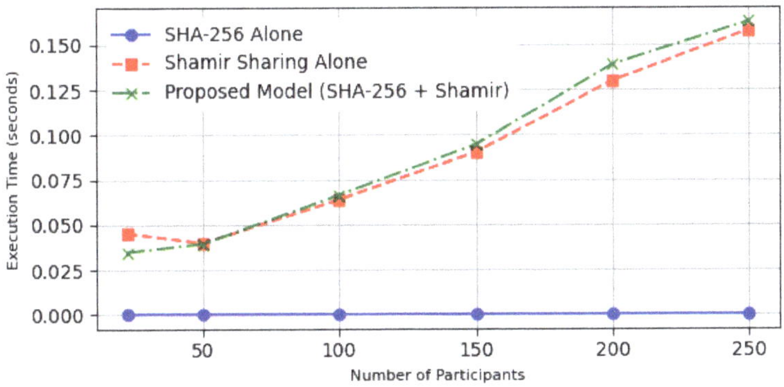

Figure 13. Scalability analysis of the proposed model [1].

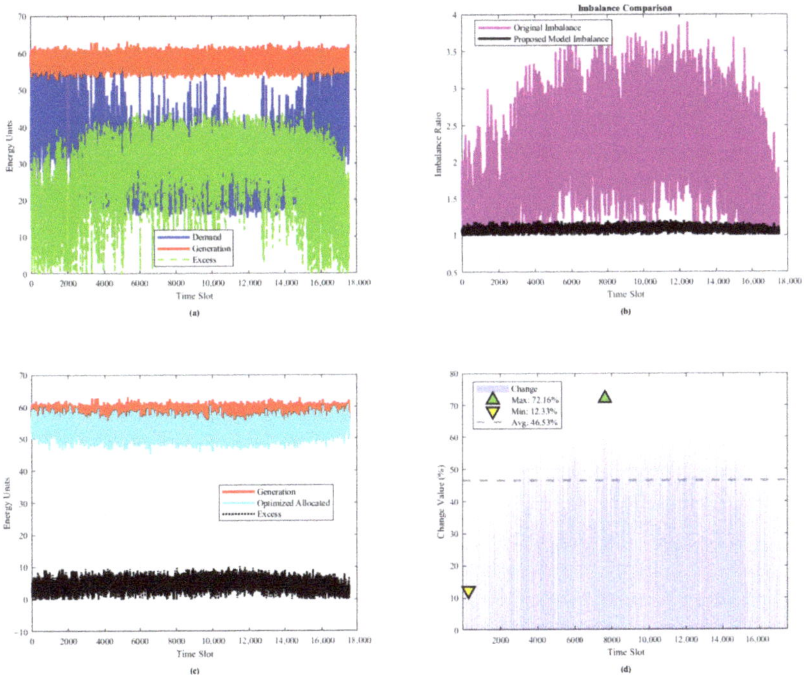

Figure 14. Impact on grid imbalance: Buyers' Mode over one-year dataset [1]. (**a**) Presents the relationship between the demand, generation and difference among them as surplus energy. (**b**) presents the original imbalance Vs Proposed model imbalance, (**c**) presents the difference in generation and optimized allocation, and (**d**) presents the percentage impact of grid imbalance state through the proposed model.

5.3. Cyberattack Detection in Smart Grid

This case study demonstrates the detection of cyberattacks in smart grids, specifically focusing on False Data Injection Attacks (FDIAs) that can manipulate participants' shared information and disrupt market operations, based on the recent study [451].

With the rapid integration of renewable energy and advanced communication technologies, smart grids have become crucial for efficient power generation and distribution. However, this interconnectivity exposes grids to cyber threats such as FDIAs, which can compromise state estimation, disrupt grid operations, and potentially cause large-scale blackouts. Traditional machine learning techniques often fail to capture temporal dependencies in smart grid time-series data, reducing detection accuracy. To address this, Ref. [451] proposes a robust detection framework using a Bidirectional Long Short-Term Memory (Bi-LSTM) network integrated with an Attention Mechanism, offering high detection accuracy, computational efficiency, and interpretability.

The architecture diagram of the proposed model is shown in Figure 15.

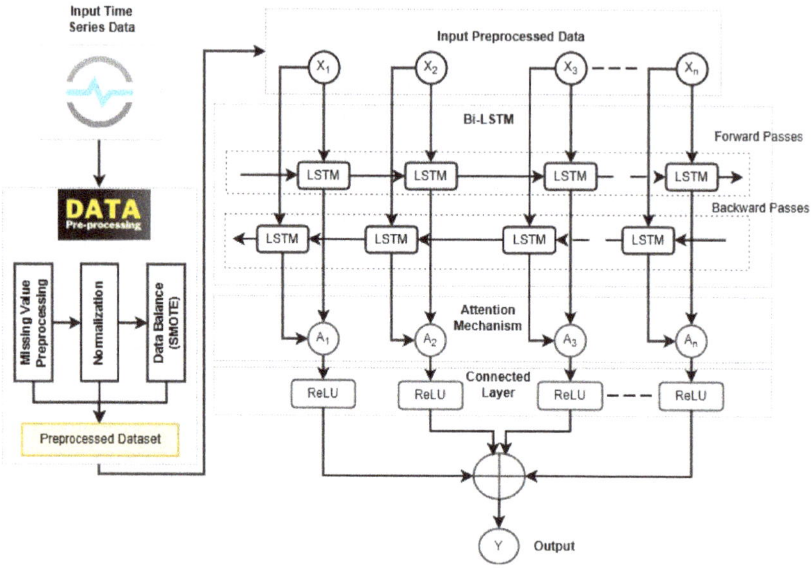

Figure 15. Architecture of the Bi-LSTM with Attention Mechanism for FDIA detection [451].

As illustrated in Figure 15, the proposed model workflow is as follows:

- Data Acquisition: Time-series data from smart meters, including participant information such as energy demand, surplus energy, offered and proposed prices, is collected.
- Data Preprocessing: The raw input data passes through a preprocessing block to clean, normalize, and structure it for use in the Bi-LSTM model.
- Bi-LSTM Network: Captures both forward and backward temporal dependencies, enhancing recognition of subtle patterns associated with potential FDIA.
- Attention Mechanism: Dynamically assigns weights to critical time steps, focusing on significant events while filtering irrelevant information, thereby improving detection accuracy and interpretability.

This integrated Bi-LSTM and Attention framework enables the smart grid system to effectively detect FDIAs, ensuring the integrity of participant data and maintaining stable energy market operations.

The proposed model is evaluated on a real-time smart grid dataset, and the reported results are summarized in Table 33. From Table 33, it can be observed that the proposed Bi-LSTM with Attention model effectively identifies potential cyberattacks in the form of False Data Injection Attacks (FDIAs), achieving an accuracy of 92.32%, which outperforms other state-of-the-art baseline models.

Table 33. Performance metric of the proposed model [451] in comparison with other state-of-the-art baseline models.

Model	Accuracy (%)	F1-Score (Attack)	F1-Score (Natural)	F1-Score (NoEvents)
Proposed Model	92.32	88.17	89.54	99.14
LSTM-CNN	79.01	66.29	73.82	96.92
LSTM-Autoencoder	90.17	86.21	87.80	97.81
Bert	80.00	68.25	74.60	98.51

6. Future Research Directions

This section briefly explores potential future research directions based on the extensive literature reviewed in the preceding sections.

6.1. Privacy-Preserving Future Research Directions

The analysis of privacy-preserving techniques in smart grids reveals several limitations that highlight essential avenues for future research. For cryptographic methods such as Homomorphic Encryption (HE), the high computational overhead and latency in processing large-scale data necessitate the development of streamlined algorithms to reduce costs and enhance feasibility for real-time applications [64,65,73]. Similarly, Zero-Knowledge Proofs (ZKPs) face challenges in computational complexity and scalability, indicating the need for optimized protocols such as zk-SNARKs and zk-STARKs for efficient deployment in large-scale networks [85,86]. Secure Multiparty Computation (SMPC) also requires advancements to mitigate high communication overhead and improve scalability in decentralized systems [93,97].

In anonymization and aggregation techniques, differential privacy (DP) demands improved optimization of noise addition to balance privacy guarantees with data utility, particularly for demand-response management [100,106]. The vulnerabilities of k-Anonymity to auxiliary information attacks call for more robust models incorporating clustering and noise injection to enhance resistance [111,117]. Data aggregation models should aim to minimize generalization losses while still supporting individualized services without compromising overall grid optimization [122,123].

Blockchain-based mechanisms, including smart contracts and decentralized access control, require solutions to scalability and transaction speed limitations to accommodate increasing network traffic [130,142]. Similarly, tokenization approaches demand improved key management to prevent misuse and ensure seamless integration into energy trading systems [156,157].

Machine learning approaches, such as federated learning, must address challenges related to data heterogeneity and communication inefficiencies to achieve better model convergence for anomaly detection [160,164]. Adversarial learning, particularly using GANs, should focus on improving the quality of synthetic data and computational efficiency to develop robust privacy-preserving models [168,169].

Table 34 summarizes the key future research directions for privacy-preserving techniques in smart grids.

Although the discussed literature presents remarkable efforts to preserve participants' privacy and ensure grid stability, their resilience against coordinated, multi-layered state-sponsored attacks as discussed in Section 2.3 remains uncertain and highly dependent on implementation rigor and adherence to established standards.

For instance, while Homomorphic Encryption ensures data confidentiality during processing, its high computational overhead may make it susceptible to resource exhaustion attacks, a common tactic in state-sponsored DDoS campaigns, unless complemented by

NIST-recommended resource management and intrusion detection systems [453]. Similarly, Zero-Knowledge Proofs, although effective for authentication without exposing data, face scalability challenges and may not withstand sustained high-volume attacks without IEC 62351-compliant key management and session integrity controls.

Blockchain-based solutions provide tamper-resistant transaction records and decentralized trust; however, their slower transaction speeds and scalability limitations (as highlighted in Tables 7 and 13) could be exploited in prolonged attacks targeting consensus mechanisms. The International Electrotechnical Commission (IEC) [454] emphasizes the necessity of resilient consensus algorithms and real-time anomaly detection to mitigate such risks.

Table 34. Future research directions in privacy preservation for smart grids.

Research Area	Future Research Direction	Refs.
Cryptographic Techniques (HE)	Streamline methods to minimize computational overhead and latency for large-scale SG networks.	[64,65,73]
Cryptographic Techniques (ZKPs)	Optimize protocols for reduced complexity and improved scalability in real-time applications.	[85,86]
Cryptographic Techniques (SMPC)	Enhance efficiency to handle high communication overhead in decentralized systems.	[93,97]
Anonymization and Aggregation (DP)	Optimize noise addition for better balance between privacy and data utility.	[100,106]
Anonymization and Aggregation (k-Anonymity)	Develop robust models resistant to auxiliary attacks using clustering and noise.	[111,117]
Anonymization and Aggregation (Data Aggregation)	Minimize data generalization losses for individualized services and grid optimization.	[122,123]
Blockchain-Based Mechanisms (Smart Contracts/Access Control)	Address scalability and transaction speed for large network traffic.	[130,142]
Blockchain-Based Mechanisms (Tokenization)	Improve key management to prevent misuse in energy trading.	[156,157]
Machine Learning Approaches (Federated Learning)	Mitigate data heterogeneity and communication issues for model convergence.	[160,164]
Machine Learning Approaches (Adversarial Learning)	Enhance synthetic data quality and reduce computational costs.	[168,169]

Although differential privacy and k-Anonymization techniques protect against data linkage attacks, they may fail under advanced adversarial inference leveraging auxiliary data, a known capability of state actors. IEC 62443-3-3 [455] recommends augmenting these techniques with continuous monitoring and adaptive noise injection to maintain utility while resisting sophisticated de-anonymization attempts.

Furthermore, the hierarchical communication architecture of smart grids (HAN/BAN, WAN) introduces multiple attack surfaces. State-sponsored actors often exploit unpatched systems, vulnerable protocols, and insider threats, as illustrated in Figure 2. NIST CSF and IEC 62351 stress the importance of defense-in-depth strategies, including (i) network segmentation and encryption across all layers, (ii) regular security audits and patch management, (iii) multi-factor authentication and strict access controls, and (iv) real-time intrusion detection and response systems.

Based on this discussion, it is recommended that future research adopt a holistic, standards-based approach integrating cryptographic methods with robust network security practices, continuous monitoring, and incident response frameworks to enhance smart grid resilience against advanced persistent threats.

6.2. Future Research Direction: Pricing Models

Future research in energy pricing mechanisms for smart grid P2P trading should focus on addressing the operational challenges and limitations observed in synchronous and asynchronous pricing models. For synchronous pricing, efforts could aim at enhancing market efficiency by mitigating manipulation in the merit order, improving congestion handling in uniform pricing, and developing more accurate grid constraint modeling for Locational Marginal Pricing (LMP) in dynamic systems with high renewable penetration. Similarly, Supply–Demand Ratio (SDR)-based pricing requires advancements to better handle fluctuating renewable energy markets through improved accuracy in supply and demand data.

Asynchronous mechanisms offer opportunities to reduce inefficiencies in pay-as-bid pricing by minimizing bid manipulation and price uncertainty, while also optimizing bilateral negotiations for greater transparency and mutual agreement on contract terms. Reserved pricing could benefit from methods to determine optimal reserve prices without hindering market clearing, and forward/futures contracts require improved models for contract pricing under market volatility.

Regarding price determination techniques, future directions include refining game-theoretic approaches to better model participant uncertainty and equilibrium calculations in large markets, as well as effectively handling non-cooperative behaviors in dynamic environments. Optimization methods should address forecasting data accuracy and computational demands for large-scale problems, particularly in systems with high renewable variability. Numerical methods can evolve to reduce computational costs in real-time simulations and improve model calibration under dynamic conditions. AI-based techniques present significant potential to overcome data quality challenges, enhance model interpretability, and reduce training time for large-scale adaptations to market dynamics.

Table 35 summarizes the future research directions for pricing models in smart grid P2P energy trading.

Table 35. Future research directions in energy pricing.

Research Area	Future Research Direction	Refs.
Synchronous Pricing Mechanisms	Mitigate price distortion and congestion issues in uniform pricing; improve grid constraint modeling for LMP in high renewable systems; enhance data accuracy for SDR in fluctuating markets.	[18,183–186,192–207]
Asynchronous Pricing Mechanisms	Reduce bid manipulation and uncertainty in pay-as-bid/Vickrey; improve transparency in bilateral negotiations; optimize reserve prices to avoid market inefficiency; handle volatility in forward/futures contracts.	[208–221]
Game-Theoretic Techniques	Address uncertainty in behavior modeling and equilibrium in large markets; handle non-cooperative behaviors in dynamic settings.	[19,241,287–289,293,301–311]
Optimization Techniques	Improve data accuracy in forecasting; reduce computational resources for large problems; model renewable variability effectively.	[313,319,321,326–339]
Numerical Method-Based Techniques	Lower computational costs for real-time simulations; enhance calibration under dynamic market conditions.	[22,340–351]
AI-Based Techniques	Overcome data quality and availability issues; improve model interpretability; reduce training time for large-scale models.	[352–356,358–365]

6.3. Future Research Directions: DSBM

The Demand–Supply Balance Program (DSBP) faces several challenges, including achieving scalability, real-time adaptability, and effective integration of strategic participant interactions. Iterative models, while simple and scalable, often exhibit slow convergence and suboptimal performance in dynamic environments [20,370]. Future research should focus on accelerating convergence through advanced synchronization techniques and enhancing robustness against communication delays.

Optimization models can achieve near-optimal solutions but are computationally intensive and struggle with high-dimensional problems [21,385]. Developing hybrid algorithms that combine heuristic methods with real-time data processing could mitigate these limitations.

Game-theoretic models capture strategic interactions effectively but face challenges in ensuring unique equilibria and preserving participant privacy [22,229]. Future research should explore decentralized game-theoretic frameworks integrated with blockchain to enable secure, privacy-preserving interactions.

AI-based models offer significant potential in managing uncertainty and large datasets but require substantial training data and can face integration challenges with legacy grid infrastructure [23,408,456]. Future efforts should focus on developing lightweight AI models, leveraging transfer learning to reduce data dependency, and ensuring seamless integration with existing systems.

Table 36 summarizes the key future research directions for DSBM.

Table 36. Future research directions for Demand–Supply Balance Program (DSBP).

Research Area	Future Research Direction	References
Iterative Models	Enhance convergence speed using advanced synchronization techniques; improve robustness against communication delays	[20,370]
Optimization Models	Develop hybrid optimization algorithms combining heuristics with real-time data processing to reduce computational intensity	[21,385]
Game-Theoretic Models	Explore decentralized frameworks with blockchain for secure, privacy-preserving strategic interactions	[22,229]
AI-Based Models	Develop lightweight AI models for seamless integration with legacy systems; leverage transfer learning to reduce data dependency	[23,408]

6.4. Future Research Directions: ESS

Energy storage systems (ESSs) play a crucial role in balancing demand and supply in smart grids, yet each battery technology presents unique challenges. Lead–acid batteries, while low-cost, suffer from short lifecycles and limited performance under adverse conditions [25,418]. Future research should focus on enhancing lifecycle performance through advanced thermal management and hybrid configurations.

Nickel-based batteries offer durability but are constrained by high self-discharge rates [420,424]. Developing low-cost, high-efficiency electrolytes and improved cell designs could mitigate this limitation. Sodium–sulfur batteries provide high energy density but require frequent maintenance [427,428]. Research should explore low-maintenance architectures and novel electrolyte additives to improve operational stability.

Sodium-ion batteries are promising for stationary storage but exhibit lower energy density [28,437]. Future work should investigate advanced cathode materials and optimized cell designs to enhance energy density while maintaining cost-effectiveness. Lithium-ion batteries, with high energy density and scalability, face challenges related to cost and degradation [442,448]. Research should emphasize second-life applications, advanced battery management systems, and recycling strategies to extend lifespan and reduce costs.

Table 37 summarizes the key future research directions for ESSs in smart grid applications.

Table 37. Future research directions for energy storage systems (ESSs).

Research Area	Future Research Direction	References
Lead–Acid Batteries	Improve lifecycle through advanced thermal management and hybrid configurations	[25,418]
Nickel-Based Batteries	Develop low-cost, high-efficiency electrolytes to reduce self-discharge rates	[420,424]
Sodium–Sulfur Batteries	Explore low-maintenance designs and novel electrolyte additives for enhanced stability	[427,428]
Sodium-Ion Batteries	Investigate advanced cathode materials to improve energy density while maintaining cost-effectiveness	[28,437]
Lithium-Ion Batteries	Focus on second-life applications and advanced battery management systems to extend lifespan and reduce costs	[442,448]

7. Conclusions

The smart grid (SG) represents a transformative platform in energy management, integrating efficiency, sustainability, and reliability through two-way communication, the Internet of Things (IoT), and Advanced Metering Infrastructure (AMI). However, the extensive transfer of sensitive data exposes SGs to privacy and security risks, as evidenced by recent global incidents including data interception, energy theft, and cyberattacks such as MITM, DoS, and replay attacks. This review systematically analyzes the SG architecture and its core components substations, concentrators, aggregators, smart appliances, and renewable energy sources highlighting associated vulnerabilities and fundamental security requirements, including confidentiality, integrity, availability, secure storage, and scalability.

Effective countermeasures include cryptographic protocols (Homomorphic Encryption, Zero-Knowledge Proofs, Secure Multiparty Computation), anonymization techniques (differential privacy, k-Anonymity), and data aggregation methods. Recent studies also demonstrate that machine learning and blockchain-based solutions can complement these defenses. Further analyses of pricing models, Demand–Supply Balance Programs (DSBPs) using optimization and game-theoretic approaches, AI-driven models, and energy storage systems across various battery chemistries underscore their crucial role in maintaining grid stability amid the stochastic nature of renewable energy sources.

Finally, this review identifies persistent research gaps, such as the computational overhead of cryptographic techniques, trade-offs in the utility of anonymized data, and network scalability challenges. Addressing these issues is critical to developing resilient SGs that safeguard user privacy, mitigate cyber threats, and facilitate widespread adoption of sustainable energy technologies.

Author Contributions: I.N., N.M. and W.A. equally contributed in all sections. All authors have read and agreed to the published version of the manuscript.

Funding: This research received no external funding.

Data Availability Statement: The original contributions presented in the study are included in the article; further inquiries can be directed to the corresponding author on request.

Conflicts of Interest: The authors declare no conflicts of interest.

Abbreviations

The following abbreviations are used in this manuscript:

SG	Smart Grid
IoT	Internet of Things
AMI	Advanced Metering Infrastructure
MITM	Man-in-the-Middle
DoS	Denial-of-Service
CCs	Control Centers
SMs	Smart Meters
RESs	Renewable Energy Sources
DSBPs	Demand–Supply Balance Programs
ESSs	Energy Storage Systems
EVs	Electric Vehicles
ICT	Information and Communication Technology
HAN	Home Area Network
BAN	Building Area Network
WAN	Wide Area Network
DDoS	Distributed Denial of Service
HE	Homomorphic Encryption
ZKPs	Zero-Knowledge Proofs
SMPC	Secure Multiparty Computation
DP	Differential Privacy
zk-SNARKs	Zero-Knowledge Succinct Non-interactive ARguments of Knowledge
zk-STARKs	Zero-Knowledge Scalable Transparent ARguments of Knowledge

References

1. Amin, W.; Huang, Q.; Li, J.; Khan, A.A.; Subramaniam, U.; Selvam, S. A secure energy management model for Peer-to-Peer smart grids with user-centric constraints. *Internet Things* **2025**, *33*, 101678. [CrossRef]
2. Dhaigude, R.; Chanda, R. Real-Time Data and Analytics on Building Management System to Control Energy Consumption. In Proceedings of the International Conference on Information and Communication Technology for Competitive Strategies, Jaipur, India, 15–17 December 2023; pp. 295–305.
3. Dutta Pramanik, P.K.; Upadhyaya, B.K.; Kushwaha, A.; Bhowmik, D. Harnessing IoT: Transforming Smart Grid Advancements. In *IoT for Smart Grid: Revolutionizing Electrical Engineering*; Wiley Online Library: Hoboken, NJ, USA, 2025; pp. 127–174.
4. Ahsan, F.; Dana, N.H.; Sarker, S.K.; Li, L.; Muyeen, S.M.; Ali, M.F.; Tasneem, Z.; Hasan, M.M.; Abhi, S.H.; Islam, M.R.; et al. Data-driven next-generation smart grid towards sustainable energy evolution: Techniques and technology review. *Prot. Control Mod. Power Syst.* **2023**, *8*, 43. [CrossRef]
5. Althobaiti, A.; Jindal, A.; Marnerides, A.K.; Roedig, U. Energy theft in smart grids: A survey on data-driven attack strategies and detection methods. *IEEE Access* **2021**, *9*, 159291–159312. [CrossRef]
6. Jain, H.; Kumar, M.; Joshi, A.M. Intelligent energy cyber physical systems (iECPS) for reliable smart grid against energy theft and false data injection. *Electr. Eng.* **2022**, *104*, 331–346. [CrossRef]
7. Said, D. A survey on information communication technologies in modern demand-side management for smart grids: Challenges, solutions, and opportunities. *IEEE Eng. Manag. Rev.* **2022**, *51*, 76–107. [CrossRef]
8. Amin, W.; Huang, Q.; Umer, K.; Zhang, Z.; Afzal, M.; Khan, A.A.; Ahmed, S.A. A motivational game-theoretic approach for peer-to-peer energy trading in islanded and grid-connected microgrid. *Int. J. Electr. Power Energy Syst.* **2020**, *123*, 106307. [CrossRef]
9. Zhao, X.; Gao, W.; Qian, F.; Ge, J. Electricity cost comparison of dynamic pricing model based on load forecasting in home energy management system. *Energy* **2021**, *229*, 120538. [CrossRef]
10. Alomari, M.A.; Al-Andoli, M.N.; Ghaleb, M.; Thabit, R.; Alkawsi, G.; Alsayaydeh, J.A.J.; Gaid, A.S.A. Security of smart grid: Cybersecurity issues, potential cyberattacks, major incidents, and future directions. *Energies* **2025**, *18*, 141. [CrossRef]
11. Amanlou, S.; Hasan, M.K.; Mokhtar, U.A.; Malik, K.M.; Islam, S.; Khan, S.; Khan, M.A.; Khan, M.A. Cybersecurity challenges in smart grid systems: Current and emerging attacks, opportunities, and recommendations. *IEEE Open J. Commun. Soc.* **2025**. [CrossRef]

12. Guo, C.; Luo, F.; Cai, Z.; Dong, Z.Y. Integrated energy systems of data centers and smart grids: State-of-the-art and future opportunities. *Appl. Energy* **2021**, *301*, 117474. [CrossRef]
13. Sevilla, F.R.S.; Liu, Y.; Barocio, E.; Korba, P.; Andrade, M.; Bellizio, F.; Bos, J.; Chaudhuri, B.; Chavez, H.; Cremer, J.; et al. State-of-the-art of data collection, analytics, and future needs of transmission utilities worldwide to account for the continuous growth of sensing data. *Int. J. Electr. Power Energy Syst.* **2022**, *137*, 107772. [CrossRef]
14. Amuthan, N.; Sathya, M.; Rani, N.C. Empowering Consumers and Utilities for a Smarter Future: The Pivotal Role of Advanced Metering Infrastructure (AMI) in Smart Meter Technology. In *Cloud Computing in Smart Energy Meter Management*; Wiley Online Library: Hoboken, NJ, USA, 2025; pp. 31–65.
15. Mubeen, M.; Arslan, M.; Anandhi, G. Strategies to Avoid Illegal Data Access. *J. Commun. Eng. Syst.* **2022**, *12*, 29–40.
16. Shi, M.; Hoffmann, A.; Wagner, A.; Huyeng, T.; Thiele, C.-D.; Rüppel, U. Using blockchain technology to implement peer-to-peer network in construction industry. In Proceedings of the 18th International Conference on Computing in Civil and Building Engineering, Sao Paulo, Brazil, 18–20 August 2020; pp. 839–849.
17. Liu, J.; Hu, H.; Yu, S.S.; Trinh, H. Electricity pricing and its role in modern smart energy system design: A review. *Designs* **2023**, *7*, 76. [CrossRef]
18. Soh, C.; Markus, M.L.; Goh, K.H. Electronic marketplaces and price transparency: Strategy, information technology, and success. *MIS Q.* **2006**, *30*, 705–723. [CrossRef]
19. González, F.; Arévalo, P.; Ramirez, L. Game Theory and Robust Predictive Control for Peer-to-Peer Energy Management: A Pathway to a Low-Carbon Economy. *Sustainability* **2025**, *17*, 1780. [CrossRef]
20. Paul, S.; Padhy, N.P. A new iterative mixed integer linear programming based real time energy efficient management of AC–DC distribution networks. *Int. J. Electr. Power Energy Syst.* **2024**, *157*, 109793. [CrossRef]
21. Zhao, Z.; Holland, N.; Nelson, J. Optimizing smart grid performance: A stochastic approach to renewable energy integration. *Sustain. Cities Soc.* **2024**, *111*, 105533. [CrossRef]
22. Cheng, L.; Huang, P.; Zhang, M.; Yang, R.; Wang, Y. Optimizing electricity markets through game-theoretical methods: Strategic and policy implications for power purchasing and generation enterprises. *Mathematics* **2025**, *13*, 373. [CrossRef]
23. Wang, X.; Wang, H.; Bhandari, B.; Cheng, L. AI-empowered methods for smart energy consumption: A review of load forecasting, anomaly detection and demand response. *Int. J. Precis. Eng. Manuf.-Green Technol.* **2024**, *11*, 963–993. [CrossRef]
24. Antonopoulos, I.; Robu, V.; Couraud, B.; Kirli, D.; Norbu, S.; Kiprakis, A.; Flynn, D.; Elizondo-Gonzalez, S.; Wattam, S. Artificial intelligence and machine learning approaches to energy demand-side response: A systematic review. *Renew. Sustain. Energy Rev.* **2020**, *130*, 109899. [CrossRef]
25. Luo, X.; Barreras, J.V.; Chambon, C.L.; Wu, B.; Batzelis, E. Hybridizing lead–acid batteries with supercapacitors: A methodology. *Energies* **2021**, *14*, 507. [CrossRef]
26. Chandra Bhowmik, K.; Rahman, M.A.; Billah, M.M.; Paul, A. From lithium-ion to sodium-ion batteries for sustainable energy storage: A comprehensive review on recent research advancements and perspectives. *Chem. Rec.* **2024**, *24*, e202400176. [CrossRef]
27. Miroshnyk, O.; Moroz, O.; Shchur, T.; Chepizhnyi, A.; Qawaqzeh, M.; Kocira, S. Investigation of smart grid operation modes with electrical energy storage system. *Energies* **2023**, *16*, 2638. [CrossRef]
28. Dorau, F.A.; Sommer, A.; Koloch, J.; Röß-Ohlenroth, R.; Schreiber, M.; Neuner, M.; Gamra, K.A.; Lin, Y.; Schoeberl, J.; Bilfinger, P.; et al. Comprehensive analysis of commercial sodium-ion batteries: Structural and electrochemical insights. *J. Electrochem. Soc.* **2024**, *171*, 090521. [CrossRef]
29. Luo, X.R.; Li, H.; Hu, Q.; Xu, H. Why individual employees commit malicious computer abuse: A routine activity theory perspective. *J. Assoc. Inf. Syst.* **2020**, *21*, 5. [CrossRef]
30. Farkas, T.; Hronyecz, E. The Risks and Danger of Smart Devices Exposing Personal Information: Dark Side of Convenience. In Proceedings of the 2024 IEEE 22nd Jubilee International Symposium on Intelligent Systems and Informatics (SISY), Pula, Croatia, 19–21 September 2024; IEEE: New York, NY, USA, 2024; pp. 39–44.
31. Nguyen, L.-H.; Nguyen, V.-L.; Hwang, R.-H.; Kuo, J.-J.; Chen, Y.-W.; Huang, C.-C.; Pan, P.-I. Towards secured smart grid 2.0: Exploring security threats, protection models, and challenges. *IEEE Commun. Surv. Tutor.* **2024**, *27*, 2581–2620. [CrossRef]
32. Peter, J.S.P.; Babu, C.R.; Esther, B.P. Cybersecurity in ICT-Enabled Smart Metering Systems: Addressing Challenges and Implementing Solutions. In *Cloud Computing in Smart Energy Meter Management*; Wiley Online Library: Hoboken, NJ, USA, 2025; pp. 263–290.
33. Vighneswari, B.D.; Kothai Andal, C. Smart Meter Security—Fraud Detection in Power Theft. In *Cloud Computing in Smart Energy Meter Management*; Wiley Online Library: Hoboken, NJ, USA, 2025; pp. 239–262.
34. Abir, S.M.A.A.; Anwar, A.; Choi, J.; Kayes, A.S. Iot-enabled smart energy grid: Applications and challenges. *IEEE Access* **2021**, *9*, 50961–50981. [CrossRef]
35. Senthilkumar, S.P.; Subramani, B. Internet of Things in Low-Power Wide Area Network and Short Range Network: A Review. *I-Manag. J. Comput. Sci.* **2023**, *10*, 33.

36. Lounis, K.; Zulkernine, M. Attacks and defenses in short-range wireless technologies for IoT. *IEEE Access* **2020**, *8*, 88892–88932. [CrossRef]
37. Suhaimy, N.; Radzi, N.A.M.; Ahmad, W.S.H.M.W.; Azmi, K.H.M.; Hannan, M.A. Current and future communication solutions for smart grids: A review. *IEEE Access* **2022**, *10*, 43639–43668. [CrossRef]
38. Raza, M.H.; Rind, Y.M.; Javed, I.; Zubair, M.; Mehmood, M.Q.; Massoud, Y. Smart meters for smart energy: A review of business intelligence applications. *IEEE Access* **2023**, *11*, 120001–120022. [CrossRef]
39. Rasheed, M.B.; Javaid, N.; Awais, M.; Khan, Z.A.; Qasim, U.; Alrajeh, N.; Iqbal, Z.; Javaid, Q. Real time information based energy management using customer preferences and dynamic pricing in smart homes. *Energies* **2016**, *9*, 542. [CrossRef]
40. Beno, M.M.; Sivakumar, N.; Saravanan, R. Demystifying Smart Meters: Powering the Next-Generation Grid. In *Cloud Computing in Smart Energy Meter Management*; Wiley Online Library: Hoboken, NJ, USA, 2025; pp. 67–97.
41. Noorazar, H.; Srivastava, A.; Pannala, S.; Sadanandan, S.K. Data-driven operation of the resilient electric grid: A case of COVID-19. *J. Eng.* **2021**, *2021*, 665–684. [CrossRef] [PubMed]
42. Lang, A.; Wang, Y.; Feng, C.; Stai, E.; Hug, G. Data aggregation point placement for smart meters in the smart grid. *IEEE Trans. Smart Grid* **2021**, *13*, 541–554. [CrossRef]
43. Liu, H.; Du, Z.; Xue, T.; Jiang, T. Enhancing smart building performance with waste heat recovery: Supply-side management, demand reduction, and peak shaving via advanced control systems. *Energy Build.* **2025**, *327*, 115070. [CrossRef]
44. Amin, W.; Xie, S.; Vasa, L.; Mentel, U. Role of land use, green energy, and water resources for food accessibility: Evidence from emerging economies in the lens of COP 28. *Land Degrad. Dev.* **2024**, *35*, 4607–4622. [CrossRef]
45. Ajayi, O.O.; Alozie, C.E.; Abieba, O.A. Enhancing cybersecurity in energy infrastructure: Strategies for safeguarding critical systems in the digital age. *Trends Renew. Energy* **2025**, *11*, 201–212. [CrossRef]
46. Sivasubramanian, S.S. Detection of Man in the Middle Attack on Fog Layer Using Intrusion Detection Systems. Ph.D. Thesis, National College of Ireland, Dublin, Ireland, 2023.
47. Cui, L.; Qu, Y.; Gao, L.; Xie, G.; Yu, S. Detecting false data attacks using machine learning techniques in smart grid: A survey. *J. Netw. Comput. Appl.* **2020**, *170*, 102808. [CrossRef]
48. Tripathi, N.; Hubballi, N. Application layer denial-of-service attacks and defense mechanisms: A survey. *ACM Comput. Surv.* **2021**, *54*, 1–33. [CrossRef]
49. Salim, M.M.; Rathore, S.; Park, J.H. Distributed denial of service attacks and its defenses in IoT: A survey. *J. Supercomput.* **2020**, *76*, 5320–5363. [CrossRef]
50. Bushigampala, B.K.; Inaganti, A.C. Threat Detection in Critical Infrastructure Using AI Models. *Int. J. Acta Inform.* **2023**, *2*, 196–208.
51. Lee, I. Analysis of insider threats in the healthcare industry: A text mining approach. *Information* **2022**, *13*, 404. [CrossRef]
52. Medcalfe, D. *Critical Infrastructure in the Face of Global Cyber Threats*; Université de Sherbrooke: Sherbrooke, QC, USA, 2024.
53. Lackenbauer, P.W. Situating the Yukon in Canadian Arctic Defence and Security. 2024. Available online: https://www.naadsn.ca/wp-content/uploads/2024/07/2024apr-Lackenbauer-Yukon-defence-security.pdf (accessed on 1 September 2025).
54. Gomila, D.; Carreras, B.A.; Reynolds-Barredo, J.-M.; Martínez-Barbeito, M.; Colet, P.; Gomis-Bellmunt, O. Reducing Blackout Risk by Segmenting European Power Grid with HVDC Lines. *J. Mod. Power Syst. Clean Energy* **2025**. [CrossRef]
55. Konstantinoudis, G.; Riou, J. Excess mortality attributable to the 2025 Iberian Peninsula blackout. *medRxiv* **2025**. [CrossRef]
56. Iftikhar, S. Cyberterrorism as a global threat: A review on repercussions and countermeasures. *PeerJ Comput. Sci.* **2024**, *10*, e1772. [CrossRef]
57. Alromih, A. Developing an Efficient and Privacy-Preserving Energy Theft Detection System for Smart Grids. Ph.D. Thesis, University of Sheffield, Sheffield, UK, 2023.
58. Liu, L.; Workman, M.; Hayes, S. Net Zero and the potential of consumer data-United Kingdom energy sector case study: The need for cross-sectoral best data practice principles. *Energy Policy* **2022**, *163*, 112803. [CrossRef]
59. Singh, S.K.; Kumar, S.; Garg, S.; Arora, S.; Sharma, S.K.; Arya, V.; Chui, K.T. Blockchain-Based Data Security in Smart Cities: Ensuring Data Integrity and Trustworthiness. In *Digital Forensics and Cyber Crime Investigation*; CRC Press: Boca Raton, FL, USA, 2025; pp. 17–41.
60. Shah, M.S.M.; Leau, Y.-B.; Anbar, M.; Bin-Salem, A.A. Security and integrity attacks in named data networking: A survey. *IEEE Access* **2023**, *11*, 7984–8004. [CrossRef]
61. Zografopoulos, I.; Hatziargyriou, N.D.; Konstantinou, C. Distributed energy resources cybersecurity outlook: Vulnerabilities, attacks, impacts, and mitigations. *IEEE Syst. J.* **2023**, *17*, 6695–6709. [CrossRef]
62. Herath, H.M.S.S.; Herath, H.M.K.K.M.B.; Madhusanka, B.G.D.A.; Guruge, L.G.P.K. Data protection challenges in the processing of sensitive data. In *Data Protection: The Wake of AI and Machine Learning*; Springer: Berlin/Heidelberg, Germany, 2024; pp. 155–179.
63. Haque, A.B.; Bhushan, B.; Dhiman, G. Conceptualizing smart city applications: Requirements, architecture, security issues, and emerging trends. *Expert Syst.* **2022**, *39*, e12753. [CrossRef]

64. Munjal, K.; Bhatia, R. A systematic review of homomorphic encryption and its contributions in healthcare industry. *Complex Intell. Syst.* **2023**, *9*, 3759–3786. [CrossRef]
65. Bozduman, H.Ç.; Afacan, E. Simulation of a homomorphic encryption system. *Appl. Math. Nonlinear Sci.* **2020**, *5*, 479–484. [CrossRef]
66. Xiao, Y.; Xu, J.; Lin, Z.; Xie, Y.; Liu, R.; Yan, L.; Feng, P. Privacy Protection Anomaly Detection in Smart Grids Based on Combined PHE and TFHE Homomorphic Encryption. *Electronics* **2025**, *14*, 2386. [CrossRef]
67. Zhao, S. Smart Grids Data Aggregation Method on Paillier Homomorphic Encryption. *Appl. Artif. Intell.* **2024**, *38*, 2327901. [CrossRef]
68. Sousa-Dias, D.; Amyot, D.; Rahimi-Kian, A.; Bashari, M.; Mylopoulos, J. Cyclic Homomorphic Encryption Aggregation (CHEA)—A Novel Approach to Data Aggregation in the Smart Grid. *Energies* **2024**, *17*, 878. [CrossRef]
69. Westrich, B. Privacy-Preserving Analytics for Smart Meter (AMI) Data: A Hybrid Approach to Comply with CPUC Privacy Regulations. *arXiv* **2025**, arXiv:2505.08237. [CrossRef]
70. Parihar, V.; Malik, A.; Bhushan, B.; Bhattacharya, P.; Shankar, A. Innovative smart grid solutions for fostering data security and effective privacy preservation. In *Data Analytics for Smart Grids Applications—A Key to Smart City Development*; Springer: Berlin/Heidelberg, Germany, 2023; pp. 351–380.
71. Syed, D.; Refaat, S.S.; Bouhali, O. Privacy preservation of data-driven models in smart grids using homomorphic encryption. *Information* **2020**, *11*, 357. [CrossRef]
72. Bakar, A.A.; Yussof, S.; Ghapar, A.A.; Sameon, S.S.; Jørgensen, B.N. A review of privacy concerns in energy-efficient smart buildings: Risks, rights, and regulations. *Energies* **2024**, *17*, 977. [CrossRef]
73. Mollakuqe, E.; Parduzi, A.; Rexhepi, S.; Dimitrova, V.; Jakupi, S.; Muharremi, R.; Hamiti, M.; Qarkaxhija, J. Applications of homomorphic encryption in secure computation. *Open Res. Eur.* **2024**, *4*, 158. [CrossRef]
74. Singh, P.; Masud, M.; Hossain, M.S.; Kaur, A. Blockchain and homomorphic encryption-based privacy-preserving data aggregation model in smart grid. *Comput. Electr. Eng.* **2021**, *93*, 107209. [CrossRef]
75. Samuel, A.J. Optimizing energy consumption through AI and cloud analytics: Addressing data privacy and security concerns. *World J. Adv. Eng. Technol. Sci.* **2024**, *13*, 789–806. [CrossRef]
76. Ali, M.; Suchismita, M.; Ali, S.S.; Choi, B.J. Privacy-Preserving Machine Learning for IoT-Integrated Smart Grids: Recent Advances, Opportunities, and Challenges. *Energies* **2025**, *18*, 2515. [CrossRef]
77. Cali, U.; Gourisetti, S.N.G.; Sebastian-Cardenas, D.J.; Catak, F.O.; Lee, A.; Zeger, L.M.; Ustun, T.S.; Dynge, M.F.; Rao, S.; Ramirez, J.E. Emerging technologies for privacy preservation in energy systems. In Proceedings of the European Interdisciplinary Cybersecurity Conference, Xanthi, Greece, 5–6 June 2024; pp. 163–170.
78. Desai, S.; Sabar, N.; Alhadad, R.; Mahmood, A.; Chilamkurti, N. *Mitigating Consumer Privacy Breach in Smart Grid Using Obfuscation-Based Generative Adversarial Network*; La Trobe: Melbourne, Australia, 2022.
79. Bamberger, K.A.; Canetti, R.; Goldwasser, S.; Wexler, R.; Zimmerman, E.J. Verification dilemmas in law and the promise of zero-knowledge proofs. *Berkeley Tech. LJ* **2022**, *37*, 1. [CrossRef]
80. Bhattacharya, S.; Seth, D.; Panyam, S.; Gangrade, P. Enhancing Digital Privacy: The Application of Zero-Knowledge Proofs in Authentication Systems. *Int. J. Comput. Trends Technol.* **2024**, *72*, 34–41. [CrossRef]
81. Liang, W.; Liu, Y.; Yang, C.; Xie, S.; Li, K.; Susilo, W. On identity, transaction, and smart contract privacy on permissioned and permissionless blockchain: A comprehensive survey. *ACM Comput. Surv.* **2024**, *56*, 1–35. [CrossRef]
82. Ballesteros-Rodríguez, A.; Sánchez-Alonso, S.; Sicilia-Urbán, M.-Á. Enhancing Privacy and Integrity in Computing Services Provisioning Using Blockchain and zk-SNARKs. *IEEE Access* **2024**, *12*, 117970–117993. [CrossRef]
83. Lodder, M. Token Based Authentication and Authorization with Zero-Knowledge Proofs for Enhancing Web API Security and Privacy. 2023. Available online: https://scholar.dsu.edu/cgi/viewcontent.cgi?article=1426&context=theses (accessed on 15 August 2025).
84. Giaconi, G.; Gündüz, D.; Poor, H.V. 10 smart meter data privacy. In *Advanced Data Analytics for Power Systems*; Cambridge University Press: Cambridge, UK, 2021; p. 230.
85. El-Hajj, M.; Oude Roelink, B.; Sarmah, D. Systematic review: Comparing zk-SNARK, zk-STARK, and bulletproof protocols for privacy-preserving authentication. *Secur. Priv.* **2024**, *7*, e401.
86. Manzoor, A. *Performance and Security Analysis of Distributed Ledger Technologies in IoT*; Oulun Yliopisto: Oulu, Finland, 2024.
87. Dehalwar, V.; Kolhe, M.L.; Deoli, S.; Jhariya, M.K. Blockchain-based trust management and authentication of devices in smart grid. *Clean. Eng. Technol.* **2022**, *8*, 100481. [CrossRef]
88. Koukaras, P.; Afentoulis, K.D.; Gkaidatzis, P.A.; Mystakidis, A.; Ioannidis, D.; Vagropoulos, S.I.; Tjortjis, C. Integrating blockchain in smart grids for enhanced demand response: Challenges, strategies, and future directions. *Energies* **2024**, *17*, 1007. [CrossRef]
89. Rahman, H. Securing Data Privacy in Blockchain Networks. In *Blockchain Technology Applications in Knowledge Management*; IGI Global Scientific Publishing: Palmdale, PA, USA, 2025; pp. 367–410.
90. Chen, X.; Wang, F.; Huang, Z.; Chen, Y.-C. Smart grid aggregation billing scheme based on blockchain. *J. Netw. Intell.* **2022**, *7*, 878–893.

91. Kumar, R.; Khan, A.A.; Kumar, J.; Golilarz, N.A.; Zhang, S.; Ting, Y.; Zheng, C.; Wang, W. Blockchain-federated-learning and deep learning models for covid-19 detection using ct imaging. *IEEE Sens. J.* **2021**, *21*, 16301–16314. [CrossRef] [PubMed]
92. Li, W. Frontiers in Blockchain for Secure Information Sharing in Next Generation Transportation Systems. Ph.D. Thesis, University of Delaware, Newark, DE, USA, 2022.
93. Rahaman, M.; Arya, V.; Orozco, S.M.; Pappachan, P. Secure multi-party computation (SMPC) protocols and privacy. In *Innovations in Modern Cryptography*; IGI Global: Palmdale, PA, USA, 2024; pp. 190–214.
94. Zhou, I.; Tofigh, F.; Piccardi, M.; Abolhasan, M.; Franklin, D.; Lipman, J. Secure multi-party computation for machine learning: A survey. *IEEE Access* **2024**, *12*, 53881–53899. [CrossRef]
95. Khan, S. Secure multi-party computation for privacy preservation in big data analytics. *J. Big Data Priv. Manag.* **2024**, *2*, 168–179.
96. Fowlie, M.; Wolfram, C.; Baylis, P.; Spurlock, C.A.; Todd-Blick, A.; Cappers, P. Default effects and follow-on behaviour: Evidence from an electricity pricing program. *Rev. Econ. Stud.* **2021**, *88*, 2886–2934. [CrossRef]
97. Fan, C.; Chen, M.; Wang, X.; Wang, J.; Huang, B. A review on data preprocessing techniques toward efficient and reliable knowledge discovery from building operational data. *Front. Energy Res.* **2021**, *9*, 652801. [CrossRef]
98. Sasikumar, K.; Nagarajan, S. Comprehensive review and analysis of cryptography techniques in cloud computing. *IEEE Access* **2024**, *12*, 52325–52351. [CrossRef]
99. Bjarghov, S.; Löschenbrand, M.; Saif, A.U.I.; Pedrero, R.A.; Pfeiffer, C.; Khadem, S.K.; Rabelhofer, M.; Revheim, F.; Farahmand, H. Developments and challenges in local electricity markets: A comprehensive review. *IEEE Access* **2021**, *9*, 58910–58943. [CrossRef]
100. Zhao, Y.; Chen, J. A survey on differential privacy for unstructured data content. *ACM Comput. Surv.* **2022**, *54*, 1–28. [CrossRef]
101. Hu, C.; Liu, Z.; Li, R.; Hu, P.; Xiang, T.; Han, M. Smart contract assisted privacy-preserving data aggregation and management scheme for smart grid. *IEEE Trans. Dependable Secur. Comput.* **2023**, *21*, 2145–2161. [CrossRef]
102. Rai, H.M.; Shukla, K.K.; Tightiz, L.; Padmanaban, S. Enhancing data security and privacy in energy applications: Integrating IoT and blockchain technologies. *Heliyon* **2024**, *10*, e38917. [CrossRef]
103. Gai, N.; Xue, K.; Zhu, B.; Yang, J.; Liu, J.; He, D. An efficient data aggregation scheme with local differential privacy in smart grid. *Digit. Commun. Netw.* **2022**, *8*, 333–342. [CrossRef]
104. Cai, Z.; Xiong, Z.; Xu, H.; Wang, P.; Li, W.; Pan, Y. Generative adversarial networks: A survey toward private and secure applications. *ACM Comput. Surv.* **2021**, *54*, 1–38. [CrossRef]
105. Chen, W.-N.; Song, D.; Ozgur, A.; Kairouz, P. Privacy amplification via compression: Achieving the optimal privacy-accuracy-communication trade-off in distributed mean estimation. *Adv. Neural Inf. Process. Syst.* **2023**, *36*, 69202–69227.
106. Gough, M.B.; Santos, S.F.; AlSkaif, T.; Javadi, M.S.; Castro, R.; Catalão, J.P.S. Preserving privacy of smart meter data in a smart grid environment. *IEEE Trans. Ind. Inform.* **2021**, *18*, 707–718. [CrossRef]
107. Razi, Q.; Piyush, R.; Chakrabarti, A.; Singh, A.; Hassija, V.; Chalapathi, G.S.S. Enhancing Data Privacy: A Comprehensive Survey of Privacy-Enabling Technologies. *IEEE Access* **2025**, *13*, 40354–40385. [CrossRef]
108. Tran, H.-Y.; Hu, J.; Pota, H.R. Smart meter data obfuscation with a hybrid privacy-preserving data publishing scheme without a trusted third party. *IEEE Internet Things J.* **2022**, *9*, 16080–16095. [CrossRef]
109. Kua, J.; Hossain, M.B.; Natgunanathan, I.; Xiang, Y. Privacy preservation in smart meters: Current status, challenges and future directions. *Sensors* **2023**, *23*, 3697. [CrossRef]
110. Li, M.; Zhu, L.; Zhang, Z.; Lal, C.; Conti, M.; Alazab, M. User-defined privacy-preserving traffic monitoring against n-by-1 jamming attack. *IEEE/ACM Trans. Netw.* **2022**, *30*, 2060–2073. [CrossRef]
111. Etezady, N.N. A survey of privacy metrics for smart homes. *Cybersecur. Pedagog. Pract. J.* **2023**, *2*, 26–37.
112. Bhattacharjee, K. Interactive Visualization Workflows for Mitigating Analytical Uncertainty. Ph.D. Thesis, New Jersey Institute of Technology, Newark, NJ, USA, 2024.
113. Sakthivel, S.; Vinotha, N. An intellectual optimization of k-anonymity model for efficient privacy preservation in cloud platform. *J. Intell. Fuzzy Syst.* **2023**, *45*, 1497–1512. [CrossRef]
114. Li, S.; Schneider, M.J.; Yu, Y.; Gupta, S. Reidentification risk in panel data: Protecting for k-anonymity. *Inf. Syst. Res.* **2023**, *34*, 1066–1088. [CrossRef]
115. Adewole, K.S.; Torra, V. DFTMicroagg: A dual-level anonymization algorithm for smart grid data. *Int. J. Inf. Secur.* **2022**, *21*, 1299–1321. [CrossRef]
116. Avraam, D.; Wilson, R.; Butters, O.; Burton, T.; Nicolaides, C.; Jones, E.; Boyd, A.; Burton, P. Privacy preserving data visualizations. *EPJ Data Sci.* **2021**, *10*, 2. [CrossRef]
117. Sun, L.; Dou, Y.; Yang, C.; Zhang, K.; Wang, J.; Yu, P.S.; He, L.; Li, B. Adversarial attack and defense on graph data: A survey. *IEEE Trans. Knowl. Data Eng.* **2022**, *35*, 7693–7711. [CrossRef]
118. Arshad, R.; Asghar, M.R. Characterisation and quantification of user privacy: Key challenges, regulations, and future directions. *IEEE Commun. Surv. Tut.* **2024**. . [CrossRef]
119. Ul Hassan, M. Efficient Privacy Preservation in Smart Grid. Ph.D. Thesis, Swinburne, Hawthorn, Australia, 2021.

120. Kapoor, A.; Kumar, D. Federated learning for urban sensing systems: A comprehensive survey on attacks, defences, incentive mechanisms, and applications. *IEEE Commun. Surv. Tut.* **2024**, *27*, 1293–1325. [CrossRef]
121. Zhan, Y.; Zhou, L.; Wang, B.; Duan, P.; Zhang, B. Efficient function queryable and privacy preserving data aggregation scheme in smart grid. *IEEE Trans. Parallel Distrib. Syst.* **2022**, *33*, 3430–3441. [CrossRef]
122. Teng, F.; Chhachhi, S.; Ge, P.; Graham, J.; Gunduz, D. *Balancing Privacy and Access to Smart Meter Data*; Energy Futures Lab briefing paper; Imperial College: London, UK, 2022.
123. Diraco, G.; Rescio, G.; Caroppo, A.; Manni, A.; Leone, A. Human action recognition in smart living services and applications: Context awareness, data availability, personalization, and privacy. *Sensors* **2023**, *23*, 6040. [CrossRef] [PubMed]
124. Aga, D.T.; Chintanippu, R.; Mowri, R.A.; Siddula, M. Exploring secure and private data aggregation techniques for the internet of things: A comprehensive review. *Discov. Internet Things* **2024**, *4*, 28. [CrossRef]
125. Xu, H.; Seng, K.P.; Ang, L.M.; Smith, J. Decentralized and distributed learning for AIoT: A comprehensive review, emerging challenges, and opportunities. *IEEE Access* **2024**, *12*, 101016–101052. [CrossRef]
126. Tong, X.; Hamzei, M.; Jafari, N. Towards Secure and Efficient Data Aggregation in Blockchain-Driven IoT Environments: A Comprehensive and Systematic Study. *Trans. Emerg. Telecommun. Technol.* **2025**, *36*, e70061. [CrossRef]
127. Behrouz, P.; Khaksari, A.; Konstantinidis, O.; Pagourtzis, A.; Spyrakou, M. Privacy-preserving methods for smart meter data-a systematization of knowledge. In Proceedings of the CP904: Institute of Engineering and Technology Conference, Singapore, October 2024; pp. 64–71. Available online: https://digital-library.theiet.org/doi/10.1049/icp.2024.4638 (accessed on 1 September 2025).
128. Le, M.; Huynh-The, T.; Do-Duy, T.; Vu, T.-H.; Hwang, W.-J.; Pham, Q.-V. Applications of distributed machine learning for the internet-of-things: A comprehensive survey. *IEEE Commun. Surv. Tut.* **2024**, *27*, 1053–1100. [CrossRef]
129. Yan, R.; Zheng, Y.; Yu, N.; Liang, C. Multi-smart meter data encryption scheme based on distributed differential privacy. *Big Data Min. Anal.* **2023**, *7*, 131–141. [CrossRef]
130. Hasan, M.K.; Alkhalifah, A.; Islam, S.; Babiker, N.B.; Habib, A.A.; Aman, A.H.M.; Hossain, M.A. Blockchain technology on smart grid, energy trading, and big data: Security issues, challenges, and recommendations. *Wirel. Commun. Mob. Comput.* **2022**, *2022*, 9065768. [CrossRef]
131. Saeed, N.; Wen, F.; Afzal, M.Z. Decentralized peer-to-peer energy trading in microgrids: Leveraging blockchain technology and smart contracts. *Energy Rep.* **2024**, *12*, 1753–1764. [CrossRef]
132. Nazir, I.; Mushtaq, N.; Ishfaq, H.; Amin, W.; Maaliw, R.R.; Shi, X.; Saleem, M.A. Blockchain Enabled Reserved Pricing and Privacy Preserving Model for Peer-to-Peer Energy Trading in Smart Grid. *IEEE Trans. Consum. Electron.* **2025**, 1. [CrossRef]
133. Alshahrani, H.; Islam, N.; Syed, D.; Sulaiman, A.; Al Reshan, M.S.; Rajab, K.; Shaikh, A.; Shuja-Uddin, J.; Soomro, A. Sustainability in blockchain: A systematic literature review on scalability and power consumption issues. *Energies* **2023**, *16*, 1510. [CrossRef]
134. Alsheavi, A.N.; Hawbani, A.; Othman, W.; Wang, X.; Qaid, G.; Zhao, L.; Al-Dubai, A.; Zhi, L.; Ismail, A.S.; Jhaveri, R. Iot authentication protocols: Challenges, and comparative analysis. *ACM Comput. Surv.* **2025**, *57*, 1–43. [CrossRef]
135. Honari, K.; Rouhani, S.; Falak, N.E.; Liu, Y.; Li, Y.; Liang, H.; Dick, S.; Miller, J. Smart contract design in distributed energy systems: A systematic review. *Energies* **2023**, *16*, 4797. [CrossRef]
136. Fatima, S.; Arshad, M.J. A Comprehensive Review of Blockchain and Machine Learning Integration for Peer-to-Peer Energy Trading in Smart Grids. *IEEE Access* **2025**, *13*, 92756–92782. [CrossRef]
137. Ishtaiwi, A.; Aldweesh, A.; al-Qerem, A.; Alkasassbeh, M. AI-Powered Smart Contracts for Energy Grid Optimization: A Hybrid Intelligence Approach. In *Blockchain Applications for the Energy and Utilities Industry*; IGI Global Scientific Publishing: Palmdale, PA, USA, 2025; pp. 235–264.
138. Sigalov, K.; Ye, X.; König, M.; Hagedorn, P.; Blum, F.; Severin, B.; Hettmer, M.; Hückinghaus, P.; Wölkerling, J.; Groß, D. Automated payment and contract management in the construction industry by integrating building information modeling and blockchain-based smart contracts. *Appl. Sci.* **2021**, *11*, 7653. [CrossRef]
139. Kalbantner, J. Security and Privacy Issues of Distributed, Complex P2P Smart Grid Marketplace Systems. Ph.D. Thesis, Royal Holloway University of London, London, UK, 2024.
140. Zaidi, S.Y.A.; Shah, M.A.; Khattak, H.A.; Maple, C.; Rauf, H.T.; El-Sherbeeny, A.M.; El-Meligy, M.A. An attribute-based access control for IoT using blockchain and smart contracts. *Sustainability* **2021**, *13*, 10556. [CrossRef]
141. AlMarshoud, M.; Sabir Kiraz, M.; Al-Bayatti, A.H. Security, privacy, and decentralized trust management in VANETs: A review of current research and future directions. *ACM Comput. Surv.* **2024**, *56*, 1–39. [CrossRef]
142. Rouhani, S.; Belchior, R.; Cruz, R.S.; Deters, R. Distributed attribute-based access control system using permissioned blockchain. *World Wide Web* **2021**, *24*, 1617–1644. [CrossRef]
143. Saidi, H.; Labraoui, N.; Ari, A.A.A.; Maglaras, L.A.; Emati, J.H.M. DSMAC: Privacy-aware Decentralized Self-Management of data Access Control based on blockchain for health data. *IEEE Access* **2022**, *10*, 101011–101028. [CrossRef]
144. Lankadasu, N.V.Y.; Pesarlanka, D.B.; Sharma, A.; Sharma, S. Security Aspects of Blockchain Technology. In *Driving Transformative Technology Trends with Cloud Computing*; IGI Global: Palmdale, PA, USA, 2024; pp. 259–277.

145. Babu, S.B.; Jothi, K.R. A secure framework for privacy-preserving analytics in healthcare records using zero-knowledge proofs and blockchain in multi-tenant cloud environments. *IEEE Access* **2024**, *13*, 8439–8455. [CrossRef]
146. Alladi, T.; Chamola, V.; Rodrigues, J.J.; Kozlov, S.A. Blockchain in smart grids: A review on different use cases. *Sensors* **2019**, *19*, 4862. [CrossRef] [PubMed]
147. Uddin, S.S.; Joysoyal, R.; Sarker, S.K.; Muyeen, S.M.; Ali, M.F.; Hasan, M.M.; Abhi, S.H.; Islam, R.; Ahamed, H.; Islam, M.; et al. Next-generation blockchain enabled smart grid: Conceptual framework, key technologies and industry practices review. *Energy AI* **2023**, *12*, 100228. [CrossRef]
148. Luo, H.; Yu, H.; Luo, J. PRAFT and RPBFT: A class of blockchain consensus algorithm and their applications in electric vehicles charging scenarios for V2G networks. *Internet Things Cyber-Phys. Syst.* **2023**, *3*, 61–70. [CrossRef]
149. Su, K.; Yu, Y.; Zhang, J. Blockchain-based smart grid power trading technology. *J. Eng. Appl. Sci.* **2024**, *71*, 220. [CrossRef]
150. Chikezie, U.; Karacolak, T.; Do Prado, J.C. Examining the applicability of blockchain to the smart grid using proof-of-authority consensus. In Proceedings of the 2021 IEEE 9th International Conference on Smart Energy Grid Engineering (SEGE), Oshawa, ON, Canada, 11–13 August 2021; pp. 19–25.
151. Khan, A.A.; Dhabi, S.; Yang, J.; Alhakami, W.; Bourouis, S.; Yee, L. B-LPoET: A middleware lightweight Proof-of-Elapsed Time (PoET) for efficient distributed transaction execution and security on Blockchain using multithreading technology. *Comput. Electr. Eng.* **2024**, *118*, 109343. [CrossRef]
152. Jain, A.K.; Gupta, N.; Gupta, B.B. A survey on scalable consensus algorithms for blockchain technology. *Cyber Secur. Appl.* **2025**, *3*, 100065. [CrossRef]
153. Zhang, X.; You, L.; Hu, G. An efficient and robust multidimensional data aggregation scheme for smart grid based on blockchain. *IEEE Trans. Netw. Serv. Manag.* **2022**, *19*, 3949–3959. [CrossRef]
154. Akiladevi, R.; Sardha, S.; Shruthi, R. Tokenization of energy assets: A multichain blockchain approach. In Proceedings of the 5th International Conference on Mobile Computing and Sustainable Informatics (ICMCSI), Lalitpur, Nepal, 18–19 January 2024; pp. 702–709.
155. Taherdoost, H. Blockchain integration and its impact on renewable energy. *Computers* **2024**, *13*, 107. [CrossRef]
156. Hu, C.; Cheng, X.; Tian, Z.; Yu, J.; Lv, W. Achieving privacy preservation and billing via delayed information release. *IEEE/ACM Trans. Netw.* **2021**, *29*, 1376–1390. [CrossRef]
157. Karaduman, Ö.; Gülhas, G. Blockchain-Enabled Supply Chain Management: A Review of Security, Traceability, and Data Integrity Amid the Evolving Systemic Demand. *Appl. Sci.* **2025**, *15*, 5168. [CrossRef]
158. Khemaissia, R. Using Blockchain to Protect Personal Data. Ph.D. Thesis, Université Echahid Cheikh Larbi-Tebessi-Tébessa, Tébessa, Algeria, 2025.
159. An, Q.; Jiang, F.; Dong, C.; Pal, S.; Li, J.; Neiat, A.G.; Yeoh, W. A blockchain-powered secure architecture for cyber marketplaces of electric vehicles. *IEEE Trans. Ind. Appl.* **2025**, *61*, 4198–4213. [CrossRef]
160. Fekri, M.N.; Grolinger, K.; Mir, S. Distributed load forecasting using smart meter data: Federated learning with Recurrent Neural Networks. *Int. J. Electr. Power Energy Syst.* **2022**, *137*, 107669. [CrossRef]
161. Su, Z.; Wang, Y.; Luan, T.H.; Zhang, N.; Li, F.; Chen, T.; Cao, H. Secure and efficient federated learning for smart grid with edge-cloud collaboration. *IEEE Trans. Ind. Inform.* **2021**, *18*, 1333–1344. [CrossRef]
162. Alshehri, A.; Badr, M.M.; Baza, M.; Alshahrani, H. Deep anomaly detection framework utilizing federated learning for electricity theft zero-day cyberattacks. *Sensors* **2024**, *24*, 3236. [CrossRef]
163. Santosh, S.H.L.; Ravi, R.M. A Enhancing Privacy in IoT Systems with Federated Learning: A Multi-Agent Solution for Smart City Networks. *SGS-Eng. Sci.* **2025**, *1*. Available online: https://spast.org/techrep/article/view/5282 (accessed on 15 August 2025).
164. Qu, L.; Zhou, Y.; Liang, P.P.; Xia, Y.; Wang, F.; Adeli, E.; Fei-Fei, L.; Rubin, D. Rethinking architecture design for tackling data heterogeneity in federated learning. In Proceeding of the IEEE/CVF Conference on Computer Vision and Pattern Recognition, New Orleans, LA, USA, 18–24 June 2022; pp. 10061–10071.
165. Khan, A.R. Federated learning for next generation intelligent applications. Ph.D. Thesis, University of Glasgow, Glasgow, UK, 2025.
166. Zhang, C.; Li, S. State-of-the-art approaches to enhancing privacy preservation of machine learning datasets: A survey. *arXiv* **2024**, arXiv:2404.16847.
167. Mohammadi, M.; Mohammadi, A. Empowering distributed solutions in renewable energy systems and grid optimization. In *Distributed Machine Learning and Computing: Theory and Applications*; Springer: Berlin/Heidelberg, Germany, 2024; pp. 141–155.
168. Adewole, K.S.; Torra, V. Privacy Protection of Synthetic Smart Grid Data Simulated via Generative Adversarial Networks. In Proceedings of the 20th International Conference on Security and Cryptography (SECRYPT), Rome, Italy, 10–12 July 2023; pp. 279–286.
169. Takiddin, A.; Ismail, M.; Serpedin, E. Robust data-driven detection of electricity theft adversarial evasion attacks in smart grids. *IEEE Trans. Smart Grid* **2022**, *14*, 663–676. [CrossRef]
170. Abdi, N.; Albaseer, A.; Abdallah, M. The role of deep learning in advancing proactive cybersecurity measures for smart grid networks: A survey. *IEEE Internet Things J.* **2024**, *11*, 16398–16421. [CrossRef]

171. Pelekis, S.; Koutroubas, T.; Blika, A.; Berdelis, A.; Karakolis, E.; Ntanos, C.; Spiliotis, E.; Askounis, D. Adversarial machine learning: A review of methods, tools, and critical industry sectors. *Artif. Intell. Rev.* **2025**, *58*, 226. [CrossRef]
172. Ramirez, J.E.; Sebastian, C.; David, J. *A Computational Review of Privacy-Preserving Mechanisms for the Smart Grid*; Pacific Northwest National Laboratory (PNNL): Richland, WA, USA, 2024. Available online: https://www.pnnl.gov/main/publications/external/technical_reports/PNNL-36774.pdf (accessed on 1 September 2025).
173. Zhang, J. Efficient Zero-Knowledge Proofs: Theory and Practice. Ph.D. Thesis, UC Berkeley, Berkeley, CA, USA, 2023. Available online: https://www2.eecs.berkeley.edu/Pubs/TechRpts/2025/EECS-2025-20.pdf (accessed on 1 September 2025).
174. Egunjobi, O.O.; Gomes, A.; Egwim, C.N.; Morais, H. A systematic review of blockchain for energy applications. *E-Prime Electr. Eng. Electron. Energy* **2024**, *9*, 100751. [CrossRef]
175. Huang, Q.; Amin, W.; Umer, K.; Gooi, H.B.; Eddy, F.Y.S.; Afzal, M.; Shahzadi, M.; Khan, A.A.; Ahmad, S.A. A review of transactive energy systems: Concept and implementation. *Energy Rep.* **2021**, *7*, 7804–7824. [CrossRef]
176. Asad, M.; Shaukat, S.; Hu, D.; Wang, Z.; Javanmardi, E.; Nakazato, J.; Tsukada, M. Limitations and future aspects of communication costs in federated learning: A survey. *Sensors* **2023**, *23*, 7358. [CrossRef] [PubMed]
177. Zhang, J.; Shi, C. Efficient Secure Data Aggregation for Real-Time Smart Grid Monitoring: A Lightweight Privacy-Preserving Approach. *IEEE Trans. Comput. Soc. Syst.* **2025**, *12*, 1563–1573. [CrossRef]
178. Yang, R.; Gao, H.; Si, F.; Wang, J. Advancing User Privacy in Virtual Power Plants: A Novel Zero-Knowledge Proof-Based Distributed Attribute Encryption Approach. *Electronics* **2024**, *13*, 1283. [CrossRef]
179. Hu, H.; Hsu, C.; Cui, J.; Harn, L.; Hou, Q.; Li, L. Advancing User Privacy in Virtual Power Plants: Provably secure and lightweight authentication protocol using PUF and blockchain for smart grids. *J. Supercomput.* **2025**, *8*, 949. [CrossRef]
180. Soto, E.A.; Bosman, L.B.; Wollega, E.; Leon-Salas, W.D. Peer-to-peer energy trading: A review of the literature. *Appl. Energy* **2021**, *283*, 116268. [CrossRef]
181. Le, H.T.; Sanseverino, E.R.; Nguyen, D.-Q.; Di Silvestre, M.L.; Favuzza, S.; Pham, M.-H. Critical assessment of feed-in tariffs and solar photovoltaic development in Vietnam. *Energies* **2022**, *15*, 556. [CrossRef]
182. Lin, Y.; Wang, J. Realizing the transactive energy future with local energy market: An overview. *Curr. Sustain. Renew. Energy Rep.* **2022**, *9*, 1–14. [CrossRef]
183. Munoz, F.D.; Wogrin, S.; Oren, S.S.; Hobbs, B.F. Economic inefficiencies of cost-based electricity market designs. *Energy J.* **2018**, *39*, 51–68. [CrossRef]
184. Wei, W.; Liu, F.; Mei, S. Energy pricing and dispatch for smart grid retailers under demand response and market price uncertainty. *IEEE Trans. Smart Grid* **2014**, *6*, 1364–1374. [CrossRef]
185. Bouloumpasis, I.; Steen, D. Congestion management using local flexibility markets: Recent development and challenges. In Proceedings of the 2019 IEEE PES Innovative Smart Grid Technologies Europe (ISGT-Europe), Bucharest, Romania, 29 September–2 October 2019; pp. 1–5.
186. Biegel, B.; Andersen, P.; Stoustrup, J.; Bendtsen, J. Congestion management in a smart grid via shadow prices. *IFAC Proc. Vol.* **2012**, *45*, 518–523. [CrossRef]
187. Armstrong, M. Price discrimination. In *The New Palgrave Dictionary of Economics*; Macmillan: London, UK, 2008.
188. Willems, B.; Yueting, Y. Bidding and Investment in Wholesale Electricity Markets: Discriminatory Versus Uniform-Price Auctions. TSE Working Paper 2023. Available online: https://www.tse-fr.eu/sites/default/files/TSE/documents/doc/wp/2023/wp_tse_1462.pdf (accessed on 15 August 2025).
189. Jaskowski, P.; Biruk, S.; Czarnigowska, A. Strategy for mark-up definition in competitive tenders for construction work. In Proceedings of the IOP Conference Series Materials Science and Engineering, Bristol, UK, 19–22 August 2019; p. 112060.
190. Beiranvand, N. The Role of Algorithmic Pricing in Modern Markets: Price Discrimination, Fairness, and Collusion. 2023. Available online: https://unitesi.unipv.it/bitstream/20.500.14239/27269/1/Collusion%20and%20Price%20Discrimination%20with%20Algorithmic%20Pricing%20Efficiency%20and%20Fairness%20Issue.pdf (accessed on 15 August 2025).
191. Damiano, E.; Li, H. Price discrimination and efficient matching. *Econ. Theory* **2007**, *30*, 243–263. [CrossRef]
192. Pollitt, M.G. Locational Marginal Prices (LMPs) for Electricity in Europe? The Untold Story. 2023.
193. Li, F. Continuous locational marginal pricing (CLMP). *IEEE Trans. Power Syst.* **2007**, *22*, 1638–1646. [CrossRef]
194. Tesfatsion, L. Locational marginal pricing: A fundamental reconsideration. *IEEE Open Access J. Power Energy* **2024**, *11*, 104–116. [CrossRef]
195. Wang, Q.; Zhang, G.; McCalley, J.D.; Zheng, T.; Litvinov, E. Risk-based locational marginal pricing and congestion management. *IEEE Trans. Power Syst.* **2014**, *29*, 2518–2528. [CrossRef]
196. Tan, Z.; Cheng, T.; Liu, Y.; Zhong, H. Extensions of the locational marginal price theory in evolving power systems: A review. *IET Gener. Transm. Distrib.* **2022**, *16*, 1277–1291. [CrossRef]
197. Liyanapathirane, U.; Khorasany, M.; Razzaghi, R. Optimization of economic efficiency in distribution grids using distribution locational marginal pricing. *IEEE Access* **2021**, *9*, 60123–60135. [CrossRef]

198. Chen, B. Real-Time Electricity Market Pricing Methods Incorporating Power System Frequency Dynamics. Ph.D. Thesis, University of British Columbia, Vancouver, BC, Canada, 2025.
199. Tang, W.; Shi, L.; Xu, P.; Luo, H.; Li, L. Assessing the Influencing Factors of Locational Marginal Price in Day-Ahead Hydro-Thermal Power Spot Market. *IEEE Access* **2025**, *13*, 47055–47067. [CrossRef]
200. Gill, S.; MacIver, C.; Bell, K. Exploring Market Change in the GB Electricity System: The Potential Impact of Locational Marginal Pricing. 2023. Available online: https://pureportal.strath.ac.uk/en/publications/exploring-market-change-in-the-gb-electricity-system-the-potentia-2 (accessed on 15 August 2025).
201. Arora, M.; Vishwanath, G.M.; Sharma, A.; Chilamkurti, N. A novel congestion based pricing model for prosumer-centric smart grid systems. *Electr. Power Syst. Res.* **2025**, *247*, 111732. [CrossRef]
202. Niu, T.; Li, H.; Chen, G.; Fang, S.; Liao, R. Pricing and Distributed Scheduling Framework of Multi-Microgrid System Based on Coupled Electricity-Carbon Market. *J. Mod. Power Syst. Clean Energy* **2024**, *13*, 1026–1039. [CrossRef]
203. Syed, A.A.; Qi, H.; Waqas, A.; Muhammad, A.; Fazal, H.; Muhammad Husnain, H. A fair and effective approach to managing distributed energy resources through peer-to-peer energy trading with load prioritization among smart homes. *Energy Rep.* **2023**, *10*, 4402–4419.
204. Alsolami, M.; Alferidi, A.; Lami, B.; Slama, S.B. Peer-to-peer trading in smart grid with demand response and grid outage using deep reinforcement learning. *Ain Shams Eng. J.* **2023**, *14*, 102466. [CrossRef]
205. Park, S.; Lee, J.; Bae, S.; Hwang, G.; Choi, J.K. Contribution-based energy-trading mechanism in microgrids for future smart grid: A game theoretic approach. *IEEE Trans. Ind. Electron.* **2016**, *63*, 4255–4265. [CrossRef]
206. Amin, W.; Ahmad, F.; Umer, K.; Khawaja, A.H.; Afzal, M.; Ahmad, S.A.; Chaitusaney, S. An effective pricing mechanism for electricity trading considering customer preference and reserved price in direct P2P electricity market under uncertainty in grid supply. *IEEE Access* **2022**, *10*, 96197–96211. [CrossRef]
207. Umar, A.; Kumar, D.; Ghose, T. A decentralized energy management system inculcating supply–demand ratio (SDR) and miners' stake through blockchain smart contract for community microgrid operation. *IEEE J. Emerg. Sel. Top. Ind. Electron.* **2023**, *5*, 369–380. [CrossRef]
208. Oprea, S.-V.; Bâra, A. An adaptive pricing mechanism framework for post-auction in the local electricity markets integrating a higher share of renewables. *Renew. Energy* **2025**, *256*, 124056. [CrossRef]
209. Caragiannis, I.; Jiang, Z.; Skoulakis, S. Rethinking Pricing in Energy Markets: Pay-as-Bid vs Pay-as-Clear. *arXiv* **2025**, arXiv:2507.06035.
210. Akbari-Dibavar, A.; Mohammadi-Ivatloo, B.; Zare, K. Electricity market pricing: Uniform pricing vs. pay-as-bid pricing. In *Electricity Markets: New Players and Pricing Uncertainties*; Springer: Berlin/Heidelberg, Germany, 2020; pp. 19–35.
211. Shahinzadeh, H.; Nasrabadi, M.N.; Nafisi, H.; Moazzami, M.; Jurado, F.; Abdelaziz, A.Y. Enhancing Transparency in the Peer-to-Peer Transactive Energy Market: Smart Contract Monitoring in Bilateral Negotiations with the Ethereum Platform. In Proceedings of the 9th International Conference Technology and Energy Management (ICTEM), Hoes Lane, Piscataway, NJ, USA, 14–15 February 2024; pp. 1–8.
212. Cretan, A.-G. Leveraging AI to revolutionize business processes. *Chall. Knowl. Soc.* **2025**, 504–517.
213. Atanasov, A. The Economics of Water Conflict: Transaction Costs, Bargaining Failure, and Sustainable Cooperation in the Nile Dispute. Available online: https://papers.ssrn.com/sol3/papers.cfm?abstract_id=5371192 (accessed on 15 August 2025).
214. Hassan, F.; Abinader, L.G.; Kavanagh, M.M. Law and power in pandemic negotiations: Policymaking via contract in South Africa's dealings with COVID-19 vaccine companies. *Glob. Public Health* **2025**, *20*, 2537688. [CrossRef]
215. Talebi, E.; Mehdinejad, M.; Mohammadi-Ivatloo, B.; Abapour, M.; Tohidi, S. A robust framework for peer-to-peer energy trading with transmission costs consideration: A fuzzy possibilistic programming model. *Appl. Energy* **2025**, *398*, 126379. [CrossRef]
216. van de Water, P.R.; Doumen, S.C.; Campfens, J.K.E.K.; Wieczorek, A.J.; Nguyen, P.H. An implementation pathway for price formation mechanisms in local electricity markets. *Energy Rep.* **2025**, *13*, 6384–6396. [CrossRef]
217. Wang, Y.; Li, T.; Li, Y.; Shao, N.; Wang, Y. Spot market clearing model and flexibility premium assessment method considering flexible regulation of virtual power plants. *IEEE Access* **2024**, *12*, 53855–53870. [CrossRef]
218. Jiang, Y.; Sioshansi, R. Optimal and incentive-compatible scheduling of flexible generation in an electricity market. *Energy J.* **2024**, 01956574251340293. [CrossRef]
219. Taslimi, M.; Khosravi, A. Flexibility and profitability in power-to-methanol plants: Integrating storage solutions and reserve market participation. *Energy* **2025**, *327*, 136438. [CrossRef]
220. Xu, Y.; Bai, Y.; Zhang, Y.; Wang, P.; Wang, X.; Guo, J.; Xie, K.; Zhao, R. Analyzing the Impact of Strategic Bidding on the Reserve Capacity via a Bi-Level Model. *arXiv* **2025**, arXiv:2506.20493. [CrossRef]
221. Johnson, A.; Fraser, A.; York, D. *Enabling Industrial Demand Flexibility: Aligning Industrial Consumer and Grid Benefits*; White Paper; ACEEE: Washington, DC, USA, 2024.
222. Cramton, P. Competitive bidding behavior in uniform-price auction markets. In Proceedings of the 37th Annual Hawaii International Conference on System Sciences Proceeding, Big Island, HI, USA, 5–8 January 2004; p. 11.

223. Vlachos, A.G.; Biskas, P.N. Balancing supply and demand under mixed pricing rules in multi-area electricity markets. *IEEE Trans. Power Syst.* **2010**, *26*, 1444–1453. [CrossRef]
224. Zhang, C.; Wang, Y.; Yang, T. Iterative auction for P2P renewable energy trading with dynamic energy storage management. *Energies* **2020**, *13*, 4963. [CrossRef]
225. Amin, W.; Huang, Q.; Afzal, M.; Khan, A.A.; Zhang, Z.; Umer, K.; Ahmed, S.A. Consumers' preference based optimal price determination model for P2P energy trading. *Electr. Power Syst. Res.* **2020**, *187*, 106488. [CrossRef]
226. Xu, S.; Zhao, Y.; Li, Y.; Zhou, Y. An iterative uniform-price auction mechanism for peer-to-peer energy trading in a community microgrid. *Appl. Energy* **2021**, *298*, 117088. [CrossRef]
227. Xu, S.-S. Pion superfluid phase transition at finite isospin chemical potential. *Eur. Phys. J. A* **2021**, *57*, 298. [CrossRef]
228. Tushar, W.; Chai, B.; Yuen, C.; Huang, S.; Smith, D.B.; Poor, H.V.; Yang, Z. Energy storage sharing in smart grid: A modified auction-based approach. *IEEE Trans. Smart Grid* **2016**, *7*, 1462–1475. [CrossRef]
229. Doan, H.T.; Cho, J.; Kim, D. Peer-to-peer energy trading in smart grid through blockchain: A double auction-based game theoretic approach. *IEEE Access* **2021**, *9*, 49206–49218. [CrossRef]
230. Angaphiwatchawal, P.; Sompoh, C.; Chaitusaney, S. A Multi-k double auction pricing mechanism for peer-to-peer energy trading market of prosumers. In Proceedings of the 2021 18th International Conference on Electrical Engineering/Electronics, Computer, Telecommunications and Information Technology (ECTI-CON), Chiang Mai, Thailand, 19–22 May 2021.
231. Zhang, Z.; Tang, H.; Ren, J.; Huang, Q.; Lee, W.-J. Strategic prosumers-based peer-to-peer energy market design for community microgrids. *IEEE Trans. Ind. Appl.* **2021**, *57*, 2048–2057. [CrossRef]
232. Salehi, Z.; Chen, Y.; Petersen, I.R.; Shi, G.; Callaway, D.S.; Ratnam, E.L. Peer-to-Peer Energy Markets with Uniform Pricing: A Dynamic Operating Envelope Approach. *arXiv* **2025**, arXiv:2506.19328.
233. Kwag, K.; Shin, H.; Oh, H.; Yun, H.; Yoon, H.; Kim, W. Quantifying the impact and policy implications of transitioning to zonal and nodal pricing in the electricity market: A South Korean case study. *Appl. Sci.* **2025**, *15*, 716. [CrossRef]
234. Hu, J.; Mitchell, J.E.; Pang, J.-S.; Yu, B. On linear programs with linear complementarity constraints. *J. Glob. Optim.* **2012**, *53*, 29–51. [CrossRef]
235. Fernández-Blanco, R.; Arroyo, J.M.; Alguacil, N. Consumer payment minimization under uniform pricing: A mixed-integer linear programming approach. *Appl. Energy* **2014**, *114*, 676–686. [CrossRef]
236. Tierney, S.; Schatzki, T.; Mukerji, R. Uniform-Pricing versus Pay-as-bid in Wholesale Electricity Markets. Analysis Group ISO New York. 2008, 24p. Available online: https://g-city.sass.org.cn/_upload/article/files/7f/9e/6effc73d4950949afcbca0df3872/0dcf8fef-1149-428d-adb8-e0c8b0f1e18d.pdf (accessed on 15 August 2025).
237. Paudel, A.; Beng, G.H. A hierarchical peer-to-peer energy trading in community microgrid distribution systems. In Proceedings of the 2018 IEEE Power Energy Soc. Gen. Meet. (PESGM), Portland, OR, USA, 5–10 August 2018; IEEE: New York, NY, USA, 2018; pp. 1–5.
238. Rasheed, M.B.; Qureshi, M.A.; Javaid, N.; Alquthami, T. Dynamic pricing mechanism with the integration of renewable energy source in smart grid. *IEEE Access* **2020**, *8*, 16876–16892. [CrossRef]
239. Fabra, N. Market power and price discrimination: Learning from changes in renewables regulation. *EnergyEcoLab* **2020**. Available online: https://repository.graduateinstitute.ch/record/319894?v=pdf (accessed on 15 August 2025).
240. El Gohary, F.; Nyström, S.; Reitsma, L.; Bartusch, C. Identifying challenges in engaging users to increase self-consumption of electricity in microgrids. *Energies* **2021**, *14*, 1257. [CrossRef]
241. Malik, S. Peer-to-Peer Energy Trading in Microgrids: A Game-Theoretic Approach. Ph.D. Thesis, University of Galway, Galway, Ireland, 2024; pp. 1–144.
242. Tushar, W.; Yuen, C.; Smith, D.B.; Poor, H.V. Price discrimination for energy trading in smart grid: A game theoretic approach. *IEEE Trans. Smart Grid* **2016**, *8*, 1790–1801. [CrossRef]
243. Tushar, W.; Yuen, C.; Chai, B.; Smith, D.B.; Poor, H.V. Feasibility of using discriminate pricing schemes for energy trading in smart grid. In Proceedings of the IEEE Global Communication Conference, Austin, TX, USA, 8–12 December 2014; pp. 3138–3144.
244. PVcase. Why Locational Marginal Pricing (LMP) Data Is Essential for Solar Development. 2023. Available online: https://pvcase.com/blog/why-locational-marginal-pricing-lmp-data-is-essential-for-solar-development (accessed on 31 July 2025).
245. SlideServe. Locational Marginal Pricing Overview. 2023. Available online: https://www.slideserve.com/shina/locational-marginal-pricing-overview (accessed on 31 July 2025).
246. Thomaßen, G.; Fuhrmanek, A. Where to build renewables in Europe? The benefits of locational auction design. *Energy Econ.* **2025**, *147*, 108579. [CrossRef]
247. Hausman, E.; Fagan, R.; White, D.; Takahashi, K.; Napoleon, A.; Economics, S.E. *LMP Electricity Markets: Market Operations, Market Power*; Synapse Energy Economics: Cambridge, MA, USA, 2006.
248. Liu, H.; Tesfatsion, L.; Chowdhury, A. Derivation of Locational Marginal Prices for Restructured Wholesale Power Markets. 2009. Available online: https://faculty.sites.iastate.edu/tesfatsi/archive/tesfatsi/DerivationLMP.HLLTAC.pdf (accessed on 15 August 2025).

249. Wu, J.; Guo, X.; Zhu, Q.; Guo, J.; Han, Y.; Zhong, L.; Liu, S. Threshold effects and supply-demand ratios should be considered in the mechanisms driving ecosystem services. *Ecol. Indic.* **2022**, *142*, 109281. [CrossRef]
250. Alfaverh, F.; Denai, M.; Sun, Y. A dynamic peer-to-peer electricity market model for a community microgrid with price-based demand response. *IEEE Trans. Smart Grid* **2023**, *14*, 3976–3991. [CrossRef]
251. Crespo-Vazquez, J.L.; AlSkaif, T.; Gonzalez-Rueda, A.M.; Gibescu, M. A community-based energy market design using decentralized decision-making under uncertainty. *IEEE Trans. Smart Grid* **2020**, *12*, 1782–1793. [CrossRef]
252. Kim, S.-J.; Giannakis, G.B. An online convex optimization approach to real-time energy pricing for demand response. *IEEE Trans. Smart Grid* **2016**, *8*, 2784–2793. [CrossRef]
253. El Rahi, G.; Etesami, S.R.; Saad, W.; Mandayam, N.B.; Poor, H.V. Managing price uncertainty in prosumer-centric energy trading: A prospect-theoretic Stackelberg game approach. *IEEE Trans. Smart Grid* **2017**, *10*, 702–713. [CrossRef]
254. Kim, J.-Y.; Brünner, T.; Skiera, B.; Natter, M. A comparison of different pay-per-bid auction formats. *Int. J. Res. Mark.* **2014**, *31*, 368–379. [CrossRef]
255. Galgana, R.; Golrezaei, N. Learning in repeated multiunit pay-as-bid auctions. *Manuf. Serv. Oper. Manag.* **2025**, *27*, 200–229. [CrossRef]
256. Carlos Carbajal, J.; Dharanan, G.V.A. GVA, Pay-as-Bid Auctions with Budget Constrained Bidders. 2025. Available online: https://www.isid.ac.in/~acegd/acegd2024/papers/GVADharanan.pdf (accessed on 15 August 2025).
257. Song, Y.; Chu, Y.; Yoon, Y.; Jin, Y. Virtual Power Plant Bidding Strategies in Pay-as-Bid and Pay-as-Clear Markets: Analysis of Imbalance Penalties and Market Operations. *Energies* **2025**, *18*, 1383. [CrossRef]
258. Federico, G.; Rahman, D. Bidding in an electricity pay-as-bid auction. *J. Regul. Econ.* **2003**, *24*, 175–211. [CrossRef]
259. Rothkopf, M.H.; Teisberg, T.J.; Kahn, E.P. Why are Vickrey auctions rare? *J. Polit. Econ.* **1990**, *98*, 94–109. [CrossRef]
260. Evans, J.R. *Determining Consumer Perceptions of and Willingness to Pay for Appalachian Grass-Fed Beef: An Experimental Economics Approach*; West Virginia University: Morgantown, MV, USA, 2007.
261. Yassine, A.; Hossain, M.S.; Muhammad, G.; Guizani, M. Double auction mechanisms for dynamic autonomous electric vehicles energy trading. *IEEE Trans. Veh. Technol.* **2019**, *68*, 7466–7476. [CrossRef]
262. Sutterer, P.; Waldherr, S.; Bichler, M. Are truthful bidders paying too much? Efficiency and revenue in display ad auctions. *ACM Trans. Manag. Inf. Syst. (TMIS)* **2019**, *10*, 1–18. [CrossRef]
263. Potfer, M.; Baudry, D.; Richard, H.; Perchet, V.; Wan, C. Improved learning rates in multi-unit uniform price auctions. *Adv. Neural Inf. Process. Syst.* **2024**, *37*, 130237–130264.
264. Lamp, S.; Samano, M.; Tiedemann, S. Firms' Bidding Behavior in a New Market: Evidence from Renewable Energy Auctions. Avail. SSRN 4780866. 2024. Available online: https://papers.ssrn.com/sol3/papers.cfm?abstract_id=4780866 (accessed on 15 August 2025).
265. [CrossRef] Bertsimas, D.; Koulouras, A.G. Adaptive Pricing in Unit Commitment Under Load and Capacity Uncertainty. *arXiv* **2023**, arXiv:2309.08162. [CrossRef]
266. Just, S.; Weber, C. Pricing of reserves: Valuing system reserve capacity against spot prices in electricity markets. *Energy Econ.* **2008**, *30*, 3198–3221. [CrossRef]
267. Hummon, M.; Denholm, P.; Jorgenson, J.; Palchak, D.; Kirby, B.; Ma, O. *Fundamental Drivers of the Cost and Price of Operating Reserves*; National Renewable Energy Laboratory (NREL): Golden, CO, USA, 2013.
268. Rosen, C.; Madlener, R. An auction design for local reserve energy markets. *Decis. Support Syst.* **2013**, *56*, 168–179. [CrossRef]
269. Cramton, P. How best to auction natural resources. In *The Taxation of Petroleum and Minerals*; Routledge: London, UK, 2010; pp. 305–332.
270. Mehrtash, M.; Hobbs, B.F.; Ela, E. Reserve and energy scarcity pricing in United States power markets: A comparative review of principles and practices. *Renew. Sustain. Energy Rev.* **2023**, *183*, 113465. [CrossRef]
271. Nazir, I.; Mushtaq, N.; Ziafat, B.; Fareed, M.F.; Amin, W. A Reserved Pricing Mechanism for Nano/Peco Smart Grid Peer-to-Peer Energy Trading. In Proceedings of the 3rd International Conference Emerging Trends Electrical, Control, and Telecommunication Engineering (ETECTE), Lahore, Pakistan, 26–27 November 2024; pp. 1–5.
272. Deman, L.; Siddiqui, A.S.; Clastres, C.; Boucher, Q. Day-ahead and reserve prices in a renewable-based power system: Adapting electricity-market design for energy storage. *Energy J.* **2025**, *46*, 67–98. [CrossRef]
273. Ribeiro, L.; Street, A.; Arroyo, J.M.; Moreno, R. A Causation-Based Framework for Pricing and Cost Allocation of Energy, Reserves, and Transmission in Modern Power Systems. *arXiv* **2025**, arXiv:2505.24159. [CrossRef]
274. Qi, N.; Pinson, P.; Almassalkhi, M.R.; Cheng, L.; Zhuang, Y. Chance-constrained generic energy storage operations under decision-dependent uncertainty. *IEEE Trans. Sustain. Energy* **2023**, *14*, 2234–2248. [CrossRef]
275. Wang, Y.; Jiang, S.; Zhou, H.; Zhu, M.; Zhang, Y. A two-stage optimization strategy for VPP trading in multi-market considering setting method and marginal revenue and expenditure of standby capacity. *J. Renew. Sustain. Energy* **2025**, *17*. [CrossRef]
276. Chow, Y.-F.; McAleer, M.; Sequeira, J. Pricing of forward and futures contracts. *J. Econ. Surv.* **2000**, *14*, 215–253. [CrossRef]
277. Injadat, E.M.M. Futures and forwards contracts from perspective of Islamic law. *J. Econ. Polit. Econ.* **2014**, *1*, 241–252.

278. Landman, D.J. Real Option Analysis of Primary Rail Contracts in Grain Shipping. Master's Thesis, North Dakota State University, Fargo, ND, USA, 2017.
279. Munné-Collado, Í.; Bullich-Massagué, E.; Aragüés-Peñalba, M.; Olivella-Rosell, P. Local and micro power markets. In *Micro and Local Power Markets*; Wiley: Hoboken, NJ, USA, 2019; pp. 37–96.
280. Schlecht, I.; Maurer, C.; Hirth, L. Financial contracts for differences: The problems with conventional CfDs in electricity markets and how forward contracts can help solve them. *Energy Policy* **2024**, *186*, 113981. [CrossRef]
281. Ordoñez, C.C.; Organero, M.M.; Ramirez-Gonzalez, G.; Corrales, J.C. Smart Contracts as a Tool to Support the Challenges of Buying and Selling Coffee Futures Contracts in Colombia. *Agriculture* **2024**, *14*, 845. [CrossRef]
282. Ogundu, P.G. The Strategic Implications of Financial Derivatives in Hedging Corporate Exposure to Global Economic volatility. *World J. Adv. Res. Rev.* **2025**, *25*, 1218–1234. [CrossRef]
283. Dahlberg, J.; Rehula, L. The Effect of Hedging with Derivatives on Firm Performance: Evidence from European Energy Firms During the Global Energy Crisis in 2021–2023. 2024. Available online: https://www.diva-portal.org/smash/get/diva2:1877521/FULLTEXT01.pdf (accessed on 15 August 2025).
284. Mohandes, N.; Bayhan, S.; Sanfilippo, A.; Abu-Rub, H. Peer-to-peer trade and the sharing economy at distribution level: A review of the literature. *IEEE Access* **2023**, *11*, 122842–122858. [CrossRef]
285. Ahmadian-Yazdi, F.; Sokhanvar, A.; Roudari, S.; Tiwari, A.K. Dynamics of the relationship between stock markets and exchange rates during quantitative easing and tightening. *Financ. Innov.* **2025**, *11*, 51. [CrossRef]
286. Bensalah, N. Challenges and Opportunities in Energy Economics: Balancing Cost, Sustainability, and Innovation in the Global Energy Transition. *Mod. Econ.* **2024**, *15*, 1147–1180. [CrossRef]
287. Karaki, A.; Al-Fagih, L. Evolutionary game theory as a catalyst in smart grids: From theoretical insights to practical strategies. *IEEE Access* **2024**, *12*, 186926–186940. [CrossRef]
288. Wan, X.; Wen, X.; Tang, B.; Sun, Q. Trading strategy for virtual power plant clusters based on a multi-subject game model. *J. Comput. Methods Sci. Eng.* **2023**, *23*, 2261–2274. [CrossRef]
289. Izanlo, A.; Sheikholeslami, A.; Gholamian, S.A.; Kazemi, M.V.; Hosseini, S.N. A combination of MILP and game theory methods for P2P energy trading by considering network constraints. *Appl. Energy* **2024**, *374*, 123916. [CrossRef]
290. Wang, B.; Guo, X. Blockchain-enabled transformation: Decentralized planning and secure peer-to-peer trading in local energy networks. *Sustain. Energy Grids Netw.* **2024**, *40*, 101556. [CrossRef]
291. Malik, S.; Thakur, S.; Duffy, M.; Breslin, J.G. Comparative double auction approach for peer-to-peer energy trading on multiple microgrids. *Smart Grids Sustain. Energy* **2023**, *8*, 21. [CrossRef]
292. Zhang, W.; Ji, C.; Liu, Y.; Hao, Y.; Song, Y.; Cao, Y.; Qi, H. Dynamic interactions of carbon trading, green certificate trading, and electricity markets: Insights from system dynamics modeling. *PLoS ONE* **2024**, *19*, e0304478. [CrossRef]
293. Abate, A.G.; Majdi, D.; Kazempour, J.; Kamgarpour, M. Learning to bid in forward electricity markets using a no-regret algorithm. *Electr. Power Syst. Res.* **2024**, *234*, 110693. [CrossRef]
294. Amin, W.; Huang, Q.; Afzal, M.; Khan, A.A.; Umer, K.; Ahmed, S.A. A converging non-cooperative & cooperative game theory approach for stabilizing peer-to-peer electricity trading. *Electr. Power Syst. Res.* **2020**, *183*, 106278.
295. Raja, A.A.; Grammatico, S. Online coalitional games for real-time payoff distribution with applications to energy markets. *IEEE Trans. Energy Mark. Policy Regul.* **2023**, *1*, 97–106. [CrossRef]
296. Raja, A.A.; Grammatico, S. Bilateral peer-to-peer energy trading via coalitional games. *IEEE Trans. Ind. Inform.* **2022**, *19*, 6814–6824. [CrossRef]
297. Mitrea, D.; Chifu, V.R.; Cioara, T.; Anghel, I.; Pop, C.B. Social Factors in P2P Energy Trading Using Hedonic Games. *IEEE Access* **2024**, *12*, 71256–71270. [CrossRef]
298. Azar, A.G.; Nazaripouya, H.; Khaki, B.; Chu, C.-C.; Gadh, R.; Jacobsen, R.H. A non-cooperative framework for coordinating a neighborhood of distributed prosumers. *IEEE Trans. Ind. Inform.* **2018**, *15*, 2523–2534. [CrossRef]
299. Bandeiras, F.; Gomes, Á.; Gomes, M.; Coelho, P. Application and challenges of coalitional game theory in power systems for sustainable energy trading communities. *Energies* **2023**, *16*, 8115. [CrossRef]
300. Gerwin, C.; Mieth, R.; Dvorkin, Y. Compensation mechanisms for double auctions in peer-to-peer local energy markets. *Curr. Sustain. Renew. Energy Rep.* **2020**, *7*, 165–175. [CrossRef]
301. Emami, I.T.; Samani, E.; Abyaneh, H.A.; Mohsenian-Rad, H.; Bakhshai, A. A conceptual analysis of equilibrium bidding strategy in a combined oligopoly and oligopsony wholesale electricity market. *IEEE Trans. Power Syst.* **2022**, *37*, 4229–4243. [CrossRef]
302. Song, H.; Liu, C.-C.; Lawarrée, J. Nash equilibrium bidding strategies in a bilateral electricity market. *IEEE Trans. Power Syst.* **2002**, *17*, 73–79. [CrossRef]
303. Cheng, H.; Li, Y.; Jiang, H.; Sun, W.; Wei, C.; Huang, X. A Multi-Type Distributed Renewable Energy Aggregators Collaborative Operation Model Based on Non-Cooperative Game Theory. *J. Phys. Conf. Ser.* **2025**, *3015*, 012011. [CrossRef]
304. Mignoni, N. *Non-Cooperative Game Theoretical Control for Green and Efficient Energy Communities*; Politecnico di Bari: Bari, Italy, 2025.

305. Lei, Z.; Liu, M.; Shen, Z.; Lu, J.; Lu, Z. A Nash–Stackelberg game approach to analyze strategic bidding for multiple DER aggregators in electricity markets. *Sustain. Energy Grids Netw.* **2023**, *35*, 101111. [CrossRef]
306. Yu, X.; Pan, D.; Zhou, Y. A Stackelberg game-based peer-to-peer energy trading market with energy management and pricing mechanism: A case study in Guangzhou. *Sol. Energy* **2024**, *270*, 112388. [CrossRef]
307. Li, Y.; Yang, Y.; Zhang, F.; Li, Y. A Stackelberg game-based approach to load aggregator bidding strategies in electricity spot markets. *J. Energy Storage* **2024**, *95*, 112509. [CrossRef]
308. Xu, N.; Hu, Y.; Zhang, J.; Hao, R. Load-Side Shared Energy Storage New Energy Consumption Strategy Based on Nash Bargaining. In Proceedings of the 7th International Conference on Renewable Energy and Power Engineering, Beijing China, 25–27 September 2024; pp. 137–142.
309. Becchetti, L.; Salustri, F. Renewable energy communities and the ecological transition: A game theoretic bargaining approach. *Util. Policy* **2025**, *96*, 102006. [CrossRef]
310. Tanis, Z.; Durusu, A. Cooperative behaviors and multienergy coupling through distributed energy storage in the peer-to-peer market mechanism. *IEEE Access* **2025**, *13*, 12081–12102. [CrossRef]
311. Wang, Y.; Zhang, B.; Ma, J.; Jin, Q. Earning while learning: An adversarial multi-armed bandit based real-time bidding scheme in deregulated electricity market. *IEEE Trans. Netw. Sci. Eng.* **2022**, *9*, 3991–4000. [CrossRef]
312. Alsalloum, H.; Merghem-Boulahia, L.; Rahim, R. A systematical analysis on the dynamic pricing strategies and optimization methods for energy trading in smart grids. *Int. Trans. Electr. Energy Syst.* **2020**, *30*, e12404. [CrossRef]
313. Umer, K.; Huang, Q.; Khorasany, M.; Amin, W.; Afzal, M. A novel prosumer-centric approach for social welfare maximization considering network voltage constraints in peer-to-peer energy markets. *Int. J. Electr. Power Energy Syst.* **2023**, *147*, 108820. [CrossRef]
314. Nekrasov, S.A. Reducing costs for integration of renewable energy sources: A way to making renewable energy more accessible. *Therm. Eng.* **2021**, *68*, 593–603. [CrossRef]
315. Mushtaq, N.; Nazir, I.; Azad, M.; Ishfaq, H.; Fareed, M.F.; Amin, W. Optimal Selection of Trading Partner in Peer-to-Peer Energy Markets. In Proceedings of the 3rd International Conference Emerging Trends Electrical, Control, and Telecommunication Engineering (ETECTE), Lahore, Pakistan, 26–27 November 2024; pp. 1–5.
316. Camelo, M.M.; de Andrade, C.F.; de Athayde Prata, B. A mixed-integer linear programming model for optimizing green hydrogen supply chain networks. *Int. J. Hydrog. Energy* **2025**, *118*, 134–145. [CrossRef]
317. Yoon, S.-J.; Ryu, K.-S.; Kim, C.; Nam, Y.-H.; Kim, D.-J.; Kim, B. Optimal Bidding Scheduling of Virtual Power Plants Using a Dual-MILP (Mixed-Integer Linear Programming) Approach under a Real-Time Energy Market. *Energies* **2024**, *17*, 3773. [CrossRef]
318. Jin, X.; Liu, B.; Liao, S.; Cheng, C.; Zhao, Z.; Zhang, Y. Robust optimization for the self-scheduling and bidding strategies of a hydroproducer considering the impacts of crossing forbidden zones. *J. Water Resour. Plan. Manag.* **2023**, *149*, 05022017. [CrossRef]
319. Zhou, W.; Zheng, J.; Wang, H.; Yin, P.; Zhang, J.; Zhang, Y. Coordinated and optimized dispatch of smart building group-energy storage power station based on multiple games. In Proceedings of the 36th Chinese Control and Decision Conference, Xi'an, China, 25–27 May 2024; pp. 2180–2185.
320. Twaisan, K.; Barışçı, N. Integrated distributed energy resources (DER) and microgrids: Modeling and optimization of DERs. *Electronics* **2022**, *11*, 2816. [CrossRef]
321. Di Persio, L.; Alruqimi, M.; Garbelli, M. Stochastic Approaches to Energy Markets: From Stochastic Differential Equations to Mean Field Games and Neural Network Modeling. *Energies* **2024**, *17*, 6106. [CrossRef]
322. Zheng, K.; Sun, Z.; Song, Y.; Zhang, C.; Zhang, C.; Chang, F.; Yang, D.; Fu, X. Stochastic scenario generation methods for uncertainty in wind and photovoltaic power outputs: A comprehensive review. *Energies* **2025**, *18*, 503. [CrossRef]
323. Tang, J.; Qian, B.; Luo, Y.; Lin, X.; Zhou, M.; Zhang, F.; Wang, H. Evolutionary Game Theory-Based Analysis of Power Producers' Carbon Emission Reduction Strategies and Multi-Group Bidding Dynamics in the Low-Carbon Electricity Market. *Processes* **2025**, *13*, 952. [CrossRef]
324. Yu, J.; Liu, S.; Zou, Y.; Wang, G.; Hu, C. Auction Theory and Game Theory Based Pricing of Edge Computing Resources: A Survey. *IEEE Internet Things J.* **2025**, *12*, 32394–32418. [CrossRef]
325. Amin, W.; Ahmad, F.; Umer, K.; Khawaja, A.H.; Afzal, M.; Ahmed, S.A.; Chaitusaney, S. P2p energy trading: An optimal solution for energy shortage in pakistan. *IEEE Access* **2022**, *10*, 89614–89633. [CrossRef]
326. Zheng, Y.; Wang, Y.; Yang, Q. Bidding strategy design for electric vehicle aggregators in the day-ahead electricity market considering price volatility: A risk-averse approach. *Energy* **2023**, *283*, 129138. [CrossRef]
327. Rousta, M.; Güsewell, J.; Eltrop, L. Integrated Production and Multi-Market Optimization of Biomethane in Germany: A Two-Step Linear Programming Approach. *Energies* **2025**, *18*, 2991. [CrossRef]
328. Pinthanon, P.; Rianmora, S.; Chanchiewvichai, W. Decision-Making Platform for Design Optimization by Integrating Linear Programming and House of Quality. *Eng. J.* **2025**, *29*, 95–122. [CrossRef]
329. Kuram, Ç.; Bilişik, Ö.N. A mixed-integer linear programming model proposal to determine material sustainability for new product development processes in production. *Comput. Chem. Eng.* **2025**, *198*, 109108. [CrossRef]

330. Liaquat, S.; Hussain, T.; Kassab, F.A.; Ali, A.; Celik, B.; Fourney, R.; Hansen, T.M. An Integrated Two-Stage Hybrid P2P-DR Transactive Energy Trading Platform Using Iterative Distributed-Mixed Integer Linear Optimisation. *IET Smart Grid* **2025**, *8*, e70027. [CrossRef]
331. Deng, J.; He, F.; Zeng, Q.; Yan, J.; Liu, R.; He, D.; Zhou, S. Coordinated Optimal Dispatch of Distribution Grids and P2P Energy Trading Markets. *Energy Sci. Eng.* **2025**, *13*, 2206–2219. [CrossRef]
332. Yaghoubi, M. Resilient Peer-to-Peer Energy Trading: Addressing Price Uncertainty and Communication Delays in Decentralized Markets. Ph.D. Thesis, École de Technologie Supérieure, Montréal, QC, Canada, 2025.
333. Akülker, H.; Alakent, B.; Aydin, E. A two-stage stochastic MINLP model to design and operate a multi-energy microgrid by addressing carbon emission regulatory policies uncertainty. *arXiv* **2025**, arXiv:2501.18988. [CrossRef]
334. Kassab, F.A. Co-Optimisation of the Sizing and Control of an Urban Microgrid. Ph.D. Thesis, Université de Technologie de Compiègne, Compiègne, France, 2024.
335. Ling, J.Q.; Tan, W.-S.; Ding, Z.Y.; Wu, Y.-K. Multi-Objective Optimization of Cloud Energy Storage Placement and Sizing in Peer-to-Peer Energy Market Under Line Congestion. In Proceedings of the 2025 IEEE Industry Applications Society Annual Meeting (IAS), Taipei, Taiwan, 15–20 June 2025; IEEE: New York, NY, USA, 2025; pp. 1–8.
336. dos Santos, S.A.B.; Coutinho, L.R.R.; Tofoli, F.L.; Barroso, G.C. Community energy management system for residential energy communities integrating demand response, distributed generation, and energy storage systems. *J. Energy Storage* **2025**, *132*, 117832. [CrossRef]
337. Ahmadi Kamarposhti, M.; Ghandour, R.; Abdel-Aty, M.; Hafez, M.; Alfiras, M.; Alkhazaleh, S.; Colak, I.; Solyman, A. Optimizing capacitor bank placement in distribution networks using a multi-objective particle swarm optimization approach for energy efficiency and cost reduction. *Sci. Rep.* **2025**, *15*, 12332. [CrossRef]
338. Vaithiyanathan, B.; Jothi, K.R.; Paramasivam, S.K. Develop a Hybrid Intelligent Fuzzy Self-Adaptive Mutated Genetic Algorithm and Deep Reinforcement Learning for Efficient Home Energy Management. *Trait. Signal* **2025**, *42*, 2. [CrossRef]
339. Zubin, J.B.; Sunitha, R.; Pathirikkat, G. Integrated Bidding and Battery Scheduling in a Microgrid for Sealed-Bid Double Auction Power Trading with Peer Microgrids under Uncertainty and Its Blockchain-Based Implementation. *IEEE Access* **2025**, *13*, 117953–117970.
340. Li, S.; Gu, N.; Wu, C. Electricity Tariff Insurance Design Under Real-Time Pricing. In Proceedings of the 7th Conference on Energy Internet & Energy System Integration, Hangzhou, China, 15–18 December 2023; pp. 3172–3177
341. Yang, A.; Li, A.; Li, Z.; Sun, Y.; Gao, J. Investment Analysis of Low-Carbon Yard Cranes: Integrating Monte Carlo Simulation and Jump Diffusion Processes with a Hybrid American–European Real Options Approach. *Energies* **2025**, *18*, 1928. [CrossRef]
342. Kluge, T. Pricing Derivatives in Stochastic Volatility Models Using the Finite Difference Method. Master's Thesis, Technische Universität Chemnitz, Chemnitz, Germany, 2002.
343. Baamonde-Seoane, M.A. Modelling, Mathematical Analysis and Numerical Simulation to Value Derivatives Related to Renewable Energy Certificates. 2021. Available online: https://ruc.udc.es/entities/publication/21889049-7496-4f83-8f44-3333e810ed59. (accessed on 15 August 2025).
344. Voronin, S.; Partanen, J. Price forecasting in the day-ahead energy market by an iterative method with separate normal price and price spike frameworks. *Energies* **2013**, *6*, 5897–5920. [CrossRef]
345. Kim, H.J.; Kim, M.K. Data-Driven Virtual Power Plant Bidding Strategy in Electricity Markets Integrating Hybrid Forecasting Model and Customized Incentive Demand Response. *IEEE Internet Things J.* **2025**, *12*, 13851–13869. [CrossRef]
346. Oprea, S.-V.; Bara, A. Two-step price mechanism using Newton-Raphson method and peer-to-peer mediation for local electricity markets. *Energy Strategy Rev.* **2025**, *59*, 101701. [CrossRef]
347. Shiltz, D.J.; Cvetković, M.; Annaswamy, A.M. An integrated dynamic market mechanism for real-time markets and frequency regulation. *IEEE Trans. Sustain. Energy* **2015**, *7*, 875–885. [CrossRef]
348. Lu, J.; Xie, Z.; Xu, H.; Liu, J. Optimizing Joint Bidding and Incentivizing Strategy for Price-Maker Load Aggregators Based on Multi-Task Multi-Agent Deep Reinforcement Learning. *IEEE Access* **2024**, *12*, 63988–164001. [CrossRef]
349. Tang, C.; Qin, Y.; Wu, F.; Tang, Z. Dynamic demand-aware power grid intelligent pricing algorithm based on deep reinforcement learning. *IEEE Access* **2024**, *12*, 75809–75817. [CrossRef]
350. Basnet, M.R.; Bryan, J.A.; Dana, S.J.; Meek, A.S.; Wang, H.; Talbot, P. Stochastic optimization and uncertainty quantification of Natrium-based nuclear-renewable energy Systems for Flexible Power Applications in deregulated markets. *Appl. Energy* **2024**, *375*, 124105. [CrossRef]
351. Wang, Y.; Lin, T. A novel deterministic probabilistic forecasting framework for gold price with a new pandemic index based on quantile regression deep learning and multi-objective optimization. *Mathematics* **2023**, *12*, 29. [CrossRef]
352. Nizharadze, N.; Farokhi Soofi, A.; Manshadi, S. Predicting the gap in the day-ahead and real-time market prices leveraging exogenous weather data. *Algorithms* **2023**, *16*, 508. [CrossRef]
353. Li, W.; Zhang, N.; Liu, Z.; Ma, S.; Ke, H.; Wang, J.; Chen, T. MLfus: A real-time forecasting architecture for low communication costs in electricity IoT based on ensemble learning. *IET Commun.* **2023**, *17*, 145–161. [CrossRef]

354. Alhosani, A. Development and Application of Forecasting Machine Learning Approach for Renewable Energy-Related Products' Prices. Ph.D. Thesis, Khalifa University of Science, Abu Dhabi, United Arab Emirates, 2023.
355. Saeed, N.; Shafi, I.; Pervez, S.; Thompson, E.B.; Castilla, A.K.; Samad, M.A.; Ashraf, I. Intelligent Decision Making for Commodities Price Prediction: Opportunities, Challenges and Future Avenues. *Comput. Econ.* **2025**, 1–59. [CrossRef]
356. Liu, F.; Li, X. Integrating deep reinforcement learning with evolutionary algorithms for advanced optimization in smart city energy management. *IEEE Access* **2024**, *12*, 177103–177118. [CrossRef]
357. Jeong, J.; Kim, S.W.; Kim, H. Deep reinforcement learning based real-time renewable energy bidding with battery control. *IEEE Trans. Energy Mark., Policy Regul.* **2023**, *1*, 85–96. [CrossRef]
358. Yadav, D.; Mekhilef, S.; Singh, B.; Rawa, M. Deep Reinforcement Learning and Energy Price Prediction. In *Artificial Intelligence-Based Smart Power Systems*; Wiley Online Library: Hoboken, NJ, USA, 2023; pp. 207–231.
359. Song, J.; Zheng, W.; Chen, M.; Qi, L.; Zou, Q. Analysis of collusive bidding characteristics in the power spot market based on multi-agent reinforcement learning simulation. *Electr. Eng.* **2025**, *107*, 10069–10090. [CrossRef]
360. Alavi, S.A.; Mehran, K.; Vahidinasab, V.; Catalão, J.P.S. Forecast-based consensus control for DC microgrids using distributed long short-term memory deep learning models. *IEEE Trans. Smart Grid* **2021**, *12*, 3718–3730. [CrossRef]
361. Aslam, S.; Ayub, N.; Farooq, U.; Alvi, M.J.; Albogamy, F.R.; Rukh, G.; Haider, S.I.; Azar, A.T.; Bukhsh, R. Towards electric price and load forecasting using cnn-based ensembler in smart grid. *Sustainability* **2021**, *13*, 12653. [CrossRef]
362. Bakir, A.; Rami, A. Enhanced electricity price forecasting in smart grids using an optimized hybrid convolutional Multi-Layer Perceptron deep network with Marine Predators Algorithm for feature selection. *Energy Sources Part B Econ. Plan. Policy* **2025**, *20*, 2456058. [CrossRef]
363. Guo, X.; Zhao, Q.; Zheng, D.; Ning, Y.; Gao, Y. A short-term load forecasting model of multi-scale CNN-LSTM hybrid neural network considering the real-time electricity price. *Energy Rep.* **2020**, *6*, 1046–1053. [CrossRef]
364. Zhu, Z. *Applications of Reinforcement Learning on Deregulated Active Distribution Networks*; Hong Kong Polytechnic University: Hong Kong, China, 2023.
365. Meydani, A.; Shahinzadeh, H.; Ramezani, A.; Moazzami, M.; Nafisi, H.; Askarian-Abyaneh, H. Comprehensive review of artificial intelligence applications in smart grid operations. In Proceedings of the 2024 9th International Conference on Technology Energy Management (ICTEM), 445 Hoes Lane, Piscataway, NJ, USA, 14–15 February 2024; pp. 1–13.
366. Richarz, J.; Henn, S.; Osterhage, T.; Müller, D. Optimal scheduling of modernization measures for typical non-residential buildings. *Energy* **2022**, *238*, 121871. [CrossRef]
367. Alghamdi, H.; Hua, L.-G.; Hafeez, G.; Murawwat, S.; Bouazzi, I.; Alghamdi, B. Optimal adaptive heuristic algorithm based energy optimization with flexible loads using demand response in smart grid. *PLoS ONE* **2024**, *19*, e0307228. [CrossRef] [PubMed]
368. Gronier, T.; Fitó, J.; Franquet, E.; Gibout, S.; Ramousse, J. Iterative sizing of solar-assisted mixed district heating network and local electrical grid integrating demand-side management. *Energy* **2022**, *238*, 121517. [CrossRef]
369. Li, J.; Mi, Z.; Yu, Y. Research on the classification and evaluation of large-scale user load patterns for grid supply–demand balance regulation demands. *Appl. Energy* **2025**, *397*, 126301. [CrossRef]
370. Yu, S.; Fan, Y.; Shi, Z.; Zhang, J.; Zhang, T.; Zhang, J.; Liu, Z. Innovative heat management method and metaheuristic algorithm optimized power supply–demand balance for PEMFC–ASHP–CHP system. *Appl. Energy* **2024**, *371*, 123778. [CrossRef]
371. Chandlekar, S.; Boroju, A.; Jain, S.; Gujar, S. A novel demand response model and method for peak reduction in smart grids–PowerTAC. *arXiv* **2023**, arXiv:2302.12520.
372. Yu, M.; Hong, S.H. Supply–demand balancing for power management in smart grid: A Stackelberg game approach. *Appl. Energy* **2016**, *164*, 702–710. [CrossRef]
373. Chen, Q.; Kumar, P.; Baskonus, H.M. Modeling and analysis of demand–supply dynamics with a collectability factor using delay differential equations in economic growth via the Caputo operator. *AIMS Math.* **2024**, *9*, 7471–7491. [CrossRef]
374. Toderean, L.; Cioara, T.; Anghel, I.; Sarmas, E.; Michalakopoulos, V.; Marinakis, V. Demand response optimization for smart grid integrated buildings: Review of technology enablers landscape and innovation challenges. *Energy Build.* **2025**, *326*, 115067. [CrossRef]
375. Rashid, A.; Biswas, P.; Nasim, M.D.; Gupta, K.D. Power plays: Unleashing machine learning magic in smart grids. *arXiv* **2024**, arXiv:2410.15423. [CrossRef]
376. Du, Y.; Wu, J.; Li, S.; Long, C.; Onori, S. Hierarchical coordination of two-time scale microgrids with supply–demand imbalance. *IEEE Trans. Smart Grid* **2020**, *11*, 3726–3736. [CrossRef]
377. Ghorbanian, M.; Dolatabadi, S.H.; Siano, P. Game theory-based energy-management method considering autonomous demand response and distributed generation interactions in smart distribution systems. *IEEE Syst. J.* **2020**, *15*, 905–914. [CrossRef]
378. Naughton, J.; Wang, H.; Cantoni, M.; Mancarella, P. Co-optimizing virtual power plant services under uncertainty: A robust scheduling and receding horizon dispatch approach. *IEEE Trans. Power Syst.* **2021**, *36*, 3960–3972. [CrossRef]

379. Cheng, L.; Wei, X.; Li, M.; Tan, C.; Yin, M.; Shen, T.; Zou, T. Integrating evolutionary game-theoretical methods and deep reinforcement learning for adaptive strategy optimization in user-side electricity markets: A comprehensive review. *Mathematics* **2024**, *12*, 3241. [CrossRef]
380. Wen, Y.; Lu, Y.; Gou, J.; Liu, F.; Tang, Q.; Wang, R. Robust transmission expansion planning of ultrahigh-voltage AC–DC hybrid grids. *IEEE Trans. Ind. Appl.* **2022**, *58*, 3294–3302. [CrossRef]
381. Lee, J.-W.; Kim, M.-K. An evolutionary game theory-based optimal scheduling strategy for multiagent distribution network operation considering voltage management. *IEEE Access* **2022**, *10*, 50227–50241. [CrossRef]
382. Zhang, X.; Cai, Z.; Yang, X.; Luo, J.; Zhang, R.; Zhang, Y. Game theory-based demand-side management for efficient energy collaboration in smart networks: A neighborhood-scale optimization framework. *Electr. Eng.* **2024**, *107*, 9071–9088. [CrossRef]
383. Chen, P.; Han, L.; Xin, G.; Zhang, A.; Ren, H.; Wang, F. Game theory based optimal pricing strategy for V2G participating in demand response. *IEEE Trans. Ind. Appl.* **2023**, *59*, 4673–4683. [CrossRef]
384. Ding, Y.; Xie, D.; Hui, H.; Xu, Y.; Siano, P. Game-theoretic demand side management of thermostatically controlled loads for smoothing tie-line power of microgrids. *IEEE Trans. Power Syst.* **2021**, *36*, 4089–4101. [CrossRef]
385. Yadav, R.K.; Bhadoria, V.S.; Hrisheekesha, P.N. Demand side management using ant colony optimization algorithm in renewable energy integrated smart grid. *J. Intell. Fuzzy Syst.* **2024**, *46*, 7627–7642. [CrossRef]
386. Nanibabu, S.; Baskaran, S.; Marimuthu, P. Realisation of demand side management in smart distribution grid with commercial loads using butterfly optimization algorithm. In Proceedings of the Third International Conference on Intelligent Techniques in Control, Optimization and Signal Processing (INCOS), Tamil Nadu, India, 14–16 March 2024; pp. 1–6.
387. Maity, S.; Chanda, S.; Rai, A.K.; Biswas, P.; De, A. A DE modified quantum PSO based demand response program for social welfare optimization with scalable renewable energy penetration in smart grid. *SAMRIDDHI A J. Phys. Sci. Eng. Technol.* **2024**, *16*, 9–19. [CrossRef]
388. Xiong, Q.; Chen, Y.; Song, L.; Huang, Z.; Dong, C.; Zou, G.; Yu, S. Research on multi-scenario source–grid–load–storage optimization and comprehensive optimization strategies considering user satisfaction and economy. In Proceedings of the 6th International Conference on Energy, 2314 Power and Grid (ICEPG), Guangzhou, China, 27–29 September 2024; pp. 1717–1723.
389. Woo, Y.-B.; Moon, I. Bilevel optimization for multi-user systems with mixed demand response programs for enhanced operational efficiency in electric power grids. *Appl. Energy* **2025**, *399*, 126507. [CrossRef]
390. Xu, W.; Zhou, D.; Huang, X.; Lou, B.; Liu, D. Optimal allocation of power supply systems in industrial parks considering multi-energy complementarity and demand response. *Appl. Energy* **2020**, *275*, 115407. [CrossRef]
391. Xinfa, T.; Jingjing, W.; Yonghua, W.; Youwei, W. The optimization of supply–demand balance dispatching and economic benefit improvement in a multi-energy virtual power plant within the Jiangxi power market. *Energies* **2024**, *17*, 4691. [CrossRef]
392. Silva, W.N.; Henrique, L.F.; Silva, A.F.P.d.C.; Dias, B.H.; Soares, T.A. Market models and optimization techniques to support the decision-making on demand response for prosumers. *Electr. Power Syst. Res.* **2022**, *210*, 108059. [CrossRef]
393. Zhang, Z.; Yang, H.; Zhang, Z.; Wang, Y. Optimization model of supply and demand matching of energy right products based on pre-listing transactions. In Proceedings of the 4th International Conference on Big Data Economy and Information Management, Tianjin, China, 27–29 October 2023; pp. 704–708.
394. Stanelyte, D.; Radziukyniene, N.; Radziukynas, V. Overview of demand-response services: A review. *Energies* **2022**, *15*, 1659. [CrossRef]
395. Honarmand, M.E.; Hosseinnezhad, V.; Hayes, B.; Shafie-Khah, M.; Siano, P. An overview of demand response: From its origins to the smart energy community. *IEEE Access* **2021**, *9*, 96851–96876. [CrossRef]
396. Yang, P.; Jiang, H.; Liu, C.; Kang, L.; Wang, C. Coordinated optimization scheduling operation of integrated energy system considering demand response and carbon trading mechanism. *Int. J. Electr. Power Energy Syst.* **2023**, *147*, 108902. [CrossRef]
397. Baldivieso Monasterios, P.R.; Verba, N.; Morris, E.A.; Morstyn, T.; Gaura, E.; McArthur, S. Incorporating forecasting and peer-to-peer negotiation frameworks into a distributed model predictive control approach for meshed electric networks. *arXiv* **2021**, arXiv:2105.14547.
398. Sampath, L.P.M.I.; Paudel, A.; Nguyen, H.D.; Foo, E.Y.S.; Gooi, H.B. Peer-to-peer energy trading enabled optimal decentralized operation of smart distribution grids. *IEEE Trans. Smart Grid* **2021**, *13*, 654–666. [CrossRef]
399. Yan, M.; Shahidehpour, M.; Paaso, A.; Zhang, L.; Alabdulwahab, A.; Abusorrah, A. Distribution network-constrained optimization of peer-to-peer transactive energy trading among multi-microgrids. *IEEE Trans. Smart Grid* **2020**, *12*, 1033–1047. [CrossRef]
400. Li, J.; Khodayar, M.E.; Wang, J.; Zhou, B. Data-driven distributionally robust co-optimization of P2P energy trading and network operation for interconnected microgrids. *IEEE Trans. Smart Grid* **2021**, *12*, 5172–5184. [CrossRef]
401. Pop, C.; Antal, M.; Cioara, T.; Anghel, I.; Salomie, I.; Bertoncini, M. A fog computing enabled virtual power plant model for delivery of frequency restoration reserve services. *Sensors* **2019**, *19*, 4688. [CrossRef]
402. Wang, W.; Chen, P.; Zeng, D.; Liu, J. Electric vehicle fleet integration in a virtual power plant with large-scale wind power. *IEEE Trans. Ind. Appl.* **2020**, *56*, 5924–5931. [CrossRef]

403. Van Summeren, L.F.M.; Wieczorek, A.J.; Bombaerts, G.J.T.; Verbong, G.P.J. Community energy meets smart grids: Reviewing goals, structure, and roles in virtual power plants in Ireland, Belgium and the Netherlands. *Energy Res. Soc. Sci.* **2020**, *63*, 101415. [CrossRef]
404. Phyo, P.-P.; Jeenanunta, C. Advanced ML-based ensemble and deep learning models for short-term load forecasting: Comparative analysis using feature engineering. *Appl. Sci.* **2022**, *12*, 4882. [CrossRef]
405. Sivakumar, S.A. Predictive analytics for demand response management with AI. *Acta Energetica* **2024**, *2*, 12–22.
406. Khan, M.A.; Saleh, A.M.; Waseem, M.; Sajjad, I.A. Artificial intelligence enabled demand response: Prospects and challenges in smart grid environment. *IEEE Access* **2022**, *11*, 1477–1505. [CrossRef]
407. Aguiar-Pérez, J.M.; Pérez-Juárez, M.A. An insight of deep learning based demand forecasting in smart grids. *Sensors* **2023**, *23*, 1467. [CrossRef]
408. Sangoleye, F.; Jao, J.; Faris, K.; Tsiropoulou, E.E.; Papavassiliou, S. Reinforcement learning-based demand response management in smart grid systems with prosumers. *IEEE Syst. J.* **2023**, *17*, 1797–1807. [CrossRef]
409. Wen, X.; Liao, J.; Niu, Q.; Shen, N.; Bao, Y. Deep learning-driven hybrid model for short-term load forecasting and smart grid information management. *Sci. Rep.* **2024**, *14*, 13720. [CrossRef]
410. El Maghraoui, A.; El Hadraoui, H.; Ledmaoui, Y.; El Bazi, N.; Guennouni, N.; Chebak, A. Revolutionizing smart grid-ready management systems: A holistic framework for optimal grid reliability. *Sustain. Energy Grids Netw.* **2024**, *39*, 101452. [CrossRef]
411. Ma, R.; Yi, Z.; Xiang, Y.; Shi, D.; Xu, C.; Wu, H. A blockchain-enabled demand management and control framework driven by deep reinforcement learning. *IEEE Trans. Ind. Electron.* **2022**, *70*, 430–440. [CrossRef]
412. Dascalu, A.; Cruden, A.J.; Sharkh, S.M. Experimental investigations into a hybrid energy storage system using directly connected lead-acid and Li-ion batteries. *Energies* **2024**, *17*, 4726. [CrossRef]
413. Szott, M.; Wermiński, S.; Jarnut, M.; Kaniewski, J.; Benysek, G. Battery energy storage system for emergency supply and improved reliability of power networks. *Energies* **2021**, *14*, 720. [CrossRef]
414. Diaz-Gonzalez, F.; Chillón-Antón, C.; Llonch-Masachs, M.; Galceran-Arellano, S.; Rull-Duran, J.; Bergas-Jane, J.; Bullich-Massagué, E. A hybrid energy storage solution based on supercapacitors and batteries for the grid integration of utility scale photovoltaic plants. *J. Energy Storage* **2022**, *51*, 104446. [CrossRef]
415. Zhang, L.; Hu, X.; Wang, Z.; Ruan, J.; Ma, C.; Song, Z.; Dorrell, D.G.; Pecht, M.G. Hybrid electrochemical energy storage systems: An overview for smart grid and electrified vehicle applications. *Renew. Sustain. Energy Rev.* **2021**, *139*, 110581. [CrossRef]
416. Wang, Z.; Tuo, X.; Zhou, J.; Xiao, G. Performance study of large capacity industrial lead-carbon battery for energy storage. *J. Energy Storage* **2022**, *55*, 105398. [CrossRef]
417. Han, K.B.; Jung, J.; Kang, B.O. Real-time load variability control using energy storage system for demand-side management in South Korea. *Energies* **2021**, *14*, 6292. [CrossRef]
418. Kebede, A.A.; Kalogiannis, T.; Van Mierlo, J.; Berecibar, M. A comprehensive review of stationary energy storage devices for large scale renewable energy sources grid integration. *Renew. Sustain. Energy Rev.* **2022**, *159*, 112213. [CrossRef]
419. Theocharis, A.; Hamanee, S. Battery storage at the secondary distribution electricity grid by investigating end-users load demand measurements. *Energies* **2022**, *15*, 2743. [CrossRef]
420. Barakat, S.; Emam, A.; Samy, M.M. Investigating grid-connected green power systems' energy storage solutions in the event of frequent blackouts. *Energy Rep.* **2022**, *8*, 5177–5191. [CrossRef]
421. Phogat, P.; Dey, S.; Wan, M. Powering the sustainable future: A review of emerging battery technologies and their environmental impact. *RSC Sustain.* **2025**, *3*, 3266–3306. [CrossRef]
422. Zeng, Y.; Li, X.; Fu, J.; Shen, P.K.; Zhang, X. Corrigendum: High-energy-density and high-rate membraneless Ni–Fe battery enabled by a 1-butyl-3-methylimidazolium bromide-based gel electrolyte. *ChemElectroChem* **2023**, *10*, e202300354. [CrossRef]
423. Sun, H.; Wang, C.; Qi, Z.; Hu, W.; Zhang, Z. Nanostructure nickel-based selenides as cathode materials for hybrid battery–supercapacitors. *Front. Chem.* **2021**, *8*, 611032. [CrossRef]
424. Salkuti, S.R. Electrochemical batteries for smart grid applications. *Int. J. Electr. Comput. Eng.* **2021**, *11*, 1849–1856. [CrossRef]
425. Li, Y.; Tan, Z.; Liu, Y.; Lei, C.; He, P.; Li, J.; He, Z.; Cheng, Y.; Wu, F.; Li, Y. Past, present and future of high-nickel materials. *Nano Energy* **2024**, *119*, 109070. [CrossRef]
426. Li, F.; Wang, D.; Liu, D.; Yang, S.; Sun, K.; Liu, Z.; Yu, H.; Qin, J. A comprehensive review on energy storage system optimal planning and benefit evaluation methods in smart grids. *Sustainability* **2023**, *15*, 9584. [CrossRef]
427. Huang, Z.; Song, B.; Zhang, H.; Feng, F.; Zhang, W.; Lu, K.; Chen, Q. High-capacity and stable sodium–sulfur battery enabled by confined electrocatalytic polysulfides full conversion. *Adv. Funct. Mater.* **2021**, *31*, 2100666. [CrossRef]
428. Wang, L.; Wang, T.; Peng, L.; Wang, Y.; Zhang, M.; Zhou, J.; Chen, M.; Cao, J.; Fei, H.; Duan, X.; et al. The promises, challenges and pathways to room-temperature sodium–sulfur batteries. *Natl. Sci. Rev.* **2022**, *9*, nwab050. [CrossRef] [PubMed]
429. Wang, Y.; Zhang, Y.; Cheng, H.; Ni, Z.; Wang, Y.; Xia, G.; Li, X.; Zeng, X. Research progress toward room temperature sodium–sulfur batteries: A review. *Molecules* **2021**, *26*, 1535. [CrossRef] [PubMed]

430. Wu, J.; Tian, Y.; Gao, Y.; Gao, Z.; Meng, Y.; Wang, Y.; Wang, X.; Zhou, D.; Kang, F.; Li, B.; et al. Rational electrolyte design toward cyclability remedy for room-temperature sodium–sulfur batteries. *Angew. Chem. Int. Ed.* **2022**, *61*, e202205416. [CrossRef]
431. Yan, Z.; Zhao, L.; Wang, Y.; Zhu, Z.; Chou, S.-L. The future for room-temperature sodium–sulfur batteries: From persisting issues to promising solutions and practical applications. *Adv. Funct. Mater.* **2022**, *32*, 2205622. [CrossRef]
432. Murugan, S.; Klostermann, S.V.; Schützendübe, P.; Richter, G.; Kästner, J.; Buchmeiser, M.R. Stable cycling of room-temperature sodium–sulfur batteries based on an in situ crosslinked gel polymer electrolyte. *Adv. Funct. Mater.* **2022**, *32*, 2201191. [CrossRef]
433. Zhang, Y.; Guo, X.; Yang, Q.; Shao, Y.; Du, Y.; Qi, J.; Zhao, M.; Shang, Z.; Hao, Y.; Tang, Y.; et al. Chemical and spatial dual-confinement engineering for stable Na–S batteries with approximately 100% capacity retention. *Proc. Natl. Acad. Sci. USA* **2023**, *120*, e2314408120. [CrossRef]
434. Basel, J.; Sapkota, N.; Parekh, M.; Rao, A.M. Electrolyte optimization for sodium–sulfur batteries. *Appl. Phys. Lett.* **2024**, *124*, 123901. [CrossRef]
435. Liu, D.; Li, Z.; Li, X.; Chen, X.; Li, Z.; Yuan, L.; Huang, Y. Stable room-temperature sodium–sulfur batteries in ether-based electrolytes enabled by the fluoroethylene carbonate additive. *ACS Appl. Mater. Interfaces* **2022**, *14*, 6658–6666. [CrossRef]
436. Zhang, H.; Gao, Y.; Liu, X.; Zhou, L.; Li, J.; Xiao, Y.; Peng, J.; Wang, J.; Chou, S.-L. Long-cycle-life cathode materials for sodium-ion batteries toward large-scale energy storage systems. *Adv. Energy Mater.* **2023**, *13*, 2300149. [CrossRef]
437. Singh, A.N.; Islam, M.; Meena, A.; Faizan, M.; Han, D.; Bathula, C.; Hajibabaei, A.; Anand, R.; Nam, K.-W. Unleashing the potential of sodium-ion batteries: Current state and future directions for sustainable energy storage. *Adv. Funct. Mater.* **2023**, *33*, 2304617. [CrossRef]
438. Zhu, Q.; Wu, J.; Li, W.; Hu, X.; Tian, N.; He, L.; Li, Y. Boosting sodium-ion battery performance by anion doping in NASICON $Na_4MnCr(PO_4)_3$ cathode. *J. Colloid Interface Sci.* **2024**, *663*, 191–202. [CrossRef]
439. Tyagaraj, H.B.; Marje, S.J.; Ranjith, K.S.; Hwang, S.-K.; Al Ghaferi, A.; Chodankar, N.R.; Huh, Y.S.; Han, Y.-K. Sodium-ion batteries: Charge storage mechanisms and recent advancements in diglyme-based electrolytes. *J. Energy Storage* **2023**, *74*, 109411. [CrossRef]
440. Pahari, D.; Verma, P.; Puravankara, S. Are Na-ion batteries nearing the energy storage tipping point?—Current status of non-aqueous, aqueous, and solid-state Na-ion battery technologies for sustainable energy storage. *J. Energy Storage* **2022**, *56*, 105961.
441. Ma, W.-L.; Zhou, Y.; Zhao, X.-W.; Cao, X.; Wu, P.; Zhu, X.-S.; Wei, S.-H.; Sun, K.; Zhou, H.; Zhou, Y.-M. Ultra-fast-charging, long-duration, and wide-temperature-range sodium storage enabled by multiwalled carbon nanotube-hybridized biphasic polyanion-type phosphate cathode materials. *ACS Appl. Mater. Interfaces* **2024**, *16*, 34819–34829. [CrossRef]
442. Parthasarathy, C.; Laaksonen, H.; Redondo-Iglesias, E.; Pelissier, S. Aging aware adaptive control of Li-ion battery energy storage system for flexibility services provision. *J. Energy Storage* **2023**, *57*, 106268. [CrossRef]
443. Rouholamini, M.; Wang, C.; Nehrir, H.; Hu, X.; Hu, Z.; Aki, H.; Zhao, B.; Miao, Z.; Strunz, K. A review of modeling, management, and applications of grid-connected Li-ion battery storage systems. *IEEE Trans. Smart Grid* **2022**, *13*, 4505–4524. [CrossRef]
444. Cheng, L.; Wan, Y.; Zhou, Y.; Gao, D.W. Operational reliability modeling and assessment of battery energy storage based on lithium-ion battery lifetime degradation. *J. Mod. Power Syst. Clean Energy* **2021**, *10*, 1738–1749. [CrossRef]
445. Iqbal, H.; Sarwar, S.; Kirli, D.; Shek, J.K.H.; Kiprakis, A.E. A survey of second-life batteries based on techno-economic perspective and applications-based analysis. *Carbon Neutrality* **2023**, *2*, 8. [CrossRef]
446. Kostenko, G.; Zaporozhets, A. Transition from electric vehicles to energy storage: Review on targeted lithium-ion battery diagnostics. *Energies* **2024**, *17*, 5132. [CrossRef]
447. Wang, W.; Liu, S.; Ma, X.-Y.; Jiang, J.; Yang, X.-G. Advancing smart lithium-ion batteries: A review on multi-physical sensing technologies for lithium-ion batteries. *Energies* **2024**, *17*, 2273. [CrossRef]
448. Nazaralizadeh, S.; Banerjee, P.; Srivastava, A.K.; Famouri, P. Battery energy storage systems: A review of energy management systems and health metrics. *Energies* **2024**, *17*, 1250. [CrossRef]
449. Solyali, D.; Safaei, B.; Zargar, O.; Aytac, G. A comprehensive state-of-the-art review of electrochemical battery storage systems for power grids. *Int. J. Energy Res.* **2022**, *46*, 17786–17812. [CrossRef]
450. Waqas, A.; Qi, H.; Jian, L.; Abdullah, A.K.; Umashankar, S.; Sivakumar, S.S. Reputation-based uniform pricing & energy distribution in peer-to-peer energy trading market. *Ain Shams Eng. J.* **2025**, *16*, 103715.
451. Ishfaq, H.; Amin, W.; Ashfaq, S.; Mushtaq, N.; Shi, X. Attention-Enhanced Bidirectional LSTM for Accurate False Data Injection Attack Detection in Smart Grid. *Trans. Emerg. Telecommun. Technol.* **2025**, *36*, 70238. [CrossRef]
452. Ahmed, S.A.; Huang, Q.; Zhang, Z.; Li, J.; Amin, W.; Afzal, M.; Hussain, J.; Hussain, F. Optimization of social welfare and mitigating privacy risks in P2P energy trading: Differential privacy for secure data reporting. *Appl. Energy* **2024**, *356*, 122403. [CrossRef]
453. Force, J.T.; Benysek, G. Security and Privacy Controls for Information Systems and Organizations. National Institute of Standards and Technology. 2017. Available online: https://nvlpubs.nist.gov/nistpubs/SpecialPublications/NIST.SP.800-53r5.pdf (accessed on 1 September 2025).

454. International Electrotechnical Commission (IEC). Power Systems Management and Associated Information Exchange—Data and Communications Security—Part 3: Communication Network and System Security—Profiles Including TCP/IP. International Electrotechnical Commission. 2014. Available online: https://cdn.standards.iteh.ai/samples/105100/b7a344cdd98f47e4a67e4a94e9df50b5/IEC-62351-3-2023.pdf (accessed on 1 September 2025).
455. Cindrić, I.; Jurčević, M.; Hadjina, T. Mapping of Industrial IoT to IEC 62443 Standards. *Sensors* **2025**, *25*, 728. [CrossRef]
456. Ishfaq, H.; Kanwal, S.; Anwar, S.; Abdussalam, M.; Amin, W. Enhancing Smart Grid Security and Efficiency: AI, Energy Routing, and T&D Innovations (A Review). *Energies* **2025**, *18*, 4747. [CrossRef]

energies

MDPI

Review

Enhancing Smart Grid Security and Efficiency: AI, Energy Routing, and T&D Innovations (A Review)

Hassam Ishfaq †, Sania Kanwal †, Sadeed Anwar †, Mubarak Abdussalam † and Waqas Amin *,†

School of Information and Control Engineering, Southwest University of Science and Technology, Mianyang 621000, China; hassamashfaqkhan@mails.swust.edu.cn (H.I.); saniakanwal@mails.swust.edu.cn (S.K.); sadeedanwar@mails.swust.edu.cn (S.A.); mubbysalam@mails.swust.edu.cn (M.A.)

* Correspondence: sheikhwaqasamin@hotmail.com

† These authors contributed equally to this work.

Abstract

This paper presents an in-depth review of cybersecurity challenges and advanced solutions in modern power-generation systems, with particular emphasis on smart grids. It examines vulnerabilities in devices such as smart meters (SMs), Phasor Measurement Units (PMUs), and Remote Terminal Units (RTUs) to cyberattacks, including False Data Injection Attacks (FDIAs), Denial of Service (DoS), and Replay Attacks (RAs). The study evaluates cutting-edge detection and mitigation techniques, such as Cluster Partition, Fuzzy Broad Learning System (CP-BLS), multimodal deep learning, and autoencoder models, achieving detection accuracies of (up to 99.99%) for FDIA identification. It explores critical aspects of power generation, including resource assessment, environmental and climatic factors, policy and regulatory frameworks, grid and storage integration, and geopolitical and social dimensions. The paper also addresses the transmission and distribution (T&D) system, emphasizing the role of smart-grid technologies and advanced energy-routing strategies that leverage Artificial Neural Networks (ANNs), Generative Adversarial Networks (GANs), and game-theoretic approaches to optimize energy flows and enhance grid stability. Future research directions include high-resolution forecasting, adaptive optimization, and the integration of quantum–AI methods to improve scalability, reliability, and resilience.

Keywords: energy generation; energy transmission & distribution; energy routing; energy utilization; smart grid; cyberattacks

1. Introduction

The global energy landscape is undergoing a significant transformation, driven by the urgent need to mitigate climate change and facilitate the transition to sustainable energy systems. Renewable energy sources, such as solar photovoltaics and wind turbines, are increasingly being integrated into smart grids to reduce carbon emissions and dependence on fossil fuels. However, the intermittent nature of these resources poses challenges for grid reliability, stability, and cybersecurity, exposing to vulnerabilities such as False Data Injection Attacks (FDIA) and Denial of Service (DoS) attacks.

Improving the functionality and adaptability of smart grids is increasingly reliant on artificial intelligence (AI) and machine learning (ML). The paper briefly describes advanced models such as the Long Short-Term Memory (LSTM) network, the Random Forest (RF) model, and the autoencoder, among others, which have been applied in demand prediction, pattern detection, and anomaly detection in energy systems. Specifically, LSTM and Gated

Recurrent Units (GRUs) do not suffer from the vanishing gradient problem, which supports their application in smart grid for tasks such as load-profile analysis and energy-routing optimization. Moreover, the Self-Organizing Maps (SOMs) can be employed to identify anomalies in consumer-load patterns and irregular sensor records, which may indicate equipment malfunctions, including defects or FDIA-related cyberattacks. These AI-based technologies enhance grid intelligence and adaptability, enabling real-time decision-making and effective use of resources to meet the demands of dynamic energy requirements.

In addition, the increasing digitalization of smart grids presents cybersecurity challenges, whereby smart meters (SMs), Phasor Measurement Units (PMUs), and Remote Terminal Units (RTUs) are vulnerable to cyber intrusions targeting their control. In this regard, this paper identifies and discusses the application of advanced solutions, such as the Cluster Partition–Fuzzy Broad Learning System (CP-FBLS) and multimodal deep learning, for the immediate and swift response to FDIA and DoS attacks in power-generation systems. The methods exploit neural network topologies as well as decentralized control and ensure effective resource management at high performance levels. The threat of the Advanced Persistent Threats (APTs), which can cripple grid functionality and cause cascading failures, is also addressed in this review study. This paper further notes that secure and locally resilient architectures are critical in safeguarding critical energy infrastructures against emerging cyberattacks, through the exploration of not only the frameworks but also the overall effects of integrating grid-level architectures into motion-control strategies that mitigate the degradation of resilient consensus.

Renewable energy intermittency and the resulting requirements to manage it pose significant demands for energy-storage systems (ESSs) and comprehensive policy frameworks necessary to support the sustainability of power systems. The paper also discusses hybrid energy-storage systems (HESSs), which consist of a combination of mechanical, electrochemical, and thermal storage technologies to improve grid stability and efficiency. The co-optimization models reviewed have reduced transmission expansion requirements by nearly 10%, thus cutting infrastructure costs and enhancing grid reliability. Moreover, the research highlights the role of policy instruments in overcoming the challenges of political opposition, financial limitations, and social acceptance. Combining econometric modeling, policy studies, and AI-driven optimization methods, the work offers a comprehensive framework for developing scalable, secure, and sustainable energy chains, capable of meeting the increasing global energy demand and ensuring environmental protection and system stability.

The smart-grid technologies, such as energy management systems (EMSs) and Advanced Metering Infrastructure (AMI), further enhance the efficiency of energy networks. EMS provides real-time visibility and control over grid operations as well as optimization, and AMI allows demand-side management based on consumption profiles derived in real time from consumption data. The paper also discusses energy routing, incorporating AI-based solutions and multi-agent systems (MASs) as approaches enabling decentralized control and peer-to-peer scheduling to enhance resilience. Nonetheless, these technologies face challenges such as heavy computational burdens and scalability limitations, underscoring the importance of future research into hybrid protocols that integrate metaheuristics, AI, and blockchain to create secure and flexible energy networks. By examining these developments in detail, this study contributes to the ongoing discourse on power system transformation, aiming to achieve sustainability, security, and efficiency in response to evolving energy needs.

This review aims to provide a holistic security and efficiency analysis for modern smart grids based on an extensive literature review, as presented in Figure 1. While existing surveys often treat cybersecurity and renewable integration in isolation, this work synthesizes these domains to highlight their interplay. Furthermore, it addresses a significant gap in the literature by dedicating substantial focus to energy-routing protocols, a strategic decision-making layer essential for managing power flow in decentralized, renewable-heavy grids. By evaluating AI, metaheuristic, and multi-agent systems for routing optimization, this review offers a distinct perspective on achieving grid resilience, going beyond conventional discussions to address the core operational challenge of future energy networks.

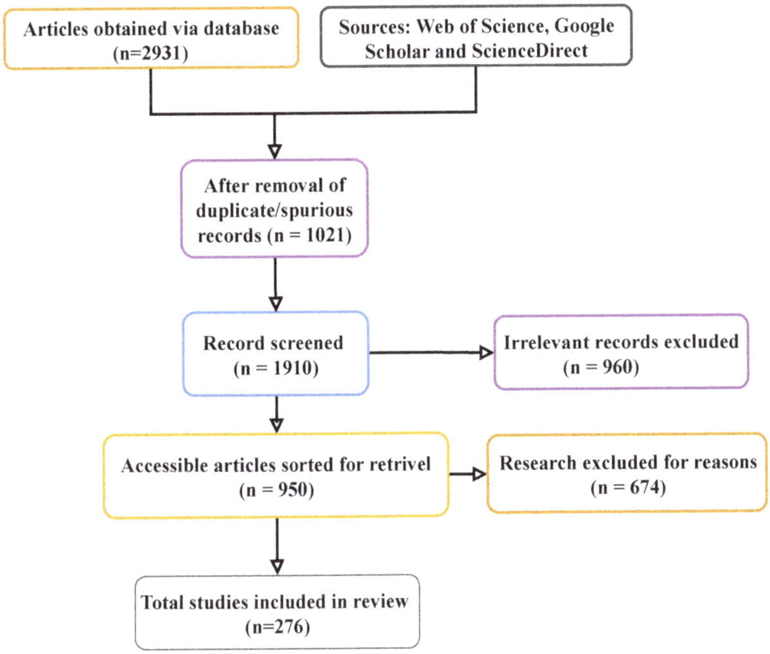

Figure 1. Flow diagram showing study selection process.

2. Energy Generation

2.1. Factors Impacting Energy Generation

The generation of energy is influenced by a combination of physical, technological, environmental, and socio-political factors. Understanding these factors is essential for planning sustainable and efficient energy systems. The key factors that shape energy generation include (1) resource availability, which determines the type and reliability of energy that can be produced; (2) technological efficiency, which affects how effectively these resources are converted into usable energy; (3) environmental and climatic conditions, which influence the performance and stability of generation systems; (4) policy and regulatory frameworks, which provide guidelines and incentives for energy production; (5) grid and storage integration, which is crucial for balancing supply and demand in real time; and (6) geopolitical and social factors, which affect energy security, market stability, and public acceptance. Each of these dimensions plays a critical role in shaping both the current and future landscape of energy generation.

The availability of resources is fundamental in determining the viability, productivity, and feasibility of generating energy. In recent years, advancements in forecasting models, geospatial analysis tools, and hybrid system design have significantly improved our ability to evaluate and utilize energy resources such as solar irradiance, wind potential, and hydropower capacity. The following are some of the most powerful contributions in this field.

One example is the study [1] which contributed by presenting a GIS-based multi-criteria decision analysis framework employing the Analytic Hierarchy Process (AHP) for solar site selection. The proposed approach integrates satellite-based solar irradiance data, topography-related parameters, land-use/-cover maps and accessibility measures to generate spatial suitability maps. In this case, the method was applied in semi-arid area to select the best location for photovoltaic (PV) deployment. The major contribution of the study lies in providing a solution for policy planners and investors, enabling them to use the method to identify high potential zones with consideration of environmental, technical, and infrastructural bottlenecks using data.

Another potential work is by [2], whose method predicted the long-term dynamic of available resources using the Earth System Models (ESMs). The authors compared climatic projections of potential solar and wind energy using multi-model collections of climatic change scenarios. They determined that, despite the predicted growth in the total amount of worldwide renewable resource potential, there will be a reduction in reliability, due to increased variability and significant occurrences in certain regions. The analysis is of great significance to future energy planning, as it indicates the need to incorporate the results of climate forecasts into plans for implementing renewables to avoid over-projecting the consistency of resources in the future.

Ref. [3] presented a comprehensive study on the use of Artificial Neural Networks (ANNs) in hybriding wind and solar energy by retrieving real-time meteorology information. The authors focused their attention on the development of a predictive model that can suggest the availability of solar radiation and wind speed to enable hybrid renewable systems at a single moment . The model was applied to adjusted forecast data from 2022 for weather and energy generation, and its performance was evaluated using standard metrics of a forecast such as the RMSE and MAE. The usefulness of the paper is enormous, as the framework was practically applied in locations where solar and wind resources were available but differed in abundance depending on weather conditions. This combined forecasting approach considers both the advantage of maximising generation efficiency and improving the planning of grid connection for hybrid systems.

Similarly, among the more recent publications, Ref. [4] integrated numerical weather prediction (NWP) models with ANNs during data training to improve the accuracy of solar irradiance forecasting. When their hybrid scheme relied on a regional collection of solar radiation observation data and high-resolution weather forecasts, significant gains in forecast accuracy were achieved over individual models. This study is crucial for solar energy systems, as accurate forecasting allows operators to optimize power generation, storage, and dispatch strategies in advance, thereby minimizing curtailment and improving system reliability.

Finally, Ref. [5] addressed a persistent challenge in the field the lack of open, high-resolution operational data by publishing a large-scale empirical dataset collected from six operational wind and solar farms in China. The dataset includes synchronized SCADA measurements, environmental conditions, and real-time power output data over two years. This openly available dataset enables researchers to validate forecasting and optimization models against real-world generation conditions. By filling a critical data gap, this study strengthens the foundation for empirical resource analysis and model development in renewable energy systems.

Table 1 presents a summary of these works along with their model deployment and main contributions.

Table 1. Summary of relevant research.

References	Model/Method Used	Dataset/Region	Main Contribution
[1]	GIS + Analytic Hierarchy Process (AHP)	Satellite solar maps, terrain and land-use data	Designed a GIS-based solar site selection framework using AHP to identify optimal PV deployment zones.
[2]	Earth System Models (ESMs)	Multi-model solar/wind projections (Global)	Reviewed long-term climate model outputs to assess how resource availability will shift under future climate scenarios.
[3]	Artificial Neural Network (ANN)	Meteorological and power data (2022)	Developed a hybrid ANN model for forecasting solar and wind availability in real time to support hybrid system planning.
[4]	Numerical Weather Prediction (NWP) + LSTM	Solar radiation and weather forecasts (Regional)	Combined deep learning with weather models to significantly improve short-term solar irradiance forecasting.
[5]	Empirical SCADA Data Publication	6 Chinese wind/solar farms (2019–2020)	Published a large, real-world dataset with synchronized SCADA and environmental data for model training/validation.

2.1.1. Technological Efficiency

Technological efficiency is crucial for optimizing energy systems, reducing costs, and enhancing sustainability. Recent advancements in computational models, machine learning (ML), and deep learning (DL) have significantly improved the efficiency of power generation from conventional plants to renewable sources. These technologies enable accurate performance forecasting, real-time optimization, and waste reduction, thereby lowering the Levelized Cost of Energy (LCoE).

For example, a study by [6] proposed hybrid ML/DL models, including XGBoost, CatBoost, LightGBM, and LSTM, to predict solar and wind power generation. Their parallel fusion approach reduced the average forecast error to 8.02%, demonstrating that combining multiple algorithms can optimize energy management and enhance prediction accuracy for variable renewable energy sources. Another study by [7] reviewed the application of AI-driven optimization techniques, such as genetic algorithms, surrogate modeling, and Bayesian optimization, in offshore wind-turbine tower design. This study emphasized that these AI tools can optimize structural design to reduce costs and improve efficiency, ultimately lowering the LCoE for offshore wind systems. In [8], the authors applied thermodynamic cycle analysis to combined cycle power plants (CCPPs). This study demonstrated that optimizing operational settings based on real plant and environmental data could significant improve efficiency, achieving a 1.8% increase in thermal efficiency, which enhances overall plant performance.

Additionally, Ref. [9] proposed a digital-twin solution for floating offshore wind turbines, specifically for the TetraSpar prototype. By integrating real-time SCADA data with aerodynamic estimations and physics-based modeling, their approach enabled improved performance monitoring and load forecasting, thereby enhancing both turbine efficiency and reliability. Similarly, the authors of [10] introduced physics-informed neural networks (PINNs) for wind-turbine power forecasting, embedding physical laws within the model architecture. This integration led to highly accurate predictions, with a Mean Absolute Error (MAE) of 3.9%, which is critical for providing reliable power-generation forecasts during real-time operations. Table 2 presents a summary of these studies and their performance metrics.

Table 2. Summary of relevant research: technological efficiency.

References	Methodology/Model Used	Dataset/Region	Main Contribution	Performance Metrics/Results
[6]	XGBoost, CatBoost, LightGBM, LSTM	Real-time solar/wind data	Developed hybrid ML/DL models to predict solar and wind power generation	Achieved an average error rate of 8.02% in power prediction
[7]	AI-based optimization techniques	-	Reviewed AI methods for optimizing offshore wind turbine towers, focusing on structural efficiency and cost reduction.	-
[8]	Thermodynamic cycle analysis	Combined-cycle power plant	Applied thermodynamic analysis to optimize efficiency in combined-cycle power plants.	Achieved a 1.8% improvement in thermal efficiency.
[9]	Digital-twin modeling	TetraSpar prototype data	Proposed a validated digital-twin solution for floating offshore wind turbines.	Estimated damage equivalent loads with 10–15% accuracy.
[10]	Physics-informed neural networks	SCADA data from 4 turbines	Implemented PINNs for accurate wind-turbine power forecasting with uncertainty quantification.	Mean Absolute Error (MAE) of 3.9% in power prediction.

2.1.2. Environmental and Climatic Conditions

Environmental and climatic conditions play a pivotal role in the efficiency, reliability, and sustainability of power-generation systems. Recent studies have highlighted how factors such as extreme weather events, temperature variations, and hydrological changes can directly affect the performance of both renewable and conventional energy sources.

For example, a study by [11] proposed a systematic framework to assess the impact of extreme summer droughts on power-generation capacity in the U.S. power grid. This study, which focused on the PJM and SERC regions, showed that during severe drought conditions, the usable capacity of hydroelectric and thermal power plants could decrease by 19–29%. These findings underscore the vulnerability of power systems to prolonged dry periods, particularly in regions dependent on hydro generation, where reduced water flow can significantly impair energy output. The study provides critical insights into planning and mitigation strategies for climate-induced disruptions in power generation.

A study by [9] examined the resilience of the Italian power system to extreme weather events such as heatwaves and cold spells. The study revealed that extreme weather conditions could severely disrupt thermal generation and hydropower systems, highlighting the increasing need for solar and wind energy integration. It is projected that Italy's power system would require an additional 5–8 GW of PV capacity by 2030 to meet renewable energy targets and mitigate the effects of extreme weather. The findings emphasize the need for adaptation strategies in energy infrastructure, especially in Mediterranean regions where climate change is expected to increase the frequency and intensity of heatwaves.

Further expanding on the role of extreme weather, Ref. [12] introduced a novel method to identify weather-induced extreme events in highly renewable energy systems. By analyzing electricity shadow prices from a European power system model based on 40 years of reanalysis data, they identified key weather patterns that lead to system stress. Their findings showed that multi-day weather events, rather than single weather shocks, were the primary cause of renewable energy shortages during extreme conditions. This study calls for interdisciplinary collaboration between energy meteorology and system modeling to better prepare for such events.

Another study, also by [11], examined how extreme summer droughts impact the eastern U.S. power system, particularly in the PJM and SERC regions. By using capacity derating models, the authors simulated the effects of prolonged droughts on both hydroelectric and thermal power generation. This study found that during periods of severe drought, power plants in these regions could experience up to a 29% reduction in usable capacity, potentially leading to power shortages. The research provides actionable insights for energy planners to develop strategies that mitigate the risk of power-generation losses during drought conditions.

Similarly, Ref. [13] explored how climate change impacts the adequacy of Europe's power system. Their study focused on temperature variations and hydrological changes, examining how these factors could alter electricity demand and hydropower generation in the future. By modifying inputs to a large-scale electricity market model, they found that while climate change may reduce winter heating demand; this decrease would likely be offset by increased challenges related to hydrological variability. These findings suggest that future energy planning must account for the dual impact of climate change in terms of both reduced demand and increased supply risks from hydropower. Table 3 summarizes the relevant studies on environmental and climatic conditions.

Table 3. Summary of relevant studies: environmental and climatic conditions.

References	Methodology	Dataset and Region	Main Contribution	Performance Metrics/Results
[11]	Capacity-derating models	PJM and SERC regions (eastern U.S.)	Developed a framework to assess the impact of extreme summer droughts on hydro and thermal power plants.	Found a 19–29% decrease in usable capacity during severe drought.
[9]	Power dispatch and capacity expansion models	Italy (future 2030 projections)	Investigated the impact of extreme weather on thermal generation and hydropower systems, highlighting the need for solar and wind integration.	Projected a 5–8 GW increase in PV capacity required by 2030 to cope with extreme weather.
[12]	Electricity shadow price analysis	European reanalysis data (40 years)	Identified weather-induced extreme events and multi-day renewable energy shortages using electricity shadow prices derived from long-term reanalysis data.	Found that multi-day weather events, rather than single events, were responsible for the majority of renewable energy shortages.
[13]	Post-processing of market models	Europe (electricity market data)	Analyzed the dual impact of climate change on electricity demand and hydropower generation in Europe.	LOLE reduction by 50% due to decreased heating demand, but increased LOLE due to hydrological changes.

In addition to environmental effects, the emergence of renewable energy sources (RES) significantly changes the energy system in terms of the cybersecurity system resilience. This transformation has a two-fold impact. On the one hand, it widens the attack surface. The network configuration composed of arrays of solar inverters, wind-farm controllers, and related sensors is a distributed system containing multiple entry points under of cyberattacks. They tend to be in a more weakened security than conventional utility assets, leaving them susceptible to attacks such as False Data Injection Attacks (FDIA) which can hack into the generation forecast or inverter parameters and short-circuit the system. Moreover, the intrinsic variability of RES can be exploited by attackers to conceal their operations, further complicating defense mechanisms. Conversely, resilience can enhanced through a diversified, decentralized system where the RES are properly integrated. Such a grid would be less susceptible to attacks targeting single points of failure by reducing dependence on large centralized generation centers. If one components of the renewable fleet is affected, others will be able to operate. This is the benefit of resilience. However, this advantage depends on the enactment of strong cybersecurity levels such as the AI-driven detection methods mentioned in this review on all the tier levels of the distributed system to ensure that the new complexity as a result of the introduction of RES is secured.

2.1.3. Policy and Regulatory Framework

The policy and regulatory framework plays a crucial role in power-generation systems, particularly in the context of the transition to renewable energy and the need for energy security in the face of climate change. These frameworks influence how energy is generated, distributed, and consumed, impacting both conventional power plants and renewable energy systems. Recent studies have explored how policy instruments, regulatory measures, and financial incentives affect the efficiency, reliability, and sustainability of power generation, with a focus on renewable energy adoption, emissions reductions, and energy security.

In a study by [14–16] the authors assessed the effects of four major climate and energy policies: carbon taxes, emissions trading systems (ETS), renewable energy standards, and energy-efficiency programs on CO_2 emissions from electricity generation. Using econometric models, the researcher were able to determine how each of the policies impacted the reduction of emissions in the power-generation industry. It is noteworthy that, as the authors reported, a combination of these policies proved to be significantly more productive than relying on any single one, much of what was required of power-generation systems that needed to decrease emissions, alongside securing reliability and coping with the growing consumer energy demand. As this work indicates, harmonizing various policy tools can help develop a more integrated framework aimed at advancing clean energy production and addressing grid stability and capacity in a more balanced manner .

Another study examined the impact of climate laws implemented in countries on energy security performance (ESP) [17]. The analysis was targeted at the positive outcomes of properly developed climate policies, i.e., the possibility to enhance the energy security due to the diversification of the energy resources supplies more resilience of the systems. These consistent energy security policies aligned and corresponding to one another in terms of energy security targets are bound to create a sustainable economy of power-generation systems that can sustain power-generation performance even in the face of environmental pressures. The paper points out that all these policies only work depending on the design and implementation of the policies, in addition to the broader political and economic context. Regulations promoting clean energy production and grid stability are a necessity in an attempt to expand power-generating capabilities and reduce vulnerability to the challenges of climate upheaval.

Similarly, Ref. [18] conceptualized the influence of financial and political aspects in the passage of climate bills that will enable the migration toward renewable energy systems. These led to the identification of the most important policy instruments, i.e., subsidies, tax incentives and renewable energy targets which have been effective in delivering renewable energy technologies, thus facilitating the shift toward more sustainable forms of power generation. Yet, barriers to policy implementation, as highlighted in the study, were found to be quite challenging, such as political resistance, economic restrictions, and social acceptability. This research stipulates that a multi-faceted policy, which involves not only subsidizing but also establishing robust regulatory procedures, is required to provide a shifting, as well as efficient, method for transitioning the energy-generation sector to renewable energies.

The Inter-American Development Bank examined the regulations and policy design in a report [19] on how to assist renewable energy in Latin America and the Caribbean. The report highlighted the significance of transparent and sound regulatory frameworks in the attraction of private investment in its renewable energy scheme, which is crucial for boosting the production of solar, wind, and hydropower resources. Public–private partnership (PPP) was presented as the effective step in scaling renewable energy projects especially in those areas that have not fully realized their potential of renewable energies. The paper emphasizes the significance of capacity building to the regulatory entities in order to allow for successful policy implementation and sustainability to renewable energy production over the long run.

The literature depicts that, for the future of power generation, policy and regulatory frameworks are essential. Renewable energy technologies can be integrated into energy-generation processes, advancing the energy-generation systems with the help of effective climate policies, financial instruments, and regulatory frameworks. The effectiveness of these policies, however, depends on how they are determined and enacted, as well as in tandem with the overall objectives of energy security and policy sustainability. Looking at the future, there is a need to have a holistic and all-encompassing policy strategy that makes it possible to have an easy and sustainable switch to clean energy creation.

For this purpose, to make a regulatory strategies paractically relevent, researchers have also explored interdisciplinary approaches. For example, authors in [20] proposed advanced econometric frameworks that allow for the quantitative assessment of carbon tax effectiveness and policy synergies in achieving energy resilience and emission reduction. Similarly, in the SG environments, multi-agent system (MAS) simulations have emerged as a powerful tool, where autonomous agents represent utilities, policy stakeholders, and consumers. The MAS-based framework of study [21] supports collaborative decision-making under regulatory uncertainty. Integrating such interdisciplinary tools provided a better way of response among stakeholders and more robust insights into policy, which ultimately leading to more effective and adaptive energy governance.

2.1.4. Grid and Storage Integration

A power grid that incorporates energy-storage systems (ESS) has now been identified as one of the essential approaches for developing the reliability, flexibility, and resilience of current power-generation systems. With the increasing presence of RES, including solar power and wind power, the intermittency of RES presents major problems when it comes to ensuring grid stability. Energy-storage technologies provide beneficial solutionsby storing energy when there is excess supply and supplying it when needed most or when there is a shortage in energy supply. This is achieved through the capability to match the supply and demand aspects required to make sure of a consistent power provision. High supply and demand, ensure a stable power supply.

The development of hybrid energy-storage systems (HESS) is one of the most significant evolutions in energy storage. These systems combine several different storage technologies to leverage their features. A paper by [22] provides an extensive overview of HESS, including their structures, planning decisions, and control measures. The study highlighted the finding that the optimal design of storage technology selection and operation under a hybrid system can have a significant impact on improving grid performance and stability. Through the combination of different storage technologies, HESS emerges as an economical and adaptable mechanism to address the drawbacks of renewable energy intermittency, optimizing electric systems as a whole. This review emphasize that the hybrid storage systems have to be optimized further in order to enable more sustainable balance between industry generation and demand on the grid.

Table 4 presents the comparative analysis of the studies relevant to policy and regulatory framework.

Table 4. Summary of relevant research: policy and regulatory framework.

References	Methodology	Dataset	Main Contribution	Findings
[14]	Econometric modeling and policy evaluation	Panel data from 109 countries over 30 years	Evaluated the static and dynamic effects of four national climate and energy policies carbon tax, ETS, renewable energy standards, and energy-efficiency programs on CO_2 emissions in the electricity generation sector.	Found that a combination of policies is more effective than individual measures. Recommended integrating various policy instruments to achieve significant emissions reductions in the power sector.
[17]	Policy analysis and comparative study	National climate laws and energy security performance data	Examined the short- and long-term impacts of national climate laws and policies on energy security performance (ESP).	Found that well-designed climate policies can enhance energy security by promoting the diversification of energy sources and improving infrastructure resilience. However, the effectiveness of these policies depends on their design, implementation, and the political and economic context in which they are applied.
[18]	Policy review and analysis	Financial and political factors aiding climate legislation	Explored how financial and political factors aid climate legislation and policies for renewable energy transition.	Identified key policy instruments, such as subsidies, tax incentives, and renewable energy targets, that have been effective in promoting the adoption of renewable energy technologies. Discussed challenges and barriers to policy implementation, including political opposition, economic constraints, and social acceptance.
[19]	Policy analysis and framework development	Energy-transition initiatives in Latin America and the Caribbean	Examined incentive mechanisms for both grid-scale and distributed generation in Latin America and the Caribbean.	Highlighted the importance of clear and stable regulatory frameworks in attracting investment in renewable energy projects. Discussed the role of public-private partnerships and the need for capacity building in regulatory institutions to effectively implement and enforce policies.

Further expanding on the integration of storage with grid systems study [23] demonstrated the value of co-optimizing the sizing of wind, solar, storage, and grid connection capacity. By employing a co-optimization model, the study found that integrating energy storage could significantly reduce the need for extensive transmission expansion, which is often required to handle growing renewable energy generation. The researchers found that co-optimizing storage and transmission planning could decrease the need for transmission expansion by around 10%, thereby lowering infrastructure costs while enhancing the efficiency and stability of the power grid. This strategy highlights how storage systems can help alleviate congestion in the grid, providing more flexibility and reliability for integrating renewables. The study underscores the economic benefits of integrating storage solutions into grid planning to optimize both energy-generation and transmission.

Long-Duration Energy Storage (LDES) has become one of the potential technologies to provide long-term solutions to grid stability. In [24,25], the authors compared the possibilities of LDES as contributing to grid operations, specifically, in reducing curtailment and increasing flexibility in operations. The researchers investigated the energy capacity requirements of LDES with respect to the benefits that would be gained in terms of electricity price setting and the achievement of decarbonization targets. The researchers discovered that LDES systems may provide seasonal energy balance, which is highly needed to support grids that rely on variable injections such as wind or solar energy. LDES has the potential to become a key element in green transition due to its ability to offer sustainable energy storage on a large scale over time, making the grid flexible enough to operate effectively in the future.

In an attempt to achieve optimal performance and ensure the longevity of energy-storage systems, as explored in [26], this article presents a novel algorithm for weather-driven priority charging. This algorithm, incorporates the real-time weather forecasting and monitoring to enables optimal charging of batteries and energy-storage systems. Using weather information, the algorithm will be able to take advantage of times when there is an oversupply of renewable energy whilst ensuring that practices of inefficient charging do not wear down the storage systems. The approach enhances the reliability and efficiency of the hybrid renewable power grids, making it an important solution to integrating the storage system effectively with the power grid. The researchers have demonstrated the way smart charging solutions can optimize energy consumption, extend battery life, and stabilize the grid. Universal charging support: The analysis emphasizes how universal charging, capable of supporting energy storage, can optimize energy use, increase battery life, and balance the grid.

Lastly, the U.S. Energy Information Administration (EIA) documented that grid-scale battery storage is growing rapidly in the United States, with the storage capacity set to rise to 30.9 GW by the end of 2024. The rise is fueled by declining battery prices, favorable governments policies and overcoming the issue of grid reliability. Large-scale battery-storage systems are gaining importance to help curb the problem posed by the growing proportion of renewable energy sources. Batteries enable the storage of surplus generated during off-peak power and also provide flexibility to release offering a stable and flexible alternative to managing renewable energy. The addition of grid-scale storage will be essential in enhancing the resilience and stability of the grid as it becomes increasingly reliant on intermittent renewable sources of energy [27]. Table 5 summarizes the few studies relevant to grid and storage integration.

Table 5. Summary of relevant research: grid and storage integration.

Study	Methodology	Contribution
[22]	Review of HESS components and control strategies	Explored the integration of multiple storage technologies
[23]	Co-optimization modeling of storage and transmission planning	Investigated the impact of storage on transmission expansion
[25]	Evaluation of LDES capacity mandates	Analyzed the benefits of LDES for grid stability
[26]	Development of priority charging algorithms	Proposed a weather-driven charging strategy for battery systems
[27]	Analysis of battery-storage deployment trends	Assessed the growth of grid-scale battery storage

2.1.5. Geopolitical and Social Factors

Geopolitical tensions and societal dynamics significantly shape the production, distribution, and adoption of power-generation systems worldwide. Recent studies reveal how national security concerns, global conflicts, public sentiment, and regulatory environments influence the path and pace of energy transitions.

A study by [28] empirically confirmed that geopolitical risks hinder energy transitions by increasing supply-chain volatility and price fluctuations. However, countries with robust regulatory frameworks and well-developed renewable industries were shown to weather these disruptions better and accelerate renewable energy deployment, reinforcing the role of resilient policy design in sustaining clean power generation.

A Germany-focused case study [29] analyzed the Energiewende, highlighting how Germany's strong political commitment to green policies has reshaped its energy mix. The research showed that national strategies balancing energy security, economic competitiveness, and diplomatic alignment led to the increased deployment of renewables. However, it also emphasized the delicate balancing act required to manage energy independence, affordability, and international cooperation.

On the social acceptance front, Ref. [30] applied machine learning (ML) to Twitter data to assess public sentiment toward wind energy in Norway. The study found an increases in negative public viewpoints during 2018–2020, demonstrating how local social opposition can impede project rollout and delay infrastructure development. This highlights the need for early engagement and transparent communication in planning renewable generation projects.

Similarly, Ref. [31] explored the geoeconomics of renewable energy, using game-theoretic models to show how countries like China strategically leverage investment, trade ties, and global policy cooperation to secure technological leadership in clean energy markets. The study concluded that geopolitical competition can drive innovation and efficiency in power generation, provided that international coordination and balanced policy frameworks are maintained.

Some researchers have also explored ML–based sentiment analysis, enabling policymakers to respond and monitor public opinion more effectively. For example, a study by [32] implemented RoBERT, a transformer-based language model, to analyse over 266,000 tweets about solar energy in the U.S., revealing temporal and regional variations in sentiment that were related to policy support. Sentiment analysis approaches guide the optimal placement of energy storage, public attitudes in real-time, grid infrastructure, or renewable installations, especially in socially sensitive areas. Similarly, authors in [33] conducted a thorough review of sentiment analysis techniques and highlighted how real-time public feedback, gathered through platforms like Twitter, guides more inclusive and socially accepted transitions toward sustainable energy systems.

Table 6 summarizes the studies' focus on geopolitical and social factors.

Table 6. Summary of relevant research: geopolitical and social factors.

Study	Methodology	Dataset/Scope	Contribution	Key Results
[28]	Econometric modeling	Cross-country energy market and policy data	Showed geopolitical risk hampers energy transition via supply-chain disruption	Robust regulatory and industry infrastructure mitigate negative impacts
[29]	Qualitative case study	Germany's energy-transition policies	Provided insights from Germany's Energiewende on policy balancing	Green policies enhance renewables but require careful geopolitical coordination
[30]	ML-based sentiment analysis	68,828 tweets on wind energy in Norway	Highlighted how social opposition can delay wind project deployment	Negative sentiment increased 2018–2020, underscoring the need for public engagement
[31]	Game-theoretic modeling	Global renewable energy trade and policy data	Illustrated the influence of geopolitical and economic strategies	Strategic investment/sprawl enhances renewables; cooperation ensures balanced growth

2.2. Impact of Cyber Attacks on Power Generation and Detection Using ML and DL Techniques

The increasing integration of digital technologies in modern power-generation systems has significantly improved operational efficiency but has also exposed these critical infrastructures to a wide range of cyber threats. As power generation forms the backbone of national energy security, any successful cyber intrusion, particularly in the form of False Data Injection (FDI), Denial of Service (DoS), or control signal manipulation, can lead to catastrophic instability and widespread outages. This growing concern has driven the research community to explore intelligent, data-driven solutions for intrusion detection and mitigation.

In this context, ML and deep learning (DL) models have emerged as powerful tools for detecting, classifying, and responding to cyber attacks within the power-generation domain. For example, a study by [34] compares k-means clustering, Auto Encoders (AE), and Graph Neural Networks (GNNs) on IEEE-68 bus simulations. The GNN model demonstrated superior performance in detecting and localizing False Data Injection Attacks (FDIAs), achieving accuracy exceeding 95%. Similarly, Ref. [35] leveraged Principal Component Analysis (PCA)-enhanced DL models on Phasor Measurement Unit (PMU) time-series data and integrated them with decision trees to distinguish between disturbances and cyber attacks in smart power systems.

Another study by [36] applied supervised ML models, including Support Vector Machine (SVM), K-Nearest Neighbors (KNN), and Artificial Neural Network (ANN), along with heuristic feature selectors on IEEE 14/57/118 bus systems. The study highlighted that optimal feature selection improves FDIA detection accuracy, with the SVM model achieving the highest accuracy of 90% with 122 selected features. In the realm of reinforcement learning, a Deep Q-Network (DQN)-based cybersecurity assessment framework was introduced by [37], which effectively identified and scored attack strategies in wind-integrated grids, outperforming conventional graph-search methods.

The vulnerability of ML models to adversarial attacks was addressed by [38], demonstrating how inputs generated using the Jacobian Saliency Map Algorithm (JSMA) can bypass RF and J48 Intrusion Detection System (IDS) models. Adversarial training was shown to restore and even enhance model robustness. Complementing this, the authors of [39] proposed stacked Convolutional Neural Network (CNN) + Long Short-Term Memory (LSTM) architectures, which significantly improved detection accuracy in Supervisory Control and Data Acquisition (SCADA) systems due to ensemble-style learning.

The authors of [40] comprehensively highlight recent DL frameworks across Industrial Control Systems (ICS), including power-generation environments. The authors identify the need for scalable and real-time defenses tailored to grid characteristics. Methodologically structured guides, such as [41], help formalize a 6-step DL pipeline for cyber–physical system (CPS) security, tested on public CPS datasets. To improve resilience against adversarial noise, Ref. [42] proposed stacked DL architectures (StackMean, StackMax, StackRF) enhanced with Synthetic Minority Over-sampling Technique (SMOTE). Their approach demonstrated improved detection in Industrial Internet of Things (IIoT) environments. However, the authors of [43] warned that AE-based models remain susceptible to evasive attacks, suggesting model-level defenses. Exploring hybrid GNN Deep Neural Network (DNN) designs, the National Renewable Energy Laboratory (NREL) proposed a state-prediction framework combining LSTM with GNN to detect FDIAs based on prediction errors in IEEE-39 bus systems [44]. Meanwhile, the authors of [45] proposed a streaming ensemble learner for real-time FDIA classification on live PMU data, showcasing adaptability to evolving grid threats. In [46], the authors proposed PowerFDNet, a spatiotemporal DL model combining CNN and GNN modules. Tested on IEEE-118 and Alternating Current (AC)-model grids, it effectively detected stealthy attacks and proved efficient for edge deployment. To reduce false alarms, an eigenvalue-residual similarity method (MERS) was applied by [47] on IEEE-39 buses, achieving over 96% accuracy.

Other anomaly-based methods include an AE approach by [48] on IEEE-118, which trains on known-good data and detects deviations via reconstruction errors. In [49], the authors applied Graph Signal Processing (GSP) to spatial grid data, detecting anomalies through k-means clustering of graph features. Similarly, transfer learning has also emerged as a promising strategy. The authors of [50] trained models on IEEE-14/118 and fine-tuned them on real-world data to improve detection in the presence of model uncertainty. Similarly, the authors of [51] proposed a graphical FDIA detector combining temporal and topological features in IEEE-39 systems. Adversarial simulation methods, such as Multi-Agent Adversarial Reinforcement Learning (MARL), were explored by [52], where agents learned attack and defense policies on inverter-rich grids.

Additionally, [53] presented Moving Target Defense (MTD) enhanced DNNs, which use randomized network topology to resist adversarial inputs across IEEE bus systems. In renewable-heavy environments, deep generative methods have been utilized. Ref. [54] combined denoising AE and Generative Adversarial Networks (GANs) to detect FDIAs under load-forecasting applications, maintaining 98% accuracy. Conditional GANs were also used in state estimation models in [55], achieving a success rate of 96%. Another study [55] proposed a state-space decomposition method using Sliding-Mode Observers (SMO) and H-resilient control to mitigate FDIA in Load Frequency Control (LFC) systems, reducing the impact of control-versus-measurement attacks and improving frequency response accuracy. To address unsupervised FDIA detection, [56] integrated LSTM and adversarial AE on IEEE-13/123 bus systems, yielding robust detection in unbalanced distributed energy networks without labeled anomaly data.

Building upon Load Frequency Control (LFC) security, the authors of [57] proposed a multi-model fusion framework that combines LSTM and AE to detect FDIAs and load-switching anomalies. The proposed model achieves 99.4% accuracy, detecting threats within 0.12 s on average. Meanwhile, a study by [58] introduced a federated LSTM-graph convolutional networks (GCN) detection system for distributed FDIA localization across IEEE-57/118/300 buses, balancing accuracy with data privacy. To address hybrid cyber–physical threats, Ref. [59] applied random matrix theory (RMT) and a singular value decomposition (SVD-CNN) classifier to identify and localize replay and FDIA attacks on IEEE-14 and IEEE-57 systems.

Expanding on control-layer resilience under cyber threats, Ref. [60] introduced a Load Frequency Control mechanism using off-the-shelf redundant communication channels, equipped with a QoS-dependent event-triggered communication (QEC) scheme. This approach maintains system stability and control accuracy even in the presence of FDIA-induced communication degradation by dynamically adjusting packet rates based on network conditions, ensuring both robust performance and efficient resource use. Another study [61] introduced a Cluster Partition–Fuzzy Broad Learning System (CP-FBLS) framework that partitions attack clusters, achieving ultra-fast FDIA detection in real-time generation networks while maintaining high classification precision.

These studies collectively reveal how ML and DL techniques have been innovatively applied to detect, localize, and mitigate cyberattacks across power-generation infrastructures. From traditional supervised learning to advanced spatiotemporal models, the literature underscores a growing trend toward hybrid, privacy-preserving, and adversarially resilient architectures. Each study contributes to addressing the multifaceted challenge of ensuring secure and stable power generation in increasingly digitized grids. Table 7 provides a comparative analysis of several studies addressing cyberattacks on power generation.

Table 7. Summary of studies addressing cyberattacks in power generation.

Study	Model	Dataset	Contribution	Performance
[34]	GNN	IEEE-68 bus	FDIA detection & localization	95% accuracy
[35]	PCA-DL + Decision Tree	PMU Time-Series	Disturbance vs. cyber-attack detection	97% accuracy
[36]	SVM	IEEE 14/57/118	Improved accuracy with feature selection	97% accuracy
[37]	DQN	Polish 2383-bus	Attack strategy learning	-
[38]	RF	ICS logs	Adversarial attack vulnerability	94% precision
[39]	Stacked CNN+LSTM	Power SCADA systems	Improved detection via ensemble	99.99% accuracy
[40]	AE	SWAT	DL approaches in ICS	0.9798
[41]	DL IDS pipeline	CPS benchmark	6-step CPS security framework	-
[42]	StackMean, StackMax, StackRF	WUSTL-IIOT-2021	Balanced & resilient detection	Accuracy: StackMean: 99.99, StackMax: 99.99, StackRF: 99.99
[43]	AE	Power SCADA systems	Adversarial evasion vulnerability	100% precision and 100% recall
[44]	LSTM + GNN	IEEE-39	FDIA via prediction errors	AUC: 0.9857
[45]	Streaming Ensemble	NY grid PMU	Real-time FDIA detection	-
[46]	PowerFDNet (CNN + GNN)	SimBench	Spatiotemporal DL-based PowerFDNet for SFDIA detection	Precision: 99.324 Recall: 99.209 F1: 99.267
[47]	MERS	IEEE-39	False alarm mitigation in AC-SE	96% accuracy
[48]	AE	IEEE-118	PMU anomaly reconstruction	TPR: 93.6%
[49]	GSP + K-means	IEEE 14-BUS systems	Real-time FDIA via graph signal processing	-
[50]	DNN (Transfer Learning)	IEEE-14/118	Robustness to modeling errors	Accuracy: 99.99%
[51]	Graph Hybrid(CNN+LSTM)	IEEE-39	Temporal and topological FDIA detection	<90%
[52]	MARL	Simulated inverter grid	MARL to defend against stealthy FDIAs	-
[53]	MTD-DNN	IEEE bus systems	Moving target defense (topology randomization)	Accuracy: 99%
[54]	SMDAE + GAN	IEEE-118 + real	Load forecasting under FDIA	~98% accuracy
[55]	CGAN	IEEE-57/118	Stealthy FDIA simulation for training	Success rate: 96%
[62]	State-space + SMO + $H\infty$ control	Renewable-rich LFC grid	Separate/mitigate control vs. measurement FDIA	~30% AE MSE reduction, improved stability
[56]	LSTM + AAE	IEEE 13 & 123	Unsupervised FDIA detection	Accuracy: 99.5% (IEEE 13), 99.6% (IEEE 123)
[63]	LSTM from partial observations	IEEE test systems	Attack localization + state reconstruction	-

Table 7. *Cont.*

Study	Model	Dataset	Contribution	Performance
[57]	Multi-model fusion (LSTM+AE)	LFC microgrid	Real-time FDIA & load-switching detection	99.4% accuracy; <0.12 s delay
[58]	Federated LSTM+GCN	IEEE 57/118/300	Decentralized FDIA detection with privacy	F1: 97–98%
[59]	RMT + SVD CNN	IEEE 14 & 57	Replay vs. FDIA differentiation	Accuracy: 100% (IEEE 14), 95.98% (IEEE 57)
[60]	Median filtering (MF)	IEEE 30	LFC resilience strategy against FDIA	-
[61]	CP-FBLS	IEEE 34 & 123	FDIA detection/localization via CP-FBLS	98.43% accuracy, 0.34 ms detection

3. Transmission and Distribution (T&D) System of Power Grid

In a traditional power grid, power flows in one direction from the utility through transmission and distribution systems to consumers. This one-way flow is a key component of traditional grid architecture, where centralized monitoring is essential to balance generation with consumer demand while respecting system limitations [64]. Utility operators manage every stage of the process to ensure both reliability and financial sustainability. However, the rapid growth in electricity demand from new devices, notably electric vehicles (EVs), has introduced vulnerabilities in traditional grids, including losses, limited monitoring and communication, poor distribution routes, and a lack of advanced automation and digital technologies [65,66].

To address these limitations and make the grid system more reliable, the smart grid has been introduced. By integrating technologies like the Internet of Things (IoT), modern digital technology, and automation, the smart grid facilitates two-way communication and allows electricity to flow in both directions [67]. These advancements enhance the monitoring of electricity transmission and distribution, support the seamless integration of electricity sources, and enable real-time monitoring and immediate response to disruptions, making the grid system more reliable and efficient. Smart-grid technology is better equipped to handle energy demands and operational issues in real time [68].

However, smart grids have complex architecture, which makes it more challenging to maintain effective power quality when dealing with a variety of load types. It is more difficult to maintain voltage stability, control harmonic distortion, and prevent other quality concerns when the network faces issues caused by diverse devices and generating sources [69].

Advanced, AI-powered grid interfaces are becoming essential tools for real-time control and adaptability [70,71]. AI-enhanced power quality-monitoring devices predict and mitigate future issues while also addressing current abnormalities. They achieve this by monitoring system conditions [72], forecasting disruptions, and automatically modifying operating conditions. This proactive approach support future innovations like direct-current distribution networks [73], which might further reduce power delivery issues in data centers, homes, and renewable energy agriculture. This enhances the functionality of connected equipment while encouraging safer, more efficient grid operation.

While integrating these IoT and modern digital techniques introduces challenges and safety risks like cyber threats, power outages, and blackouts [74], an overview of the smart grid is presented in Figure 2. To address these problems, much research has been conducted in recent years.

Utkarsh et al. [75] introduced a Self-Organizing Map (SOM) method for resilience assessment and restoration control in a distribution system during severe events. The study proposes a temporal resilience index to support resource allocation and operational dispatch. The proposed SomRes framework detects weak states and promotes pre-emptive load shedding, thereby improving coordination of T&D system responses during contingencies.

A fuzzy SOM-based third-party approach for electricity distribution networks was presented by Ovidiu Ivanov. Under an affinity control policy, consumer classification through load-profile shapes was performed for the rapid development of a Typical Load Profiles (TLPs) database, which is essential for effective T&D planning. The approach demonstrated good strength and stability of segmentation under load.

Wu et al. [76] proposed Gridtopo-GAN, a GAN-based method which uses topology-preserved node embeddings to correctly identify meshed and radial patterns in large-scale networks and performs effectively on both IEEE benchmark systems and real-world networks.

Yan et al. [77] proposed UG-GAN, a model that synthesizes three-phase unbalanced distribution systems. By observing real network random walks, it emulates local structures and enriches the synthetic grids with realistic nodal demands and components. UG-GAN was tested on smart meter data to balance data privacy and structural realism.

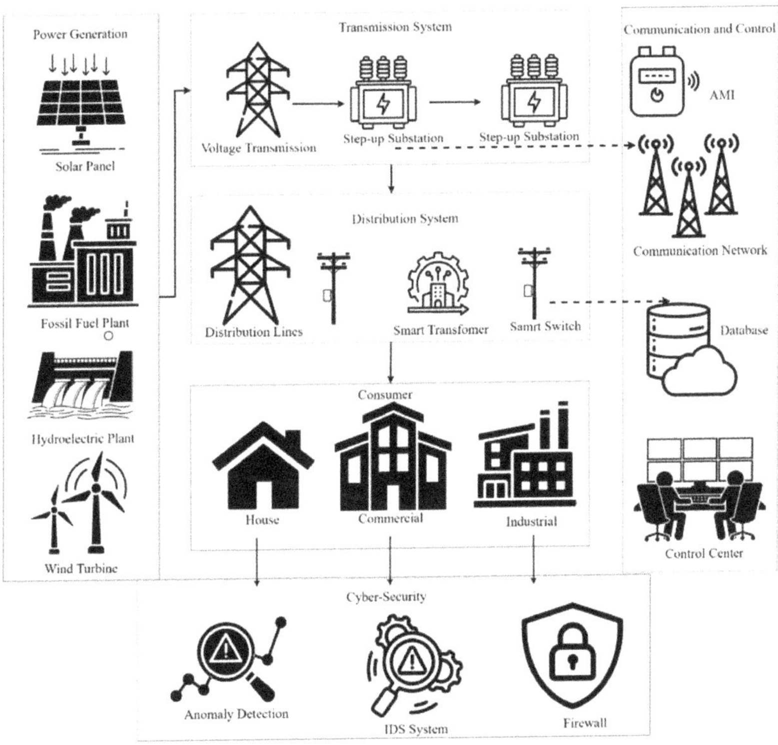

Figure 2. Overview of smart grid.

Venkatraman et al. [78] developed a CoTDS model with static/induction loads, reactive shunts, and DG inverters. The PI control and small-signal model serve as the foundation for analyzing mutual influences between the T&D systems. Simulations highlight the necessity of coordinated grid control under high DG penetration.

Mohammadabadi et al. [79] introduced a GenAI-based communication model that employs pre-trained CNNs to generate synthetic time-series data. Local agents process the data, reducing transmission load and preserving privacy. This method supports scalable, distributed learning for tasks such as anomaly detection and energy optimization in smart grids.

Shankar et al. [80] developed a deep learning OPA (dLOPA) model based on LSTM for smart-grid contingency analysis. It incorporates a pre-processing step that improves training efficiency, achieving 93% accuracy while maintaining low false-positive and false-negative rates. The dLOPA model does not target T&D systems specifically but contributes to the reliability of the grid.

Gonzalez et al. [81] introduced an RNN approach for demand prediction in distribution network planning. By incorporating confidence intervals, the model successfully respects explicit transformer thermal boundaries and enhances planning when accounting for load variations and distributed generation.

Shrestha et al. [58] proposed a federated learning LSTM-AE model for anomaly detection. The model achieved a 99% F1-score, maintained privacy, and outperformed conventional techniques.

Liu et al. [82] introduced MDRAE, an attention-based autoencoder for missing data recovery in transmission and distribution systems, that outperformed comparison models and enhanced both data quality and grid stability.

Raghuvamsi et al. [83] proposed a TCDAE for topology detection in distribution systems with missing measurements that outperformed other types of DAE variants.

Li et al. [84] proposed GTCA, a mixed T&D contingency analysis application that integrates the Global Power Flow and DC models. The approach is more effective than traditional methods, especially for looped networks.

3.1. Important Factors of Transmission & Distribution (T&D) System

There are several important factors of the T&D system in the smart grid, which are presented in Figure 3.

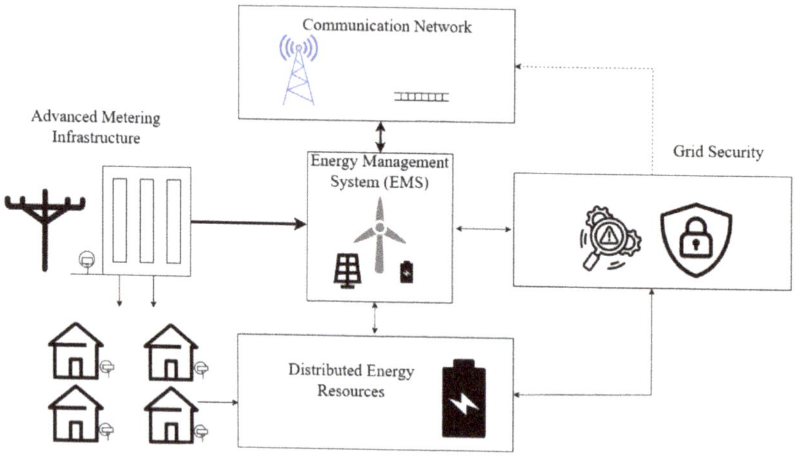

Figure 3. Important factors of transmission and distribution.

3.2. Advanced Metering Infrastructure (AMI)

AMI is an advanced digital system that automatically measures energy consumption in real time [85]. In contrast, traditional metering relies on analog devices requiring manual readings. AMI enables two-way communication between consumers and prosumers, providing accurate consumption monitoring and billing based on exact energy usage. This technology also facilitates demand response programs and allows consumers to track their real-time energy usage, helping them to reduce consumption during peak hours and optimize costs [86].

3.2.1. Communication Networks

Communication networks enable frequent status checks, the sharing of data, and two-way communication to enable accurate T&D data which, in consideration of smart-grid networks, requires a high level of security. With constant monitoring, prosumers can track grid status and identify potential problems that might occur in order to enhance distribution of energy properly. It is also remotely managed through automated control systems that operates electronic devices such as routers and electric vehicles. This enhances the efficiency of transfer data and communication channels between prosumers and consumers more efficiently [87].

3.2.2. Distributed Energy Resources (DER)

Distributed energy resources (DER) include facilities located closer to consumption points than conventional generation facilities. These systems comprise small energy sources like solar arrays, batteries, wind generators, and electric cars that produce and store energy close to the households and enterprise premises. Implementing DER on the smart grid increases reliability and flexibility while reducing dependences on conventional power plants. Moreover, DER expands the utilization of green energy and reduces carbon emission. Moreover, self-sufficient energy systems, known as microgrids, can operate independently and control multiple processes with the smart grids [88,89].

3.3. Energy Management Systems (EMS)

An energy management system (EMS) is a computer-based system designed to monitor, manage, and optimize the operations of the smart power grid. It also assists utilities in generating, distributing, and managing electricity consumption in real time [90]. The overall goal of EMS is to enhance grid stability through effective management of the power resources, energy supplies, and loss as well as enhance efficiency in general. Information collected by EMS in actual time varies in different sectors of the smart-grid tools in order to carry out functions such as forecasting about the energy consumption and regulation of the amount of needed power. Machine learning methods are also applied in EMS to increase system performance, enhance prediction accuracy, and enable autonomous systems to work [91,92].

Grid Security

Grid security aims to protect the smart-grid infrastructure from cyberattacks, which can disrupt operations and cause major losses, such as data theft. To prevent such disruptions, smart grids employ security systems that ensure secure and reliable operation [93]. Cyberattacks primarily target smart grid-connected systems containing critical information. To prevent these attacks, grid security employs powerful encryption, secure authentication methods, and real-time monitoring for immediate threat identification and prevention. In addition, grids implement physical safeguards, such as fences, cameras, and guards, to prevent unauthorized physical access that could lead to equipment damage or cyber intrusions [94,95].

3.4. Detecting Methods

To ensure grid security and sustainability, several detection methods have been proposed in the literature, as summarized in Table 8.

Table 8. Summary of studies addressing grid security.

Method	Used Techniques	Description	References
Supervised Learning	Classification, regression	Learns by harnessing labeled input-output data to fit models that make predictions. Typical of email filtering, demand forecasting, and diagnostics.	[96,97]
Unsupervised Learning	Clustering, association	Tests the patterns and groupings in data sets with no predefined labels. Commonly used in anomaly detection and data segmentation.	[98,99]
Reinforcement Learning	Sequential decision making	Learns optimal actions with interactions with an environment. Tries to garner as much mutual gains in the long run, usually has a Markov Decision Processes model.	[100–102]
Dynamic Programming	Derivation of optimal policy	Divides complex tasks into solvable subproblems that are solved recursively. It is appropriate in cases where models of the systems are clear and where policies have to be optimized.	[103,104]
Monte Carlo Methods	Value-estimation methods	By random sampling, approximates returns/outcomes. Performs well on simulation-based settings but prone to the quality and overfitting concerns in samples.	[105,106]
Temporal Difference Methods	Online learning, partial feedback	Learns online without waiting until the end of the situation to know an outcome, or the outcome of learning is incomplete. Composes Monte Carlo ideas with dynamic programming to carry out adaptive predictions.	[107,108]
Deep Q-Network (DQN)	High-dimensional state control	Applies the principles of deep learning to generalize Q-values in large or continuous spaces and allow for adaptation in complex environments, making decisions when facing extreme phenomena.	[109,110]
Bayesian Methods	Probabilistic modeling of uncertainty, decision uncertainty	Applies Bayesian probability in estimating certain uncertainties and, therefore, decision making, particularly when dealing with limited or noisy data.	[111,112]
Support Vector Machine (SVM)	Classification, regression	Seeks the optimal hyperplane between the classes, which supports linear and nonlinear classification by the use of kernel methods.	
Decision Tree	Interpretable rule-based modeling	Tree-based decisions are constructed as a sequence of rule-based splits. Simple and interpretable, normally used in classification.	[113,114]

3.4.1. Optimization Models

Optimization models are mathematical and computational tools used to enhance the performance and efficiency of electricity transmission and distribution. These models allow smart-grid systems to make intelligent decisions and determine the optimal operational strategy in various scenarios. They also help reduce energy loss and improve real-time efficiency. Recent investigations demonstrate that these approaches are effective in both simulations and real-world situations [115].

Barajas et al. [116] introduced optimization models to enhance the generation, transmission, and distribution of electricity in a microscopic system using the energy hub concept. The first linear programming model addressed hourly operations , while a mixed-integer linear programming (MILP) model was applied for 24 h planning with 21,773 constraints. Results from Mexico's national grid show that the integrated-cycle plant generates 50% of power, while the hydroelectric plant provides 18%. The proposed model achieved a 30% energy objective using the existing architecture, demonstrating its suitability for large-scale energy management.

Tianqi et al. [117] introduced an optimization model for smart-grid investment planning during T&D tariff reform. The proposed model integrates indices for planning efficiency and benefits, while also considering constraints from previous projects and the relative significance of different projects. A risk index model was additionally developed to address the uncertainty that can affect planning decisions. The results show that the proposed models outperform traditional models by effectively responding to changing user load profiles under peak–valley time-of-use pricing schemes and enhancing planning efficiency.

In their study, Ogbogu et al. [118] used an optimal power-flow model to reduce investment costs, transmission losses, and maintenance requirements, as well as optimize power distribution. The proposed model is subject to constraints, such as generator power limits, bus voltages, transformer tap applications, and transmission line capability. These limitations are determined by the physical and operational principles of the power system.

3.4.2. Numerical Methods

Numerical models refer to mathematical tools that deal with physical systems by dexterously representing a system with numbers and an algorithm to simulate and analyze real-life problems. These models are critical for monitoring, planning, and optimization in the system of transmission and distribution. They are perfectly applicable in the calculation of solutions, more so in complex systems beyond the scope of analytical or manual calculation methods [119].Bragin et al. [120] proposed a surrogate Lagrangian relaxation (SLR) model to address the same issue in the smart grid by solving the coordinated operation of the entire T&D system. Such work aims to improve the efficiency of operationsand resource utilization. Convergence is compared with conventional methods like the sub-gradient model, where the results indicate that the model is effective and gives an added advantage to the previously available methods in enhancing the coordinated decision making.

A model where the electricity market equilibrium was presented as a transmission and distribution system in the work of Saukh et al. [121], using electricity circuit theory, Kirchhoff's laws, and a quadratic loss function. By using the measurement factor, this technique assists in making a more realistic simulation of the market and efficiently.

An OpenDSS model of the smart-grid planning using GIS data and CIM files was developed by Cordova et al. [122]. This strategy can achieve better results of simulating and enhancing decision-making within the distribution system, thus supporting effective open-source smart-grid planning.

Moradi et al. [123] proposed a coordinated mechanism of improving the performance of the T&D power system in smart-grid networks. The approach uses a mathematical step-by-step model, which considers grid risks that include the varying electricity demand, production, and behavior of electric vehicles. The proposed model is employed in battery storage, wind farms, switchable feeders, distributed generation and demand–response programs. There are also EV charging stations across different grid levels to cope with short-term uncertainty. The efficiency of the model was validated using with the IEEE RTS and test feeder of 33 nodes, validating the test experiment.

3.4.3. Artificial Intelligence Model (AI)

AI-based models are computer programs that use data for classification, prediction, and pattern recognition, encompassing a wide range of methods, such as Machine Learning (ML), Deep Learning (DL), data mining, evolutionary computation, and fuzzy logic [124]. One of the most well-known subsets of AI is ML, which focuses on identifying patterns in large datasets [125]. ML analyzes input features to develop a model that can be used for tasks like prediction, classification, and clustering.

There are three main categories of ML: Supervised Learning [126], Unsupervised Learning [127], and Reinforcement Learning (RL) [128]. Supervised learning models are trained on labeled datasets with known input–output pairs. The aim of supervised learning is to build a model that can accurately map the input to output. Common models include linear regression, Support Vector Machines, and Random Forests.

In contrast, unsupervised learning is used for unlabeled data. The primary objective is to identify hidden patterns and structures within the data. This method is primarily used for clustering and pattern recognition with the help of models like k-means and autoencoders.

Reinforcement learning uses a goal-driven approach where agents learn by interacting with the environment to optimize long-term cumulative rewards based on prior experience. A key principle of this learning method is the Markov Decision Process (MDP), which helps the agent determine the most effective policy. RL also identifies and manages the trade-off between exploration (trying new actions to discover outcomes) and exploitation (using known-good actions).

3.4.4. Random Forest (RF)

RF is an ML model used for prediction, which enhances accuracy, performance, and model stability by integrating the outputs of multiple decision trees [129]. RF commonly employs historical data from sensors, SCADA systems, and meteorological input to forecast a range of operational parameters. RF is also known for its ability to manage high-dimensional, nonlinear datasets that are typical of complex grid operations, and its resilience to overfitting, making it essential for applications such as load forecasting, fault detection, and optimizing power flow within smart grids [130,131].

3.4.5. Support Vector Machine (SVM)

SVM is a supervised learning method used for complex tasks such as classification and regression [132]. To address these two functions, it is defined as either Support Vector Regression (SVR) for continuous value prediction or Support Vector Classification (SVC) for classification.

SVR is used to predict continuous values. The primary objective of SVR is to identify a hyperplane that accurately fits the training data, while minimizing the error for data points that fall outside a predefined margin of tolerance [133,134].

3.4.6. XGBoost

Extreme Gradient Boosting, or XGBoost, is a powerful technique for building predictive models. The process entails sequentially adding decision trees, where each new tree corrects the errors of its predecessors. It uses internal regularization controls to prevent over-complexity and thus reduces the risk of overfitting. Due to its speed and accuracy, XGBoost is a popular tool in a variety of fields. It is designed to process big data quickly and efficiently by performing several tasks in parallel. Because of these benefits, it is commonly used to predict energy consumption, identify equipment anomalies, and detect abnormal trends in modern power-grid systems for energy distribution and delivery [135].

3.4.7. Artificial Neural Network (ANN)

ANNs are computational networks modeled after the human brain [136] and are composed of hierarchies of interconnected neurons. Each neuron performs a weighted summation of its inputs, followed by an activation function [137].

3.4.8. Long Short-Term Memory (LSTM)

Long Short-Term Memory (LSTM) networks are a variant of Recurrent Neural Networks (RNNs) designed to capture long-term dependencies to address the vanishing gradient problem. LSTMs also incorporate memory cells and a gating mechanism comprising an input gate, a forget gate, and an output gate to control the flow of information [138].

LSTMs are deployed in smart-grid transmission and distribution systems, where they have been applied to load forecasting, anomaly detection, and identifying cyberattacks. This is because they can effectively model temporal dependencies within time-series data from sensors and meters. This capability enables earlier detection of errors or attacks, enhancing grid stability and security [139,140].

3.4.9. Convolutional Neural Network (CNN)

CNNs are a type of deep learning model designed to capture spatial and temporal features hierarchies from input data. In smart-grid transmission and distribution, CNNs are used to identify anomalies, cyberattacks, and system faults using real-time data streaming from devices such as Phasor Measurement Units (PMUs), Remote Terminal Units (RTUs), and smart meters. The building block of a CNN is the convolutional layer, which applies learned filters to the input data to extract informative features [141].

In the context of the smart grid, CNNs are fundamental to many applications. They enhance cybersecurity by identifying anomalies resulting from infiltrations such as False Data Injection Attacks (FDIA), Denial of Service (DoS), and Load Redistribution Attacks. They also support in fault diagnosis by detecting and localizing line faults, transformer failures, and voltage instability. Finally, they enhance short-term load forecasting by learning temporal patterns from historical data [142]. CNNs play an essential role in improving the efficiency, reliability, and robustness of smart-grid T&D networks, as they can automatically extract spatial–temporal dependencies.

3.4.10. Recurrent Neural Network (RNN)

Recurrent Neural Networks (RNNs) are deep learning models used to process time-series and sequential data. For smart-grid transmission and distribution systems, RNNs are useful in applications with time-dependent characteristics, such as load forecasting, anomaly detection, and state estimation. Unlike feedforward networks, RNNs do not maintain a separate representation for time steps; instead, information persists through repeating loops of weights [140].

Due to their ability to model temporal dynamics, RNNs are suitable for forecasting future power load based on historical data. They can also detect patterns that indicate imminent failure or identify anomalies (e.g., cyberattacks like FDIA and DoS attacks) [143]. For fault diagnosis, RNNs can process voltage and current waveforms to detect abnormalities by comparing them with normal behavioral trends. In cybersecurity, the RNN can examine communication or measurement sequences for unusual patterns.

RNN-based algorithms also have potential for load forecasting, as they can extract daily and seasonal variations, thus supporting utilities in generation and distribution planning. More advanced versions, such as LSTM and GRUs, overcome shortcomings caused by the vanishing gradients problem, making RNNs more useful for smart-grid applications [144]. Overall, RNNs play an important role in enhancing the intelligence, adaptivity, and security of smart-grid operations.

3.4.11. Autoencoder Models

Autoencoders are unsupervised neural network architectures that learn compressed representation of the input data. They are consist of two parts: the encoder, which compresses the input into a latent-space representation, and the decoder, which reconstructs the input from this representation. The purpose of the autoencoder is to reconstruct the input as accurately as possible, with the latent space capturing the most salient features of the data [145].

In smart-grid transmission and distribution systems, autoencoders can be used for anomaly detection, fault diagnosis, and data denoising. As an inherent feature of their learning process, autoencoders that have been trained on non-anomalous operational data are highly sensitive to deviations that result from False Data Injection Attacks (FDIA), Load Redistribution Attacks, and sensor failures [146]. By examining the reconstruction error, the model can determine when a data point is not consistent with the expected norm. For instance, a sudden voltage collapse or abnormal load generation can be flagged as suspicious if the reconstruction loss exceeds a predefined threshold.

Autoencoders can also be employed for dimensionality reduction of high-dimensional observations from multiple smart-grid sensors, retaining the most relevant features for use as input for forecasting or classification. Furthermore, models such as sparse autoencoders, denoising autoencoders, and variational autoencoders (VAEs) enhance performance through improved feature extraction, noise robustness, and probabilistic interpretation [147]. In this context, autoencoders represent a valuable tool for the reliable identification of disturbances and efficient representation of data, thus contributing to the security, resilience, and efficiency of smart-grid systems.

3.4.12. Self-Organizing Map (SOM)

The self-organizing map (SOM) is an unsupervised neural network model suggested by Teuvo Kohonen, known for clustering and visualizing high-dimensional data by projecting it into a lower-dimensional grid (usually 2D) [148].

In smart-grid transmission and distribution systems, SOMs are especially beneficial for load profiling, anomaly detection, consumer behavior analysis, and fault-type detection. The SOM is consists of neurons in a 2D lattice, each of which is linked to a weight vector of the same size as the input [149,150]. During training, the SOM undergoes competitive learning, reorganizing itself so that similar input patterns will activate neurons that are nearby each other on the map.

In the context of the smart grid, SOMs are employed to identify patterns in consumer-load profiles to group similar consumption behavior and assist with demand-side management. They can also detect abnormal sensor or measurement data that could signal a failure or equipment defect. In addition, SOMs can help classify power quality events (PQEs) or cluster different parts of a network according to their operation [151]. As SOMs can be trained in an unsupervised manner, they are well-suited for situations where data labeling is costly or impossible. Ultimately, SOMs enhance smart-grid operations, imparting the abilities of pattern discovery, anomaly detection, and dataset visualization.

3.4.13. Generative Adversarial Networks (GANs)

Generative Adversarial Networks (GANs), a family of deep learning models proposed in 2014 by Ian Goodfellow, operate on a generate-and-test basis. A GAN consists of two neural networks: a generator, tasked with producing synthetic data that appears realistic, and a discriminator, which acts as a classifier to determine whether the data is realistic or unrealistic [152,153]. The generator strives to produce more realistic data to deceive the discriminator, while the discriminator works to improve its classification accuracy. With each iteration, the generator gets better at deceiving the discriminator, and over time, it produces high-quality synthetic data that closely matches the real data distribution.

With regard to smart-grid transmission and distribution systems, GANs have several impactful applications. They can be used for data augmentation, particularly in datasets with imbalanced representation where incidents like cyberattacks (e.g., False Data Injection Attacks, DoS) are rare. By creating synthetic data of these rare incidents, GANs enhance the effectiveness of classifiers trained to detect them.

GANs are also used for anomaly detection, where they learn the characteristics of normal operational data and flag data points with high discriminator loss as anomalies. Additionally, GANs can enable privacy-preserving grid data sharing, enabling secure risk-sharing partnerships among utilities. In predictive maintenance, GANs can be used to model the effects of hypothetical fault conditions and overloads under future operating conditions [154]. Other variants, such as Conditional GANs (CGANs) and Wasserstein GANs (WGANs), improve training stability and enable conditional generation based on labels like grid zones, time windows, or attack types. In general, GANs can make the grid smarter and more secure by promoting realistic simulation, data augmentation, and anomaly detection in complex T&D networks [155].

3.5. Hybrid Models

The methodology in [156] employs multivariable regression methods, including linear, polynomial, and exponential regressions, for predicting peak loads. When these models were applied to Jordan's electricity grid, achieved an accuracy of nearly 90%, with a performance comparable to regular exponential regression techniques. As discussed in [157], the aim is to examine issues of power consumption at the national level through active long-term demand forecasting for a deregulated electricity market. The study also underlines the importance of activities like energy procurement, infrastructure expansion, and contract management.

[158] highlight a deficiency in prior research regarding hybrid deep learning. They note that limited work has explored combine multiple deep learning structures for large-scale smart-grid load forecasting. According to their findings, combining models improves predictive accuracy and provides insights into the relationships between various models and complex data patterns. In [159], a hybrid algorithm was proposed that combines evolutionary deep learning with LSTM and a genetic algorithm to optimize parameters as lag windows and the number of neurons. This mixed method proved more accurate and flexible than traditional forecasting methods.

In [160], the authors examined two recurrent neural network sequence-to-sequence (S2S) implementations to predict building-level energy consumption using GRU and LSTM layers. They also compared these hybrid models with five conventional forecasting techniques: multiple linear regression, stochastic time-series analysis, exponential smoothing, Kalman filtering, and state-space modeling, using an identical dataset for the comparison. An LSTM-based approach proposed by Yong et al. [161] was specifically designed to forecast the same-day energy demand of buildings.

3.6. Cyberattacks in Transmission and Distribution

Cyberattacks are malicious activities designed to damage networks, steal data, and disrupt normal processes in the smart grid. Their primary goal is to compromise transmission and distribution, potentially leading to blackouts and physical damage. Several types of cyberattacks target T&D systems, with each having unique features and potential outcomes. For instance, a Man-in-the-Middle (MiTM) attack targets communication between two devices or users. Similarly, Replay Attacks (RA) and Time Synchronization Attacks (TSA) can replay or alter measurement data and signals.

TSA and False Data Injection Attacks (FDIA) are designed to manipulate smart meter readings in order to manipulate state estimation and evade bad-data detection systems. Additionally, Denial of Service (DoS) attacks prevent information from reaching its intended destination. Countermeasures are introduced to detect these attacks, protect the grid, and prioritize cybersecurity efforts and resources. Some of these cyberattacks pose a greater threat than others due to their potential impact on grid stability and security [162]. Table 9 provides a comparative analysis of several studies focusing on cyberattacks in T&D.

Table 9. Comparative analysis of model in detection of cyberattacks (T&D).

Models	Attack Type	Learning Objective	Strengths (+)
DQDN [163]	False Data Injection	Derive optimal defense using sliding window over observation space	(+) Performs better than existing models in IEEE bus systems
DPDG [164]	DDoS	Monitor and manage network traffic to retain bandwidth	(+) Effective traffic analysis and filtering strategy
DPDG [165]	False Data Injection	Determine re-close timing after FDI attacks	(+) Considers generator rotor behavior changes post-attack
Actor-Critic NN [166]	Multiple Attack Types	Learn dynamic defense by observing CPS states in real-time	(+) Enables real-time operation (+) Learns optimal defense and attack strategies
POMDP [167]	DoS, FDI, Jamming	Minimize false alarms and detection delay by optimal action selection	(+) Operates online (+) Model-free approach
Q-learning [168]	Multiple ZigBee Attacks	Identify ideal actions to prevent ZigBee network threats	(+) Evaluated across six attack types (+) Combines detection and prevention using ML
Inverse RL [169]	Back-off Attack	General defense using Generative Adversarial Imitation Learning	(+) Works in both offline and live settings (+) Combines attack detection and prevention
DQN [170]	Jamming	Learn frequency patterns to optimize spectrum selection	(+) SINR and switching cost-aware (+) Reduces detection time and false alarms
Q-learning [171]	DoS Amplification	Alleviate DNS congestion post-amplification attacks	(+) Based on real DNS amplification scenarios
POMDP [172]	False Data Injection	Maintain optimal power flow and improve grid resilience	(+) Uses Maximum Likelihood Estimation (MLE) for effective recovery

3.6.1. Man-in-the-Middle Attack (MiTM)

A Man-in-the-Middle (MiTM) attack occurs when an attacker intercepts communication between two parties or devices and impersonates one of them, making it seem as though the information is being transmitted normally [173]. Kulkarni et al. [174] investigated the potential security threats related to MiTM attacks in power systems, focusing on vulnerabilities in the Modbus TCP/IP protocol used for communication.

Fritz et al. [175] developed a prototype of a MiTM attack on a smart-grid emulation platform. The authors also demonstrated a technique for compromising the integrity and validity of IEEE Synchrophasor protocol packets. Detecting packet interception is difficult because of the physical distance between the PMU and the Phasor Data Concentrator (PDC). Additionally, the PDC acquires data with minimal time for encryption, authentication, and integrity testing.

3.6.2. Replay Attack (RA) and Time-Delay Attack (TDA)

Accurate timing of control signals is essential for maintaining proper system operation. Time-Delay Attacks (TDAs) compromise this by introducing random delays in data transmission and reception, which can disrupt timing-dependent control components [175]. In contrast, Replay Attacks (R) are used to capture sensor measurements over a set period and then replace real-time data with these previous measurements. Additionally, they may involve retransmitting past recorded control signals from the operator to the actuator [176,177]. These two attacks use legitimate past data to manipulate system behavior, resulting in a system state that differs from the intended scenario. This can lead to major outages by disrupting and damaging the power grid [178].

3.6.3. False Data Injection Attack (FDIA)

Smart grids represent a radical departure from conventional electrical networks, aiming to promote interconnection among operators, suppliers, and users to create a tightly linked environment. This greater interconnection introduces significant challenges, particularly concerning consumer data privacy and the safe operation of the grid. A major concern is the vulnerability of real-time system operations, including generator synchrophasormeasurements, to cyber threats like FDIAs.

FDIAs pose a significant threat to smart-grid cybersecurity. First proposed by Liu et al. [179], compromises the state estimation (SE) process at the control center by injecting false values into measurement data. Although bad-data detection (BDD) tools, which use the l2-norm, are employed to identify inconsistencies, it is possible to design attack strategies that manipulate these systems. By injecting errors stealthily, attackers can bypass the BDD tools.

The FDIA is a hidden attack due to the subtle modification of the attack vector in the measurement data. Attackers are typically modeled as either requiring knowledge of the grid topology to launch an attack (FC-attackers) relying solely on observed data without topology information. FDIA behavior is also affected by application-specific vulnerabilities. For example, open communication links make Wireless Sensor Networks (WSNs) more susceptible. In power systems, attacks are more difficult to execute because of the complexity of modeling real grid parameters. Attackers may conduct broad or targeted attacks depending on their objectives and the available system information. Crucially, these attacks may threaten the integrity, availability, or confidentiality of the system. One example is providing false meter readings, which can lead to incorrect system operation, service outages, and private customer data leakage, as recorded in the AMI [180].

3.6.4. Load Redistribution Attack (LRA)

LRA is a type of FDIA attack proposed by Yuan et al. [181] and collaborators, which aims to disrupt the operation of the power system by targeting the Economic Dispatch (ED) process. ED aims to minimize generation costs and optimize power distribution to meet load demands. When LRA manipulates the estimated system state, it can lead to incorrect ED decisions, resulting in grid instable and inefficiency.

LRA has two primary objectives: immediate effects or delayed effects. The immediate effect aims to significantly increase the operating cost right after the attack. In contrast, the delayed effect aims to gradually overload power lines, which can ultimately cause physical damage to the power grid.

3.6.5. Denial of Service (DoS)

In the smart grid, there are numerous measurement devices, such as smart meters (SMs), data aggregators, Phasor Measurement Units (PMUs), Remote Terminal Units (RTUs), Intelligent Electronic Devices (IEDs), and Programmable Logic Controllers (PLCs). These devices exhibit vulnerabilities that can be exploited for Denial of Service (DoS) attacks.

A DoS attack in a power system makes measurement data unavailable to both users and the control system, disrupting communication and causing network instability. The attack also hinders the smart-grid ability to capture events during its occurrence. A DoS attack can be caused by flooding a device or communication channel with high volumes of data or exploiting system weaknesses such as jamming or routing vulnerabilities [182–184]. A puppet attack is a newer form of DoS attack that blocks communication in the Advanced

Metering Infrastructure (AMI) network [185]. The attacker selects normal nodes to send attack data. When those nodes receive the data, the attacker continues to send more, overloading the system and blocking communication.

4. Energy Utilization

The objectives of energy T&D models are to effectively transfer energy from generation plants to consumers. Energy utilization has become a critical concern due to the pressing need to reduce consumption, combat climate change, and enhance the sustainability of energy systems. While the oil and gas industry has long driven global economic growth, the increasing link between national economies and global energy demands necessitates maximizing energy efficiency to harmonize economic growth with environmental protection. Buildings, which account for about 34% of global energy consumption and carbon emissions, are a major source of energy use and therefore a priority for efficiency enhancement, as discussed in [186]. This specific study, however, did not use a dedicated model to examine energy utilization. In contrast, studies by [187,188] used optimization models to quantify the impacts of energy use and its utilization.

Conventional building energy management strategies have traditionally been based on fixed power grids and conventional energy systems, which are often inefficient, costly, and environmentally harmful [189–192]. Over the last few years, there has increasingly shifted toward smarter energy management systems through the integration of renewable energy sources (RES) into built environments [193–196]. These systems, which include solar photovoltaic (PV) systems, wind turbines (WTs), and energy-storage systems (ESS), aim to reduce dependence on the grid, lower energy costs, and minimize building's carbon footprints [197–199].

The concept of energy utilization optimization in buildings encompasses a diverse range of strategies, including demand-side management (DSM), smart grids, energy storage, and real-time monitoring of consumption patterns [200–204]. These studies highlight that optimization techniques such as genetic algorithms and demand-side strategies are pivotal in addressing the dynamic challenges of balancing energy supply and demand, managing uncertainty, and coordinating resources. The ability to track energy utilization in real time enables adaptive building operations to optimize energy use, minimize losses, and coordinate consumption with periods of high renewable energy production [205–207]. These studies also emphasize the evolution of solar technologies, from first-generation silicon-based cells to advanced PV/T and quantum dot cells, aiming for improved efficiency and lower costs. Such developments, supported by policy incentives and innovation, are critical for maximizing solar energy utilization and achieving global decarbonization goals.

With increasing urgency in energy conservation efforts, the advantages of combining smart meters with building energy management systems (BEMS) are abundant [191,208]. These systems can automate energy optimization processes, respond to changes in energy generation, and enhance overall efficiency. Smart meters facilitate flexible pricing, enable effective peak demand management, and improve grid stability. They also empower occupants to monitor and control their energy usage. At the same time, rapid cost reductions in solar, wind, and battery-storage technologies, along with climate policies and carbon pricing, have accelerated a global shift toward electricity-dominated energy systems [209,210]. Recent research on wind energy emphasizes the optimization of turbine performance, control systems, and offshore infrastructure to enhance efficiency, reliability, and integration into future hydrogen economies and applications.

This section discusses the importance of smart meters in maximizing energy utilization in buildings, exploring how they increase energy efficiency, decrease operational expenses, and facilitate the integration of renewable sources. By investigating new developments and trends in smart meter technology, this section provides a comprehensive overview of how these systems transform energy use in the built environment and contribute to the broader goal of sustainability. Figure 4 provides a complete overview of energy utilization in the smart grid.

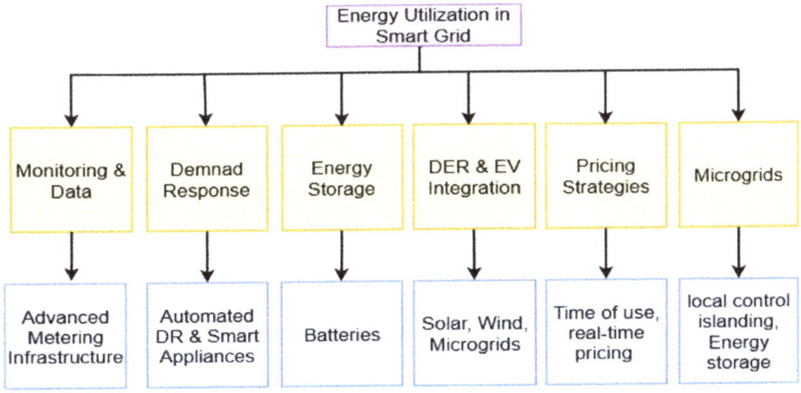

Figure 4. Energy utilization strategies in smart grid.

4.1. Technologies for Energy Utilization in Smart Grid

The integration of smart grids has transformed energy systems by introducing advanced technologies designed to enhance efficiency, reliability, and sustainability in energy use. For example, policy-driven demand-side management in industrial sectors further improves energy efficiency and support broader sustainability goals [68,211]. Smart grids are characterized by their ability to provide real-time monitoring, data analytics, and automated control, which collectively enable better energy management and optimization. As discussed by [212,213], demand-side management plays a pivotal role in enhancing energy efficiency, reducing emissions, and facilitating smart-grid integration through optimized consumption strategies and the use of hybrid algorithms.

Several major technologies are key to the effective operation of smart grids [214–216]. These include renewable energy integration, Advanced Metering Infrastructure (AMI), advances in energy-storage technologies, and demand-side management tools. The key technologies that enable optimal energy use within smart grids and how they affect efficient grid operation are presented in Table 10.

Table 10. Comparison table of energy utilization strategies in smart grids.

Technology	Description	Components/Tools	Advantages	Limitations
Advanced Metering Infrastructure (AMI)	Real-time monitoring of energy consumption and grid conditions with two-way communication	Smart meters, communication networks, data management systems	Granular consumption insights, remote meter reading, improved billing accuracy	Privacy concerns, high deployment costs, data management complexity
Renewable Energy Integration	Incorporation of variable renewable sources (solar PV, wind, hydropower) into the grid	Solar PVs, wind turbines, hydropower, forecasting, data analytics, machine learning	Reduces fossil fuel reliance, supports distributed energy resources, optimizes grid with forecasting	Variability and intermittency, grid balancing, forecasting accuracy

Table 10. *Cont.*

Technology	Description	Components/Tools	Advantages	Limitations
Energy-Storage Systems (ESS)	Storing excess energy for balancing supply and demand	Batteries (Li-ion, flow), pumped hydro storage	Grid stability, peak load management, reduces need for peaking plants	High capital costs, energy losses during storage, lifespan limits
Demand-Side Management (DSM)	Strategies to influence consumer energy use for efficiency and peak load reduction	Dynamic pricing, smart appliances, incentives	Peak load reduction, energy efficiency, demand flexibility	Requires consumer participation, behavioral change, technology adoption
Artificial Intelligence & Machine Learning (AI/ML)	Advanced analytics and real-time optimization for grid management	Predictive analytics, fault-detection algorithms, energy dispatch optimization	Improved forecasting, fault prediction, optimized resource dispatch	Data quality, algorithm transparency, computational resources

4.1.1. Advanced Metering Infrastructure (AMI)

Smart grids rely on Advanced Metering Infrastructure (AMI), which is an integrated system of smart meters, communication networks, and data management platfoms. AMI systems gather real-time data on energy usage, voltage, and power quality for analysis and monitoring by utilities. AMI systems support self-configuration, algorithm integration, and scalable deployment, paving the way for intelligent energy market participation and system resilience [215,217].

These systems also enable utilities to monitor energy consumption in detail, providing valuable insights into usage patterns. AMI helps facilities gather real-time energy usage data, enabling instant access for both consumers and utilities. Furthermore, AMI facilitates two-way communication that supports remote meter control and troubleshooting for both consumers and utilities. Table 11 presents a comparative analysis of several studies and their key findings.

Table 11. Summary of recent AMI research and market trends.

Study	Year	Focus/Scope	Key Findings/Highlights	Market/Tech Trends	Key Utilities/Regions Covered
[218]	2025	Market forecasts and technology trends	Global AMI market projected to reach USD 29.63B in 2025 at 11.23% CAGR; smart-grid integration drives adoption; AI-enabled data analytics and automation improving operational efficiency; government policies promote AMI adoption	Rise of smart-grid solutions; real-time data analytics; AI integration; automation for outage management	U.S., Germany, China leading smart-city initiatives
[219]	2025	Global survey and case studies of AMI implementations	AMI enables major utility operation changes beyond meter-to-cash; mature, standardized technologies; detailed cost/benefit and implementation data from utilities worldwide	Standardized AMI functionalities; impact on utility operations and resource management	Utilities in U.S., Italy (ENEL), Brazil (Electrobras), India, Australia

Table 11. Cont.

Study	Year	Focus/Scope	Key Findings/Highlights	Market/Tech Trends	Key Utilities/Regions Covered
[220]	2025	Market size and strategic insights	Market estimated at USD 21.98B in 2024, growing to USD 24.89B in 2025, projected USD 45.48B by 2030; emphasizes regulatory and digital transformation drivers	Digital transformation; regulatory compliance; tariff impacts on AMI deployment	Global overview with emphasis on U.S. tariff impact
[221]	2025	Technical differentiation and solutions	AMI provides two-way, real-time communication beyond AMR; features include outage detection, demand response; improves operational efficiency and reduces labor	Advanced communication networks; real-time data collection and remote monitoring	General market focus with technology comparisons

4.1.2. Renewable Energy Integration

Smart grids are crucial for integrating renewable energy systems, such as solar photovoltaics (PV), wind turbines (WTs), and hydropower, to overcome the constraints of conventional fossil-fuel-based energy systems. Advanced optimization techniques and smart-grid technologies are key to managing renewable variability and ensuring reliable, efficient, and sustainable energy supply [222,223].

Integrating a large number of renewable energy sources into the grid introduces variability and uncertainty. Smart grids handle this through advanced forecasting, energy storage, and demand–response techniques. The integration of PVs and electric vehicles (EVs) requires coordinated operation models and simulation-based techniques to mitigate grid instability, optimize load scheduling, and reduce energy costs. In parallel, modern power systems leverage AI, machine learning, and dynamic modeling to enhance grid resilience, enabling real-time control and supporting the seamless integration of renewable energy sources [224,225]. Moreover, intelligent grids allow for the efficient use of distributed energy resources (DERs), including rooftop solar panels and small-scale wind turbines, to provide local energy. Smart grids also employ weather forecasting, data mining, and machine learning to predict the availability of renewable energy, thereby streamlining grid operations, especially during periods of fluctuating energy production [226,227].

Table 12 presents a comparative analysis of several studies, highlighting their limitations.

Table 12. Comparison of existing studies on renewable energy integration in energy utilization.

Paper/Source	Focus/Scope	Key Findings/Highlights	Limitations/Challenges Identified by Paper
[228]	Optimization of grid reliability and profitability with renewables and energy-storage systems (ESS)	Introduces reliable efficiency index (REI); improves grid reliability and profitability by ∼38% in tests; environmental benefits demonstrated	Focuses on simulation results (IEEE 33-bus); real-world validation needed; complexity of calculations limits immediate practical adoption
[229]	Review of challenges using energy storage to mitigate renewables intermittency	Highlights critical role of ESS in enhancing grid flexibility and stability; discusses market and technical barriers	Lacks new experimental data; relies on review of existing literature; economic barriers remain substantial

Table 12. *Cont.*

Paper/Source	Focus/Scope	Key Findings/Highlights	Limitations/Challenges Identified by Paper
[230]	Proposes energy hub framework optimizing multi-carrier energy systems and managing renewable uncertainty	Framework optimizes renewables integration with demand–response and storage; manages variability effectively	Model assumptions may oversimplify real energy systems; uncertainty quantification depends on input data quality
[231]	Examines demand-side flexibility impact on renewable integration and decarbonization in regional power grids	Demonstrates demand flexibility as key to enabling higher renewable penetration and supporting low-carbon goals	Case study limited to West Inner Mongolia, may not generalize globally; data availability constraints noted
[232]	Develops optimization models balancing cost, emissions, and grid stability during integration	Incorporates uncertainty, storage, and demand response; provides robust planning tools for renewable deployment	Computational complexity limiting scalability; real-time control integration not addressed

4.1.3. Energy-Storage Systems (ESS)

One of the key elements of smart grids is the energy-storage systems (ESS), which includes batteries and pumped hydro storage facilities. These systems store energy surpluses generated by renewable sources during periods of high production and release them during times of low generation to maintain the balance between supply and demand [233]. Accurate load forecasting using advanced deep learning models, such as the IntDEM framework, enables optimized charging and discharging schedules for ESS in IoT-enabled smart grids [234].

ESS ensures grid stability by storing energy during low-demand periods and discharging it during peak demand, thus preventing blackouts and maintaining a constant supply. By discharging stored energy during peak load periods, ESS helps reduce the need for costly peaking power plants and reduce electricity prices [235]. An integrated demand–response framework that combines medium-term forecasting and an optimized battery ESS can reduce costs and emissions while improving energy efficiency [236]. Furthermore, ESS enables greater integration of intermittent renewable energy sources by compensating for fluctuations in generation, thereby ensuring a continuous and reliable energy supply.

The following Table 13 compares several studies in the domain of energy utilization by highlighting their key findings and limitations.

Table 13. Comparison of existing studies: energy-storage systems in energy utilization.

Paper/Source	Focus	Key Findings	Limitations
[237]	Integration of solar cells with energy-storage devices (batteries, supercapacitors)	Reviews latest advances in material composition, system construction, and performance; addresses intermittent solar generation by integrated storage	Emphasis on lab-scale and prototype systems; practical large-scale deployment and cost challenges remain open
[238]	Comprehensive categorization and description of mechanical, electrochemical, thermal ESS	Provides systemic classification, explains working principles, and discusses suitability for grid and renewable integration	Lacks in-depth economic analysis or real-world validation of some emerging technologies

Table 13. Cont.

Paper/Source	Focus	Key Findings	Limitations
[239]	Overview of technological advancements in high-power ESS including lithium-ion batteries	Highlights latest improvements in power density, cycle life, and applications in grid and transport sectors	Focuses mostly on lithium-ion technologies; emerging alternatives less emphasized; safety concerns noted
[240]	Explores ESS role in enhancing grid flexibility and renewable energy integration	Demonstrates grid services improvements with ESS deployment; discusses technical integration and policy frameworks	Implementation barriers include high costs, regulatory uncertainty, and interoperability challenges
[241]	Reviews advances in battery ESS technologies, focusing on Li-ion and alternative chemistries	Details technological progress in energy density, lifespan, sustainability, and safety; commercial trends tracked	Battery degradation, recycling challenges, and raw material supply risks highlighted

4.1.4. Demand-Side Management (DSM)

Demand-side management (DSM) refers to strategies employed to influence consumers' energy usage in order to improve efficiency and reduce grid stress. DSM systems enable utilities to actively manage consumption patterns through incentives, dynamic pricing, and direct control of appliances. For instance, one study used conceptual frameworks like intersectionality and social license to examine social barriers to DSM adoption, while another employed statistical and machine learning models to optimize DSM implementation through dynamic pricing and load-control strategies [242,243].

Smart grids enable dynamic pricing schemes that encourage the use of energy during off-peak periods, thereby reducing peak load demand and promoting a more balanced energy distribution [244]. Furthermore, appliances equipped with smart technologies, such as smart thermostats, washing machines, and refrigerators, can be controlled remotely to optimize energy consumption based on grid conditions and pricing signals. The key findings and limitations of several previous studies are presented in Table 14.

Table 14. Comparison of existing studies addressing demand-side management in energy utilization.

Paper/Source	Focus	Key Findings	Limitations
[245]	Development of DSM model for Pakistan analyzing multiple DSM scenarios (energy efficiency, conservation, load management)	DSM can reduce electricity demand by ∼26.4% and significantly decrease GHG emissions; DSM needs institutional policy and regulatory support for effectiveness	Model based on projections; practical implementation and regulatory framework lacking in Pakistan; regional focus limits generalizability
[246]	Integration of electric vehicles (EVs) with DSM to optimize peak load shifting and distributed resource allocation	Combining DSM with EVs effectively shifts peak load and improves grid flexibility; optimal resource allocation enhances cost savings	Limited by assumption of EV penetration rates and grid conditions; complexities of large-scale EV integration not fully addressed
[247]	Large-scale DSM control using discrete deep reinforcement learning (DRL) for residential energy management	DRL enables scalable, adaptive load control, reducing peak demand and smoothing load profiles in smart residential districts	DRL model requires extensive data and computational resources; transferability to other regions with different load patterns not tested

Table 14. *Cont.*

Paper/Source	Focus	Key Findings	Limitations
[248]	Economic analysis of DSM system profitability under dynamic European electricity markets	DSM systems can be profitable under flexible pricing and market conditions; financial incentives are key to DSM adoption	Profitability sensitive to market conditions and tariff structures; analyses rely on market forecasts which can be volatile
[249]	Examines combined DSM strategies and their effect on microgrid cost control and reliability	Multiple DSM strategies help maintain power balance and reduce operational costs without compromising reliability	Microgrid-scale focus limits applicability to larger grids; strategy coordination complexity may increase operational overhead

4.1.5. Artificial Intelligence and Machine Learning

Artificial intelligence (AI) and machine learning (ML) play an increasingly important role in smart-grid technology by enabling advanced analytics, real-time decision-making, and optimization. For example, an AI-driven framework using Deep Reinforcement Learning, blockchain, and federated learning has demonstrated enhance virtual power plant and smart-grid performance by optimizing renewable integration and ensuring secure, scalable operation [250,251].

ML algorithms can predict energy demand patterns, renewable energy generation, and potential faults in the grid, thereby allowing for proactive management and reduced downtime. Furthermore, AI-powered systems optimize energy dispatch from various resources, including renewable energy, storage, and conventional power plants, ensuring the most efficient and cost-effective energy mix. AI algorithms can analyze sensor data to detect anomalies, predict equipment failures, and initiate preventive maintenance, thereby improving grid reliability [252].

A comparison of several studies highlights the importance of AI in the energy system, as presented in Table 15.

Table 15. Comparative analysis of recent research on AI in energy systems.

Paper/Source	Focus	Key Findings	Limitations
[253]	Cross-sector review of AI's role in energy efficiency and grid management	AI reduces energy use in buildings (9-30%), communication networks (up to 40%), and manufacturing; supports predictive maintenance, smart grid, and renewable integration	Highlights large, uncertain, and rising energy demand of AI systems and data centers; variability in projected energy savings; data privacy and ethical concerns
[254]	State-of-the-art and trends in AI for energy production/distribution, smart grid, demand response, and asset management	AI optimizes forecasting, grid management, energy efficiency, storage, and asset maintenance; advances in ML, digital twins, and predictive controls	Key challenges include data scarcity, model explainability, energy/resource footprint of AI, trust, and regulatory/licensing complexity
[255]	Review of AI and ML for building energy forecasting and system optimization	AI models outperform conventional approaches in accuracy and adaptivity; Hybrid AI-IoT methods enable real-time management	Demands large, high-quality datasets and significant computational resources; Generalization to varied building types is challenging

Table 15. *Cont.*

Paper/Source	Focus	Key Findings	Limitations
[256]	Comprehensive review of AI/ML in energy generation, forecasting, grid stability, predictive maintenance	AI/ML algorithms improve grid reliability, enable predictive maintenance, optimize energy storage, and reduce waste in renewable integration	Integration is challenged by intermittency, requirement for robust real-time data, and transferability across regions and technologies
[257]	Scoping review of AI for emission reduction and energy-efficiency measures	AI enables targeted, real-time emission reduction, improved energy management, and dynamic grid balancing	Access to reliable emissions data is limited; implementation of AI at nationwide scale is complex; cybersecurity and transparency limitations

4.2. Cyber Attacks on Energy Utilization in Smart Grids

The integration of advanced technologies in smart grids has clearly transformed energy management into a more efficient, sustainable, and adaptive system. Nevertheless, this revolution has introduced new vulnerabilities, especially with respect to the security of digital infrastructure. As detailed in a comprehensive review by [258,259], the rising threat of cyberattacks is a major challenge due to the widespread integration of IoT in modern energy systems. These studies examine smart-grid architecture, categorize cyber threats, and review past attack incidents. As smart grids rely on sophisticated networks of communication and data management and interconnected devices, they are prime targets for cybercriminals who exploit potential flaws.

The effects of such cyberattacks on energy systems can be devastating, leading to shortages of energy supply, misrepresentation of consumption data, loss of revenue, and threats to national security. This section discusses different types of cyber threats to smart grids, the severity of these attacks, and strategies for risk mitigation.

Figure 5 illustrates electricity consumption throughout the life cycle of an AI model. For example, when implementing AI models in the energy market, electricity is consumed during model development, training, and deployment. These values are detailed in a study [253]. Furthermore, the energy sector continues to be heavily impacted by ransomware attacks, with critical infrastructure facing increasing vulnerabilities. According to a report by Sophos [260], the primary root causes of these attacks were exploited vulnerabilities (35%), compromised credentials (32%), and phishing emails (27%), as shown in Figure 6, while misconfiguration and other causes contributed marginally.

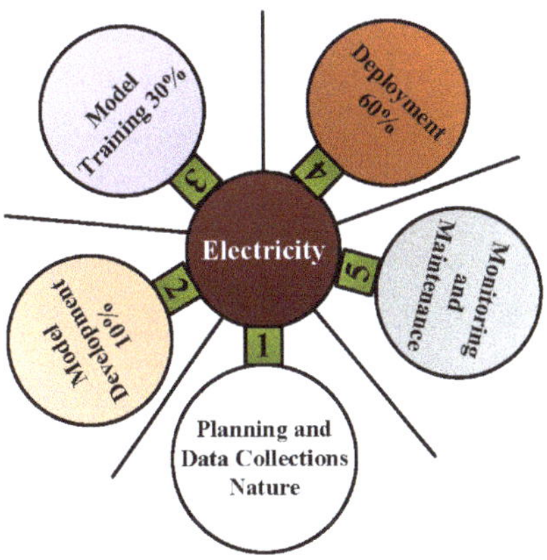

Figure 5. Electricity utilization in AI life cycle.

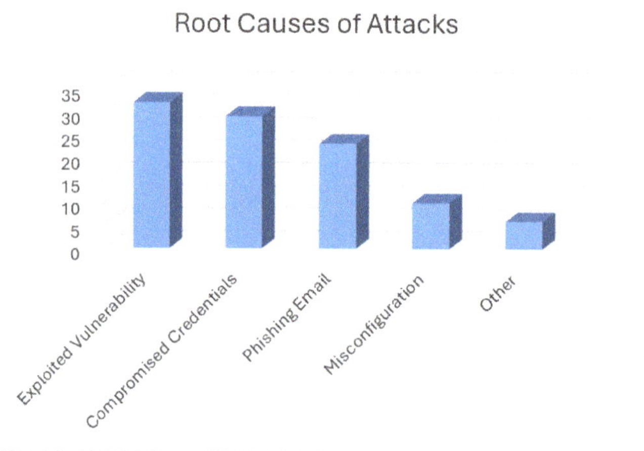

Figure 6. Root causes of attacks on energy utilization.

Figure 7 highlights the recovery methods, showing a strong reliance on backups, with the majority of organizations using backups to restore data after an attack. A smaller fraction reported paying ransom, recreating data, or using a combination of these methods.

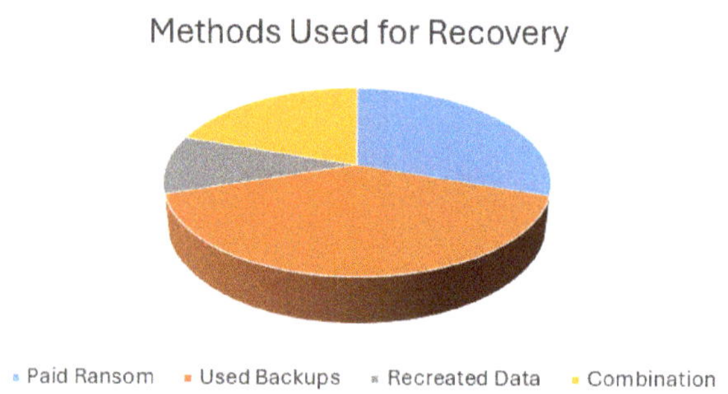

Figure 7. Methods used for recovery in energy utilization.

Furthermore, an infection range analysis, shown in Figure 8, revealed that in 30% of incidents, over 75% of the organization's devices were impacted, demonstrating the devastating reach of such attacks. Approximately 22% of cases involved infection rates below 25%, suggesting some success in containment strategies.

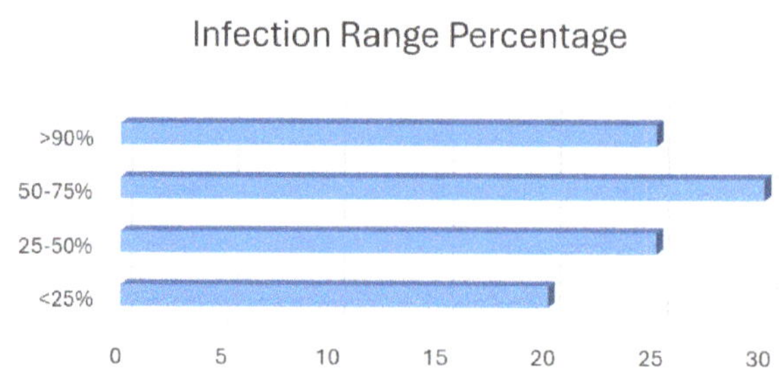

Figure 8. Infection Range Percentage in Energy Utilization.

4.3. Types of Cyberattacks on Energy Systems

Energy grids are typically targeted by cyberattacks that exploit the digital infrastructure supporting the generation, distribution, and consumption of energy. Such attacks vary in size, complexity, and purpose and can be classified into the following categories.

Denial of Service (DoS) Attacks

When energy management systems are attacked, an overload of traffic can cause a disruption in services. In the smart grid, a DoS attack can target communication networks, leading to delays or a complete shutdown of energy-distribution systems. Studies by [66] and [261] discuss cybersecurity in energy systems using advanced models. The first paper proposes a resilient consensus control strategy embedded with reduced-order dynamic gain filters to identify DoS attacks in multi-agent systems, thereby improving energy efficiency and resilience. The second paper presents a new DDoS attack model for electric vehicle charging stations that employs a time-variant Poisson process and Ornstein–Uhlenbeck dynamics, though it does not provide specific datasets or quantitative accuracy metrics.

4.4. False Data Injection Attack (FDIA)

In a False Data Injection Attack (FDIA), an adversary selectively manipulates measurement data or sends misleading information to the grid's communication system. This is particularly dangerous in IoT-driven smart grids, where a large number of sensors and interconnected devices including AMI gateways, smart meters, distributed energy resources (DERs) controllers, smart inverters, Phasor Measurement Units (PMUs), Remote Terminal Units (RTUs), smart thermostats, and grid-connected sensors constantly monitor and transmit data. By vulnerabilities in these IoT devices and their messages, attackers can induce errors in state estimation, tamper with electricity consumption reports, conceal illegal activity, interfere with automatic control signals, or cause false alarms at the control center. Within the framework of smart grids, FDIAs have the potential to destabilize the grid, improperly trigger demand–response functions, change dynamic prices, and cause significant losses for both utilities and consumers.

The sophisticated nature of FDIAs makes them difficult to detect, as the injected data can closely mimic legitimate measurements and bypass common anomaly detection [262]. The study by [263] briefly highlights the importance and outcomes of FDIA attacks and the associated challenges. Researchers have addressed these challenges; for example, one study [264] proposed a multimodal deep learning model, including a variational graph auto-encoder (VGAE), a temporal convolutional network (TCN), and a gated recurrent unit (GRU), for the identification and classification of FDIAs under varied power-system topologies. Tested on the IEEE-14 and IEEE-118 bus systems, the model achieved an accuracy of 91.2%. To reduce the impact of FDIAs, a CNN-based model incorporating noise elimination and state estimation was implemented in [265]. In [266,267] researchers suggested a model based on variational mode decomposition (VMD), Fast Independent Component Analysis (FastICA), and an XGBoost classifier optimized with Particle Swarm Optimization (PSO). When tested on the IEEE-14 bus system, this model achieved an accuracy of 99.84%. The sophisticated nature of FDIAs makes them difficult to identify when attackers exploit flaws in communication networks. A common technique used to facilitate such data manipulation is the Man-in-the-Middle (MiTM) attack.

4.4.1. Man-in-the-Middle Attack (MiTM)

In a MiTM attack, an attacker secretly intercepts the traffic between two parties and may alter the data. By placing themselves in the communication process, they can sabotage the confidentiality and integrity of the information. MiTM attacks pose a significant threat to IoT-based smart grids, where data interactions must be secure and trustworthy to ensure reliable operations. To counter these threats, researchers have proposed several countermeasures. For example, Ref. [268] proposed an improved intrusion detection system (IDS) that integrates a machine-learning-based detection, a Quick Random Disturbance Algorithm, and a feature-selection technique. The model was validated on the IEEE-39 and IEEE-118 bus systems with an accuracy rate of 99.99%. In a different study, Ref. [269] proposed a BP-based hybrid deep learning model (AEXB) that integrates an autoencoder with XGBoost to detect and prevent MiTM attacks in real time. Similarly, in [270], a hybrid deep learning algorithm combining a Convolutional Neural Network (CNN) and Long Short-Term Memory (LSTM) was developed, tested on network traffic data, and achieved an accuracy of 97.85%.

4.4.2. Ransomware Attack

Besides MiTM and FDIA, ransomware attacks have emerged as a significant cybersecurity threat in smart grids [271]. Ransomware refers to malicious software which encrypts vital system information to extort money from victims in exchange for data recovery. Ransomware's rapid spread necessitates timely detection. To address these issues, various researchers have proposed diverse solutions. For example, Ref. [272] presented a ransomware detection model based on a CNN, which converts binary executable files into 2D grayscale images. This approach achieved an accuracy of 96.22%. In [273], the authors presented a twofold taxonomy of ransomware in cyber–physical systems (CPS), discussing infection vectors, targets, and objectives. They also analyzed real-life cases of ransomware against CPS to identify major security weaknesses. In a separate study, Ref. [274] developed a deep-learning-based ransomware detection framework for SCADA-automated vehicle charging station. Appraising the performance of three deep learning models a deep neural network (DNN), a convolutional neural network (CNN), and a long short-term memory (LSTM) the authors reported that their approach had an average accuracy of 98%.

4.4.3. Advanced Persistent Threats (APTs)

These represent long-term and strategic attacks intended to gain unauthorized entry into critical energy infrastructure. The actors are typically sophisticated and/or state-sponsored. APTs can be difficult to detect, as they are covert and designed to accumulate intelligence or cause destruction over an extended period. APTs may tamper with energy grid operations, steal secrets, or disrupt renewable energy systems [275,276]. These studies also investigate the growing threat of APTs in energy systems, using the MITRE ATTACK framework to identify weaknesses in current mitigation strategies.

Table 16 presents a comparative analysis of several studies, identifying the key findings and limitations.

Table 16. Comparative analysis of recent research on cyberattack detection in energy systems.

Paper/Source	Focus	Dataset(s) Used	Model(s) Proposed	Key Findings	Limitations
[277]	Deep learning for cyberattack detection in energy systems	SCADA system logs (imbalanced), simulated attack data	Hybrid deep learning w/feature reduction	Hybrid DL model improves threat detection in SCADA systems on imbalanced data; addresses spatial/temporal/structural patterns via feature reduction	Performance relies on quality of log data; practical deployment tested only on synthetic/simulated data
[278]	ML/Explainable AI for SCADA intrusion detection	SCADA wind farm operational data (real-world & simulated)	CatBoost, XGBRegressor, RNN, SHAP, LIME	XAI improves trust and transparency; RNNs outperform baseline; CatBoost/XGBRegressor excel in anomaly detection	Model explainability can add computational overhead; results may not generalize across SCADA types

Table 16. *Cont.*

Paper/Source	Focus	Dataset(s) Used	Model(s) Proposed	Key Findings	Limitations
[279]	ML for cyber event classification in grids	Mississippi State Univ./Oak Ridge National Lab datasets	XGBoost Classifier	Achieves 85% accuracy, F1 = 0.91 across cyber/physical event subdatasets	Dataset scope limited to North American grid scenarios; data labeling subjectivity
[280]	Privacy-preserving anomaly detection (federated ML)	Collaborative smart-grid substation operational logs	Federated Unsupervised Cyberattack Detection (UCAD)	UCAD model reliably detects attacks without centralizing raw data, preserving privacy	Requires homogeneous data distribution across substations; performance degrades with noisy data inputs
[281]	Cyber-robust data-driven decision support	Big data from smart renewables (various, not specified)	Data-driven Decision Tree (DT-DD) with security	DT-DD framework enhances environmental/operational security for renewables; detects data manipulation attacks	Concrete dataset not openly specified; focus is primarily on environmental rather than real-time attack scenarios
[282]	Hybrid ensemble learning for SCADA cybersecurity	UNSW-NB15, MSU water system intrusion dataset	Hybrid Ensemble Learning Model (HELM)	HELM using network sensors boosts breach detection over single-model baselines	Heterogeneity between network sensor types can affect harmonization and detection accuracy
[283]	Explainable DL for attack detection (EV charging)	CIC EV Charger Attack Dataset 2024 (CICEVSE2024)	Custom Explainable Deep Learning Model	Novel DL model with explainability outperforms traditional methods on EV charging cyberattack scenarios	Only tested for EV chargers; explainability adds computational cost; limited dataset diversity
[284]	Attack patterns in blockchain comms for energy	Blockchain-based comm-net dataset (Solana, open blockchains)	Statistical pattern analysis, descriptive ML	Presents labeled dataset for SQL Injection, Spoofing, MitM; highlights blockchain's vulnerabilities	Data is open, but scenarios rely more on simulation than threats observed in field deployments
[285]	AI/ML for anomaly detection, testbeds in energy	Emulated energy cyber–physical models and real testbeds	ML models (deep learning, XAI), anomaly detectors	Compares XAI approaches, underlines need for transparent AI; testbeds bridge lab-to-field knowledge gap	Testbed realism, transfer of lab results to real-world utilities remains challenging; heavy computational demands

4.5. Potential Impact of Cyberattacks on Energy Utilization

The implications of cyberattacks on smart grids can have significant consequences, not just for energy distribution but also for public safety, financial markets, and the environment. These are some of the potential effects:

1. Grid Disruptions and Blackouts: Cyberattacks on critical infrastructure can lead to widespread energy outages. A successful attack on the controllers of power stations or substations can result in blackouts, leaving millions of people without power and disrupting daily life. A coordinated attack on multiple grid nodes can make recovery efforts take days or even weeks.
2. Financial Losses: Cyberattacks can result in significant financial losses for energy utilities and consumers. Downtime, system recovery costs, and legal liabilities can be extremely expensive for utilities. Data manipulation can lead to unnecessarily high bills for consumers, who may even face penalties for alleged over-utilization of energy due to tampered smart meters.
3. Energy Efficiency and Sustainability Setbacks: Cyberattacks can sabotage energy-efficiency programs by disrupting smart data and compromising information in smart meters and grid control systems. For example, tampering with energy consumption data might lead to inefficient demand–response programs, which undermines the goals of energy use optimization and carbon reduction.
4. Loss of Consumer Trust: A cybersecurity incident can erode public confidence in the energy system and the smart grid. When consumers believe their energy or personal information is at risk, they may be unwilling to accept smart meters or participate in energy-efficiency programs. This lack of public trust can make it significantly harder to implement smart-grid projects.

5. Energy Routing and Energy Internet

Energy Internet (EI) is a emerging innovation that enables peer-to-peer (P2P) energy trading and facilitates highly efficient energy distribution. According to [286], EI can be understood as a smart-grid implementation in which energy flows over the Internet similarly to data packets. The study in [287] highlights the advantages of EI, noting that it is open, robust, and reliable. While the concept of EI is widely discussed in China and the United States, analogous frameworks have been identified globally under different terminologies. For example, in Japan, the EI concept is commonly referred to as the Digital Grid.

The Digital Grid in Japan represents a fully decentralized energy system [288], enabling energy transactions for P2P, stabilizing and reinforcing the grid, and supporting new on-demand energy markets. Routers within the Digital Grid facilitate the transfer of energy from point to point, analogous to the movement of data packets on the Internet. This innovation is transforming energy delivery and market dynamics, allowing for more versatile and efficient energy flow. The EI architecture can be structured into seven layers, as illustrated in Figure 9, which are based on the Open Systems Interconnection (OSI) model.

The foundational layer is the physical layer, which consists of energy cells. The energy link layer interconnects these energy cells to enable P2P trading within an Energy Intranet. Energy Routers (ERs) play a central role in establishing and maintaining these connections. The communication layer ensures seamless information exchange among the system, stakeholders, and other entities. The business layer manages financial transactions, while the transmission layer facilitates the transfer of power between various Energy Intranets. Finally, the consumption layer regulates energy usage in the energy cells.

Recent research, such as [289], describes the EI as a transformative network system comprising three primary layers: the physical, information, and value layers. The physical layer interconnects diverse energy sources, including heat, electricity, gas, and cooling to facilitate demand–response and energy sharing among distributed energy resources (DERs). The information layer collects data from the physical layer, enabling real-time coordination and decentralized energy management. The value layer leverages this information to create new business models.

Figure 9. Energy Internet architecture.

The authors in [290,291] presented an EI model as a decentralized process for trading energy, where resources can exchange energy without the intervention of a centralized operator. System stability is maintained through the operation of an Energy Internet

Service Provider (E-ISP), which ensures secure and stable energy exchange by regulating transaction volumes in real-time and managing centralized resources when necessary. The EI architecture is designed to be structurally compatible with traditional Internet protocols, incorporating EI Cards (analogous to MAC addresses) and Energy IP addresses to enable mobility tracking across networks. Additionally, a dedicated energy transport layer protocol ensures reliable delivery of energy communications, while an application layer facilitates standardized energy communication exchanges.

Energy Routers (ERs) play a pivotal role in integration resources within the EI, by enabling the routing of energy across geographically dispersed resources. These devices manage power dispatch, facilitate information exchange, and perfrom transmission scheduling, thereby supporting a dynamic grid and flexible grid structure. ERs are essential for regulating energy flows along interconnected lines, ensuring both stability and operational efficiency of the EI. A typical ER consists of input–output ports that interface with energy sources, electrical loads, and other ERs, as well as a core module comprising power converters and a controller. The literature presents several ER architectures with a generalized structure, illustrated in Figure 10.

Microgrids (MGs) play a vital role within the EI by integrating intelligent devices, storage systems, distributed energy sources, and communication networks to optimize the flow of energy. As the network architecture evolves, traditional routers are being enhanced to support energy-routing functionalities, enabling bidirectional energy transfer and advance routing operations. Energy-routing challenges include consumer matching, congestion mitigation, and failure prevention through transmission scheduling, and determination of optimal energy paths. To address these issues, routing algorithms must be embedded within ERs to distribute energy from a source to a destination. Optimized energy routing not only minimizes transmission losses and maximizes energy delivery but also reduces greenhouse gas emissions and fossil fuel reliance. Furthermore, efficient routing supports renewable energy integration, enhances grid reliability, and stabilizes electricity flow. P2P energy trading and decentralized energy markets heavily rely on these routing protocols, as they facilitate cleaner energy consumption and reduce reliance on fossil-fuel-based power plants.

This section is motivated by the existing gaps in the literature on energy-routing protocols. Existing reviews [292–297] exhibit notable limitations: they often fail to provide a systematic classification of routing methods based on their characteristics and functionalities, and lack comprehensive comparative analyses, and offer limited guidance for future research directions. Additionally, these studies do not thoroughly address the diverse methodologies and terminologies employed to tackle energy-routing challenges, which impedes the development of robust and innovative solutions. Energy routing in modern energy networks is a rapidly evolving area, presenting challenges such as dynamic operation, security, and computational efficiency. Therefore, there is a pressing need for an in-depth comparative framework. Such a framework should evaluate routing protocols in terms of their strengths limitations, and suitability across different application environments.

This section addresses these gaps by preenting a comprehensive discussion of energy routing protocols, with particular emphasis on artificial intelligence (AI), Multi-Agent Systems (MASs), and optimization models, including metaheuristics. It profferes an in-depth overview and a structured classification of existing methods, highlighting their strengths and limitations. Furthermore, the proposed framework integrates metaheuristic optimization for minimizing energy losses, AI-driven predictive routing for adaptive decision making, and MAS to enable decentralized control in dynamic, renewable energy networks. The paper also identifies critical challenges such as reducing energy losses, ensuring scalability, integrating renewable energy sources, and responding to fluctuating demand. The future research directions identified herein are intended to guide the development of efficient and sustainable energy routing, ultimately optimizing energy flow within intelligent and decentralized power networks.

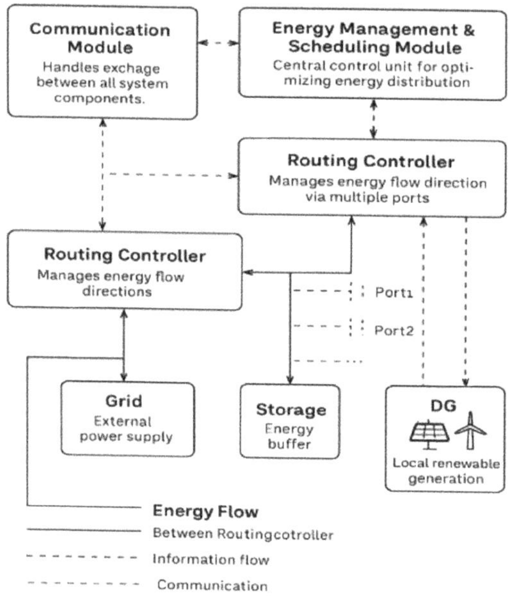

Figure 10. Energy Router general architecture.

5.1. Energy Network Architecture and Routing Constraints

The energy network architecture comprises two layers: the energy transmission layer and the communication layer, interconnected through multiple Energy Routers (ERs) [298–304]. While the architecture of ERs remains largely consistent across existing studies [305–308], variations emerge in terms of implementation details and control design strategies.

Energy-routing algorithms, implemented within the ER's routing controller as shown in Figure 10, adapt their routing decisions dynamically based on information received from the communication layer. It is important to distinguish the energy-routing problem from the Optimal Energy Flow (OEF) problem. The OEF problem primarily addresses system-level optimization, such as determining generation levels and reduce transmission costs. In contrast, the energy-routing problem focuses on real-time energy routing within decentralized networks that involving ERs, distributed energy resources (DERs), and peer-to-peer (P2P) energy trading. It aims to ensure effective energy distribution by determining the best routes for source–load pairs while minimizing transmission losses and adhering to energy constraints.

The (EI) model can be represented as a graph, where of Energy Routers (ERs) represent the vertices and energy transmission lines from the edges, as shown in Figure 11. The routing of energy should be subjected to the following requirements:

- The total energy loss across all paths within a network, $\sum P_{loss}$, in Equation (1), must be less than the total transmitted energy P_{TX}. Each path consists of one or more ERs that transfer energy packet between the destination and source.

$$\sum P_{loss} < P_{TX} \tag{1}$$

- In Equation (2), the transmitted energy P_{TX} must not exceed the maximum capacity of the selected path. This capacity, denoted as C_{path}, is determined by the minimum ER interface capacity C_{int}, and the minimum energy line capacity, C_{line}.

$$P_{TX} \leq C_{path} = \min(C_{int}, C_{line}) \tag{2}$$

- In Equation (3), the total energy transmitted through an energy line must not exceed its capacity, C_{line}.

$$P_{line} \leq C_{line} \tag{3}$$

- In Equation (4), the energy entering a given ER interface must not exceed its interface capacity, C_{int}.

$$P_{in\_int} \leq C_{int} \tag{4}$$

- Minimizing transmission losses TL to identify optimal energy-routing paths is a central focus in the literature [309–318]. The losses in a DC lines connecting two ERs depend on the line resistance R_{ij}, voltage V_{ij}, prior energy P_{ij}, and the transmitted energy P_{TX}. These parameters are incorporated into Equation (5) to compute TL for the line L_{ij}.

$$P_{loss\_line} = \frac{P_{TX}^2 R_{ij}}{V_{ij}^2} \tag{5}$$

- The energy loss within an ER (i), as described in Equation (6), depends on the electronic converter efficiency η_i of that ER.

$$P_{loss\_ER_i} = P_{in\_ER_i}(1 - \eta_i) \tag{6}$$

- The total energy loss along a transmission path between a consumer and a producer as expressed in Equation (7), is the sum of all losses in the ERs and the connecting energy lines along that path.

$$TL = \sum P_{loss\_ER} + \sum P_{loss\_line} \tag{7}$$

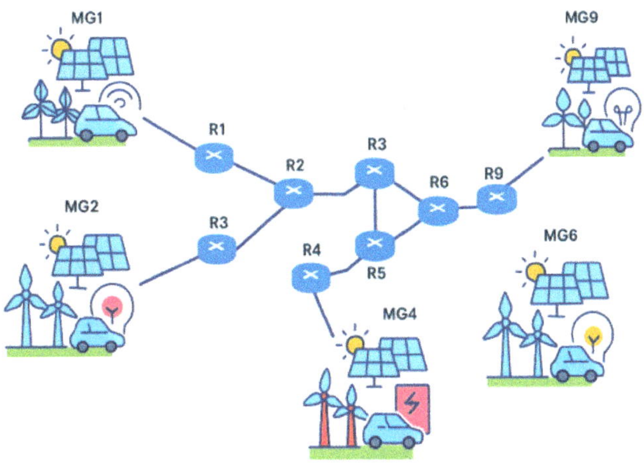

Figure 11. Energy Internet model with Energy Routers between microgrids.

5.2. Existing Research on Energy Routing

Numerous studies have investigated the challenges of energy routing in decentralized networks, aiming to enhance efficiency and reliability. These works propose a variety of routing algorithms designed to achieve objectives, such as minimizing energy loss, ensuring scalability and adaptability , and responding to dynamic network conditions. As illustrated in Figure 12, key strategies include Multi-Agent (MA) architectures for decentralized decision-making, metaheuristic optimization techniques for determining optimal energy paths, and artificial intelligence (AI) methods for predictive and adaptive routing. Emerging approaches based on graph theory and game theory are also being explored to further advance energy routing, with comprehensive reviews of these methodologies addressed in separate publications.

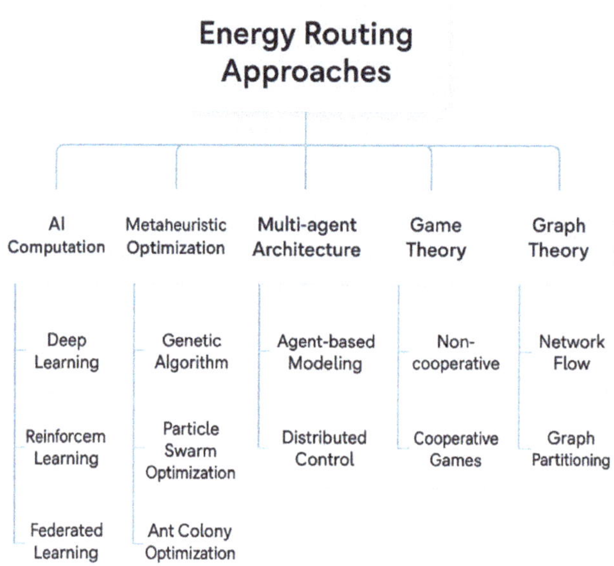

Figure 12. Energy-routing approaches.

5.2.1. AI-Based Approach in Energy Routing

The application of Artificial Intelligence (AI) models and algorithms has gained prominence in automating the complexities of energy management in modern energy systems [319–321]. This section provides an in-depth analysis of AI-based computation, highligting its roles, advantages, limitations, and applicability in enhancing energy routing within decentralized energy networks. A comprehensive overview for researchers and practitioners is presented in Table 17.

An example of this approach is a model-free Deep Reinforcement Learning (DRL) algorithm that optimizes energy management by controlling Energy Routers (ERs) and Distributed Generators (DGs) within sub-grids. The algorithm leverages real-time observations of the system state to make energy allocation decisions [322]. While evaluating on a nine-node network revealed limitations, including high computational demands, the requirement for large dataset, and challenges in hyper-parameter tuning.

In studies by [323,324], neural network-based reinforcement learning methods, including actor–critic and Q-learning algorithms, are employed to enable adaptive energy routing and demand prediction in large-scale energy networks. While these frameworks improve efficiency and resilience, they need for large training datasets, and limited scalability under dynamic conditions.

AI techniques, including DRL and neural network-based approaches, enhance energy routing by enabling adaptive, efficient, and cost-effective management in decentralized networks. However, their high computational demands, dependence on large datasets, and challenges in hyper-parameter tuning underscore the need for further research to improve scalability and support practical deployment. Table 17 presents a comparative analysis of selected studies that apply AI to energy routing.

Table 17. Energy-routing strategies based on AI (limitations, applicability, and advantages).

References	Limitations	Applicability	Advantages
[322]	Involves heavy computation, substantial data input, and meticulous hyperparameter optimization.	Ideal for energy management in real-time within compact to moderately sized networks.	Ensures stable power regulation for ERs and DGs across subgrids, fitting for networks with dynamic behavior.
[323]	Relies on large-scale data and exhibits significant computational load, which may hinder real-time functionality in dynamic environments.	Ideal for real-time operations in controlled or limited-scale systems, but performance may degrade in complex, evolving environments.	Achieves holistic optimization by disseminating local data, cutting down the amount of communication required between ERs.
[324]	Based solely on the training dataset, with no emphasis on generalization, and involves tuning of hyperparameters	Struggles in scalable systems, particularly when data is lacking and adaptability is crucial	Real-time optimization of energy flow routes

5.2.2. Metaheuristic Approach in Energy Routing

Metaheuristic algorithms plays a crucial role in optimizing energy-routing protocols by improving path selection, resource allocation, and load balancing, thereby contributing to robust and efficient energy-distribution networks. Their key characteristics, limitations, applicability, and advantages are summarized in Table 18.

Firefly Algorithm (FA) The Firefly Algorithm (FA), inspired by the bioluminescent behavior, is employed in [325,326] to address the subscriber-matching problem in energy networks. In [325], FA is utilized to match consumers with producers based on Euclidean distance and energy price, aiming to minimize overall costs, transmission losses, and distances. The algorithm operates in three phases: (i) parameter initialization(energy quantities, pricing, positions), (ii) fitness evaluation considering demand, supply, losses, and price, and (iii) energy quantity updates to efficiently satisfy consumer requirements. Validated on a 17-node Energy Internet (EI) network, this approach offers a decentralized solution but overlooks critical factors such as Energy Router (ER) impacts, voltage drops, and link impedance, which could lead to increased losses in smart grids. The study in [326] extends FA to optimize energy-efficient routing paths and is tested on an 11-node network with four consumers and seven producers. While both approaches show promise, they could benefit from incorporating dynamic real-time data, enhancing scalability for larger networks, and adopting more accurate loss models. Future research may focus on developing hybrid FA-based models for real-time congestion management and power flowoptimization.

Genetic Algorithms (GAs): Genetic algorithms (GAs) are widely applied in energy optimization,as demonstrated in [327,328]. In [328], GA is employed for subscriber matching by encoding solutions as chromosomes that represent producer–consumer links, aiming to minimize energy losses and overall costs in a 17-node EI network. Similarly, Refs. [327,329] utilize GA for energy-efficient path selection, leveraging crossover and mutation operators to identify low-loss routing paths. These methods, validated on 7- and 17-node networks, reported a 7.75% efficiency improvement over Ant Colony Optimization (ACO). Furthermore, Ref. [328] integrates GA with graph traversal and the U-NSGA-III algorithm for multi-objective optimization, balancing carbon emissions and energy losses using a Parent-Centric Crossover (PCX) to enhance exploration. This hybrid approach was tested on an 11-node network. Despite these advances, GA methods face limitations such as scalability challenges, reliance on centralized control, and limited adaptability to dynamic conditions like network failures or demand fluctuations.

FA and GA provide robust solutions for energy routing, with FA excelling in decentralized subscriber matching and GA offering flexibility for path selection and multi-objective optimization. However, both methods face scalability and adaptability challenges, highlighting the need for hybrid and distributed approaches to improve real-time performance in complex energy networks. Additionally, bio-inspired metaheuristic algorithms, such as Discrete–Artificial Bee Colony (D-ABC) and Particle Swarm Optimization (PSO), demonstrated significant potential for optimizing energy routing in smart grid (SG) environments by effectively addressing challenges related to pathfinding, energy allocation, and operational efficiency within the Energy Internet (EI).

Discrete–Artificial Bee Colony (D-ABC): The D-ABC algorithm, as proposed in [309], enhances the traditional Artificial Bee Colony (ABC) method for efficient energy pathfinding within constrained-capacity EI networks. By introducing mutation and crossover operations during the employed and onlooker bee phases, D-ABC achieves faster convergence and mitigates suboptimal solutions through diversified search strategies. Experimental validated in smart-grid environments demonstrates its effectiveness in addressing complex routing challenges. However, the algorithm exhibits limitations, including sensitivity to parameter tuning, memory-intensive processes, and high computational requirements for large-scale networks. Future research could explore adaptive mechanisms and leverage advanced computational resources to enhance scalability and dynamic adaptability in evolving EI systems.

Particle Swarm Optimization (PSO): Particle Swarm Optimization (PSO) has been widely applied for energy routing in a smart-grid environment [330,331].In [330], PSO facilitates optimal subscriber matching by selecting producers and allocating energy to meet consumer demand efficiently. Similarly, Ref. [331] employs PSO within a multi-objective framework, optimizing energy paths based on metrics such as transmission latency, hop count, distance, and cost, with validation on a 10-node EI network. Furthermore, a discrete PSO variant, proposed in [332] addresses discontinuous routing issues in a 7-ER network derived from a 30-bus system, improving energy efficiency and minimizing transmission losses. Although PSO offers rapid convergence and high accuracy, it suffers from premature convergence and limited adaptability under dynamic conditions. Future reserach could focus on hybrid approaches and adaptive strategies to enhance scalability and robustness in real-world EI scenarios .

D-ABC and PSO represent powerful bio-inspired approaches for energy routing, with D-ABC excelling in efficient pathfinding and PSO offering versatile optimization for subscriber matching and path selection. Nevertheless, both face challenges related to computational complexity, parameter sensitivity, and scalability in dynamic EI environments. Future research should focus on adaptive and hybrid approaches to enhance robustness, real-time performance, and applicability in large-scale, heterogeneous networks.

Table 18. Overview of energy-routing optimization methods and their characteristics.

References	Limitations	Applicability	Advantages
[325]	Operates under the belief that reducing Euclidean distance directly reduces losses.	Useful for producer identification in less complex, ideal smart-grid frameworks.	Optimally identifies the best-fit producers in smart-grid networks.
[326]	Employs reduced-complexity models.	Improvement is necessary for real-world grid adaptation, along with upgrades to handle larger-scale and more complex environments.	Handle the problem of finding optimal energy paths and matching subscribers effectively.
[327]	Based on centralized control and complete network visibility, yet it fails to address subscriber matching, scheduling, and has minimal capacity for managing congestion and system failures.	Capable of efficient energy routing in orderly networks, but lacks scalability without distributed system designs.	Enhances energy-routing efficiency by minimizing power losses, particularly across extended transmission distances, and shows strong adaptability in complex network environments.
[328]	Centralized approach incurs heavy computational costs and needs enhanced responsiveness to real-time environmental changes.	Useful in energy-routing scenarios requiring multi-objective optimization, enabling a balance between sustainability and system performance.	Uses a multi-objective optimization framework that evaluates both power loss and carbon footprint, while introducing advanced exploration techniques to bypass suboptimal results.
[309]	Requires considerable time for parameter optimization and large memory resources for population storage, limiting scalability in extensive networks.	Optimizes energy routing effectively in compact networks and supports energy-efficient path selection under EI capacity constraints.	Utilizes common node-based crossover and mutation to strengthen exploration and accelerate convergence in local optimization.
[330]	Lacks scalability and adaptability, and involves substantial computational load and parameter optimization.	Designed for stable, medium-scale grids, but can be extended to larger, evolving grids using hybrid techniques.	Capable of optimizing different elements of energy routing within medium-scale grid systems.
[331]	Early convergence may occur, hindering global optimality exploration, and evaluation steps contribute to high computational overhead.	Designed for decentralized energy coordination in dynamic systems with numerous energy sources and users, and is highly applicable to real-world contexts requiring multi-objective optimization and distributed control.	Maintains equilibrium between energy supply and demand by applying multi-objective optimization, relying on multiple criteria for effective decisions. Decentralized functionality enhances both scalability and adaptability in changing smart-grid environments.

5.2.3. Multi-Agent Systems for Energy-Routing Optimization

Multi-Agent Systems (MAS) facilitate decentralized decision-making in energy networks, enabling autonomous agents to a manage energy flows, optimize routing paths, and allocate resources dynamically. By promoting distributed intelligence and local adaptability, MAS enhance efficiency, scalability, and resilience in modern energy management and routing applications.

A novel fully distributed peer-to-peer (P2P) control strategy of Networked Renewable Energy Resources (NRERs) is proposed in [333]. The proposed approach integrates MAS with IoT technologies, featuring a dual-layer architecture comprising a primary droop control layer for local energy sharing and a secondary distributed diffusion algorithm layer for network-side optimization. Experimental validation on a modified IEEE 9-node feeder demonstrates a 52.34% reduction in energy losses. However, challenges such as communication latency, system complexity, and security vulnerabilities persist, necessitating advanced strategies for enhanced ER placement and loss minimization.

The study in [334] introduces a MAS-based coordinated scheduling model for Active Distribution Networks (ADNs). The proposed model optimizes Distributed Generation (DG) outputs and minimizes operating costs through iterative bid adjustments and market clearing mechanisms. Simulation on a test comprising four suppliers, three buyers, and four ADNs demonstrates effective energy dispatch and improved system stability. However, the model primarily focuses on market coordination and does not comprehensively routing functions such as dynamic path discovery.

The work in [335] presents a MAS-based distributed electricity trading framework enabling P2P energy sharing among prosumers. The system leverages coalition formation techniques alongside blockchain to ensure secure and transparent transactions, thereby optimizing trading costs and improving scalability. However, the framework does not address energy-routing functions such as transmission-path discovery, highlighting an opportunity for future research to extend its applicability to multi-energy router networks.

References [336,337] explore the integration of MAS with reinforcement learning (RL) for energy management. In [336], a Multi-Agent Reinforcement Learning (MARL) protocol employing Q-learning optimizes energy routing across a 9-ER network, effectively minimizing losses during system failures. Similarly [337] applies Multi-Agent Deep Reinforcement Learning (MADRL) to a 16-node multi-microgrid system, using Markov games to enhance cost efficiency and network stability. However, this framework relies on switches instead of ERs, limiting its direct applicability to fully decentralized EI networks.

MAS enhances energy routing by enabling decentralized control, facilitating P2P energy sharing, ADN scheduling, and RL-based optimization to improve efficiency and resilience. Despite these advantages, challenges such as communication latency, limited scalability, and incomplete routing functionalities highlight the need for future research on adaptive ER strategies and hybrid RL-MAS approaches.

5.3. Characteristics and Key Functions of Energy-Routing Protocol

Designing a robust energy-routing protocol for the Energy Internet (EI) requires addressing key characteristics and core functionalities to ensure efficient, secure, and adaptable energy distribution in dynamic energy networks.

Energy-Routing Schemes: Energy-routing approaches can be categorized asdistributed, centralized, and semi-centralized approaches. Distributed routing allows independent path selection, which reduces losses but may not achieve global optimization. Centralized routing enforces network constraints and minimizes transmission losses but depends on a central controller, both strategies employing a central controller, potentially reducing reliability and privacy. Semi-centralized routing combines both strategies, employing a central controller for constraint management while leveraging distributed algorithms to select efficient paths, thereby balancing computational efficiency and flexibility.

Algorithm Complexity and Computation Time: Algorithm complexity, measured in terms of memory usage and execution time, depends on network size and the employed techniques. More specialized algorithms typically increase complexity, potentially slowing computations and hindering real-time performance in large-scale EI networks.

Energy-Routing Constraints: Routing is constrained by factors such as limited energy supply, Energy Router (ER) capabilities, and link capacities. These constraints require overflow prevention and secure energy delivery, directly affecting path selection and overall system stability.

Congestion Management: Rising energy demand causes congestion, which delays delivery and increases losses. Effective management requires assessing link capacities and employing multi-packet routing to meet consumer needs through multiple sources or paths, thereby mitigating delays and reducing losses.

Failure Handling and Topology Changes: Protocols must adapt dynamically adapt to ER or link failures and topology changes to ensure reliable energy delivery, thereby maintaining network resilience during disruptions.

Security: Security measures including authentication, data integrity, and confidentiality safeguard energy networks against cyber threats, ensuring the secure transmission of energy packets.

Scenarios: Protocols must accommodate diverse scenarios including single-source single-load (SSSL), single-source multi-load (SSML), multi-source multi-load (MSML), and multi-source single-load (MSSL) ensuring flexible operation across varying source and load configurations.

Energy-Routing Functions: The functions of energy routing can be summarized into three main types:

1. Energy-Efficient Path (EEP): EEP aims to minimize energy losses during transmission by accounting factors such as voltage drops, congestion, and link impedance. Optimal paths must respect ER and link capacities to prevent failures and ensure efficient energy delivery.
2. Subscriber Matching (SM): SM enables demand-driven peer-to-peer (P2P) energy trading by pairing consumers with suppliers based on delivery time, energy quantity, and price. ERs perform SM to identify optimal suppliers, supporting both one-to-one and one-to-many configurations for efficient fulfillment of critical load demands.
3. Transmission Scheduling (TS): TS addresses network congestion by scheduling optimal transmission paths according to energy demand and source availability. It mitigates risks associated with bidirectional flows, voltage fluctuations, and intermittent renewable generation, ensuring stable and efficient energy distribution.

5.4. Discussion and Future Directions for Energy-Routing Protocols

Metaheuristics, AI, and MAS architectures provide complementary strengths for energy routing. Metaheuristics excel at optimizing stable networks, AI facilitates dynamic adaptation, and MAS delivers decentralized resilience. Hybrid approaches, augmented with data-efficient AI techniques and security measures such as blockchain, can overcome scalability and computational challenges, fostering sustainable and efficient energy networks. Table 19 presents a concise comparison of energy-routing protocols, highlighting their strengths, limitations, and ideal applications in decentralized energy systems.

Metaheuristic algorithms excel at optimizing energy paths and minimizing transmission losses, making them highly effective for stable networks. However, their high computational demands and sensitivity to parameter settings limit real-time deployment. Integrating reinforcement learning for adaptive parameter tuning, hybridizing with machine learning models, and leveraging GPU-based parallel processing can significantly improve convergence speed, accuracy, and scalability in dynamic energy networks.

AI methods, such as Graph Neural Networks and network flow optimization, enable adaptive, real-time decision-making in dynamic energy networks by predicting and responding to changing conditions. However, their reliance on large datasets and high computational costs poses challenges, which can be mitigated through edge computing and data-efficient algorithms such as few-shot learning.

MAS facilitate decentralized decision-making, providing scalability and flexibility for dynamic topologies and effective congestion management. Coordination challenges in large networks can be mitigated through agent clustering and reinforcement learning, enhancing overall autonomy.

The integration of AI and metaheuristics introduces challenges such as data scarcity and high computational complexity. Energy routing can become more robust and adaptive through efficient data-driven approaches, including federated learning, transfer learning, and decentralized AI systems.

Hybrid protocols that combine metaheuristics, AI, and MAS offer resilient and scalable energy routing. AI can predict demand fluctuations, MAS frameworks enhance decentralized control, and blockchain ensures security. Future research should focus on integrating these approaches to optimize adaptability, security, and sustainability in decentralized energy networks.

Table 19. Comparison of optimization methods in energy systems.

Methods	Applications	Weaknesses	Strengths
AI Computing Method	Adaptive response to real-time network changes, with support for predictive maintenance and fault handling.	Requires substantial computing resources and a high volume of training data.	Strengthens real-time decision capabilities, enables autonomous responses to network changes, and increases scalability and robustness.
Metaheuristic Optimization	Reliable/low-variability networks	Consumes substantial processing energy, offers limited responsiveness in real-time scenarios, and has poor scalability in broad network deployments.	Capable of identifying optimal routes and lowering energy losses, best applied in stable network scenarios.
MA Architecture	Energy networks based on decentralized architecture, featuring real-time routing and traffic management.	As network size increases, agent coordination becomes more complicated, possibly leading to slower decision-making under heavy loads.	Facilitates autonomous decisions across the network, reducing the need for centralized control. It is highly scalable and flexible, well-suited to address congestion and shifting topologies

5.5. Real-World SCADA Testbed Using Protocol-Specific Traffic: A Case Study

In the literature discussed above, we have extensively analyzed various DL/ML, and hybrid models for detecting cyberattacks a on smart grid. However, a major limitation in existing smart-grid cybersecurity research is that most studies rely on datasets not collected from real-world SCADA devices. Many of these datasets also lack integration with smart grid-specific communication protocols, such as Modbus. Additionally, evaluations are often performed on balanced datasets, which do not accurately reflect the operational condition of real-world smart grids. In real SG infrastructure, a large number of heterogeneous SCADA devices continuously communicating using SCADA-specific protocols. Furthermore, under normal operating conditions, routine communication between devices are far more frequent than malicious activities, such as cyberattacks.

To address these challenges, a recent study conducted by [338] proposed a model that closely replicates real-world smart-grid SCADA deployments. In this study, the authors implemented a fully simulated yet realistic SG network topology, where heterogeneous SCADA devices including distributed energy resources (DERs), DER controllers, smart meters, RTU, field sensors, g gateways, and aggregator platforms communicate using the Modbus protocol, one of the most widely used protocols in SG SCADA system. Figure 13 illustrates the communication among these SG SCADA devices and the deployment scenarios for a potential cyberattack. In this study, the authors collected raw data from this network topology in PCAP format, capturing real-time packet-level exchanges between these devices. From the PCAP files, they extracted 29 Modbus-specific features, including packet-level metrics (such as TCP flags, source/destination IPs, and window sizes) and flow-based features (e.g., flow duration, inter-arrival times, packet entropy, retransmission counts). The resulting Modbus DDoS attack (MDDA) dataset provides one of the most protocol-relevant resources currently available for smart-grid cybersecurity research.

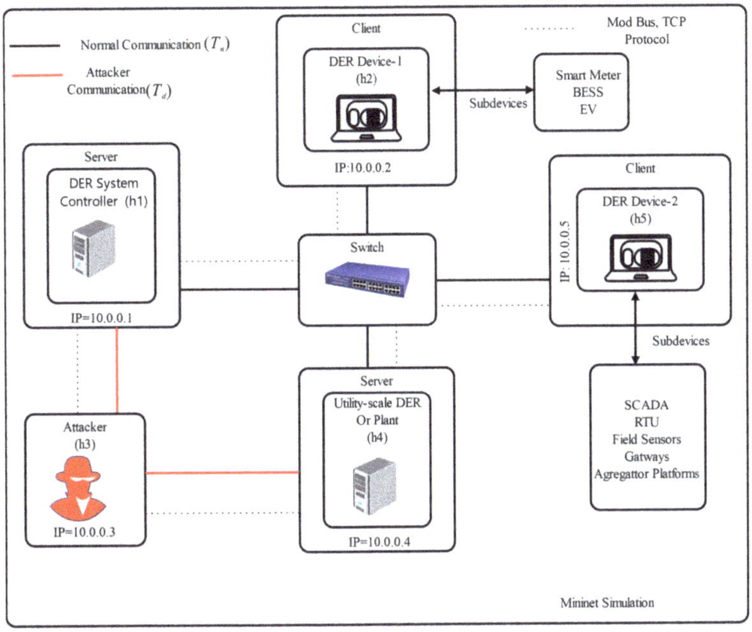

Figure 13. Attack deployment in smart grid [338].

To evaluate model performance under realistic conditions, the authors applied several ML and DL techniques, with a particular focus on a sparse autoencoder (SAE) model. The experimental setup also incorporated Gaussian noise into the network traffic and varied the intensity and durations of attacks to emulate the unpredictability of real-world SG environments.

A key feature of this study is that the authors preserved the natural class imbalance in the dataset, with 662,750 normal samples and 233,891 DDoS samples, rather than applying oversampling or balancing techniques. This approach accurately reflects the actual operational behavior of smart-grid systems, where benign traffic is far more prevalent than attack traffic. The dataset is divided into four training and testing sets to maintain temporal consistency, and the sample distribution is presented in Table 20 of the study.

Table 20. Number of normal and DDoS attack samples in each split of datasets, and total size of datasets [338].

MDDA Dataset	Train Test Splits	Normal Traffic	DDoS Traffic	Total Samples
D1	Training Set 1	93,143	63,768	156,911
	Test Set 1	32,769	34,479	67,248
D2	Training Set 2	132,792	24,119	156,911
	Test Set 2	38,566	28,682	67,248
D3	Training Set 3	128,346	28,565	156,911
	Test Set 3	45,018	22,230	67,248
D4	Training Set 4	132,143	24,771	156,914
	Test Set 4	59,973	7277	67,250
Total Samples		662,750	233,891	896,641

This real-world case study demonstrates that Modbus-specific attack detection can be effectively achieved using an SAE model, attaining up to 76% detection accuracy, which surpasses baseline models such as Random Forest, CNN, and GRU, as presented in Table 21.

Table 21. Comparison of several models' performance on the MDDA dataset [338].

Model	Accuracy	Model	Accuracy
SAE	**0.76**	Bagging	0.53
DAE	0.29	Boosting	0.44
MDAE	0.29	Stacking	0.75
ELMAE	0.29	RF	0.57
CNN-BiLSTM	0.73	NB	0.41
Claasic SVM	0.34	J48	0.52
QSVM	0.61	CNN-GRU	0.71
DT	0.59	ARIMA-LWCNN	0.69
RF	0.52	RF	0.73
QDA	0.34	NB	0.41
NB	0.41	SVM	0.66
XGB	0.56	XGB	0.55

Moreover, it underscores the importance of evaluating models on realistic, protocol-aware, and imbalanced dataset, rather than on generic benchmark datasets like CICIDS2017 or NSL-KDD, which fail to capture the communication structures and traffic characteristics of smart-grid environments.

6. Future Research Directions

6.1. Energy Generation

The comprehensive analysis of factors impacting energy generation, as discussed in the preceding sections, underscores significant advancements in resource assessment, technological efficiency, environmental considerations, policy frameworks, grid integration, geopolitical influences, and cybersecurity. Nevertheless, several gaps and opportunities remain for future research aimed at enhancing the sustainability, reliability, and security of power-generation systems. Below, we outline key directions for future investigation, building on the findings of the referenced studies, to address these challenges and foster innovation in the energy sector.

Considering advanced resource assessment and forecasting, the studies by [1–5] demonstrate significant progress in leveraging GIS-based frameworks, Earth System Models (ESMs), and Artificial Neural Networks (ANNs) for resource assessment and forecasting. However, the growing variability of renewable resources due to climate change necessitates further refinement of these models. Future research should prioritize the development of high-resolution, real-time forecasting models that integrate multi-source data, such as satellite imagery, IoT sensor networks, and advanced climate models, to improve the accuracy of solar and wind resource predictions. Furthermore, expanding the open-access dataset initiatives, as illustrated in [5], to encompass diverse geographical regions and extended temporal scales would enhance model validation and generalizability. Research should also investigate hybrid forecasting models that combine physics-based approaches with machine learning, addressing the limitations of standalone models, as suggested by [4].

Enhancing technological efficiency advancements in machine learning and digital-twin technologies, as discussed by [6–10], have significantly enhanced the efficiency of energy-generation systems. Future research should prioritize the developing adaptive, real-time optimization algorithms capable of dynamically adjusting operational parameters in response to fluctuating environmental conditions and grid demands. For example, extending the digital-twin framework proposed by [9] to other renewable technologies, such as solar farms or hybrid systems, could facilitate more and accurate performance monitoring and predictive maintenance. Moreover, integrating physics-informed neural networks (PINNs), as demonstrated in [10], with generative AI models could further improve forecasting accuracy and system resilience, particularly for offshore and remote installations.

Considering climate-resilient energy systems, the studies discussed by [9,11–13] emphasize the vulnerability of energy infrastructure to climate-induced disruptions, such as droughts and extreme weather events. Future research should prioritize developing climate-resilient energy infrastructure by integrating long-term climate projections into system design and operation. This includes creating adaptive-capacity models that account for multi-day weather events, as highlighted by [12], and investigating modular, decentralized energy systems to mitigate the impact of regional climate variability. Additionally, research should examine the feasibility of hybrid renewable systems that combine solar, wind, and hydropower to buffer against climate-induced resource shortages, building on the hybrid system insights from [3].

To address challenges and implement policy and regulatory innovations, the studies discussed by [14,17–19] emphasize the critical role of integrated policy frameworks in accelerating renewable energy adoption. Future research should investigate dynamic policy models that adapt to evolving technological and geopolitical landscapes. For instance, developing simulation-based policy evaluation tools that combine econometric models, as demonstrated by [14], with real-time energy market data could assist policymakers in designing more responsive and effective regulations. Furthermore, research should aim on overcome barriers to policy implementation, such as political resistance and public acceptance, by leveraging insights from [18] and exploring innovative engagement strategies, including gamified policy communication platforms.

The integration of hybrid energy-storage systems (HESS) and Long-Duration Energy Storage (LDES), as discussed by [22,25], is crucial for ensuring grid stability. Future research should optimizing HESS configurations using advanced control algorithms that account for real-time grid dynamics and renewable intermittency. Building on [23], research should investigate co-optimization models that integrate storage, generation, and transmission planning at a global scale, especially in regions with underdeveloped grid infrastructure. Additionally, the weather-driven charging algorithm proposed by [26] could be extended to include predictive maintenance strategies, thereby enhancing battery longevity and grid reliability in systems with high renewable penetration.

To address the geopolitical and social dynamics issues that influence grid and storage optimization, the studies discussed by [28–31] highlight the complex interplay of geopolitical risks and social acceptance in energy transitions. Future research should develop predictive models that quantify the impact of geopolitical events on energy supply chains, building on the econometric approach of [28]. Additionally, leveraging machine learning for real-time sentiment analysis, as demonstrated by [30], could be expanded to other social media platforms and regions to better understand and address public opposition to renewable projects. Research should also investigate game-theoretic models, as in [31], to design cooperative international frameworks that balance geopolitical competition with sustainable energy goals.

To ensure the security and stability of energy-generation networks, cybersecurity studies discussed by [34–38] and others highlight the growing sophistication of ML and DL techniques for detecting cyber threats. Nevertheless, challenges such as adversarial attacks and scalability persist. Future research should prioritize development of adversarially resilient models, building on [38], by incorporating Generative Adversarial Networks (GANs) to simulate and counter evasive attacks, as suggested by [43]. Additionally, research should investigate federated learning frameworks, as in [58], to enable privacy-preserving, decentralized intrusion detection systems for distributed energy networks. Real-time, scalable detection systems that integrate streaming data, as proposed by [45], should also be emphasized to address evolving cyber threats in smart grids.

A critical gap in the current literature is the lack of interdisciplinary approaches that integrate resource assessment, technological advancements, climate resilience, policy design, grid optimization, and cybersecurity. Future research should prioritize developing holistic frameworks that combine these dimensions. For example, integrating the GIS-based site selection models of [1] with the cybersecurity frameworks of [34] could facilitate the design of secure, resource-efficient renewable energy systems. Similarly, combining the climate resilience strategies of [11] with the policy insights of [14] could guide the development of adaptive, climate-resilient energy policies. Such integrated approaches would ensure that energy systems are sustainable, secure, and resilient to both environmental and geopolitical challenges. Table 22 summarizes the future research directions relevant to energy generation.

Table 22. Future research directions in energy generation.

Research Area	Future Research Focus	Relevant Studies
Resource Assessment	Develop high-resolution, multi-source forecasting models and expand open-access datasets	[1–5]
Technological Efficiency	Create adaptive optimization algorithms and extend digital-twin frameworks to diverse renewable systems	[6–10]
Climate Resilience	Design climate-resilient infrastructure and modular hybrid systems	[9,11–13]
Policy Innovation	Develop dynamic policy models and innovative public engagement strategies	[14,17–19]
Grid and Storage	Optimize HESS configurations and integrate predictive maintenance in charging algorithms	[22,23,25,26]
Geopolitical and Social	Quantify geopolitical impacts and expand real-time sentiment analysis	[28–31]
Cybersecurity	Develop adversarially resilient and scalable detection systems	[34–38,43,45,58]
Interdisciplinary Approaches	Integrate resource, technological, climate, policy, and cybersecurity frameworks	[1,11,14,34]

6.2. Future Research Directions—T&D

The evolution of smart-grid transmission and distribution (T&D) systems, as detailed in the preceding sections, highlights significant advancements through the integration of IoT, AI, and advanced automation. Nevertheless, the complexity of smart grids introduces challenges such as cybersecurity risks, power quality issues, and the need for scalable, privacy-preserving models. Based on a critical analysis of the current state of research, several directions emerge for future investigation to enhance the reliability, security, and efficiency of T&D systems.

First, the integration of distributed energy resources (DERs) and microgrids presents opportunities for improving grid resilience and reducing reliance on traditional power plants [88,89]. Future research should prioritize developing scalable optimization models for DER integration that account for dynamic load profiles and the intermittent nature of renewable energy sources. Specifically, enhancing the coordination between DERs and energy management systems (EMS) using AI-driven predictive models could improve real-time decision-making and energy distribution efficiency [90,91]. For example, extending the work of Moradi-Sepahvand [123] on integrated T&D models to include adaptive forecasting for EV charging and renewable energy fluctuations could mitigate short-term unpredictability.

Second, cybersecurity remains a critical concern due to the increasing prevalence of cyberattacks such as False Data Injection Attacks (FDIA), Load Redistribution Attacks (LRA), and Denial of Service (DoS) attacks [162,179,181]. While models like DQDN, POMDP, and Q-learning have shown promise in detecting and mitigating these threats [163,167,168], future research should prioritize developing hybrid AI models that integrate supervised, unsupervised, and reinforcement learning to detect sophisticated, multi-vector attacks. Additionally, investigating privacy-preserving techniques, including federated learning [58] and GAN-based synthetic data generation [79], could facilitate secure data sharing among utilities while safeguarding consumer privacy.

Third, the application of advanced AI models, such as Convolutional Neural Networks (CNNs), Long Short-Term Memory (LSTM) networks, and Generative Adversarial Networks (GANs), has demonstrated significant potential in anomaly detection, load forecasting, and fault diagnosis [138,141,152]. Nevertheless, the scalability of these models in large-scale, real-world T&D systems remains underexplored. Future research should examine the computational efficiency and robustness of these models under noisy or incomplete data conditions, building on the work of Raghuvamsi [83] and Liu [82]. Furthermore, integrating quantum computing with AI models could improve the processing of high-dimensional datasets, enabling faster and more accurate predictions for complex grid operations.

Fourth, the development of direct-current (DC) distribution networks, as proposed by Javid [73], presents a promising approach for reducing power delivery losses in data centers, residential buildings, and renewable energy applications. Future research could focus on designing hybrid AC-DC smart grids with AI-powered control interfaces to optimize power flow and minimize harmonic distortions. This direction could build on the findings of Venkatraman [78] regarding coordinated T&D control under high distributed generation (DG) penetration.

Fifth, the role of advanced metering infrastructure (AMI) and communication networks in enabling two-way communication and real-time monitoring is well-established [85,87]. Nevertheless, the resilience of these systems against cyber–physical attacks, such as Man-in-the-Middle (MiTM) and Time-Delay Attacks (TDA), requires further investigation [174,175]. Future research should focus on adaptive encryption and authentication protocols tailored to the low-latency requirements of synchrophasor protocols and AMI networks. Additionally, leveraging blockchain technology for secure data transmission in smart grids could further enhance grid security and transparency.

Finally, the use of hybrid models combine deep learning with traditional methods, as explored by Almalaq [158,159], demonstrates significant potential for improving forecasting accuracy and operational efficiency. Future studies should investigate the integration of hybrid models with real-time optimization techniques, such as those proposed by Barajas-Villarruel [116] and Tianqi [117], to address multi-objective optimization challenges in T&D systems. These challenges include balancing energy efficiency, cost reduction, and environmental sustainability under varying operational constraints.

The following Table 23 summarizes these future research directions, highlighting the key areas, proposed approaches, and relevant references from the current work.

Table 23. Future research directions for smart-grid T&D systems.

Research Area	Proposed Approach	Relevant References
DER and Microgrid Integration	Develop scalable AI-driven optimization models for dynamic load and renewable energy integration	[88,89,123]
Cybersecurity Enhancements	Combine hybrid AI models and privacy-preserving techniques for multi-vector attack detection	[58,79,162,163]
Scalable AI Models	Improve computational efficiency of CNNs, LSTMs, and GANs for large-scale T&D applications	[82,83,138,141,152]
DC Distribution Networks	Design hybrid AC-DC grids with AI-powered control interfaces	[73,78]
AMI and Communication Resilience	Develop adaptive encryption and blockchain-based protocols for secure data transmission	[85,87,174,175]
Hybrid Model Optimization	Integrate deep learning with real-time optimization for multi-objective T&D challenges	[116,117,158,159]

6.3. Future Research Directions—Energy Utilization

The exploration of energy utilization within smart grids, as detailed in the preceding section, highlights significant advancements in technologies such as Advanced Metering Infrastructure (AMI), renewable energy integration, energy-storage systems (ESSs), Demand-side management (DSM), artificial intelligence (AI) and machine learning (ML). Nevertheless, the paper also identifies critical gaps and challenges, particularly in scalability, real-world validation, cybersecurity, and the integration of emerging technologies. These gaps present opportunities for future research to improve the efficiency, reliability, and security of smart-grid systems, supporting global sustainability objectives [186–188].

One prominent area for future research is the scalability and practical deployment of the proposed models and frameworks. For example, studies such as [228,232] rely heavily on simulation-based results (e.g., IEEE 33-bus and IEEE-118 bus systems), which lack real-world validation. Future research should prioritize conducting large-scale, real-world pilot studies to assess the applicability of optimization models for renewable energy integration and ESS. These studies could address the computational complexity and data-quality challenges noted in [230,232], ensuring that models remain robust across diverse grid conditions and geographical regions [231].

Cybersecurity remains as critical concern, with thehighlighting vulnerabilities to Denial of Service (DoS), False Data Injection Attacks (FDIA), Man-in-the-Middle (MiTM), ransomware, and Advanced Persistent Threats (APTs) [258,259]. While models such as those in [264,266,268] demonstrate promising accuracy (e.g., 91.2–99.99%) in detecting cyberattacks, their reliance on specific datasets or simulated environments limits generalizability. Future research should prioritize developing universal, adaptive cybersecurity frameworks capable of operating across heterogeneous grid systems while incorporating real-time, privacy-preserving techniques, as suggested by [280]. Additionally, exploring blockchain-based solutions for secure data transmission in smart grids could mitigate MiTM and FDIA risks, building on the preliminary work of [250,284].

The integration of AI and ML in smart grids, as discussed in [250,253,256], offers considerable potential but faces challenges related to data scarcity, model explainability, and the energy footprint of AI systems. Future research should focus on developing lightweight, energy-efficient AI algorithms that reduce computational demands while maintaining high accuracy in forecasting and anomaly detection [254,255]. Hybrid AI-IoT frameworks, as proposed in [255], could be extended to incorporate federated learning for decentralized data processing, addressing privacy concerns highligted in [253]. Furthermore, research should examine the environmental impact of AI-driven energy systems, particularly the energy consumption associated with model training and deployment, as illustrated in Figure 5 [253].

The variability and intermittency of renewable energy sources, as highlighted in [222,223], remain significant barriers to large-scale integration. Future research should focus on developing advanced forecasting models that leverage real-time weather data and machine learning to enhance the accuracy of renewable energy-generation predictions [226,227]. Additionally, the high capital costs and lifespan limitations of ESS, as discussed in [233,241], underscore the need for research into cost-effective, sustainable battery technologies, such as next-generation lithium-ion alternatives or solid-state batteries. Studies could also investigate the integration of ESS with hybrid renewable systems to optimize load scheduling and strengthen grid stability [236].

Demand-side management (DSM) strategies, as explored in [245,248], demonstrate significant potential for reducing peak loads and emissions but face challenges related to consumer adoption and regional applicability. Future research should focus on developing user-centric DSM frameworks that incorporate behavioral economics and gamification to encourage consumer participation [242,243]. Additionally, cross-regional studies could address the generalizability issues highlighted in [245], ensuring that DSM models remain adaptable to diverse economic and regulatory environments.

This section highlights the role of policy incentives and carbon pricing in accelerating the adoption of smart-grid technologies [209,210]. Future research should examine the impact of dynamic policy frameworks on technology adoption, particularly in regions with underdeveloped regulatory systems, as emphasized in [245]. Economic analyses could also investigate the long-term cost-benefit trade-offs of deploying AMI, ESS, and renewable integration systems, addressing the high deployment costs and economic barriers identified in [218,240]. Table 24 summarizes these future research directions for energy utilization.

Table 24. Future research directions for smart-grid energy utilization.

Research Area	Key Focus	Potential Approaches	Relevant Studies
Scalability and Validation	Real-world deployment and scalability of optimization models	Large-scale pilot studies, cross-regional validation, simplified computational models	[228,231,232]
Cybersecurity	Adaptive, universal cybersecurity frameworks	Blockchain-based security, federated learning, real-time intrusion detection	[258,264,268,280]
AI and ML Optimization	Energy-efficient, explainable AI models	Lightweight algorithms, federated learning, environmental impact assessments	[250,253,255]
Renewable Integration	Improved forecasting and integration of renewables	Advanced weather forecasting, ML-based prediction models, hybrid renewable systems	[222,226,227]
Energy Storage Systems	Cost-effective, sustainable ESS technologies	Next-generation batteries, hybrid ESS integration, lifecycle analysis	[233,236,241]
Demand-Side Management	User-centric DSM frameworks	Behavioral economics, gamification, cross-regional DSM models	[242,243,245]
Policy and Economics	Dynamic policy and economic frameworks	Cost-benefit analyses, regulatory impact studies, cross-regional policy models	[209,210,245]

6.4. Future Research Directions—Energy Routing

The exploration of energy-routing protocols in the Energy Internet (EI) highlights their significant potential to enhance energy distribution efficiency, scalability, and sustainability. However, as highlighted in [292–297], the existing literature often lacks comprehensive classifications, in-depth comparative analyses, and clear guidance for future research. Building on insights from discussed AI-based, metaheuristic, and Multi-Agent System (MAS) approaches, several critical research directions emerge to address challenges related to dynamic operation, scalability, security, and computational efficiency in decentralized energy networks.

A key research direction involves developing hybrid protocols that integrate metaheuristics, AI, and MAS to leverage their complementary strengths. Metaheuristic algorithms, such as the Firefly Algorithm (FA) [325,326] and genetic algorithms (GAs) [326–328], are effective for optimizing stable networks but often face challenges in real-time adaptability and scalability. Combining these with AI techniques, such as Deep Reinforcement Learning (DRL) [322–324], can enhance dynamic adaptability by predicting demand fluctuations and network changes. Meanwhile, MAS approaches, as demonstrated in [333–337], facilitating decentralized decision-making, improving resilience and scalability. Hybrid frameworks could employ AI for predictive routing, MAS for autonomous coordination, and metaheuristics for optimal path selection, effectively addressing limitations such as high computational demands and parameter sensitivity.

Scalability remains a critical challenge, particularly for AI-based approaches require extensive datasets and high computational resources [322–324], as well as for metaheuristics such as Discrete–Artificial Bee Colony (D-ABC) [309] and Particle Swarm Optimization (PSO) [330–332]. Future research should focus on incorportaing edge computing and data-efficient learning algorithms, including few-shot learning or federated learning, to reduce computational overhead and enable real-time performance in large-scale EI networks. Furthermore, leveraging GPU-based parallel processing can significantly accelerate the convergence of metaheuristics, such as D-ABC [309], thereby enhancing their applicability in dynamic resource-constrained environments.

Security remains a critical concern, as decentralized energy networks are increasingly vulnerable to cyberattacks and data breaches [333]. Blockchain technology, as demonstrated in [335], offers a promising solution for ensuring secure and transparent P2P energy trading. Future work should prioritize integrating blockchain with MAS and AI-based frameworks to provide robust authentication, data integrity, and confidentiality in energy-routing protocols. Additionally, the development of lightweight cryptographic schemes optimized for resource-constrained Energy Routers (ERs) is essential to maintain high security standards without introducing significant computational overhead.

Dynamic network conditions, such as Energy Router (ER) or link failures and frequsent topology changes, necessitate routing protocols capable of real-time adaptability [336,337]. Future research should focus on developing adaptive ER strategies that integrate MAS with Reinforcement Learning (RL) to effectively manage congestion, mitigate failures, and address renewable energy intermittency. For example, extending MARL approaches, as proposed in [336], to multi-ER environments could significantly enhance path discovery and minimize energy losses. Furthermore, incorporating Graph Neural Networks (GNNs) to model the EI as a graph as suggested in [313,318], can enable predictive and topology-aware routing, thereby improving the networks' resilience and adaptability under dynamic conditions.

The integration of renewable energy sources introduces significant challenges due to their intermittent and unpredictable nature [289,290]. Future research should focus on designing routing protocols that optimize energy flow from distributed energy resources (DERs) by leveraging AI-based demand forecasting in conjunction with metaheuristic path optimization. For instance, extending the multi-objective optimization framework proposed in [328] to explicitly account for renewable energy variability could help balance carbon emissions, minimize energy losses, and maintain grid stability. Moreover, enhancing hybrid PSO variants [330,331] to prioritize renewable sources in real-time routing decisions would further improve energy utilization and sustainability in dynamic Energy Internet environments.

Congestion management and transmission scheduling (TS) are essential for ensuring efficient and reliable energy delivery within the Energy Internet. Future research should prioritize the development of advanced TS algorithms that utilize AI-driven models to anticipate congestion and dynamically allocate transmission paths, as proposed in [331]. Integrating these approaches with MAS-based distributed control strategies [334] could significantly reduce delays and transmission losses, particular under high-demand conditions. Furthermore, optimizing TS mechanisms for diverse operational scenarios including SSSL, MSML, and MSSL will enhance system resilience and adaptability in large-scale decentralized energy networks.

To address the absence of standardized methodologies highlighted in [292,293], future research should focus on developing a unified framework for energy-routing protocols. This framework should define standardized components such as EI Cards, Energy IP addresses, and transport layer protocols, as suggested in [290], to ensure compatibility with existing Internet architectures. Establishing such standards would enable seamless interoperability across diverse Energy Internet deployments, including large-scale initiatives like the Digital Grid in Japan [288], thereby promoting scalability, reliability, and global integration of decentralized energy systems.

In summary, future research should prioritize the development of hybrid energy-routing protocols that integrate AI, metaheuristics optimization, and MAS. These protocols should leverage enabling technologies such as edge computing, blockchain, and GNNs to ensure scalability, security, and adaptability in dynamic environments. Table 25 provides an overview of these research directions, outlining their objectives, challenges, and potential approaches.

Table 25. Future research directions for energy-routing protocols.

Research Direction	Objectives	Potential Approaches
Hybrid Protocol Development	Combine strengths of AI, metaheuristics, and MAS for adaptive, scalable routing	Integrate DRL [322], FA [326], and MARL [336] for predictive, decentralized optimization
Scalability and Computational Efficiency	Reduce computational overhead for large-scale EI networks	Use edge computing, few-shot learning, and GPU-based parallel processing [309]
Security Enhancements	Ensure secure energy packet transmission	Implement blockchain [335] and lightweight cryptography for ERs
Dynamic Network Adaptation	Handle topology changes and failures in real-time	Combine MAS with GNNs [313] and RL [336] for adaptive routing
Renewable Energy Integration	Optimize routing for intermittent renewable sources	Extend multi-objective PSO [330] and GA [328] to prioritize DERs
Congestion and Scheduling Optimization	Mitigate delays and losses in high-demand scenarios	Develop AI-based TS algorithms [331] with MAS control [334]
Standardized Framework	Ensure interoperability across EI implementations	Standardize EI Cards and Energy IP protocols [290]

6.5. Future Research Directions for Quantum-Based AI Models

Several AI, ML, DL, and hybrid models have been applied to address smart-grid challenges. These include Long Short-Term Memory (LSTM), Random Forest (RF), autoencoders, Artificial Neural Networks (ANNs), Generative Adversarial Networks (GANs), and hybrid models such as XGBoost, CatBoost, LightGBM, and Cluster Partition–Fuzzy Broad Learning System (CP-FBLS). While these models demonstrate strong performance in applications such as False Data Injection Attack (FDIA) detection, demand forecasting, and energy routing, they also exhibit theoretical and practical limitations that restrict their scalability, adaptability, and efficiency in complex and dynamic smart-grid environments.

For example, LSTM and Gated Recurrent Units (GRUs) are widely used for load-profile analysis and energy-routing optimization due to their capability to process sequential data while mitigating vanishing gradient issues [161]. However, these models are computationally intensive, requiring substantial training time and resources, particularly when applied to large-scale smart-grid datasets with high temporal resolution. Their performance is further constrained by sensitivity to noisy or incomplete data, which is prevalent in real-world grid operations [4]. Moreover, LSTM-based models often struggle with capturing long-term dependencies in highly variable renewable energy data, thereby reducing forecasting accuracy under extreme climatic conditions [2].

Random Forest (RF) models are frequently applied to anomaly detection and demand forecasting in smart grids due to their robustness against overfitting and ability to handle high-dimensional datasets [3]. However, RF lacks inherent mechanisms for modeling sequential dependencies, which restricts its capability to capture complex temporal patterns in time-series data. Consequently, its effectiveness diminishes in dynamic smart-grid environments where real-time adaptability is critical [6].

Autoencoders have demonstrated high accuracy (up to 99.99%) in detecting FDIA attacks by effectively modeling normal operational patterns [339]. However, their performance is highly sensitive to hyperparameter tuning and the quality of training data. They often fail to generalize across diverse attack scenarios, particularly against sophisticated Advanced Persistent Threats (APTs) that exploit subtle data manipulations [275]. Additionally, their reliance on unsupervised learning increases susceptibility to false positives in noisy grid environments.

ANNs are widely employed for hybrid renewable energy forecasting and energy-routing optimization [3,324]. Despite their predictive capabilities, ANNs require large labeled datasets for effective training a constraint in smart-grid applications due to data privacy concerns and limited availability of real-world operational data [5]. GANs, have emerged as a powerful tool for generating synthetic data to enhance cybersecurity training in the smart grids. However, they face significant challenges such as training instability and mode collapse, which hinder their ability to capture the full diversity of attack patterns [340].

Hybrid models such as XGBoost, CatBoost, and LightGBM have demonstrated superior predictive performance in smart-grid applications, achieving notably low forecast errors (e.g., 8.02% in [6]). Similarly, CP-FBLS exhibits strong performance in detecting FDIA attacks [162]. Despite these advantages, these models are computationally intensive, demanding substantial resources for both training and real-time deployment. Moreover, their inherent black-box nature compromises interpretability, posing a significant barrier for grid operators who require explainable decision-making processes [283]. Additionally, CP-FBLS's dependency on fuzzy logic introduces sensitivity to parameter tuning, potentially reducing robustness in dynamic and uncertain grid environments.

MAS-based frameworks have been widely adopted for decentralized energy routing and control in smart grids [334,335] However, they encounter significant scalability challenges in large-scale enviornments due to high communication overhead and increased coordination complexity. Furthermore, these systems remain vulnerable to cyberattacks that exploit inter-agent communication channels, posing security risks [338].

Collectively, these limitations computational complexity, sensitivity to data quality, limited generalization capability, and lack of interpretability underscore the necessity for innovative approaches that enhance scalability, robustness, and efficiency of AI models in smart-grid applications.

Quantum–AI integration presents a promising avenue to overcome the inherent limitations of classical AI, ML, DL, and hybrid models. By leveraging fundamental quantum principles such as superposition, entanglement, and quantum parallelism, quantum computing enables significantly accelerated computations, making it particularly suitable for complex optimization and pattern-recognition tasks in smart grids [341].

Compared to classical models such as LSTM, ANNs, and hybrid models (e.g., [6,161]), which are computationally intensive, particularly for large-scale grid datasets, quantum algorithms offer significant speed advantages. Algorithms like the Quantum Approximate Optimization Algorithm (QAOA) and Variational Quantum Eigensolver (VQE) exploit quantum parallelism to accelerate complex tasks [342]. For example, QAOA can enhance ANN training by efficiently solving combinatorial optimization problems such as hyperparameter tuning, at speeds exponentially faster than classical gradient-based methods [343]. This capability facilitates real-time energy routing and FDIA attack detection in dynamic and large-scale grid environments.

Specifically, autoencoders and GANs face challenges in generalizing across diverse attack scenarios and handling noisy data [339,340]. Quantum Neural Networks (QNNs), by leveraging quantum entanglement, can more effectively model complex, high-dimensional data distributions [344]. For instance, QNNs can enhance FDIA detection by learning subtle patterns in noisy SCADA datasets, thereby improving robustness against APTs [275]. Additionally, quantum-enhanced feature-selection techniques can reduce input dimensionality, mitigating the negative effects of incomplete or low-quality datasets [345].

Furthermore, the black-box nature of hybrid models such as CP-FBLS and XGBoost limits their interpretability [283]. Quantum-inspired algorithms, including quantum kernel methods, can generate probabilistic outputs compatible with explainable AI frameworks, providing grid operators with clearer insights into decision-making processes [343]. For example, quantum kernel-based classifiers can be integrated with MAS to deliver interpretable energy-routing decisions, thereby enhancing trust in decentralized control systems [334].

Furthermore, MAS-based energy-routing protocols encounter scalability challenges due to high communication overhead [335]. Quantum communication techniques, such as Quantum Key Distribution (QKD), can secure agent interactions against cyberattacks, while quantum optimization algorithms can enhance coordination efficiency in large-scale grids [342]. For instance, quantum-enhanced reinforcement learning (QRL) can optimize Multi-Agent energy routing by efficiently solving high-dimensional Markov Decision Processes, outperforming classical methods [336].

Current quantum hardware, including noisy intermediate-scale quantum (NISQ) devices, suffers from limited qubit counts and high error rates, making the implementation of QNNs or QAOA computationally intensive for large-scale smart-grid applications [342]. These limitations constrain their immediate deployment in resource-constrained environments. Additionally, quantum computers require specialized infrastructure, such as cryogenic cooling, which increases deployment costs relative to classical systems [343], posing a barrier to widespread adoption in smart grids. Quantum algorithms for smart-grid applications remain in the early stages of development, with limited real-world validation compared to established classical models like LSTM or RF [3,161,345]. Furthermore, implementing Quantum–AI solutions necessitates expertise in quantum computing, which is currently scarce among grid operators and researchers [346].

Advancements in quantum hardware and algorithms are expected to address these challenges in the near future. By 2030, improvements in qubit coherence times and error correction are projected to enable stable NISQ devices capable of supporting large-scale smart-grid optimization tasks [342]. Cloud-based quantum computing platforms will reduce infrastructure costs, making Quantum–AI solutions more accessible to utilities [347]. Furthermore, hybrid quantum–classical frameworks will facilitate incremental adoption by bridging current classical models with emerging quantum systems [345]. As quantum algorithm libraries mature, their integration with smart-grid applications will become increasingly seamless, allowing quantum-enhanced AI models to operate effectively for real-time forecasting, cybersecurity, and energy routing.

7. Conclusions

The increasing digitization of smart grids has transformed power generation while simultaneously introducing significant cybersecurity risks, necessitating advanced detection and mitigation strategies. This paper synthesizes state-of-the-art approaches, including CP-BLS, multimodal deep learning, and autoencoder models, demonstrating their effectiveness in countering cyber threats such as FDIA, DoS, and Replay Attacks. It further highlights the critical role of AI-driven solutions, including ANNs and GANs, in optimizing transmission and distribution systems and energy-routing protocols. The analysis encompasses resource assessment, environmental impacts, policy frameworks, grid integration, and geopolitical influences, underscoring their effect on energy security and sustainability. Despite these advances, challenges related to computational complexity, scalability, and real-time adaptability in T&D systems remain significant. Future research should prioritize hybrid routing protocols, edge computing, low-shot learning, and Quantum–AI integration to enhance grid resilience and operational efficiency. These developments are essential for aligning smart-grid systems with global sustainability objectives and ensuring robust, secure energy infrastructure.

Author Contributions: W.A., S.K. and S.A. focused on the Generation, Transmission and Distribution (T&D) sections of the work, where they provided in-depth analysis and insights into the optimization and challenges associated with these critical components of the power system. They also contributed significantly to outlining the future work related to enhancing T&D systems, particularly in the context of integrating renewable energy sources and improving grid stability. H.I. and M.A. concentrated on the Utilization, Routing, and Future Work sections. Their work emphasized the effective utilization of energy, the optimization of energy-routing strategies, and explored advanced solutions for integrating energy-storage systems. They also proposed several avenues for future work, focusing on the potential applications of AI and machine learning in enhancing the efficiency of energy-routing systems. All authors have read and agreed to the published version of the manuscript

Funding: This research received no external funding.

Data Availability Statement: The original contributions presented in the study are included in the article, further inquiries can be directed to the corresponding author on request.

Conflicts of Interest: The authors declare no conflicts of interest.

Abbreviations

The following abbreviations are used in this manuscript:

AMI	Advanced Metering Infrastructure
ANN	Artificial Neural Networks
APTs	Advanced Persistent Threats
BEMS	Building Energy Management Systems
CP-BLS	Cluster Partition Fuzzy Broad Learning System
CNN	Convolutional Neural Network
D-ABC	Discrete-Artificial Bee Colony
DER	Distributed Energy Resources
DoS	Denial of Service

DSM	Demand-Side Management
ESS	Energy Storage Systems
EMS	Energy Management Systems
FDIA	False Data Injection Attack
GAN	Generative Adversarial Networks
GNNs	Graph Neural Networks
LRA	Load Redistribution Attack
LSTM	Long Short-Term Memory
MAS	Multi-Agent Systems
MiTM	Man-in-the-Middle
ML	Machine Learning
P2P	Peer-to-Peer
PMU	Phasor Measurement Unit
PSO	Particle Swarm Optimization
Q-learning	Q-Learning (Reinforcement Learning)
RA	Replay Attack
RNN	Recurrent Neural Network
RL	Reinforcement Learning
RTU	Remote Terminal Unit
SG	Smart Grid
T&D	Transmission and Distribution
TDA	Time-Delay Attack
TS	Transmission Scheduling
XGBoost	Extreme Gradient Boosting

References

1. Jiang, W.; Zhang, S.; Wang, T.; Zhang, Y.; Sha, A.; Xiao, J.; Yuan, D. Evaluation method for the availability of solar energy resources in road areas before route corridor planning. *Appl. Energy* **2024**, *356*, 122260. [CrossRef]
2. Chen, G.; Ji, Z. A review of solar and wind energy resource projection based on the Earth system model. *Sustainability* **2024**, *16*, 3339. [CrossRef]
3. Shafi, I.; Khan, H.; Farooq, M.S.; Diez, I.d.l.T.; Miró, Y.; Galán, J.C.; Ashraf, I. An Artificial Neural Network-Based Approach for Real-Time Hybrid Wind–Solar Resource Assessment and Power Estimation. *Energies* **2023**, *16*, 4171. [CrossRef]
4. Pereira, S.; Abreu, E.F.M.; Iakunin, M.; Cavaco, A.; Salgado, R.; Canhoto, P. Method for solar resource assessment using numerical weather prediction and artificial neural network models based on typical meteorological data: Application to the south of Portugal. *Sol. Energy* **2022**, *236*, 225–238. [CrossRef]
5. Chen, Y.; Xu, J. Solar and wind power data from the Chinese state grid renewable energy generation forecasting competition. *Sci. Data* **2022**, *9*, 577. [CrossRef] [PubMed]
6. Abubakar, M.; Che, Y.; Zafar, A.; Al-Khasawneh, M.A.; Bhutta, M.S. Optimization of solar and wind power plants production through a parallel fusion approach with modified hybrid machine and deep learning models. *Intell. Data Anal.* **2025**, *29*, 808–830. [CrossRef]
7. Jiang, Z.; Li, H.; Yang, H.; Wu, H.; Liu, W.; Chen, Z. Review of Artificial Intelligence-Based Design Optimization of Wind Power Systems. *Wind* **2025**, *5*, 18. [CrossRef]
8. Méndez-Cruz, L.E.; Salazar-Pereyra, M.; Lugo-Leyte, R.; Sales-Cruz, M.; Torres-Aldaco, A.; Lugo-Méndez, H.D. Thermodynamic Evaluation of the Hybrid Combined Cycle Power Plant in the Valley of Mexico. *Energies* **2025**, *18*, 1901. [CrossRef]
9. Branlard, E.; Jonkman, J.; Brown, C.; Zhang, J. A digital twin solution for floating offshore wind turbines validated using a full-scale prototype. *Wind Energy Sci.* **2024**, *9*, 1–24. [CrossRef]
10. Gijón, A.; Pujana-Goitia, A.; Perea, E.; Molina-Solana, M.; Gómez-Romero, J. Prediction of wind turbines power with physics-informed neural networks and evidential uncertainty quantification. *arXiv* **2023**, arXiv:2307.14675. [CrossRef]
11. Shuai, H.; Li, F.; Zhu, J.; Tingen Ii, W.J.; Mukherjee, S. Modeling the impact of extreme summer drought on conventional and renewable generation capacity: Methods and a case study on the Eastern US power system. *Appl. Energy* **2024**, *363*, 122977. [CrossRef]
12. Grochowicz, A.; Van Greevenbroek, K.; Bloomfield, H.C. Using power system modelling outputs to identify weather-induced extreme events in highly renewable systems. *Environ. Res. Lett.* **2024**, *19*, 054038. [CrossRef]

13. Harang, I.; Heymann, F.; Stoop, L.P. Incorporating climate change effects into the European power system adequacy assessment using a post-processing method. *Sustain. Energy Grids Netw.* **2020**, *24*, 100403. [CrossRef]
14. Tello, W.P. Policy interactions and electricity generation sector CO2 emissions: A quasi-experimental analysis. *Energy Policy* **2025**, *198*, 114434. [CrossRef]
15. Nazir, I.; Mushtaq, N.; Ziafat, B.; Fareed, M.F.; Amin, W. A Reserved Pricing Mechanism for Nano/Peco Smart Grid Peer-to-Peer Energy Trading. In Proceedings of the 2024 3rd International Conference on Emerging Trends in Electrical, Control, and Telecommunication Engineering (ETECTE), Lahore, Pakistan, 26–27 November 2024; pp. 1–5.
16. Mushtaq, N.; Nazir, I.; Azad, M.; Ishfaq, H.; Fareed, M.F.; Amin, W. Optimal Selection of Trading Partner in Peer-to-Peer Energy Markets. In Proceedings of the 2024 3rd International Conference on Emerging Trends in Electrical, Control, and Telecommunication Engineering (ETECTE), Lahore, Pakistan, 26–27 November 2024; pp. 1–5.
17. Hu, G.; Yang, J.; Li, J.; Cheng, X.; Feng, C. National climate legislation and policymaking and energy security: International evidence. *Appl. Energy* **2024**, *375*, 124157. [CrossRef]
18. Omri, A.; Jabeur, S.B. Climate policies and legislation for renewable energy transition: The roles of financial sector and political institutions. *Technol. Forecast. Soc. Change* **2024**, *203*, 123347. [CrossRef]
19. Lopez Soto, D.; Mejdalani, A.; Nogales, A.; Tolmasquim, M.; Hallack, M. *Advancing the Policy Design and Regulatory Framework for Renewable Energies in Latin America and the Caribbean for Grid-Scale and Distributed Generation;* IDB: Washington, DC, USA, 2019.
20. Feng, K.; Yang, Z.; Zhuo, Y.; Jiao, L.; Wang, B.; Liu, Z. Impact of carbon tax on renewable energy development and environmental–economic synergies. *Energies* **2024**, *17*, 5347. [CrossRef]
21. Izmirlioglu, Y.; Pham, L.; Son, T.C.; Pontelli, E. A survey of multi-agent systems for smartgrids. *Energies* **2024**, *17*, 3620. [CrossRef]
22. Adeyinka, A.M.; Esan, O.C.; Ijaola, A.O.; Farayibi, P.K. Advancements in hybrid energy storage systems for enhancing renewable energy-to-grid integration. *Sustain. Energy Res.* **2024**, *11*, 26. [CrossRef]
23. Manocha, A.; Mantegna, G.; Patankar, N.; Jenkins, J.D. Reducing transmission expansion by co-optimizing sizing of wind, solar, storage and grid connection capacity. *Environ. Res. Energy* **2025**, *2*, 015011. [CrossRef]
24. Amin, W.; Ahmad, F.; Umer, K.; Khawaja, A.H.; Afzal, M.; Ahmed, S.A.; Chaitusaney, S. P2p energy trading: An optimal solution for energy shortage in pakistan. *IEEE Access* **2022**, *10*, 89614–89633. [CrossRef]
25. Staadecker, M.; Szinai, J.; Sánchez-Pérez, P.A.; Kurtz, S.; Hidalgo-Gonzalez, P. The value of long-duration energy storage under various grid conditions in a zero-emissions future. *Nat. Commun.* **2024**, *15*, 9501. [CrossRef]
26. Bhatt, D.; Penumatsa, S.; Singhal, N. Weather-Driven Priority Charging for Battery Storage Systems in Hybrid Renewable Energy Grids. In Proceedings of the 2025 9th International Conference on Green Energy and Applications (ICGEA), Singapore, 7–9 March 2025; pp. 170–175.
27. U.S. Energy Information Administration. Today in Energy. Available online: https://www.eia.gov/todayinenergy/detail.php?id=64586 (accessed on 1 August 2025. on).
28. Zhu, Z.; Hunjra, A.I.; Alharbi, S.S.; Zhao, S. Global energy transition under geopolitical risks: An empirical investigation. *Energy Econ.* **2025**, *145*, 108495. [CrossRef]
29. Mary, B.J. The Geopolitics of Energy Transition: Lessons from Germany's Green Policies and Their Implications for Global Sustainability. 2025. https://www.researchgate.net/publication/388952243_The_Geopolitics_of_Energy_Transition_Lessons_from_Germany's_Green_Policies_and_Their_Implications_for_Global_Sustainability (accessed on 1 August 2025)
30. Vågerö, O.; Bråte, A.; Wittemann, A.; Robinson, J.Y.; Zeyringer, M.; Sirotko-Sibirskaya, N. Machine learning of public sentiments towards wind energy in Norway. *Wind Energy* **2024**, *27*, 583–611. [CrossRef]
31. Zhao, X.; Huang, W. Global geopolitical changes and new/renewable energy game. *Energies* **2024**, *17*, 4115. [CrossRef]
32. Kim, S.Y.; Ganesan, K.; Dickens, P.; Panda, S. Public sentiment toward solar energy opinion mining of twitter using a transformer-based language model. *Sustainability* **2021**, *13*, 2673. [CrossRef]
33. Ren, J.T. Harnessing public sentiment: A literature review of sentiment analysis in energy research. *Renew. Sustain. Energy Rev.* **2025**, *219*, 115739. [CrossRef]
34. Yin, T.; Naqvi, S.A.R.; Nandanoori, S.P.; Kundu, S. Advancing cyber-attack detection in power systems: A comparative study of machine learning and graph neural network approaches. In Proceedings of the 2024 Resilience Week (RWS), Austin, TX, USA, 3–5 December 2024; pp. 1–9.
35. Almalaq, A.; Albadran, S.; Mohamed, M.A. Deep machine learning model-based cyber-attacks detection in smart power systems. *Mathematics* **2022**, *10*, 2574. [CrossRef]
36. Sakhnini, J.; Karimipour, H.; Dehghantanha, A. Smart grid cyber attacks detection using supervised learning and heuristic feature selection. In Proceedings of the 2019 IEEE 7th International Conference on Smart Energy Grid Engineering (SEGE), Oshawa, ON, Canada, 12–14 August 2019; pp. 108–112.
37. Liu, X.; Ospina, J.; Konstantinou, C. Deep Reinforcement Learning for Cybersecurity Assessment of Wind Integrated Power Systems. *IEEE Access* **2020**, *8*, 208378–208394. [CrossRef]

38. Anthi, E.; Williams, L.; Rhode, M.; Burnap, P.; Wedgbury, A. Adversarial attacks on machine learning cybersecurity defences in industrial control systems. *J. Inf. Secur. Appl.* **2021**, *58*, 102717. [CrossRef]
39. Wang, W.; Harrou, F.; Bouyeddou, B.; Senouci, S.-M.; Sun, Y. A stacked deep learning approach to cyber-attacks detection in industrial systems: Application to power system and gas pipeline systems. *Clust. Comput.* **2022**, *25*, 561–578. [CrossRef]
40. Aslam, M.M.; Tufail, A.; Irshad, M.N. Survey of Deep Learning Approaches for Securing Industrial Control Systems: A Comparative Analysis. *Cyber Secur. Appl.* **2025**, *3*, 100096. [CrossRef]
41. Zhang, J.; Pan, L.; Han, Q.-L.; Chen, C.; Wen, S.; Xiang, Y. Deep learning based attack detection for cyber-physical system cybersecurity: A survey. *IEEE/CAA J. Autom. Sin.* **2021**, *9*, 377–391. [CrossRef]
42. Wu, W.; Harrou, F.; Bouyeddo, B.; Senouci, S.-M.; Sun, Y. Deep learning-based stacked models for cyber-attack detection in industrial internet of things. *Neural Comput. Appl.* **2025**, *37*, 19617–19651 [CrossRef]
43. Khaw, Y.M.; Jahromi, A.A.; Arani, M.; Kundur, D. Evasive attacks against autoencoder-based cyberattack detection systems in power systems. *Energy AI* **2024**, *17*, 100381. [CrossRef]
44. Sahu, A.; Nguyen, T.; Chen, K.; Zhang, X.; Hassanaly, M. Detection of False Data Injection Attacks (FDIA) on Power Dynamical Systems With a State Prediction Method. *IEEE Access* **2024**, 13, 12411–12426. [CrossRef]
45. Hallaji, E.; Razavi-Far, R.; Wang, M.; Saif, M.; Fardanesh, B. A stream learning approach for real-time identification of false data injection attacks in cyber-physical power systems. *IEEE Trans. Inf. Forensics Secur.* **2022**, *17*, 3934–3945. [CrossRef]
46. Yin, X.; Zhu, Y.; Xie, Y.; Hu, J. PowerFDNet: Deep learning-based stealthy false data injection attack detection for AC-model transmission systems. *IEEE Open J. Comput. Soc.* **2022**, *3*, 149–161. [CrossRef]
47. Zhu, Y.; Liu, R.; Chang, D.; Guo, H. Detection of false data injection attacks on power systems based on measurement-eigenvalue residual similarity test. *Front. Energy Res.* **2023**, *11*, 1285317. [CrossRef]
48. Wang, C.; Tindemans, S.; Pan, K.; Palensky, P. Detection of false data injection attacks using the autoencoder approach. In Proceedings of the 2020 International Conference on Probabilistic Methods Applied to Power Systems (PMAPS), Liège, Belgium, 18–21 August 2020; pp. 1–6.
49. Drayer, E.; Routtenberg, T. Detection of false data injection attacks in smart grids based on graph signal processing. *IEEE Syst. J.* **2019**, *14*, 1886–1896. [CrossRef]
50. Xu, B.; Guo, F.; Wen, C.; Deng, R.; Zhang, W.-A. Detecting false data injection attacks in smart grids with modeling errors: A deep transfer learning based approach. *arXiv* **2021**, arXiv:2104.06307. [CrossRef]
51. Niu, X.; Li, J.; Sun, J.; Tomsovic, K. Dynamic detection of false data injection attack in smart grid using deep learning. In Proceedings of the 2019 IEEE Power & Energy Society Innovative Smart Grid Technologies Conference (ISGT), Washington, DC, USA, 18–21 February 2019; pp. 1–6.
52. Chen, K.; Nguyen, T.; Hassanaly, M. Adversarial Multi-Agent Reinforcement Learning for Proactive False Data Injection Detection. *arXiv* **2024**, arXiv:2411.12130. [CrossRef]
53. Chen, Y.; Lakshminarayana, S.; Poor, H.V. Moving Target Defense Against Adversarial False Data Injection Attacks In Power Grids. *IEEE Internet Things J.* **2025**, *12*, 26315–26327. [CrossRef]
54. Mahmoudnezhad, F.; Moradzadeh, A.; Mohammadi-Ivatloo, B.; Zare, K.; Ghorbani, R. Electric load forecasting under false data injection attacks via denoising deep learning and generative adversarial networks. *IET Gener. Transm. Distrib.* **2024**, *18*, 3247–3263. [CrossRef]
55. Wang, Q.; Zhang, J.; Hu, J.; Wu, S.; Hou, S.; Tang, Y. A CGAN-based adversarial attack method for data-driven state estimation. *Int. J. Electr. Power Energy Syst.* **2025**, *170*, 110878. [CrossRef]
56. Zideh, M.J.; Khalghani, M.R.; Solanki, S.K. An unsupervised adversarial autoencoder for cyber attack detection in power distribution grids. *Electr. Power Syst. Res.* **2024**, *232*, 110407. [CrossRef]
57. Zheng, F.; Li, W.; Li, H.; Yang, L.; Sun, Z. Research on load frequency control system attack detection method based on multi-model fusion. *Energy Inform.* **2025**, *8*, 72. [CrossRef]
58. Shrestha, R.; Mohammadi, M.; Sinaei, S.; Salcines, A.; Pampliega, D.; Clemente, R.; Sanz, A.L.; Nowroozi, E.; Lindgren, A. Anomaly detection based on lstm and autoencoders using federated learning in smart electric grid. *J. Parallel Distrib. Comput.* **2024**, *193*, 104951. [CrossRef]
59. Shen, Y.; Qin, Z. Detection, differentiation and localization of replay attack and false data injection attack based on random matrix. *Sci. Rep.* **2024**, *14*, 2758. [CrossRef]
60. Huang, C.; Deng, S.; Ge, H. Resilient load frequency control of cyber-physical power systems with off-the-shelf redundant communication channels under FDI attacks. *Meas. Energy* **2025**, 100053. [CrossRef]
61. An, H.; Xing, Y.; Zhang, G.; Bamisile, O.; Li, J.; Huang, Q. Cluster partition-fuzzy broad learning-based fast detection and localization framework for false data injection attack in smart distribution networks. *Sustain. Energy Grids Netw.* **2024**, *40*, 101534. [CrossRef]
62. Hong, C.; Liang, Z.; Yang, Y.; Li, P.; Chen, L.; Bi, L.; Zhang, Y. Research on detection and defense methods for false data injection attacks in power systems based on state-space decomposition. *Discov. Appl. Sci.* **2025**, *7*, 760. [CrossRef]

63. Zhai, Z.-M.; Moradi, M.; Lai, Y.-C. Detecting Attacks and Estimating States of Power Grids from Partial Observations with Machine Learning. *PRX Energy* **2025**, *4*, 013003. [CrossRef]
64. Abir, S.M.A.A.; Anwar, A.; Choi, J.; Kayes, A. IoT-enabled smart energy grid: Applications and challenges. *IEEE Access* **2021**, *9*, 50961–50981. [CrossRef]
65. Amin, M. A smart self-healing grid: In pursuit of a more reliable and resilient system [in my view]. *IEEE Power Energy Mag.* **2013**, *12*, 110–112. [CrossRef]
66. Kumar, K.P.; Saravanan, B.; Swarup, K.S. Optimization of renewable energy sources in a microgrid using artificial fish swarm algorithm. *Energy Procedia* **2016**, *90*, 107–113. [CrossRef]
67. Qi, F.; Yu, P.; Chen, B.; Li, W.; Zhang, Q.; Jin, D.; Zhang, G.; Wang, Y. Optimal planning of smart grid communication network for interregional wide-area monitoring protection and control system. In Proceedings of the 2018 IEEE International Conference on Energy Internet (ICEI), Beijing, China, 21–25 May 2018; pp. 190–195.
68. Cavus, M. Advancing Power Systems with Renewable Energy and Intelligent Technologies: A Comprehensive Review on Grid Transformation and Integration. *Electronics* **2025**, *14*, 1159. [CrossRef]
69. Tabassum, S.; Babu, A.R.V.; Dheer, D.K. A comprehensive exploration of IoT-enabled smart grid systems: Power quality issues, solutions, and challenges. *Sci. Technol. Energy Transit.* **2024**, *79*, 62. [CrossRef]
70. Shah, S.N.H. IoT enabled smart grid integration with edge computing method. In Proceedings of the 2023 International Conference on Communication, Computing and Digital Systems (C-CODE), Islamabad, Pakistan, 17–18 May 2023; pp. 1–6.
71. Judge, M.A.; Khan, A.; Manzoor, A.; Khattak, H.A. Overview of smart grid implementation: Frameworks, impact, performance and challenges. *J. Energy Storage* **2022**, *49*, 104056. [CrossRef]
72. Tabassum, S.; Babu, A.R.V.; Dheer, D.K.; Pasha, M.M. Inspection and surveillance of energy consumption in IoT-smart grid using wireless sensor network. In Proceedings of the 2022 IEEE 6th International Conference on Condition Assessment Techniques in Electrical Systems (CATCON), Durgapur, India, 17–19 December 2022; pp. 308–312.
73. Javid, Z.; Kocar, I.; Holderbaum, W.; Karaagac, U. Future distribution networks: A review. *Energies* **2024**, *17*, 1822. [CrossRef]
74. Rezaei, N.; Tarimoradi, H.; Deihimi, M. A coordinated management scheme for power quality and load consumption improvement in smart grids based on sustainable energy exchange based model. *Sustain. Energy Technol. Assess.* **2022**, *51*, 101903. [CrossRef]
75. Utkarsh, K.; Ding, F. Self-organizing map-based resilience quantification and resilient control of distribution systems under extreme events. *IEEE Trans. Smart Grid* **2022**, *13*, 1923–1937. [CrossRef]
76. Wu, H.; Xu, Z.; Zhao, J.; Chai, S. Gridtopo-GAN for distribution system topology identification. *IEEE Trans. Ind. Inf.* **2022**, *19*, 5356–5366. [CrossRef]
77. Yan, R.; Yuan, Y.; Wang, Z.; Geng, G.; Jiang, Q. Synthetic Active Distribution System Generation via Unbalanced Graph Generative Adversarial Network. *arXiv* **2021**, arXiv:2108.00599. [CrossRef]
78. Venkatraman, R.; Khaitan, S.K.; Ajjarapu, V. Dynamic co-simulation methods for combined transmission-distribution system with integration time step impact on convergence. *IEEE Trans. Power Syst.* **2018**, *34*, 1171–1181. [CrossRef]
79. Mohammadabadi, S.M.S.; Entezami, M.; Moghaddam, A.K.; Orangian, M.; Nejadshamsi, S. Generative artificial intelligence for distributed learning to enhance smart grid communication. *Int. J. Intell. Netw.* **2024**, *5*, 267–274.
80. Udhaya Shankar, C.; Ram Inkollu, S.; Nithyadevi, N. Deep Learning Based Effective Technique for Smart Grid Contingency Analysis Using RNN with LSTM. *Electr. Power Compon. Syst.* **2023**, 1–18. . [CrossRef]
81. Saldaña-González, A.E.; Aragüés-Peñalba, M.; Sumper, A. Distribution network planning method: Integration of a recurrent neural network model for the prediction of scenarios. *Electr. Power Syst. Res.* **2024**, *229*, 110125. [CrossRef]
82. Liu, G.; Bai, Y.; Liu, J.; Zhao, H.; Liang, G.; Zhao, J. MDRAE: An Attention Mechanism-Based Autoencoder for Missing Data Recovery in Smart Grids. In Proceedings of the 2023 IEEE 7th Conference on Energy Internet and Energy System Integration (EI2), Hangzhou, China, 15–18 December 2023; pp. 5190–5195.
83. Raghuvamsi, Y.; Teeparthi, K.; Kosana, V. Denoising autoencoder based topology identification in distribution systems with missing measurements. *Int. J. Electr. Power Energy Syst.* **2023**, *154*, 109464. [CrossRef]
84. Li, Z.; Wang, J.; Sun, H.; Guo, Q. Transmission contingency analysis based on integrated transmission and distribution power flow in smart grid. *IEEE Trans. Power Syst.* **2015**, *30*, 3356–3367. [CrossRef]
85. Fang, X.; Misra, S.; Xue, G.; Yang, D. Smart Grid The New and Improved Power Grid: A Survey. *IEEE Commun. Surv. Tutor.* **2012**, *14*, 944–980. [CrossRef]
86. Kiasari, M.; Ghaffari, M.; Aly, H.H. A Comprehensive Review of the Current Status of Smart Grid Technologies for Renewable Energies Integration and Future Trends: The Role of Machine Learning and Energy Storage Systems. *Energies* **2024**, *17*, 4128. [CrossRef]
87. Jung, O. The Increasingly Critical Role of Communication Networks in Enhancing Power Grid Resilience Under Climate Change. In Proceedings of the 14th International Conference on Smart Cities and Green ICT Systems, Porto, Portugal, 2–3 April 2025; pp. 180–187.

88. Babayomi, O.; Zhang, Z.; Dragicevic, T.; Hu, J.; Rodriguez, J. Smart grid evolution: Predictive control of distributed energy resources A review. *Int. J. Electr. Power Energy Syst.* **2023**, *147*, 108812. [CrossRef]
89. Shi, J.; Ma, L.; Li, C.; Liu, N.; Zhang, J. A comprehensive review of standards for distributed energy resource grid-integration and microgrid. *Renew. Sustain. Energy Rev.* **2022**, *170*, 112957. [CrossRef]
90. Khan, N.; Shahid, Z.; Alam, M.M.; Bakar Sajak, A.A.; Mazliham, M.; Khan, T.A.; Ali Rizvi, S.S. Energy management systems using smart grids: An exhaustive parametric comprehensive analysis of existing trends, significance, opportunities, and challenges. *Int. Trans. Electr. Energy Syst.* **2022**, *2022*, 3358795. [CrossRef]
91. Hernandez-Matheus, A.; Löschenbrand, M.; Berg, K.; Fuchs, I.; Aragüés-Peñalba, M.; Bullich-Massague, E.; Sumper, A. A systematic review of machine learning techniques related to local energy communities. *Renew. Sustain. Energy Rev.* **2022**, *170*, 112651. [CrossRef]
92. Ahmad, T.; Madonski, R.; Zhang, D.; Huang, C.; Mujeeb, A. Data-driven probabilistic machine learning in sustainable smart energy/smart energy systems: Key developments, challenges, and future research opportunities in the context of smart grid paradigm. *Renew. Sustain. Energy Rev.* **2022**, *160*, 112128. [CrossRef]
93. Alsuwian, T.; Shahid Butt, A.; Amin, A.A. Smart grid cyber security enhancement: Challenges and solutions A review. *Sustainability* **2022**, *14*, 14226. [CrossRef]
94. Hasan, M.K.; Habib, A.A.; Shukur, Z.; Ibrahim, F.; Islam, S.; Razzaque, M.A. Review on cyber-physical and cyber-security system in smart grid: Standards, protocols, constraints, and recommendations. *J. Netw. Comput. Appl.* **2023**, *209*, 103540. [CrossRef]
95. Ghiasi, M.; Niknam, T.; Wang, Z.; Mehrandezh, M.; Dehghani, M.; Ghadimi, N. A comprehensive review of cyber-attacks and defense mechanisms for improving security in smart grid energy systems: Past, present and future. *Electr. Power Syst. Res.* **2023**, *215*, 108975. [CrossRef]
96. Krivohlava, Z.; Chren, S.; Rossi, B. Failure and fault classification for smart grids. *Energy Inform.* **2022**, *5*, 33. [CrossRef]
97. Yip, S.-C.; Wong, K.; Hew, W.-P.; Gan, M.-T.; Phan, R.C.-W.; Tan, S.-W. Detection of energy theft and defective smart meters in smart grids using linear regression. *Int. J. Electr. Power Energy Syst.* **2017**, *91*, 230–240. [CrossRef]
98. Si, C.; Xu, S.; Wan, C.; Chen, D.; Cui, W.; Zhao, J. Electric load clustering in smart grid: Methodologies, applications, and future trends. *J. Mod. Power Syst. Clean Energy* **2021**, *9*, 237–252. [CrossRef]
99. Priya, B.; Malhotra, J. An Intelligent User-RAT association for 5G enabled Smart Grid. In Proceedings of the 2020 IEEE International Conference on Computing, Power and Communication Technologies (GUCON), Greater Noida, India, 2–4 October 2020; pp. 300–304.
100. Li, D.; Jayaweera, S.K. Distributed smart-home decision-making in a hierarchical interactive smart grid architecture. *IEEE Trans. Parallel Distrib. Syst.* **2014**, *26*, 75–84. [CrossRef]
101. Zhang, X.; Nakhai, M.R.; Zheng, G.; Lambotharan, S.; Chambers, J.A. Distributed foresighted energy management in smart-grid-powered cellular networks. *IEEE Trans. Veh. Technol.* **2019**, *68*, 4064–4068. [CrossRef]
102. Garcia, E.D.; Pereira, P.R.; Canha, L.N.; Popov, V.A. Grid functional blocks methodology to dynamic operation and decision making in Smart Grids. *Int. J. Electr. Power Energy Syst.* **2018**, *103*, 267–276. [CrossRef]
103. Li, D.; Jayaweera, S.K. Machine-learning aided optimal customer decisions for an interactive smart grid. *IEEE Syst. J.* **2014**, *9*, 1529–1540. [CrossRef]
104. Zhang, W.; Xu, Y.; Li, S.; Zhou, M.; Liu, W.; Xu, Y. A distributed dynamic programming-based solution for load management in smart grids. *IEEE Syst. J.* **2016**, *12*, 402–413. [CrossRef]
105. Abud, T.P.; Augusto, A.A.; Fortes, M.Z.; Maciel, R.S.; Borba, B.S.M.C. State of the art Monte Carlo method applied to power system analysis with distributed generation. *Energies* **2022**, *16*, 394. [CrossRef]
106. Khan, Z.A.; Jayaweera, D. Approach for smart meter load profiling in Monte Carlo simulation applications. *IET Gener. Transm. Distrib.* **2017**, *11*, 1856–1864. [CrossRef]
107. Zangirolami, V.; Borrotti, M. Dealing with uncertainty: Balancing exploration and exploitation in deep recurrent reinforcement learning. *Knowl. Based Syst.* **2024**, *293*, 111663. [CrossRef]
108. Yu, R.; Zhong, W.; Xie, S.; Zhang, Y.; Zhang, Y. QoS differential scheduling in cognitive-radio-based smart grid networks: An adaptive dynamic programming approach. *IEEE Trans. Neural Netw. Learn. Syst.* **2015**, *27*, 435–443. [CrossRef]
109. Wang, H.; Ruan, J.; Wang, G.; Zhou, B.; Liu, Y.; Fu, X.; Peng, J. Deep learning-based interval state estimation of AC smart grids against sparse cyber attacks. *IEEE Trans. Ind. Inf.* **2018**, *14*, 4766–4778. [CrossRef]
110. Zhao, Z.; Keerthisinghe, C. A fast and optimal smart home energy management system: State-space approximate dynamic programming. *IEEE Access* **2020**, *8*, 184151–184159. [CrossRef]
111. Bassamzadeh, N.; Ghanem, R. Multiscale stochastic prediction of electricity demand in smart grids using Bayesian networks. *Appl. Energy* **2017**, *193*, 369–380. [CrossRef]
112. Kaur, D.; Anwar, A.; Kamwa, I.; Islam, S.; Muyeen, S.M.; Hosseinzadeh, N. A Bayesian deep learning approach with convolutional feature engineering to discriminate cyber-physical intrusions in smart grid systems. *IEEE Access* **2023**, *11*, 18910–18920. [CrossRef]

113. Jindal, A.; Dua, A.; Kaur, K.; Singh, M.; Kumar, N.; Mishra, S. Decision tree and SVM-based data analytics for theft detection in smart grid. *IEEE Trans. Ind. Inf.* **2016**, *12*, 1005–1016. [CrossRef]
114. Wang, Y.; Kong, T. Air quality predictive modeling based on an improved decision tree in a weather-smart grid. *IEEE Access* **2019**, *7*, 172892–172901. [CrossRef]
115. Benti, N.E.; Chaka, M.D.; Semie, A.G. Forecasting renewable energy generation with machine learning and deep learning: Current advances and future prospects. *Sustainability* **2023**, *15*, 7087. [CrossRef]
116. Barajas-Villarruel, L.R.; Rico-Ramirez, V.; Castrejón-González, E.O. Optimization models for the generation, distribution and transmission of electricity in a macroscopic system: Representation of operational regions as energy hubs. *Chem. Eng. Res. Des.* **2022**, *187*, 645–661. [CrossRef]
117. Tianqi, L.; Guochen, Y.; Jing, G.; Xueyan, P.; Bo, Y. Research on Optimization of Power Grid Planning for Comprehensive Transmission and Distribution Tariff Reform. In Proceedings of the 2023 3rd International Signal Processing, Communications and Engineering Management Conference (ISPCEM), Montreal, QC, Canada, 25–27 November 2023; pp. 675–681.
118. Ogbogu, N.; Anaemeje, G. Electric Power Transmission Optimization. *J. Res. Natl. Dev.* **2011**, *9*, 1–9.
119. Amin, W.; Huang, Q.; Li, J.; Khan, A.A.; Subramaniam, U.; Selvam, S. A secure energy management model for Peer-to-Peer smart grids with user-centric constraints. *Internet Things* **2025**, *33*, 101678. [CrossRef]
120. Bragin, M.A.; Dvorkin, Y.; Darvishi, A. Toward Coordinated Transmission and Distribution Operations. In Proceedings of the 2018 IEEE Power & Energy Society General Meeting (PESGM), Portland, OR, USA, 5–10 August 2018; pp. 1–5.
121. Saukh, S.; Borysenko, A. Representation of transmission and distribution networks in the mathematical model of the electricity market equilibrium. In Proceedings of the 2019 IEEE 20th International Conference on Computational Problems of Electrical Engineering (CPEE), Lviv, Ukraine, 15–18 September 2019; pp. 1–4.
122. Cordova, S.; Moreira, A.; Heleno, M. Netload Range Cost Curves for a Transmission-Aware Distribution System Planning under DER Growth Uncertainty. *arXiv* **2024**, arXiv:2402.07428. [CrossRef]
123. Moradi-Sepahvand, M.; Amraee, T.; Aminifar, F.; Akbari, A. Coordinated expansion planning of transmission and distribution systems integrated with smart grid technologies. *Int. J. Electr. Power Energy Syst.* **2023**, *147*, 108859. [CrossRef]
124. Ghulam, A.; Sikander, R.; Ali, F. AI and Machine Learning-based practices in various domains: A Survey. *VAWKUM Trans. Comput. Sci.* **2022**, *10*, 21–41. [CrossRef]
125. Bashir, A.K.; Khan, S.; Prabadevi, B.; Deepa, N.; Alnumay, W.S.; Gadekallu, T.R.; Maddikunta, P.K.R. Comparative analysis of machine learning algorithms for prediction of smart grid stability. *Int. Trans. Electr. Energy Syst.* **2021**, *31*, e12706. [CrossRef]
126. Rangel-Martinez, D.; Nigam, K.; Ricardez-Sandoval, L.A. Machine learning on sustainable energy: A review and outlook on renewable energy systems, catalysis, smart grid and energy storage. *Chem. Eng. Res. Des.* **2021**, *174*, 414–441. [CrossRef]
127. Viloria, A.; Zelaya, N.A.L.; Varela, N. Unsupervised learning algorithms applied to grouping problems. *Procedia Comput. Sci.* **2020**, *175*, 677–682. [CrossRef]
128. Sivamayil, K.; Rajasekar, E.; Aljafari, B.; Nikolovski, S.; Vairavasundaram, S.; Vairavasundaram, I. A systematic study on reinforcement learning based applications. *Energies* **2023**, *16*, 1512. [CrossRef]
129. Arrieta-Prieto, M.; Schell, K.R. Spatially transferable machine learning wind power prediction models: v-logit random forests. *Renew. Energy* **2024**, *223*, 120066. [CrossRef]
130. Ponkumar, G.; Jayaprakash, S.; Kanagarathinam, K. Advanced machine learning techniques for accurate very-short-term wind power forecasting in wind energy systems using historical data analysis. *Energies* **2023**, *16*, 5459. [CrossRef]
131. Karaman, Ö. A. Prediction of wind power with machine learning models. *Appl. Sci.* **2023**, *13*, 11455. [CrossRef]
132. Pradhan, A. Support vector machine-a survey. *Int. J. Emerg. Technol. Adv. Eng.* **2012**, *2*, 82–85.
133. Awad, M.; Khanna, R. *Efficient Learning Machines: Theories, Concepts, and Applications for Engineers and System Designers*; Springer Nature: Berlin/Heidelberg, Germany, 2015.
134. Anand, P.; Rastogi, R.; Chandra, S. A class of new support vector regression models. *Appl. Soft Comput.* **2020**, *94*, 106446. [CrossRef]
135. Alshboul, O.; Shehadeh, A.; Almasabha, G.; Almuflih, A.S. Extreme gradient boosting-based machine learning approach for green building cost prediction. *Sustainability* **2022**, *14*, 6651. [CrossRef]
136. McCulloch, W.S.; Pitts, W. A logical calculus of the ideas immanent in nervous activity. *Bull. Math. Biophys.* **1943**, *5*, 115–133. [CrossRef]
137. Zakaria, M.; Mabrouka, A.S.; Sarhan, S. Artificial neural network: A brief overview. *Neural Netw.* **2014**, *1*, 2.
138. McClellan, J.; Parks, T.W.; Rabiner, L. A computer program for designing optimum FIR linear phase digital filters. *IEEE Trans. Audio Electroacoust.* **2003**, *21*, 506–526. [CrossRef]
139. Yamamoto, Y.; Anderson, B.D.O.; Nagahara, M.; Koyanagi, Y. Optimizing FIR approximation for discrete-time IIR filters. *IEEE Signal Process. Lett.* **2003**, *10*, 273–276. [CrossRef]
140. Sherstinsky, A. Fundamentals of recurrent neural network (RNN) and long short-term memory (LSTM) network. *Phys. D Nonlinear Phenom.* **2020**, *404*, 132306. [CrossRef]

141. Aghdam, H.H.; Heravi, E.J. *Guide to Convolutional Neural Networks;* Springer: New York, NY, USA, 2017; Volume 10, p. 51.
142. O'shea, K.; Nash, R. An introduction to convolutional neural networks. *arXiv* **2015**, arXiv:1511.08458. [CrossRef]
143. Wang, Y. A new concept using LSTM Neural Networks for dynamic system identification. In Proceedings of the 2017 American Control Conference (ACC), Seattle, WA, USA, 24–26 May 2017; pp. 5324–5329.
144. Grossberg, S. Recurrent neural networks. *Scholarpedia* **2013**, *8*, 1888. [CrossRef]
145. Zhai, J.; Zhang, S.; Chen, J.; He, Q. Autoencoder and its various variants. In Proceedings of the 2018 IEEE International Conference on Systems, Man, and Cybernetics (SMC), Miyazaki, Japan, 7–10 October 2018; pp. 415–419.
146. Sondak, D.; Protopapas, P. Learning a reduced basis of dynamical systems using an autoencoder. *Phys. Rev. E* **2021**, *104*, 034202. [CrossRef]
147. Chen, S.; Guo, W. Auto-encoders in deep learning a review with new perspectives. *Mathematics* **2023**, *11*, 1777. [CrossRef]
148. Kohonen, T. *Self-Organization and Associative Memory;* Springer Science & Business Media: Berlin/Heidelberg, Germany, 2012; Volume 8.
149. Vesanto, J.; Alhoniemi, E. Clustering of the self-organizing map. *IEEE Trans. Neural Netw.* **2000**, *11*, 586–600. [CrossRef] [PubMed]
150. Nazir, I.; Mushtaq, N.; Ishfaq, H.; Amin, W.; Maaliw, R.R.; Shi, X.; Saleem, M.A. Blockchain Enabled Reserved Pricing and Privacy Preserving Model for Peer-to-Peer Energy Trading In Smart Grid. *IEEE Trans. Consum. Electron.* **2025**, *71*, 1. [CrossRef]
151. Liu, Y.; Weisberg, R.H. A review of self-organizing map applications in meteorology and oceanography. In *Self-Organizing Maps: Applications and Novel Algorithm Design;* InTech: Rijeka, Croatia, 2011; Volume 1, pp. 253–272.
152. Saxena, D.; Cao, J. Generative adversarial networks (GANs) challenges, solutions, and future directions. *ACM Comput. Surv.* **2021**, *54*, 1–42. [CrossRef]
153. Mao, X.; Li, Q.; Xie, H.; Lau, R.Y.; Wang, Z.; Paul Smolley, S. Least squares generative adversarial networks. In Proceedings of the IEEE International Conference on Computer Vision, Venice, Italy, 22–29 October 2017; pp. 2794–2802.
154. Husz'ar, F. How (not) to train your generative model: Scheduled sampling, likelihood, adversary? *arXiv* **2015**, arXiv:1511.05101. [CrossRef]
155. Theis, L.; Oord, A.v.d.; Bethge, M. A note on the evaluation of generative models. *arXiv* **2015**, arXiv:1511.01844.
156. Amran, Y.A.; Amran, Y.M.; Alyousef, R.; Alabduljabbar, H. Renewable and sustainable energy production in Saudi Arabia according to Saudi Vision 2030; Current status and future prospects. *J. Clean. Prod.* **2020**, *247*, 119602. [CrossRef]
157. Bedi, J.; Toshniwal, D. Deep learning framework to forecast electricity demand. *Appl. Energy* **2019**, *238*, 1312–1326. [CrossRef]
158. Almalaq, A.; Zhang, J.J. Deep learning application: Load forecasting in big data of smart grids. In *Deep Learning: Algorithms and Applications;* Springer: Berlin/Heidelberg, Germany, 2019; pp. 103–128.
159. Almalaq, A.; Zhang, J.J. Evolutionary deep learning-based energy consumption prediction for buildings. *IEEE Access* **2018**, *7*, 1520–1531. [CrossRef]
160. Sehovac, L.; Nesen, C.; Grolinger, K. Forecasting building energy consumption with deep learning: A sequence to sequence approach. In Proceedings of the 2019 IEEE International Congress on Internet of Things (ICIOT), Milan, Italy, 8–13 July 2019; pp. 108–116.
161. Zhang, Y.; Yang, X.; Fang, C.; Cai, P.; Chen, B.; Li, T. Short-term building load forecasting based on similar day selection and LSTM network. In Proceedings of the 2018 2nd IEEE Conference on Energy Internet and Energy System Integration (EI2), Beijing, China, 20–22 October 2018; pp. 1–5.
162. Achaal, B.; Adda, M.; Berger, M.; Ibrahim, H.; Awde, A. Study of smart grid cyber-security, examining architectures, communication networks, cyber-attacks, countermeasure techniques, and challenges. *Cybersecurity* **2024**, *7*, 10. [CrossRef]
163. An, D.; Yang, Q.; Liu, W.; Zhang, Y. Defending against data integrity attacks in smart grid: A deep reinforcement learning-based approach. *IEEE Access* **2019**, *7*, 110835–110845. [CrossRef]
164. Liu, Y.; Dong, M.; Ota, K.; Li, J.; Wu, J. Deep reinforcement learning based smart mitigation of DDoS flooding in software-defined networks. In Proceedings of the 2018 IEEE 23rd International Workshop on Computer Aided Modeling and Design of Communication Links and Networks (CAMAD), Barcelona, Spain, 17–19 September 2018; pp. 1–6.
165. Wei, F.; Wan, Z.; He, H. Cyber-attack recovery strategy for smart grid based on deep reinforcement learning. *IEEE Trans. Smart Grid* **2019**, *11*, 2476–2486. [CrossRef]
166. Feng, M.; Xu, H. Deep reinforecement learning based optimal defense for cyber-physical system in presence of unknown cyber-attack. In Proceedings of the 2017 IEEE Symposium Series on Computational Intelligence (SSCI), Honolulu, HI, USA, 27 November–1 December 2017; pp. 1–8.
167. Necip Kurt, M.; Ogundijo, O.; Li, C.; Wang, X. Online cyber-attack detection in smart grid: A reinforcement learning approach. *arXiv* **2018**, arXiv:1809.05258. [CrossRef]
168. Jokar, P.; Leung, V.C.M. Intrusion detection and prevention for ZigBee-based home area networks in smart grids. *IEEE Trans. Smart Grid* **2016**, *9*, 1800–1811. [CrossRef]
169. Parras, J.; Almod'ovar, A.; Apell'aniz, P.A.; Zazo, S. Inverse reinforcement learning: A new framework to mitigate an Intelligent Backoff Attack. *IEEE Internet Things J.* **2022**, *9*, 24790–24799. [CrossRef]

170. Liu, X.; Xu, Y.; Jia, L.; Wu, Q.; Anpalagan, A. Anti-jamming communications using spectrum waterfall: A deep reinforcement learning approach. *IEEE Commun. Lett.* **2018**, *22*, 998–1001. [CrossRef]
171. Zhang, Y.; Cheng, Y. An amplification DDoS attack defence mechanism using reinforcement learning. In Proceedings of the 2019 IEEE SmartWorld, Ubiquitous Intelligence & Computing, Advanced & Trusted Computing, Scalable Computing & Communications, Cloud & Big Data Computing, Internet of People and Smart City Innovation (SmartWorld/SCALCOM/UIC/ATC/CBDCom/IOP/SCI), Leicester, UK, 19–23 August 2019; pp. 634–639.
172. Chen, Y.; Huang, S.; Liu, F.; Wang, Z.; Sun, X. Evaluation of reinforcement learning-based false data injection attack to automatic voltage control. *IEEE Trans. Smart Grid* **2018**, *10*, 2158–2169. [CrossRef]
173. Conti, M.; Dragoni, N.; Lesyk, V. A survey of man in the middle attacks. *IEEE Commun. Surv. Tutor.* **2016**, *18*, 2027–2051. [CrossRef]
174. Kulkarni, S.; Rahul, R.K.; Shreyas, R.; Nagasundari, S.; Honnavalli, P.B. MITM intrusion analysis for advanced metering infrastructure communication in a smart grid environment. In *Trends in Computational Intelligence, Security and Internet of Things, Proceedings of the Third International Conference, ICCISIoT 2020, Tripura, India, 29–30 December 2020*; Springer: Berlin/Heidelberg, Germany, 2020; pp. 256–267.
175. Fritz, J.J.; Sagisi, J.; James, J.; Leger, A.S.; King, K.; Duncan, K.J. Simulation of man in the middle attack on smart grid testbed. In Proceedings of the 2019 SoutheastCon, Huntsville, AL, USA, 11–14 April 2019; pp. 1–6.
176. Wu, Y.; Weng, J.; Qiu, B.; Wei, Z.; Qian, F.; Deng, R.H. Random delay attack and its applications on load frequency control of power systems. In Proceedings of the 2019 IEEE Conference on Dependable and Secure Computing (DSC), Hangzhou, China, 18–20 November 2019; pp. 1–8.
177. Mo, Y.; Sinopoli, B. Secure control against replay attacks. In Proceedings of the 2009 47th Annual Allerton Conference on Communication, Control, and Computing (Allerton), Monticello, IL, USA, 30 September–2 October 2009; pp. 911–918.
178. Zhu, M.; Martinez, S. On the performance analysis of resilient networked control systems under replay attacks. *IEEE Trans. Autom. Control* **2013**, *59*, 804–808. [CrossRef]
179. Liu, Y.; Ning, P.; Reiter, M.K. False data injection attacks against state estimation in electric power grids. *ACM Trans. Inf. Syst. Secur.* **2011**, *14*, 1–33. [CrossRef]
180. Chen, J.; Liang, G.; Cai, Z.; Hu, C.; Xu, Y.; Luo, F.; Zhao, J. Impact analysis of false data injection attacks on power system static security assessment. *J. Mod. Power Syst. Clean Energy* **2016**, *4*, 496–505. [CrossRef]
181. Yuan, Y.; Li, Z.; Ren, K. Modeling load redistribution attacks in power systems. *IEEE Trans. Smart Grid* **2011**, *2*, 382–390. [CrossRef]
182. Jhaveri, R.H.; Patel, S.J.; Jinwala, D.C. DoS attacks in mobile ad hoc networks: A survey. In Proceedings of the 2012 Second International Conference on Advanced Computing & Communication Technologies, Rohtak, India, 7–8 January 2012; pp. 535–541.
183. Liang, L.; Zheng, K.; Sheng, Q.; Huang, X. A denial of service attack method for an iot system. In Proceedings of the 2016 8th International Conference on Information Technology in Medicine and Education (ITME), Fuzhou, China, 23–25 December 2016; pp. 360–364.
184. Xu, W.; Ma, K.; Trappe, W.; Zhang, Y. Jamming sensor networks: Attack and defense strategies. *IEEE Netw.* **2006**, *20*, 41–47. [CrossRef]
185. Yi, P.; Zhu, T.; Zhang, Q.; Wu, Y.; Pan, L. Puppet attack: A denial of service attack in advanced metering infrastructure network. *J. Netw. Comput. Appl.* **2016**, *59*, 325–332. [CrossRef]
186. Fuller, J.; Obiomon, P.; Abood, S.I. *Power System Operation, Utilization, and Control*; CRC Press: Boca Raton, FL, USA, 2022.
187. Guo, Z.; Wei, W.; Chen, L.; Dong, Z.Y.; Mei, S. Impact of energy storage on renewable energy utilization: A geometric description. *IEEE Trans. Sustain. Energy* **2020**, *12*, 874–885. [CrossRef]
188. Wang, Z.; Li, M.; Cao, N.; Liao, P.; Ji, Y.; Han, F. Enhancing renewable energy utilization and energy management strategies for new energy yachts. *Heliyon* **2024**, *10*, e018. [CrossRef]
189. Lü, X.; Wu, Y.; Lian, J.; Zhang, Y.; Chen, C.; Wang, P.; Meng, L. Energy management of hybrid electric vehicles: A review of energy optimization of fuel cell hybrid power system based on genetic algorithm. *Energy Convers. Manag.* **2020**, *205*, 112474. [CrossRef]
190. Capehart, B.L.; Turner, W.C.; Kennedy, W.J. *Guide to Energy Management*; River Publishers: Gistrup, Denmark, 2020.
191. Rathor, S.K.; Saxena, D. Energy management system for smart grid: An overview and key issues. *Int. J. Energy Res.* **2020**, *44*, 4067–4109. [CrossRef]
192. Meliani, M.; Barkany, A.E.; Abbassi, I.E.; Darcherif, A.M.; Mahmoudi, M. Energy management in the smart grid: State-of-the-art and future trends. *Int. J. Eng. Bus. Manag.* **2021**, *13*, 18479790211032920. [CrossRef]
193. Dambhare, M.V.; Butey, B.; Moharil, S.V. Solar photovoltaic technology: A review of different types of solar cells and its future trends. *J. Phys. Conf. Ser.* **2021**, *1913*, 012053. [CrossRef]
194. Victoria, M.; Haegel, N.; Peters, I.M.; Sinton, R.; Jäger-Waldau, A.; Del Cañizo, C.; Breyer, C.; Stocks, M.; Blakers, A.; Kaizuka, I.; et al. Solar photovoltaics is ready to power a sustainable future. *Joule* **2021**, *5*, 1041–1056. [CrossRef]
195. Porté-Agel, F.; Bastankhah, M.; Shamsoddin, S. Wind-turbine and wind-farm flows: A review. *Bound.-Layer Meteorol.* **2020**, *174*, 1–59. [CrossRef] [PubMed]

196. Bošnjaković, M.; Katinić, M.; Santa, R.; Marić, D. Wind turbine technology trends. *Appl. Sci.* **2022**, *12*, 8653. [CrossRef]
197. Schaffarczyk, A.P. *Introduction to Wind Turbine Aerodynamics*; Springer Nature: Berlin/Heidelberg, Germany, 2024.
198. Luderer, G.; Madeddu, S.; Merfort, L.; Ueckerdt, F.; Pehl, M.; Pietzcker, R.; Rottoli, M.; Schreyer, F.; Bauer, N.; Baumstark, L.; et al. Impact of declining renewable energy costs on electrification in low-emission scenarios. *Nat. Energy* **2022**, *7*, 32–42. [CrossRef]
199. Balali, Y.; Stegen, S. Review of energy storage systems for vehicles based on technology, environmental impacts, and costs. *Renew. Sustain. Energy Rev.* **2021**, *135*, 110185. [CrossRef]
200. Panigone, S.; Sandri, F.; Ferri, R.; Volpato, A.; Nudo, E.; Nicolini, G. Environmental impact of inhalers for respiratory diseases: Decreasing the carbon footprint while preserving patient-tailored treatment. *BMJ Open Respir. Res.* **2020**, *7*, e000571. [CrossRef]
201. Oluokun, O.A.; Akinsooto, O.; Ogundipe, O.B.; Ikemba, S. Optimizing Demand Side Management (DSM) in industrial sectors: A policy-driven approach. *J. Ind. Energy Effic.* **2024**, *12*, 240–260. [CrossRef]
202. Sarker, E.; Halder, P.; Seyedmahmoudian, M.; Jamei, E.; Horan, B.; Mekhilef, S.; Stojcevski, A. Progress on the demand side management in smart grid and optimization approaches. *Int. J. Energy Res.* **2021**, *45*, 36–64. [CrossRef]
203. Rezaei, N.; Ahmadi, A.; Deihimi, M. A comprehensive review of demand-side management based on analysis of productivity: Techniques and applications. *Energies* **2022**, *15*, 7614. [CrossRef]
204. Aoun, A.; Ghandour, M.; Ilinca, A.; Ibrahim, H. Demand-side management. In *Hybrid Renewable Energy Systems and Microgrids*; Elsevier: Amsterdam, The Netherlands, 2021; pp. 463–490.
205. Moreno Escobar, J.J.; Morales Matamoros, O.; Tejeida Padilla, R.; Lina Reyes, I.; Quintana Espinosa, H. A comprehensive review on smart grids: Challenges and opportunities. *Sensors* **2021**, *21*, 6978. [CrossRef] [PubMed]
206. Chen, Z.; Amani, A.M.; Yu, X.; Jalili, M. Control and optimisation of power grids using smart meter data: A review. *Sensors* **2023**, *23*, 2118. [CrossRef] [PubMed]
207. Dewangan, F.; Abdelaziz, A.Y.; Biswal, M. Load forecasting models in smart grid using smart meter information: A review. *Energies* **2023**, *16*, 1404. [CrossRef]
208. Orlando, M.; Estebsari, A.; Pons, E.; Pau, M.; Quer, S.; Poncino, M.; Bottaccioli, L.; Patti, E. A smart meter infrastructure for smart grid IoT applications. *IEEE Internet Things J.* **2021**, *9*, 12529–12541. [CrossRef]
209. Yang, Y.; Bremner, S.; Menictas, C.; Kay, M. Modelling and optimal energy management for battery energy storage systems in renewable energy systems: A review. *Renew. Sustain. Energy Rev.* **2022**, *167*, 112671. [CrossRef]
210. Stephanie, F.; Karl, L. Incorporating Renewable Energy Systems for a New Era of Grid Stability. *Fusion Multidiscip. Res. Int. J.* **2020**, *1*, 37–49. [CrossRef]
211. Tavakoli, A.; Saha, S.; Arif, M.T.; Haque, M.E.; Mendis, N.; Oo, A.M.T. Impacts of grid integration of solar PV and electric vehicle on grid stability, power quality and energy economics: A review. *IET Energy Syst. Integr.* **2020**, *2*, 243–260. [CrossRef]
212. Khan, S.; Mazhar, T.; Shahzad, T.; Khan, M.A.; Rehman, A.U.; Hamam, H. Integration of smart grid with industry 5.0: Applications, challenges and solutions. *Meas. Energy* **2025**, *5*, 100031. [CrossRef]
213. Wang, Z.; Liao, P.; Liu, S.; Ji, Y.; Han, F. Scenario-based energy management optimization of hydrogen-electric-thermal systems in sustainable shipping. *Int. J. Hydrogen Energy* **2025**, *99*, 566–578. [CrossRef]
214. Ganesh, P.M.J.; Sundaram, B.M.; Balachandran, P.K.; Mohammad, G.B. IntDEM: An intelligent deep optimized energy management system for IoT-enabled smart grid applications. *Electr. Eng.* **2025**, *107*, 1925–1947. [CrossRef]
215. Amuthan, N.; Sathya, M.; Rani, N.C. Empowering Consumers and Utilities for a Smarter Future: The Pivotal Role of Advanced Metering Infrastructure (AMI) in Smart Meter Technology. In *Cloud Computing in Smart Energy Meter Management*; Wiley Online Library: Hoboken, NJ, USA, 2025; pp. 31–65.
216. Nasruddin, M.; Ramli, S.; Abdullah, M. Malaysia's Advanced Metering Infrastructure (AMI): A Regulatory Review. *IOP Conf. Ser. Earth Environ. Sci.* **2025**, *1511*, 012018. [CrossRef]
217. Azar, J.; Al Saleh, M.; Couturier, R.; Noura, H. Text Mining and Unsupervised Deep Learning for Intrusion Detection in Smart-Grid Communication Networks. *IoT* **2025**, *6*, 22. [CrossRef]
218. WKI Information. *Advanced Metering Infrastructure Market: Q1 and Q2 2025 Predictions*; WKI Information: San Francisco, CA, USA, 2025; Forecast Market Size of USD 29.63 Billion in 2025 and CAGR of 11.23% from 2025 to 2033. Available online: https://www.wkinformation.com/market-reports/advanced-metering-infrastructure-market/ (accessed on 1 August 2025).
219. World Bank. *Survey of International Experience in Advanced Metering Infrastructure and Its Implementation*; International Bank for Reconstruction and Development, The World Bank Group: Washington, DC, USA, 2018; Report. Global AMI Best Practices in Deployment, Governance, Procurement, Security, and Organization. Available online: https://documents1.worldbank.org/curated/en/957331569246407856/pdf/Survey-of-International-Experience-in-Advanced-Metering-Infrastructure-and-its-Implementation.pdf (accessed on 1 August 2025).
220. 360iResearch. *Advanced Metering Infrastructure Market: Global Forecast 2025–2030*; 360iResearch: Pune, India, 2025; Industry Report. Projected to Grow from USD 24.89 Billion in 2025 to USD 45.48 Billion by 2030 at a CAGR of 12.88%. Available online: https://www.360iresearch.com/library/intelligence/advanced-metering-infrastructure (accessed on 1 August 2025).

221. Şimşek, H.H. *Advanced Metering Infrastructure: Top 10+ Solutions [2025]*; AIMultiple: Tallinn, Estonia, 2025; Industry Review. Qualitative Analysis of AMI Solutions, Architecture Components, and Complementary Tools for Utilities. Available online: https://research.aimultiple.com/advanced-metering-infrastructure/ (accessed on 1 August 2025).
222. Lamb, M. Advanced Metering Infrastructure: Continued Evolution and Opportunities to Deliver Greater Value. *Clim. Energy* **2025**, *41*, 14–20. [CrossRef]
223. Shahveran, E.; Yousefi, H. Replacing fossil fuel-based power plants with renewables to meet Iran's environmental commitments in the electricity sector. *Renew. Sustain. Energy Transit.* **2025**, *7*, 100102. [CrossRef]
224. Meenatchi, K.; Veena, P.; Gowtham, A.; Sabarichandru, S.R.; Elavarasan, T.; Sanjay, V. AI Powered Renewable Load Sharing: A Sustainable Alternative to Fossil Fuel-Based System for Efficiency and Low Emissions. In Proceedings of the 2025 8th International Conference on Trends in Electronics and Informatics (ICOEI), Tirunelveli, India, 24–26 April 2025; pp. 227–232.
225. Li, R.; Tu, Q.; Feng, H.; Zou, Z. Demand response based battery energy storage systems design and operation optimization. *Energy Build.* **2025**, *338*, 115738. [CrossRef]
226. Ojo, O.T.; Salman, M.B.; Agbanusi, I.L.; Azubuike, C.H.; Olaleye, T.G.; Oyesanya, A.; Ajiboye, J.A. Enhancing Power Grid Resilience Through Energy Storage and Demand Response. *Path Sci.* **2025**, *11*, 8023–8029. [CrossRef]
227. Lehning, M.; Dujardin, J.; Cintas, A. The value of weather forecasts for optimally placed wind and solar electricity generators. *Front. Energy Res.* **2025**, *13*, 1533778. [CrossRef]
228. Alizadeh, H.; Khajehzadeh, A.; Eslami, M.; Shahbazzadeh, M.J. A holistic approach to enhance reliability and profitability through reliable efficiency index optimization and renewable energy integration. *Sci. Rep.* **2025**, *15*, 11853. [CrossRef]
229. Ergun, S.; Dik, A.; Boukhanouf, R.; Omer, S. Large-scale renewable energy integration: Tackling technical obstacles and exploring energy storage innovations. *Sustainability* **2025**, *17*, 1311. [CrossRef]
230. Alizadeh Aliabadi, M.; Pradhan, R.; Sattar, R.; Mahmood Fatah, A.N.; Abbas, J.K.; Kunamnei, R.; Salih, N.; Hussien Alawadi, A.; Kumar, A. Integration of renewable energy using the concept of energy hub to reduce the effect of uncertainty. *J. Chin. Inst. Eng.* **2025**, *48*, 214–226. [CrossRef]
231. Wang, H.; Zhang, J.; Wang, P.; Zhang, N. The role of demand-side flexibilities on low-carbon transition in power system: A case study of West Inner Mongolia, China. *Renew. Energy* **2025**, *242*, 122478. [CrossRef]
232. Rasool, M.H.; Taylan, O.; Perwez, U.; Batunlu, C. Comparative assessment of multi-objective optimization of hybrid energy storage system considering grid balancing. *Renew. Energy* **2023**, *216*, 119107. [CrossRef]
233. Iqbal, M.A.; Gil, J.-M.; Kim, S.K. Attention-driven hybrid ensemble approach with bayesian optimization for accurate energy forecasting in jeju island's renewable energy system. *IEEE Access* **2025**, *13*, 7986–8010. [CrossRef]
234. Alshareef, R.S.; Maghrabie, H.M. Building-integrated photovoltaics with energy storage systems—A comprehensive review. *J. Energy Storage* **2025**, *116*, 115916. [CrossRef]
235. Mahadevan, V.; Raja S, S.; Rusho, M.A.; Yishak, S. Critical review of Energy Storage Systems: A comparative assessment of Mechanisms, Advantages, Challenges, and Integration with Renewable Energy. *Results Eng.* **2025**, *26*, 105589. [CrossRef]
236. Manoharan, Y.; Olson, K.; Headley, A.J. Sensitivity of energy storage system optimization program to the source of renewable energy in the presence of demand side management: A behind-the-meter case study. *Appl. Energy* **2025**, *388*, 125557. [CrossRef]
237. Lu, Y.; Chen, M.; Zhu, G.; Zhang, Y. Recent progress in the study of integrated solar cell-energy storage systems. *Nanoscale* **2024**, *16*, 8778–8790. [CrossRef]
238. Kandhasamy, M.; Duvaragan, B.K.; Kamaraj, S.; Shanmugam, G. An Overview on Classification of Energy Storage Systems. In *Materials for Boosting Energy Storage. Volume 2: Advances in Sustainable Energy Technologies*; ACS Publications: Washington, DC, USA, 2024; pp. 1–25.
239. Aghmadi, A.; Mohammed, O.A. Energy storage systems: Technologies and high-power applications. *Batteries* **2024**, *10*, 141. [CrossRef]
240. Razmjoo, A.; Ghazanfari, A.; Østergaard, P.A.; Jahangiri, M.; Sumper, A.; Ahmadzadeh, S.; Eslamipoor, R. Moving toward the expansion of energy storage systems in renewable energy systems a techno-institutional investigation with artificial intelligence consideration. *Sustainability* **2024**, *16*, 9926. [CrossRef]
241. Njema, G.G.; Ouma, R.B.O.; Kibet, J.K. A review on the recent advances in battery development and energy storage technologies. *J. Renew. Energy* **2024**, *2024*, 2329261. [CrossRef]
242. Basmadjian, R. Demand-Side Management. In *Demand-Side Energy and Power Management in Data Centers: Are Data Centers Modern Societies' Double-Edged Sword?*; Springer: Berlin/Heidelberg, Germany, 2025; pp. 27–60.
243. Henriksen, I.M.; Strömberg, H.; Branlat, J.; Diamond, L.; Garzon, G.; Kuch, D.; Yilmaz, S.; Motnikar, L. The role of gender, age, and income in demand-side management acceptance: A literature review. *Energy Effic.* **2025**, *18*, 17. [CrossRef]
244. Zhu, Z.; Wang, D.; Tang, B.; Zhao, X.; Gu, G.; Yang, P.; Lai, X. Co-optimization of distributed generation, flexible load, and energy storage for promoting renewable energy consumption and power balancing in distribution networks. *Electr. Power Syst. Res.* **2025**, *238*, 111124. [CrossRef]

245. Chughtai, A.; Uqaili, M.A.; Hussain Mirjat, N.; Shaikh, F.; Khatri, S.A. Demand-side management scenario analysis for the energy-efficient future of Pakistan: Bridging the gap between market interests and national priorities. *Front. Energy Res.* **2024**, *12*, 1391973. [CrossRef]
246. Pang, C.; Kezunovic, M.; Ehsani, M. Demand side management by using electric vehicles as distributed energy resources. In Proceedings of the 2012 IEEE International Electric Vehicle Conference, Greenville, SC, USA, 4–8 March 2012; pp. 1–7.
247. Pinto, G.; Deltetto, D.; Capozzoli, A. Data-driven district energy management with surrogate models and deep reinforcement learning. *Appl. Energy* **2021**, *304*, 117642. [CrossRef]
248. Lindroos, T.J.; Ikäheimo, J. *Profitability of Demand Side Management Systems Under Growing Shares of Wind and Solar in Power Systems*; Taylor & Francis: Abingdon, UK, 2024.
249. Elazab, R.; Abdelnaby, A.T.; Ali, A.A. Impacts of multiple demand-side management strategies on microgrids planning: A literature survey. *Clean Energy* **2024**, *8*, 36–54. [CrossRef]
250. Tang, X.; Wang, J. Deep Reinforcement Learning-Based Multi-Objective Optimization for Virtual Power Plants and Smart Grids: Maximizing Renewable Energy Integration and Grid Efficiency. *Processes* **2025**, *13*, 1809. [CrossRef]
251. Cao, J.; Spulbar, C.; Eti, S.; Horobet, A.; Yüksel, S.; Dincer, H. Innovative approaches to green digital twin technologies of sustainable smart cities using a novel hybrid decision-making system. *J. Innov. Knowl.* **2025**, *10*, 100651. [CrossRef]
252. Al-Qarni, A.; Al-Shehri, M.; Rahman, A.-u.; Iqbal, T.; Bakry, A.; Abdulqader, M. Integrating Smart Grids in Smart Cities for Sustainable Future. *Int. J. Comput. Sci. Netw. Secur.* **2025**, *25*, 227–237.
253. World Economic Forum and Accenture. *Artificial Intelligence's Energy Paradox: Balancing Challenges and Opportunities*; World Economic Forum: Geneva, Switzerland, 2025; White Paper. An Analysis of AI's Rising Energy Demand, Strategies for Mitigation, and Its Dual Role in Accelerating Energy Transition. Available online: https://reports.weforum.org/docs/WEF_Artificial_Intelligences_Energy_Paradox_2025.pdf (accessed on 1 August 2025).
254. Amoresano, A.; Andresen, C.; Borragan, G.; Celino, M.; Chinnici, M.; Hoffmann, V.; Malek, K.; Malerba, L.; Mayo-Garcia, R.; Paltrinieri, N.; et al. An analysis of the implementation of artificial intelligence in the energy sector. In *An Analysis of the Implementation of Artificial Intelligence in the Energy Sector*; EERA: Berlin, Germany, 2025; pp. 18–20.
255. Shakir, M.A.; Appana, A.R.; Khalid, W.; Rajamanickam PJK, S.; Ahmad, M.I. Advancing insights into building electrical energy consumption: A comprehensive review of conventional and artificial intelligence methods. *Adv. Build. Energy Res.* **2025**, 1–44. . [CrossRef]
256. Amini, M.; Baradaran Rohani, M. The Role of Machine Learning and Artificial Intelligence in Enhancing Renewable Energy through Data Science. *World J. Technol. Sci. Res.* **2024**, *12*, 2341–2365. [CrossRef]
257. Alatalo, J.; Heilimo, E.; Rantonen, M.; Väänänen, O.; Sipola, T. Reducing emissions using artificial intelligence in the energy sector: A scoping review. *Appl. Sci.* **2025**, *15*, 999. [CrossRef]
258. Alomari, M.A.; Al-Andoli, M.N.; Ghaleb, M.; Thabit, R.; Alkawsi, G.; Alsayaydeh, J.A.J.; Gaid, A.S.A. Security of smart grid: Cybersecurity issues, potential cyberattacks, major incidents, and future directions. *Energies* **2025**, *18*, 141. [CrossRef]
259. Paul, B.; Sarker, A.; Abhi, S.H.; Das, S.K.; Ali, M.F.; Islam, M.M.; Islam, M.R.; Moyeen, S.I.; Badal, M.F.R.; Ahamed, M.H.; et al. Potential smart grid vulnerabilities to cyber attacks: Current threats and existing mitigation strategies. *Heliyon* **2024**, *10*, e019. [CrossRef] [PubMed]
260. Sophos. *State of Ransomware in Critical Infrastructure 2024*; Sophos Ltd.: Abingdon, UK, 2024. Available online: https://assets.sophos.com/X24WTUEQ/at/75tnw38cqsnrrv56wpwc78k/sophos-state-of-ransomware-critical-infrastructure-2024.pdf (accessed on 29 July 2025).
261. Aljohani, T.; Almutairi, A. Modeling time-varying wide-scale distributed denial of service attacks on electric vehicle charging Stations. *Ain Shams Eng. J.* **2024**, *15*, 102860. [CrossRef]
262. Behdadnia, T.; Thoelen, K.; Deconinck, G. Early-Stage Anomaly Detection in IT-OT Power Grid Communication Networks Using Wavelet Transform and Hybrid Deep Learning. *IEEE Transactions on Smart Grid.* **2025**, *16*, 3305–3322. [CrossRef]
263. Nasir, Q.; Talib, M.A.; Arshad, M.A.; Ishak, T.; Berrim, R.; Alsaid, B.; Badway, Y.; Waraga, O.A. Comparison of deep learning algorithms for site detection of false data injection attacks in smart grids. *Energy Inform.* **2024**, *7*, 71. [CrossRef]
264. Wang, J.; Chen, H.; Si, Y.; Zhu, Y.; Zhu, T.; Yin, S.; Liu, B. FDIA localization and classification detection in smart grids using multi-modal data and deep learning technique. *Comput. Electr. Eng.* **2024**, *119*, 109572. [CrossRef]
265. Priyadarsini, M.; Sonekar, N. A CNN-based approach for anomaly detection in smart grid systems. *Electr. Power Syst. Res.* **2025**, *238*, 111077. [CrossRef]
266. Zhang, J.; Zhou, K.; Wang, L.; Yang, Y.; Li, Y. Detection framework of FDIA for smart grid based on mode decomposition and gradient boosting decision. *J. Electr. Eng. Technol.* **2025**, *20*, 3601–3623. [CrossRef]
267. Anwar, S.; Kanwal, S.; Essa, S.; Abbas, S.; Meisam, K.M.; Amin, W. Detecting False Data Injection Attacks in Smart Grids Using Machine Learning: A Comparative Study. In Proceedings of the International Conference on Smart Electrical Grid and Renewable Energy, Suzhou, China, 9–12 August 2024; pp. 509–515.

268. Gao, Z.; Wang, J.; Illindala, M. Intrusion detection system with updated structure and feature selection algorithm for man-in-the-middle attacks in the smart grid. *IEEE Trans. Ind. Appl.* **2025**, *61*, 8150–8157. [CrossRef]
269. Kandasamy, V.; Roseline, A.A. Harnessing advanced hybrid deep learning model for real-time detection and prevention of man-in-the-middle cyber attacks. *Sci. Rep.* **2025**, *15*, 1697. [CrossRef] [PubMed]
270. Qiao, Y.; Chen, D.; Sun, Q.Z.; Tian, G.; Wang, W. Unveiling stealthy man-in-the-middle cyber-attacks on energy performance in grid-interactive smart buildings. *Energy Convers. Manag.* **2024**, *319*, 118949. [CrossRef]
271. Ahn, B.; Jenkins, A.M.; Kim, T.; Zeng, J.; McLauchlan, L.; Park, S.-w. Exploring ransomware attacks on smart inverters. In Proceedings of the 2023 IEEE Energy Conversion Congress and Exposition (ECCE), Nashville, TN, USA, 29 October–2 November 2023; pp. 1567-1573.
272. Alvee, S.R.B.; Ahn, B.; Kim, T.; Su, Y.; Youn, Y.-W.; Ryu, M.-H. Ransomware attack modeling and artificial intelligence-based ransomware detection for digital substations. In Proceedings of the 2021 6th IEEE Workshop on the Electronic Grid (eGRID), New Orleans, LA, USA, 8–10 November 2021; pp. 01-05.
273. Benmalek, M. Ransomware on cyber-physical systems: Taxonomies, case studies, security gaps, and open challenges. *Internet Things Cyber-Phys. Syst.* **2024**, *4*, 186–202. [CrossRef]
274. Basnet, M.; Poudyal, S.; Ali, M.H.; Dasgupta, D. Ransomware detection using deep learning in the SCADA system of electric vehicle charging station. In Proceedings of the 2021 IEEE PES Innovative Smart Grid Technologies Conference—Latin America (ISGT Latin America), Lima, Peru, 15–17 September 2021; pp 1–5.
275. Bao, K.; Chau, S.C.-K.; Elbez, G.; Liu, Q.; Hagenmeyer, V. Advanced persistent threats on consumer energy resources in decentralized energy systems. In Proceedings of the 16th ACM International Conference on Future and Sustainable Energy Systems, Rotterdam, The Netherlands, 17–20 June 2025; pp. 838–845.
276. Kanjow, S. Advanced Persistent Threats (APT) in the Energy Sector. 2025. https://www.diva-portal.org/smash/get/diva2:1972349/FULLTEXT01.pdf (accessed on 1 August 2025).
277. Aydın, Z. Detecting cybersecurity threats in digital energy systems using deep learning for imbalanced datasets. *Int. J. Energy Econ. Policy* **2025**, *15*, 614–628. [CrossRef]
278. Idima, S.; Nwaga, P.; Evah, P. Comprehensive analysis of SCADA system data for intrusion detection using machine learning. *Glob. J. Eng. Technol. Adv.* **2025**, *22*, 64–89. [CrossRef]
279. Jeje, M.O. Cybersecurity assessment of smart grid exposure using a machine learning based approach. *arXiv* **2025**, arXiv:2501.14175.
280. Zheng, H.; Li, X.; Li, F. Unsupervised cyberattack detection in smart grids: A novel approach integrating horizontal federated learning for the control center and substations. *Reliab. Eng. Syst. Saf.* **2025**, *264*, 111444. [CrossRef]
281. Ahmed, I.S.; Ahmed, A.L.; Kamalrudin, M.; Musa, M. Integrating Cybersecurity into Renewable Energy Development: A Data-Driven Decision Tree Approach for Environmental Protection. *J. Cybersecur. Inf. Manag.* **2025**, *15*, 2.
282. Almalawi, A.; Hassan, S.; Fahad, A.; Iqbal, A.; Khan, A.I. Hybrid cybersecurity for asymmetric threats: Intrusion detection and SCADA system protection innovations. *Symmetry* **2025**, *17*, 616. [CrossRef]
283. Rahman, M.M.; Chayan, M.M.H.; Mehrin, K.; Sultana, A.; Hamed, M.M. Explainable deep learning for cyber attack detection in electric vehicle charging stations. In Proceedings of the 11th International Conference on Networking, Systems, and Security, Khulna, Bangladesh, 19–21 December 2024; pp. 1–7.
284. Faheem, M.; Al-Khasawneh, M.A.; Khan, A.A.; Madni, S.H.H. Cyberattack patterns in blockchain-based communication networks for distributed renewable energy systems: A study on big datasets. *Data Brief* **2024**, *53*, 110212. [CrossRef] [PubMed]
285. Trivedi, R.; Patra, S.; Khadem, S. Data-centric explainable artificial intelligence techniques for cyber-attack detection in microgrid networks. *Energy Rep.* **2025**, *13*, 217–229. [CrossRef]
286. Sun, H.; Guo, Q.; Shen, X.; Xue, Y.; Shahidehpour, M.; Hatziargyriou, N. Energy internet: Redefinition and categories. *Energy Internet* **2024**, *1*, 3–8. [CrossRef]
287. Fawaz, A.; Mougharbel, I.; Kanaan, H.Y. Power routing algorithms: A comparative study. In Proceedings of the 2023 6th International Conference on Renewable Energy for Developing Countries (REDEC), Zouk Mosbeh, Lebanon, 5–7 July 2023; pp. 139–144.
288. Rikiya, A.; Tanaka, K.; Van Triet, N. Digital Grid in Japan. In *The Energy Internet*; Elsevier: Amsterdam, The Netherlands, 2019; pp. 241–264.
289. Ma, J.; Song, L.; Li, Y. Optimal power dispatching for local area packetized power network. *IEEE Trans. Smart Grid* **2017**, *9*, 4765–4776. [CrossRef]
290. Guo, Y.; Lin, H.; Sun, H. Energy Internet: A Standardization-Based Blueprint Design. *arXiv* **2024**, arXiv:2409.05189. [CrossRef]
291. Mubarak, A.; Ishfaq, H.; Amin, W.; Shi, X.; Duvvuri, S.S.S.S.; Saleem, M.A.; Raju, K.N. FPGA-Driven Secure Energy Routing and Attack Detection in Decentralized Microgrids With SREM-Based Dynamic Path Optimization. *IEEE Trans. Consum. Electron.* **2025**, *71*, 1. [CrossRef]

292. Tsoukalas, L.H.; Gao, R. From smart grids to an energy internet: Assumptions, architectures and requirements. In Proceedings of the 2008 Third International Conference on Electric Utility Deregulation and Restructuring and Power Technologies, Nanjing, China, 6–9 April 2008; pp. 94–98.
293. Hebal, S.; Mechta, D.; Harous, S.; Louail, L. A comparative study of energy routing algorithms to optimize energy transmission in energy internet. *J. Supercomput.* **2024**, *80*, 20583–20639. [CrossRef]
294. Jiang, X.; Sun, C.; Cao, L.; Law, N.; Loo, K.H. Review and comparative study of energy routing algorithms for energy internet. In Proceedings of the 2021 IEEE 5th Conference on Energy Internet and Energy System Integration (EI2), Taiyuan, China, 22–24 October 2021; pp. 567–572.
295. Abdella, J.; Shuaib, K.; Harous, S. Energy routing algorithms for the energy internet. In Proceedings of the 2018 International Conference on Intelligent Systems (IS), Madeira, Portugal, 25–27 September 2018; pp. 80–86.
296. Hebal, S.; Harous, S.; Mechta, D. Energy routing challenges and protocols in energy internet: A survey. *J. Electr. Eng. Technol.* **2021**, *16*, 3197–3212. [CrossRef]
297. Joseph, A.; Balachandra, P. Energy internet, the future electricity system: Overview, concept, model structure, and mechanism. *Energies* **2020**, *13*, 4242. [CrossRef]
298. Ma, J.; Zhang, N.; Shen, X. Elastic energy distribution of local area packetized power networks to mitigate distribution level load fluctuation. *IEEE Access* **2018**, *6*, 8219–8231. [CrossRef]
299. Chen, M.; Xia, M.; Chen, Q. Hierarchical structure design and primary energy dispatching strategy of net energy router. *Electr. Power Syst. Res.* **2021**, *201*, 107539. [CrossRef]
300. Chen, M.; Xia, M.; Chen, Q. Planning model for energy routers considering peer-to-peer energy transactions among multiple distribution networks. *CSEE J. Power Energy Syst.* **2022**, *10*, 1026–1043.
301. Abe, R. *Digital Grid: New Internet-Like Multi-Directional Power Supply*; Springer: Berlin/Heidelberg, Germany, 2023.
302. Schneider, K.; Wiegel, F.; Schulz, D.; Hagenmeyer, V.; Hiller, M.; Bless, R.; Zitterbart, M. Designing the interplay of energy plane and communication plane in the energy packet grid. In Proceedings of the 2021 IEEE 46th Conference on Local Computer Networks (LCN), Edmonton, AB, Canada, 4–7 October 2021; pp. 331–334.
303. Schulz, D.; Schneider, K.; Weßbecher, M.; Hagenmeyer, V.; Zitterbart, M.; Hiller, M. Composing power flow patterns through coordinated dispatch of energy packets. In Proceedings of the 2023 8th IEEE Workshop on the Electronic Grid (eGRID), Karlsruhe, Germany, 16–18 October 2023; pp. 1–6.
304. Schneider, K.; Weßbecher, M.; Schulz, D.; Hagenmeyer, V.; Hiller, M.; Zitterbart, M. Modes of energy packets in the energy packet grid. In Proceedings of the 2023 8th IEEE Workshop on the Electronic Grid (eGRID), Karlsruhe, Germany, 16–18 October 2023; pp. 1–7.
305. Schulz, D.; Schneider, K.; Weßbecher, M.; Hagenmeyer, V.; Zitterbart, M.; Hiller, M. Hardware realization of participants in an energy packet-based power grid. In Proceedings of the 2022 IEEE 13th International Symposium on Power Electronics for Distributed Generation Systems (PEDG), Kiel, Germany, 26–29 June 2022; pp. 1–6.
306. Liu, B.; Wu, W.; Zhou, C.; Mao, C.; Wang, D.; Duan, Q.; Sha, G. An AC–DC hybrid multi-port energy router with coordinated control and energy management strategies. *IEEE Access* **2019**, *7*, 109069–109082. [CrossRef]
307. Dudko, A.; Endrjukaite, T.; Roose, L.R. Open routed energy distribution network based on a concept of energy router in smart grid. In Proceedings of the 21st International Conference on Information Integration and Web-Based Applications & Services, Munich, Germany, 2–4 December 2019; pp. 483–491.
308. Zhou, J.; Wang, J. Research review on multi-port energy routers adapted to renewable energy access. *Electronics* **2024**, *13*, 1493. [CrossRef]
309. Fawaz, A.; Mougharbel, I.; Al-Haddad, K.; Kanaan, H.Y. Energy routing protocols for Energy Internet: A review on multi-agent systems, metaheuristics, and Artificial Intelligence approaches. *IEEE Access* **2025**, *13*, 41625–41643. [CrossRef]
310. Jiang, X.; Sun, C.; Cao, L.; Law, N.; Loo, K.H. Semi-decentralized energy routing algorithm for minimum-loss transmission in community energy internet. *Int. J. Electr. Power Energy Syst.* **2022**, *135*, 107547. [CrossRef]
311. Shahzad, K.; Iqbal, S. A reliable graph-based routing algorithm in residential multi-microgrid systems. In Proceedings of the 2022 International Conference on Frontiers of Information Technology (FIT), Islamabad, Pakistan, 12–13 December 2022; pp. 71–76.
312. Razi, R.; Pham, M.-C.; Hably, A.; Bacha, S.; Tran, Q.-T.; Iman-Eini, H. A novel graph-based routing algorithm in residential multimicrogrid systems. *IEEE Trans. Ind. Inf.* **2020**, *17*, 1774–1784.
313. Wang, R.; Wu, J.; Qian, Z.; Lin, Z.; He, X. A graph theory based energy routing algorithm in energy local area network. *IEEE Trans. Ind. Inf.* **2017**, *13*, 3275–3285. [CrossRef]
314. Hussain, S.S.; Aftab, M.A.; Nadeem, F.; Ali, I.; Ustun, T.S. Optimal energy routing in microgrids with IEC 61850 based energy routers. *IEEE Trans. Ind. Electron.* **2019**, *67*, 5161–5169. [CrossRef]
315. Wang, Y.; Zhang, Y. Energy routing strategy based on minimum loss and transaction price. In Proceedings of the 2022 41st Chinese Control Conference (CCC), Hefei, China, 25–27 July 2022; pp. 6073–6078.

316. Koottappillil, D.P.; Naidoo, R.M.; Mbungu, N.T.; Bansal, R.C. Distribution of renewable energy through the energy internet: A routing algorithm for energy routers. *Energy Rep.* **2022**, *8*, 355–363. [CrossRef]
317. Guo, H.; Wang, F.; Li, L.; Zhang, L.; Luo, J. A minimum loss routing algorithm based on real-time transaction in energy internet. *IEEE Trans. Ind. Inf.* **2019**, *15*, 6446–6456. [CrossRef]
318. Guo, H.; Wang, F.; James, G.; Zhang, L.; Luo, J. Graph theory based topology design and energy routing control of the energy internet. *IET Gener. Transm. Distrib.* **2018**, *12*, 4507–4514. [CrossRef]
319. Forootan, M.M.; Larki, I.; Zahedi, R.; Ahmadi, A. Machine learning and deep learning in energy systems: A review. *Sustainability* **2022**, *14*, 4832. [CrossRef]
320. Perera, A.T.D.; Kamalaruban, P. Applications of reinforcement learning in energy systems. *Renew. Sustain. Energy Rev.* **2021**, *137*, 110618. [CrossRef]
321. Fawaz, A.; Mougharbel, I.; Kanaan, H.Y. Efficient low-cost method for the estimation of clouds shading rate on PV farms-real-time reconfiguration application. In Proceedings of the 2021 IEEE 30th International Symposium on Industrial Electronics (ISIE), Kyoto, Japan, 20–23 June 2021; pp. 1–7.
322. Hua, H.; Qin, Y.; Hao, C.; Cao, J. Optimal energy management strategies for energy Internet via deep reinforcement learning approach. *Appl. Energy* **2019**, *239*, 598–609. [CrossRef]
323. Yang, G.; Cao, J.; Hua, H.; Zhou, Z. Deep learning-based distributed optimal control for wide area energy Internet. In Proceedings of the 2018 IEEE International Conference on Energy Internet (ICEI), Beijing, China, 21–25 May 2018; pp. 292–297.
324. Wang, D.-L.; Sun, Q.-Y.; Li, Y.-Y.; Liu, X.-R. Optimal energy routing design in energy internet with multiple energy routing centers using artificial neural network-based reinforcement learning method. *Appl. Sci.* **2019**, *9*, 520. [CrossRef]
325. Benchikh, L.; Louail, L.; Mechta, D. Subscriber matching in energy internet using the firefly algorithm. In *Proceedings of the Second International Conference on Innovations in Computing Research (ICR'23)*; Springer: Berlin/Heidelberg, Germany, 2023; pp. 418–432.
326. Benchikh, L.; Louail, L.; Mechta, D.; Nacef, A. Optimal path selection and producer allocation for energy distribution in energy Internet. In Proceedings of the 2023 International Conference on Networking and Advanced Systems (ICNAS), Algiers, Algeria, 21–23 October 2023; pp. 1–6.
327. Nacef, A.; Mechta, D.; Louail, L.; Harous, S. Genetic algorithm-based path selection for power-efficient energy routing in smart grids and energy internet. In Proceedings of the 2023 2nd International Engineering Conference on Electrical, Energy, and Artificial Intelligence (EICEEAI), Zarqa, Jordan, 27–28 December 2023; pp. 1–6.
328. Hua, H.; Shi, J.; Chen, X.; Qin, Y.; Wang, B.; Yu, K.; Naidoo, P. Carbon emission flow based energy routing strategy in energy Internet. *IEEE Trans. Ind. Inf.* **2023**, *20*, 3974–3985. [CrossRef]
329. Mubarak, A.; Amin, W.; Shi, X. Energy Efficiency Optimization in Vehicle Communication Network Using Hybrid-Genetic Algorithm. *Trans. Emerg. Telecommun. Technol.* **2025**, *36*, e70146. [CrossRef]
330. Hebal, S.; Mechta, D.; Harous, S.; Dhriyyef, M. Hybrid energy routing approach for energy internet. *Energies* **2021**, *14*, 2579. [CrossRef]
331. Mechta, D.; Harous, S.; Hebal, S. Energy-efficient path-aware routing Protocol based on PSO for Smart Grids. In Proceedings of the 2020 IEEE International Conference on Electro Information Technology (EIT), Chicago, IL, USA, 31 July–1 August 2020; pp. 093–097.
332. Ye, H.; Wen, S.; Zhu, M.; Lin, A.; Pan, A.; Liu, Z.; Dong, Z.; Liu, J. Virtual Energy Routing Strategy for Energy Internet with Renewable Energy. In Proceedings of the 2021 IEEE 12th Energy Conversion Congress & Exposition-Asia (ECCE-Asia), Singapore, 24–27 May 2021; pp. 1459–1463.
333. Alhasnawi, B.N.; Jasim, B.H.; Rahman, Z.-A.S.A.; Guerrero, J.M.; Esteban, M.D. A novel internet of energy based optimal multi-agent control scheme for microgrid including renewable energy resources. *Int. J. Environ. Res. Public Health* **2021**, *18*, 8146. [CrossRef]
334. Hu, J.; Cong, H.; Jiang, C. Coordinated scheduling model of power system with active distribution networks based on multi-agent system. *J. Mod. Power Syst. Clean Energy* **2018**, *6*, 521–531. [CrossRef]
335. Luo, F.; Dong, Z.Y.; Murata, J.; Xu, Z. A distributed trading system in active distribution networks based on multi-agent coalition and Blockchain. *IEEE Trans. Power Syst.* **2018**, *34*, 4097–4108. [CrossRef]
336. Fawaz, A.; Mougharbel, I.; Kanaan, H.Y. Q-Learning Routing Protocol Applied to Energy Internet for Power Distribution. In Proceedings of the 2023 IEEE 17th International Conference on Compatibility, Power Electronics and Power Engineering (CPE-POWERENG), Tallinn, Estonia, 14–16 June 2023; pp. 1–7.
337. Li, S.; Cao, D.; Hu, W.; Huang, Q.; Chen, Z.; Blaabjerg, F. Multi-energy management of interconnected multi-microgrid system using multi-agent deep reinforcement learning. *J. Mod. Power Syst. Clean Energy* **2023**, *11*, 1606–1617. [CrossRef]
338. Kanwal, S.; Amin, W.; Khan, A.A.; Rafique, B.; Huang, Q.; Jian, L.; Batool, I. Enhanced cybersecurity for smart grids: Detecting protocol-specific DDoS attacks on Modbus networks. *Comput. Electr. Eng.* **2025**, *127*, 110629. [CrossRef]
339. Cui, L.; Qu, Y.; Gao, L.; Xie, G.; Yu, S. Detecting false data attacks using machine learning techniques in smart grid: A survey. *J. Netw. Comput. Appl.* **2020**, *170*, 102808. [CrossRef]

340. Sayghe, A.; Hu, Y.; Zografopoulos, I.; Liu, X.; Dutta, R.G.; Jin, Y.; Konstantinou, C. Survey of machine learning methods for detecting false data injection attacks in power systems. *IET Smart Grid* **2020**, *3*, 581–595. [CrossRef]
341. Olaitan, O.F.; Ayeni, S.O.; Olosunde, A.; Okeke, F.C.; Okonkwo, U.U.; Ochieze, C.G.; Chukwujama, O.V.; Akatakpo, O.N. Quantum Computing in Artificial Intelligence: A Review of Quantum Machine Learning Algorithms. *Path Sci.* **2025**, *11*, 7001–7009. [CrossRef]
342. He, Z.; Amaro, D.; Shaydulin, R.; Pistoia, M. Performance of quantum approximate optimization with quantum error detection. *Commun. Phys.* **2025**, *8*, 217. [CrossRef]
343. Kottahachchi Kankanamge Don, A.; Khalil, I. Q-SupCon: Quantum-enhanced supervised contrastive learning architecture within the representation learning framework. *ACM Trans. Quantum Comput.* **2025**, *6*, 1–24. [CrossRef]
344. Ahanger, T.A.; Bhatia, M.; Aldaej, A. Quantum informative analysis in smart power distribution. *ACM Trans. Intell. Syst. Technol.* **2024**, *15*, 1–18. [CrossRef]
345. Siddi Moreau, G.; Pisani, L.; Profir, M.; Podda, C.; Leoni, L.; Cao, G. Quantum Artificial Intelligence Scalability in the NISQ Era: Pathways to Quantum Utility. *Adv. Quantum Technol.* **2024**, *7*, 2400716. [CrossRef]
346. How, M.-L.; Cheah, S.-M. Forging the future: Strategic approaches to quantum AI integration for industry transformation. *AI* **2024**, *5*, 290–323. [CrossRef]
347. Golec, M.; Hatay, E.S.; Golec, M.; Uyar, M.; Golec, M.; Gill, S.S. Quantum cloud computing: Trends and challenges. *J. Econ. Technol.* **2024**, *2*, 190–199. [CrossRef]

energies

Article

Towards Net Zero in Poland: A Novel Approach to Power Grid Balance with Centralized Hydrogen Production Units

Dariusz Bradło [1], Witold Żukowski [1,*], Jan Porzuczek [2], Małgorzata Olek [2] and Gabriela Berkowicz-Płatek [1]

1 Faculty of Chemical Engineering and Technology, Cracow University of Technology, Warszawska 24, 31-155 Cracow, Poland; dariusz.bradlo@pk.edu.pl (D.B.); gabriela.berkowicz@pk.edu.pl (G.B.-P.)
2 Faculty of Environmental Engineering and Energy, Cracow University of Technology, Warszawska 24, 31-155 Cracow, Poland; jan.porzuczek@pk.edu.pl (J.P.); malgorzata.olek@pk.edu.pl (M.O.)
* Correspondence: witold.zukowski@pk.edu.pl

Abstract: The net zero emissions policy represents a crucial component of the global initiative to address climate change. The European Union has set a target of achieving net zero greenhouse gas emissions by 2050. This study assesses Poland's feasibility of achieving net zero emissions. Currently, Poland relies on fossil fuels for approximately 71% of its electricity generation, with electricity accounting for only approximately 16% of the country's total final energy consumption. Accordingly, the transition to net zero carbon emissions will necessitate significant modifications to the energy system, particularly in the industrial, transport, and heating sectors. As this is a long-term process, this article demonstrates how the development of renewable energy sources will progressively necessitate the utilisation of electrolysers in line with the ongoing industrial transformation. A new framework for the energy system up to 2060 is presented, with transition phases in 2030, 2040, and 2050. This study demonstrates that it is feasible to attain a sustainable, zero-emission, and stable energy system despite reliance on uncontrolled and weather-dependent energy sources. Preparing the electricity grid to transmit almost three times the current amount represents a significant challenge. The resulting simulation capacities, comprising 64 GW of onshore wind, 33 GW of offshore wind, 136 GW of photovoltaic, 10 GW of nuclear, and 22 GW of electrolysers, enable a positive net energy balance to be achieved under the weather conditions observed between 2015 and 2023. To guarantee system stability, electrolysers must operate within a centralised framework, functioning as centrally controlled dispatchable load units.

Keywords: energy transition; modern energy systems; electrolysis; hydrogen; renewable energy

1. Introduction

The net zero policy is a cornerstone of global efforts to mitigate climate change. By 2050, the European Union (EU) aims to achieve zero greenhouse gas (GHG) emissions, a target formalised with the adoption of the European Green Deal [1]. An intermediate milestone requires a reduction in GHG emissions by at least 55% by 2030 compared to 1990 levels. In 2022, the total GHG emissions in the EU reached 3.14 Gt $CO_{2\text{-eq.}}$ [2], while Poland's emissions amounted to 380.5 Mt $CO_{2\text{-eq.}}$, including 315.4 Mt CO_2 [3]. Meanwhile, China, responsible for 30% of global CO_2 emissions in 2023 (11.9 Gt CO_2) [4], has committed to achieving net zero emissions by 2060 [5]. Major companies like Microsoft and IKEA also align with net zero policies [6].

The scientific community is working to identify the best path towards the elimination of GHGs. Climate models predict changes and assess mitigation strategies, emphasising the importance of achieving negative CO_2 emissions [7,8]. Various methods, including carbon dioxide capture [9] and storage technologies [10], are being developed alongside advancements in energy storage, such as hydrogen [11], battery systems [12], pumped hydro storage (PHS) [13], or compressed air energy storage (CAES) [14]. Innovations in renewable energy technologies—solar [15], wind [16], and geothermal energy [17]—are also progressing. However, detailed research is lacking on systems that can efficiently manage energy generated from intermittent sources, such as wind and solar power, especially in the absence of demand-side regulations. In energy systems dominated by weather-dependent sources, balancing demand without imposing consumer restrictions is the primary challenge. Research is needed to develop models and strategies for efficient energy supply and demand management while maintaining reliability without imposing restrictions on consumers.

Currently, Poland's power system operates with complete consumer freedom in electricity usage. Demand is balanced by Centrally Dispatched Generating Units (CDGU), primarily thermal power plants fuelled by coal, lignite, and natural gas [18]. Transitioning to a zero-emission system requires replacing these units with alternative solutions. Sole reliance on nuclear power is not viable, as these plants operate at a constant output with limited flexibility in adjusting power levels. CDGU must quickly adapt their production to match the current demand in the power system. Furthermore, nuclear power plants are costly and time-consuming to construct compared to renewable energy sources (RESs) [19].

Adapting the power system to weather-dependent energy sources necessitates new management principles. A critical step involves transferring the responsibility for system balance to energy consumers. Additionally, post-transition, industrial sectors will require significant hydrogen resources. In 2022, the production and consumption of hydrogen in Polish industries (primarily from natural gas) totalled 1 Mt [20]. Hydrogen is an important industrial feedstock, for which there are often no substitutes (e.g., fertiliser production). In this study, electrolysis was selected as the method for zero-emission hydrogen production, as other methods of hydrogen production are not at such a high level of readiness for implementation or are based on thermochemical biomass processes. In this study, it is assumed that 100% of the domestic biomass that can be used for biogas production is, in fact, used to produce biomethane, which, by powering electricity generators, supplements the shortfall in production from RESs. Integrating hydrogen production electrolysers into the power system could address the dual needs of system balancing and hydrogen demand. To achieve this, electrolysers must operate as part of the centralised power system, aligning their use with system requirements, especially for managing electricity surpluses.

The aim of this article is to present and analyse concepts for creating a sustainable and stable power system in Poland based on weather-dependent energy sources without imposing energy consumption restrictions on individual consumers and industries. This analysis covers 2024–2060 and shows the final result of net zero emissions in Poland achieved through power grid balance with centralised hydrogen production units.

2. The Current State of the Power System in Poland

In 2023, Poland's energy supply was primarily based on fossil fuels: 36% from coal, 33% from oil, and 16% from natural gas. At the same time, about 71% of electricity was generated from fossil fuels (around 60% from coal and about 11% from natural gas). The share of electricity in total final energy consumption was only about 16% in 2022. Decarbonising key sectors—transport, heating, and industry—will necessitate substantial investments and the development of complementary solutions, such as hydrogen utilisation, advanced energy storage systems, and transmission infrastructure modernisation. Poland is gradually transitioning toward renewable energy, despite significant reliance on domestic coal resources [21]. The photovoltaic (PV) market is among the fastest-growing in the EU, with increasing adoption of heat pumps and ambitious plans for offshore wind energy. By 2030, Poland aims to achieve at least 3.4 GW of offshore wind capacity [22].

At the end of 2023, Poland's installed power capacity totalled 67.8 GW, with 40.6 GW coming from conventional power plants and 27.2 GW from RES. The average annual demand was 23.1 GW, peaking at 27.3 GW [23]. Peak demand occurs during winter (from November to February), while the lowest demand is recorded between May and August (Figure 1).

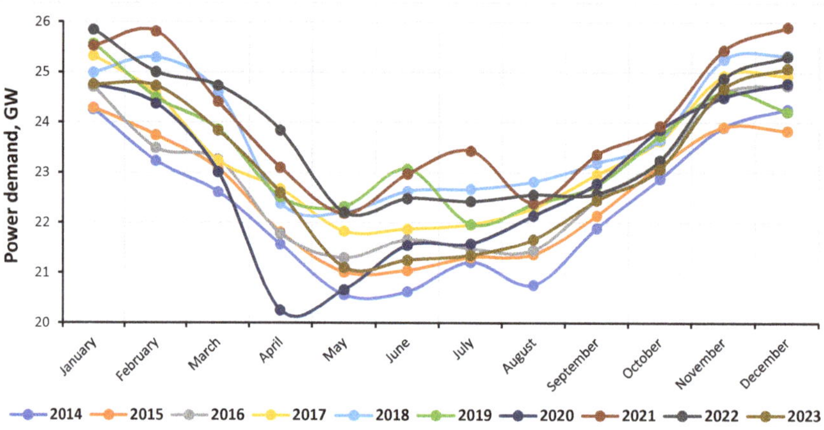

Figure 1. Average monthly power demand during daily peak load working days in 2014–2023 in Poland [23].

Electricity production in 2023 reached 163.6 TWh, a 7% decrease compared to 2022 [23]. The changes in national electricity production between 2014 and 2023 are shown in Figure 2. In 2023, the production from conventional thermal power plants returned to 2020 levels and amounted to 128.4 TWh.

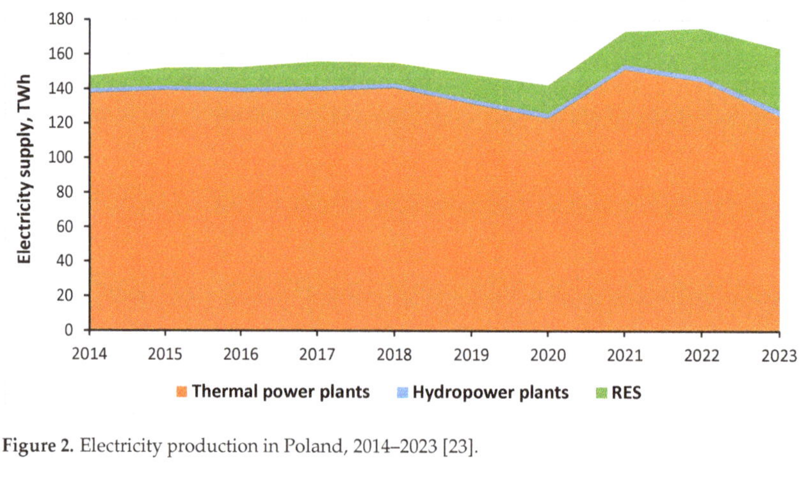

Figure 2. Electricity production in Poland, 2014–2023 [23].

At the beginning of 2024, wind power capacity reached 9.43 GW, with annual growth exceeding 1 GW in recent years (Table 1). In March 2024, Poland had 1400 onshore wind installations, with large wind farms (>30 MW) contributing approximately 7 GW [24]. The total potential of the Polish part of the Baltic Sea is estimated to be 33 GW [25]. The first offshore wind farms are expected to begin production in 2026. By 2040, offshore wind capacity could provide 11 GW to the national grid [26].

Table 1. Annual increase in installed RES capacity in 2014–2023 in Poland, GW/y [27,28].

Year	2014	2015	2016	2017	2018	2019	2020	2021	2022	2023
Wind onshore	0.44	0.75	1.23	0.04	0.01	0.06	0.43	0.77	1.18	1.13
PV	-	0.06	0.10	0.08	0.24	0.96	2.44	3.87	4.65	4.66

By early 2024, Poland's PV capacity had reached 17.1 GW. Micro-installations (<50 kW) accounted for 64%, small installations (50–1000 kW) for 25%, and large farms (>1 MW) for 11%, equating to approximately 1.6 GW. By March 2024, the total RES capacity exceeded 29 GW, with PV installations contributing 60% [27]. In recent years, PV capacity has grown annually by over 4.5 GW (Table 1).

Currently, Poland has no operational nuclear power plants, similar to 14 other EU countries [29]. The formulation of plans is currently in progress to build six pressurised water reactor (PWR) units, with a total capacity of 6–9 GW, by 2040 [30,31]. An alternative could be small modular reactors (SMRs), with powers of around 300 MW, which offer a flexible alternative for regions with high energy demand and limited RES potential [32]. Poland's grid development plan includes three SMR units, totalling 0.84 GW, by 2034 [33].

In Poland, biogas and biomethane are produced at landfill sites or are generated from sewage sludge, agricultural and municipal waste, slurry, and manure [34,35]. In 2021, Poland produced 622.98 million m^3 of biogas from landfill gas collection systems (107.43 million m^3), sewage sludge fermentation systems at 100 wastewater treatment plants (153.74 million m^3), 128 installations processing 4.9 million tons of agricultural waste (342.91 million m^3), and from municipal waste composting and fermentation systems (15.65 million m^3). This allowed for the production of 1345.58 GWh of electricity and 5383.47 TJ of heat [34]. Despite ranking fifth in biogas electricity production among EU-27 countries [36], Poland's output is far below Germany's, which generates 23 times more electricity with comparable substrate potential. Based on biogas efficiency studies conducted on nearly 3500 different substrates and analyses of the potential of the agricultural sector, Poland's biogas production potential is estimated at 13.5 billion m^3 (equivalent to 7.8 billion m^3 of biomethane) [37]. Another 10.8 billion m^3 of biogas could be produced following Germany's example and using 5% of the national farmland for energy crop production. Unlocking this potential requires changes in waste logistics, biogas purification technologies, and alternative methods such as biomethanisation or cryogenic separation [35]. Legal reforms, public awareness campaigns, and financial support are also essential [37]. Regarding the environmental impact of biomethane production, it is worth noting that it is primarily based on the use of biodegradable waste materials. The controlled management of these materials is likely to have a positive environmental effect.

3. The Methods of Simulation

Calculations were conducted using an energy system simulator developed by the Polish governmental institution, the National Centre for Research and Development (NCBR) [38,39]. The model's input parameters were calculated using MS Excel and MATLAB R2020b for various simulation time horizons, namely 2030, 2040, 2050, and 2060.

In the process, three different scenarios for achieving a net zero state by 2050 or 2060 were designed.

The simulator used in the study is based on balancing the electricity system. It encompasses three energy system components: generation, storage, and consumption. The simulation is performed for an annual time horizon, with a one-hour time step, and the individual runs are aggregated over one year. The simulator also accounts for weather conditions throughout the year with a one-hour resolution. These conditions include solar radiation, wind speeds, and ambient temperature. These data are used to calculate the efficiency of weather-dependent energy sources, the heating demand (via degree-hours), and the efficiency of heat pumps. The source data necessary for the simulations are obtained from publicly available tools and databases (PVGIS [40], ENTSO-E [41], Open-meteo [42], and Instrat [43]). For simulation purposes, it is possible to select the year for which the weather data will be considered, covering the period from 2015 to 2023.

According to the simulator's assumptions, electricity is generated by selected renewable energy sources (onshore wind, offshore wind, and PV) and nuclear power plants. Any instantaneous surplus of generated energy is sequentially sent to electrical energy storage systems (battery storage, electric vehicle batteries, and pumped hydroelectric storage), industrial heat storage, residential heat storage, and electrolysers, which are the final recipients of excess energy. Instantaneous energy shortages in the system are sequentially compensated by residential heat storage, industrial heat storage, electrical energy storage, and dispatchable generation sources powered by biomethane. The simulator determines, among other factors, the demand for biomethane to power dispatchable gas power plants and their required power levels on an hourly basis.

The starting year for this analysis is 2024. This year, Poland's power generation capacity in the electricity system primarily consists of thermal power plants fuelled by fossil fuels (hard coal, lignite, and natural gas) [21], onshore wind turbines, and photovoltaic panels. Transportation and heating are predominantly based on fossil fuels.

The new energy system model presented in this publication, designed up to 2060, is based on a net zero policy. It anticipates an increase in electricity demand resulting from GDP growth, the complete electrification of transportation, and the electrification of industrial and residential heating. Moreover, the model assumes a transformation of the industry toward reliance on green hydrogen. No fossil fuel-powered units will remain in the energy system by 2050 (or 2060, depending on the simulation variant). Instead, onshore and offshore wind farms will be expanded, and an adequate number of nuclear power plant units will be built. Power plants or generators with gas turbines running on biomethane will complement these generation facilities. Using biomethane as a fuel ensures zero emissions from these energy sources. This study assumes that electricity, biomethane, and hydrogen will be exclusively sourced from domestic production and will not involve cross-border exchanges.

In the presented model, the role of stabilising generation units in the electricity system, currently fulfilled by coal-fired power plants (referred to as centrally dispatched generation units or CDGUs), will be taken over by the following:

(a) On the generation side, generators powered by biomethane (CDGUs);
(b) On the consumption side, hydrogen-producing units, which become centrally dispatched consumption units (CDCUs).

This means that, in the new model, shortfalls in energy production are balanced in the same way as today, with the difference that biomethane-fuelled generators will be used instead of coal-fired power plants. In turn, the management of excess production using a system of centralised electrolysers would be a new element in the Polish electricity system. Papers related to the issue of management and optimisation of hydrogen production based on an integrated RES system are known [44–49]. The adopted model does not take into account the specific management of the system, but only indicates the effect of its operation, considering the specific value of the installed electrolyser's capacity. The integration of RES systems is necessary to achieve net zero policy targets, as it ensures the stability of electricity transmission and distribution across the country. The interconnection of systems also ensures energy security, as it reduces the need to import energy in case of local system balancing issues.

According to the authors, hydrogen must be produced exclusively from surpluses of renewable energy and stored or further processed.

The first step of the analysis involved determining the values of input parameter required for the model to achieve a net zero emission economy. The key input parameters included installed capacity in PV systems, onshore wind farms, offshore wind farms, nuclear power plants, energy storage capacity, and electrolyser capacity. Depending on the selected simulation variant, these parameters describe the electricity system in 2060 or 2050. Using these parameter values, simulations were conducted for all weather data available in the simulator. This approach provides insights into the variability of the energy balance depending on the weather conditions in a given year and allows for selecting input parameters to ensure adequate electricity production even under unfavourable weather conditions over the year. The model parameter selection ensures a balance between supply and demand in the system not only annually, but also momentarily, thanks to the anticipated storage capabilities of battery systems, hydrogen storage, and biomethane storage.

Subsequently, the pace of achieving this state and the intensity of the necessary investments were determined. The time horizons analysed were 2030, 2040, 2050, and 2060. The objective functions were calculated as the levels of CO_2 emission reductions in the four main areas of the economy included in the simulator: electricity production, transportation, heating, and industry. Zero-emission criteria were met when all of these areas collectively achieved net zero emissions. The pace of reaching net zero was analysed for three alternative scenarios:

Scenario I assumes that net zero will be achieved in 2060. This is the most gradual strategy, requiring the least investment intensity. The year 2060 is the net zero target set by the People's Republic of China [50].

Scenario II aligns with the European Commission's proposed European Green Deal [1], which aims to maintain net zero carbon emissions by 2050 and maintain this state through 2060.

Scenario III envisions a 90% reduction in CO_2 emissions by 2040, achieving net zero emissions by 2050 and maintaining this state through 2060 [51].

To streamline the presentation of results, the final simulation outcomes were reported based on 2023 weather conditions, which were determined during the simulation process to represent average conditions for the operation of the entire electricity system.

Additionally, a long-term simulation was conducted to analyse the storage states of green hydrogen ($Accum_{H_2}(\tau)$) and biomethane ($Accum_{CH_4}(\tau)$), the demand for power from professional gas power plants, and the effective capacity of electrolysers. The long-term simulation, which lasted nine consecutive years, was carried out using weather conditions from 2015 to 2023, while energy supply and demand values were assumed to reflect 2060. The following relationships were used to calculate hydrogen and biomethane accumulation in storage:

$$Accum_{H_2}(\tau) = \int_{\tau_0}^{\tau} Supply_{H_2}(\tau) \cdot d\tau - \int_{\tau_0}^{\tau} Demand_{H_2}(\tau) \cdot d\tau \quad (1)$$

$$Supply_{H_2}(\tau) = \frac{P_{el}(\tau)[\mathrm{GW}]}{1000} \cdot \alpha_{H_2} \left[\frac{\mathrm{mln\ m^3H_2}}{\mathrm{TWh}}\right] \quad (2)$$

$$Demand_{H_2}(\tau) = \frac{D_{H_2}[\mathrm{TWh}]}{a} \cdot \alpha_{H_2} \left[\frac{\mathrm{mln\ m^3H_2}}{\mathrm{TWh}}\right] \quad (3)$$

$$Accum_{CH_4}(\tau) = \int_{\tau_0}^{\tau} Supply_{CH_4}(\tau) \cdot d\tau - \int_{\tau_0}^{\tau} Demand_{CH_4}(\tau) \cdot d\tau \quad (4)$$

$$Supply\ CH_4 = \frac{S_{CH_4}\left[\mathrm{mln\ m^3CH_4}\right]}{a} \quad (5)$$

$$Demand_n\ CH_4 = \frac{P_g(\tau)[\mathrm{GW}]}{1000} \cdot \alpha_{CH_4} \left[\frac{\mathrm{mln\ m^3CH_4}}{\mathrm{TWh}}\right] \quad (6)$$

where:

τ_0 = 1 January 2015;
$P_{el}(\tau)$ is the power of electrolysers in time τ;
$\alpha_{H_2} = 254.5\left[\frac{\mathrm{mln\ m^3H_2}}{\mathrm{TWh}}\right]$ is the volume of hydrogen produced from 1 TWh of energy
D_{H_2} is the annual hydrogen demand;
a is the number of hours in the year;
$a = \begin{cases} 365 \cdot 24 \\ 366 \cdot 24 \end{cases} \left(\begin{matrix} for\ 2015,\ 2017,\ 2018,\ 2019,\ 2021,\ 2022\ and\ 2023 \\ for\ 2016\ and\ 2023 \end{matrix} \right)$
S_{CH_4} is the annual biomethane supply;
$P_g(\tau)$ is the power of gas generators in time τ;
$\alpha_{CH_4} = 201.6\left[\frac{\mathrm{mln\ m^3CH_4}}{\mathrm{TWh}}\right]$ is the volume of biomethene required to produce 1 TWh of energy.

In line with the assumptions described further, it was assumed that $D_{H_2} = 66.7$ [TWh] and $S_{CH_4} = 7800$ $\left[\mathrm{mln\ m^3CH_4}\right]$. For both hydrogen and biomethane, it was assumed that the storage facilities were empty at the beginning of the year.

4. Results and Discussion

The generation capacities and the amounts of electricity produced and consumed for the baseline year and the end of the simulation are shown in Table 2. According to the assumptions and calculations, the minimum renewable energy capacities required to ensure a positive overall energy balance under the weather conditions observed between 2015 and 2023 are 64 GW for onshore wind, 33 GW for offshore wind, and 136 GW for PV systems. While these installed capacities appear substantial, considering the long-time horizon for achieving them, these values are not excessive. Regarding wind energy, the potential onshore installed capacity in Poland ranges from 63.4 GW to 118.3 GW, depending on the minimum legally required distance of wind farms from residential areas (700 m or 500 m, respectively) [25]. For the simulation, the 700 m distance was assumed, corresponding to approximately 64 GW. The potential capacity for offshore wind farms in Poland's Baltic Sea region is 33 GW [25], which was adopted in the simulation. The potential for PV capacity in Poland is less clearly defined, but estimates for agri-PV alone indicate a capacity of 118.8 GW [52]. In this context, the calculated target capacity of 136 GW does not appear to be overstated. The resulting PV power is the lowest value that provides a positive balance to the electricity system under weather conditions corresponding to 2015–2023. However, challenges with connecting new PV and wind installations to the grid remain significant. In 2023 alone, connection conditions were denied for PV installations totalling 83.6 GW due to an unprepared electrical grid [27]. Similarly, in 2022, connection conditions were denied for wind installations totalling 51.1 GW [28]. These figures highlight Poland's substantial renewable energy potential, contingent upon appropriate grid expansion and modernisation investments.

Another key element in balancing the electricity system is nuclear power plants. The simulation showed that the installed nuclear capacity in 2060 (10 GW) could not be reduced while maintaining a positive energy balance under the adopted assumptions. According to estimates by the Ministry of Climate, the economic feasibility and rationale for developing nuclear power with a total capacity of 9.9 GW by 2050 have been confirmed [53]. Thus, the simulation's target capacity of 10 GW is ambitious, but realistic.

In the proposed model, professional power plants fuelled by biomethane are the last resort for supplying energy during production shortages from renewable energy sources and nuclear power. It was assumed that biomethane would be entirely produced and consumed domestically. The projected biomethane production of 7.8 billion m^3/year in 2060 aligns with national reports [37,54], based on the current total technical potential. This production level balances supply and demand during temporary shortages when nuclear power plants and energy storage facilities are insufficient.

In alignment with net zero policies, zero emissions also apply to the industrial sector, which currently uses hydrogen primarily as a feedstock in the fertiliser and petrochemical industries. The annual demand for green hydrogen (66.7 TWh, Table 2) was calculated based on recent studies [20] and is summarised in Table 3. The study assumes the complete electrification of transportation, excluding hydrogen demand for refining processes related to conventional vehicle fuels. The primary hydrogen consumers are the fertiliser and non-fuel refinery sectors, with a combined annual demand of approximately 451 ktH_2/year. Additionally, the study assumes the complete transformation of the steel industry (297 ktH_2/year) and high-temperature heat sectors, including non-ferrous metal production, glassworks, ceramics, and cement plants (10 ktH_2/year). Furthermore, aviation and maritime transport are assumed to rely entirely on synthetic fuels derived from hydrogen and carbon dioxide, with the annual hydrogen demand estimated accordingly. A surplus of 5 TWh of green hydrogen was also included for unaccounted sectors to meet EU directives, such as AFiR (1% of road transport using hydrogen) and RED III (1% of conventional fuels from hydrogen). Based on the calculated hydrogen demand, the minimum capacity of electrolysers (22 GW, Table 2) was determined to ensure a positive hydrogen balance under the weather conditions observed between 2015 and 2023.

The current electrical capacity of heat pumps (0.5 GW, Table 2) was estimated based on the following assumptions: 220,000 units, an average thermal capacity of 8 kW, and an average SCOP (Seasonal Coefficient Of Performance) of 3.5. The calculated capacity for heat pumps in 2060 corresponds to their projected 90% share in heating (with the remaining 10% provided by electric heating and cogeneration).

The target capacity of energy storage in electric vehicles (230 GWh, Table 2) assumes 23 million electric vehicles by 2060 with average battery capacities of 50 kWh, and 20% of these batteries are used as energy storage systems. Simulations revealed that large-scale electricity storage capacities, comparable to those in electric vehicles, are essential for balancing the electricity system. The target capacity for pumped storage hydropower (PSH) was stablished at 30 GWh, with the total electricity storage capacity reaching 400 GWh. The capacities of residential and industrial heat storage were set at 35 GWh and 70 GWh, respectively, due to their limited suitability as centrally dispatchable balancing components.

Table 2. Comparison of the electricity system parameter values used to carry out simulations for 2024 and 2060 (energy values are the result of simulations).

	Production					Heat Pumps	Storage					Demand				
	Wind Onshore	Wind Offshore	PV	Nuclear	Bio-CH_4		Batteries	EV Batteries	PSH	Heat Storage (Industry)	Heat Storage (Household)	Transportation	Industry Heat	Household Heat	Electricity	Electrolysers/H_2
	GW				mld m^3	GW_{el}	GWh					Electrification—Share of Current Demand				GW
2024	9	0	17	0	0	0.5	0	0	0	0	0	0%	0%	0%	100%	0
2060	64	33	136	10	7.8	9.8	140	230	30	35	70	100%	100%	100%	120%	22
	Energy, TWh															
2024	26.7	0.0	17.3	0.0	0.0							0	0	0	167.5	0
2060	189.7	139.1	138.4	70.7	38.7							102.8	70.8	66.8	174.5	66.7

Table 3. Annual industrial green hydrogen demand to achieve total decarbonisation based on [20].

Industry Sector	Hydrogen Demand, t	Energy, TWh
Nitrogen plants, refineries	451,318	19.8
Metallurgy, steel production	307,000	13.4
Maritime transport	354,667	15.5
Aviation	295,571	12.9
Other	-	5.0
SUM		66.7

All simulations conducted indicate that the balance of biomethane and hydrogen is strongly dependent on weather conditions (Figure 3).

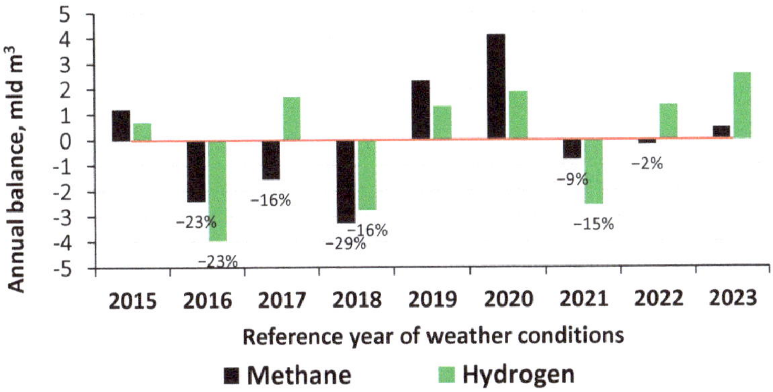

Figure 3. Balances of biomethane and green hydrogen as functions of weather conditions (in parentheses, the missing quantities are given as a percentage of the demand).

For instance, using weather data from 2018, there was a shortfall of approximately 3.3 billion m^3 of biomethane in the balance, representing 29% of the annual demand. In contrast, using weather data from 2019, a positive balance of 2.3 billion m^3 of biomethane was achieved. Weather data from 2023 were selected for further simulations because this year showed a positive biomethane balance (0.47 billion m^3) close to the average value for 2015–2023. The weather conditions in 2023 allowed for the production of a significant surplus of hydrogen (2.6 billion m^3). The capacity of installed electrolysers was calculated, and it was set to the minimum value to ensure a positive cumulative H_2 balance across all analysed years.

The energy system simulator allows for obtaining a range of valuable partial data. For example, it provides detailed energy balance components on both the demand and supply sides. Additionally, detailed data on the amounts of energy supplied to and drawn from heat and electricity storage facilities can be accessed. This enabled the creation of a comprehensive energy system balance for Poland in 2060 (Table 4). The primary annual energy sources in this system are wind farms and photovoltaic (PV) installations. The demand for biomethane to power conventional power plants during renewable energy shortages constitutes 5.9% of the total supply. On the demand side, the largest consumer in the system is electricity, with a 28.5% share. It is followed by transport with a 16.8% share and unutilised energy with a 12.8% share. This unutilised energy represents the surplus that cannot be stored or directed to electrolysers due to their limited capacities and power. The primary consumer of surplus energy is electrolysers, which produce green hydrogen for industrial applications, as detailed in Table 3. Green hydrogen accounts for 12.5% of the total demand. In contrast, the combined share of all electricity and heat storage facilities is almost twice as small at approximately 6.7%. This indicates that, despite the substantial total storage capacities (400 GWh), they are not the primary consumers of surplus energy. Instead, electrolysers take on this role, even with a moderate power level (22 GW). According to the assumptions, electrolysers operate with power adjusted to the real-time difference between available power and demand. If the hydrogen produced is not immediately used, it is stored. The simulation shows that the energy storage systems remain fully charged for most of the year, which prevents them from absorbing the surplus energy generated by renewable sources.

Table 4. Energy system balance in Poland in 2060 (weather conditions as of 2023).

Supply	TWh	Demand	TWh
Wind onshore	189.7	Electricity	174.5
Wind offshore	139.1	Industry heat	70.83
PV	138.4	Household heat	66.84
Nuclear	70.74	Transportation	102.8
Methane	36.35	Hydrogen	76.80
Cogeneration	6.120	Unused	78.24
Batteries	30.99	Batteries–charging	36.93
Heat storage (industry)	1.410	Heat storage (industry)—charging	1.444
Heat storage (household)	0.6166	Heat storage (household)—charging	2.559
		Transmission losses of methane	2.413
SUM	613.5		613.5

It is worth noting that, in 2024, the total electricity demand was 167.5 TWh, while in the target year 2060, it will rise to 481.8 TWh (Table 2), representing a 2.9-fold increase. This implies that the current electricity grid must be upgraded to ensure the stable transmission of significantly increased electricity volumes.

The geographical location of Poland in Central Europe results in distinct seasonal variability, which significantly impacts energy demand and the performance of various components in the energy mix. To illustrate these seasonal differences, Figure 4A presents the instantaneous power outputs of different energy sources during the first months of each quarter. The height of the areas under the demand curve indicates which energy sources are used to meet the demand. Additionally, a line representing the total instantaneous demand has been added, allowing for the easy identification of periods with surplus production relative to demand and times when gas turbines are required to maintain the stability of the energy system. Figure 4B, on the other hand, shows the instantaneous power demand across various sectors of the economy, the operation of electrolysers, and the real-time values of unused energy.

In January, representative of the winter season, wind energy, mainly offshore, plays a significant role. Due to short days and limited sunlight, the contribution of photovoltaics is negligible. During windless days in January, biomethane production, cogeneration, and energy storage facilities for electricity and heat stabilise the energy system. Nuclear energy provides a constant and stable production level, independent of weather conditions. In January 2060, the maximum power demand reaches 73.1 GW. During surplus power, storage facilities and electrolysers are activated; however, part of the excess power remains unutilised. A characteristic feature of this month is that periods of gas turbine operation and periods of surplus energy last for several days.

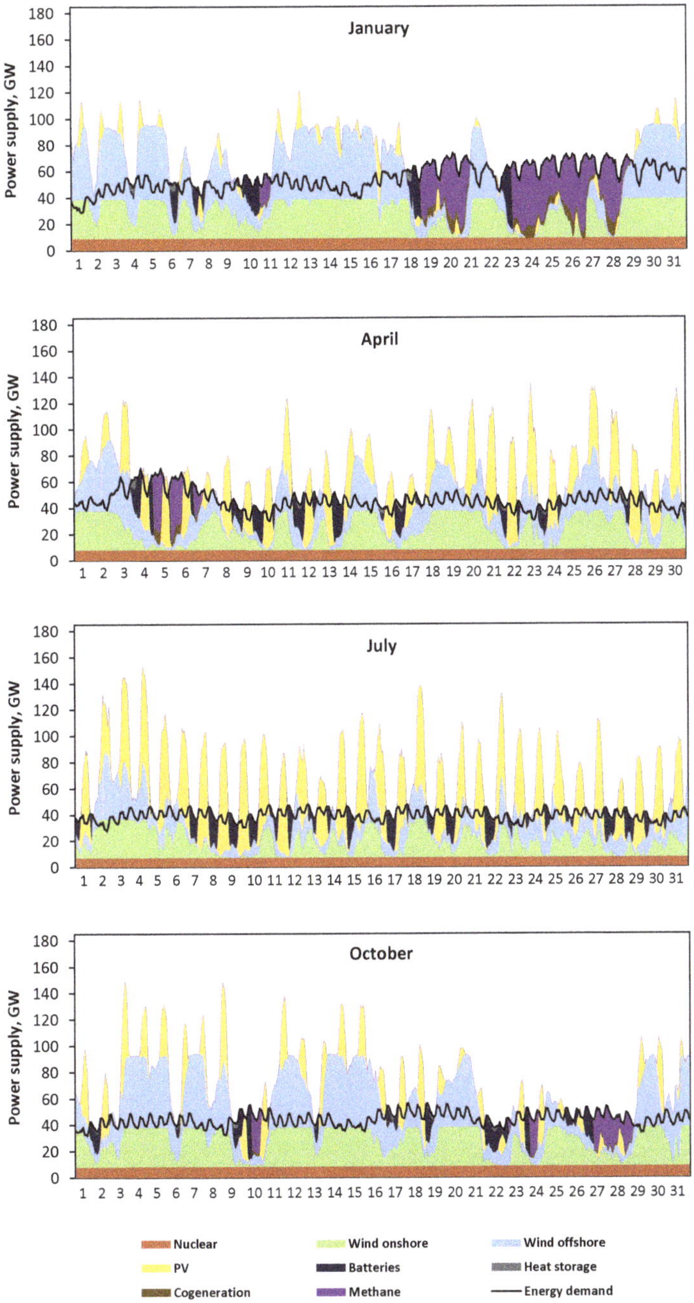

(A)

Figure 4. *Cont.*

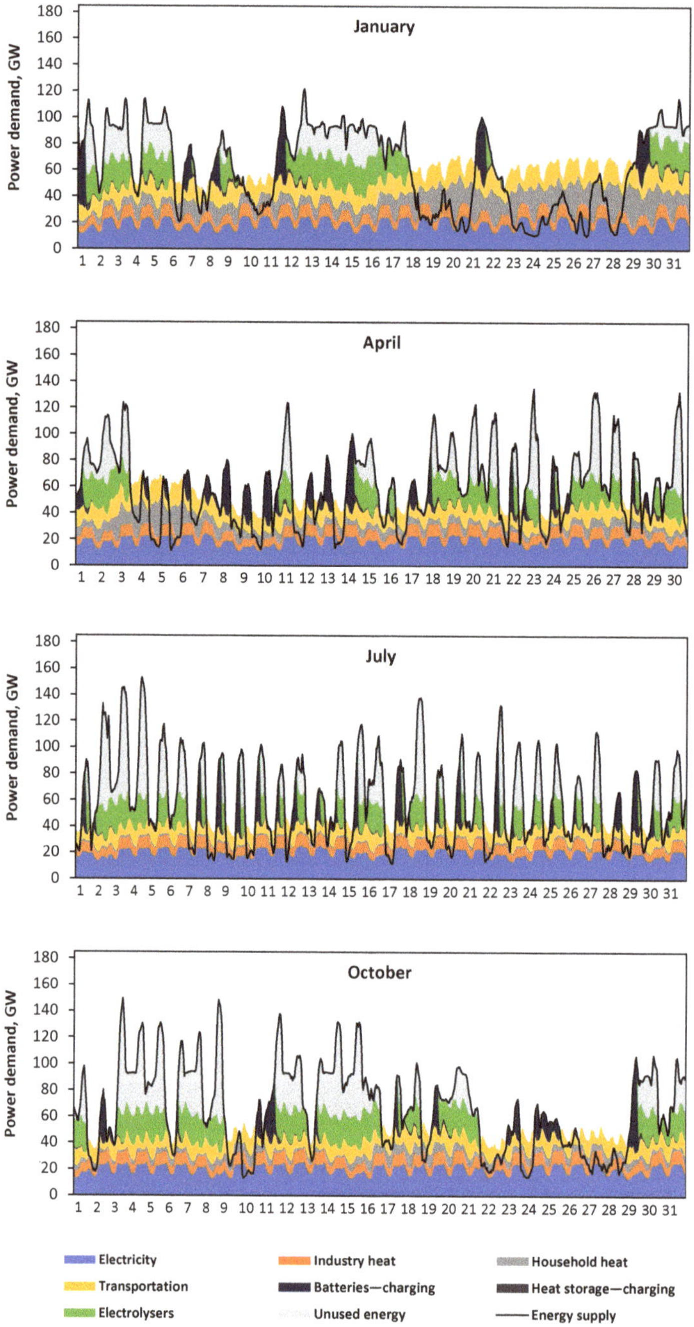

(B)

Figure 4. (**A**) The balance of the electricity system in Poland on the following days in January, April, July, and October 2060 (weather conditions as of 2023)—power supply. (**B**) The balance of the electricity system in Poland on the following days in January, April, July, and October 2060 (weather conditions as of 2023)—power demand.

As days grow longer in April, the share of photovoltaic (PV) energy increases, while wind power generation remains high. PV generation sometimes helps to meet demand gaps, but occasionally results in unused energy. During this period, biomethane consumption decreases compared to winter months, and there are periodic surpluses and deficits in renewable energy production. In the simulated scenario, early April experiences surplus renewable energy production (mainly from wind), allowing for storage facilities to be filled, hydrogen to be produced, and significant amounts of unutilised energy to be observed. On days with unfavourable weather conditions for wind energy generation, demand is supplemented by energy stored in batteries and heat storage or by activating conventional power plants fuelled by biomethane. Due to higher temperatures, the maximum power demand in April is lower than in January, reaching 70.5 GW. Electrolysers are utilised more extensively because surplus power arises intermittently from both PV and wind power.

In July, the variability in energy production is primarily driven by PV output. The peak energy generation from PV typically exceeds energy demand. The instantaneous maximum power demand reaches 47.6 GW, while PV alone can generate up to 84.7 GW at its peak. In October, Poland experiences various weather conditions, resulting in varying contributions from PV and wind farms. However, the shorter daylight hours and lower sun angles reduce PV's overall potential. Although wind energy production increases, windless days necessitate system balancing through battery storage and biomethane production. October exhibits energy production variability similar to that of January. However, the maximum power demand in October is lower, reaching 56.1 GW.

The following research stage presents three approaches to achieving net zero emissions for Poland's entire economy by 2060 (Scenario I) or by 2050 while maintaining this state through 2060 (Scenarios II and III). Table 5 outlines the parameters describing the power system and energy demand in sectors such as electricity, heating, transport, and hydrogen production, not only for the final simulation year, but also for the transition years (2030, 2040, and 2050). Depending on the adopted approach, these scenarios consider additional conditions for decarbonisation levels. For transitional periods, it was assumed that changes in the numerical parameters of the generation side would follow a steady rate of change wherever possible, considering additional constraints. Only for the initial period (2024–2030) was a slower rate of change adopted compared to subsequent periods (2030+) due to the gradual intensification of investment processes. This slower start reflects delays in regulatory changes and the necessary shifts in awareness among decision-makers and investors. Solving an optimisation problem requires selecting one of the many known methods. A heuristic approach was used instead of a strict mathematical optimisation algorithm to determine the system parameters for each scenario. As is known, heuristic methods do not guarantee finding an optimal solution. However, compared to methods such as linear programming, they offer several advantages, including resilience to potential nonlinearities, robustness to imperfect and uncertain data, and, in many cases, fast convergence to a suboptimal solution. Additionally, heuristic methods can be supported by expert knowledge (expert-in-the-loop approach) or data from long-term energy development programmes (e.g., nuclear energy planning). This method was supported by analysing various sources that describe the economic conditions for developing a zero-emission energy sector. On the energy consumption side, a gradual and consistent increase in parameters was assumed, including energy and heat storage capacities, the development of electromobility, the transition to heat pumps in heating, and the production capacities for hydrogen. Even the relatively moderate Scenario I requires simultaneous, intensive growth in renewable energy capacity, nuclear power plants, electrolysers, heat pumps, and energy storage facilities, as well as biomethane production from biomass. The most critical aspect is to compare the intensity of development of specific energy sources in the power system, depending on the scenario, and the necessary pace of changes in energy demand, as outlined in Table 6. The calculated rates of change in the system are the most achievable (and thus the most realistic) in Scenario I. They demonstrate the minimum investments needed across all areas of electricity production and other sectors responsible for greenhouse gas emissions to achieve a climate-neutral economy by 2060 at the latest. As seen in the data for Scenarios II and III, achieving climate neutrality earlier requires a significant increase in these investments. The results in Table 6 can serve as guidelines for a detailed action plan throughout the transition period from 2024 to 2060.

Applying the energy system simulator also enables the estimation of transformation costs. Given that energy transition scenarios, analysed in this paper, are long-term processes, economic analysis is inherently subject to high prediction uncertainty. Our primary objective was to conduct a simplified cost assessment solely to compare the different scenarios. The cost balance was calculated using the energy simulator [38] and includes the construction, maintenance, and operation costs of individual system components, excluding biomethane production. The simulator also does not account for other transformation costs, such as the expansion and modernisation of the power grid or the costs of transporting and storing biomethane and green hydrogen. Nevertheless, the obtained data allow for a direct comparison of the selected scenarios. Regardless of the chosen scenario, the total costs amount to approximately EUR 496.8 billion by 2060 (Figure 5). While this figure may seem substantial, it averages around EUR 13.8 billion annually. The annual costs peak at EUR 33.4 billion during 2030–2040 in Scenario III, whereas in the most moderate Scenario I, the maximum annual costs reach EUR 18.4 billion during 2050–2060. When these figures are compared to Poland's net energy and fuel import costs (exceeding EUR 25 billion annually in 2021–2023 [55]), it becomes apparent that the costs of building the new energy system components do not require additional financial resources. The analysis revealed the challenges posed by Scenario III, which aims for a 90% reduction in CO_2 emissions by 2040 and net zero emissions by 2050. In this scenario, the investments during 2030–2040 would need to be 165% higher compared to Scenario I. Additionally, the assumed growth rates for installed capacities—3 GW/year for onshore wind, 7 GW/year for PV, and 7–16 GWh/year for battery storage—would present significant challenges and appear unlikely to be achievable.

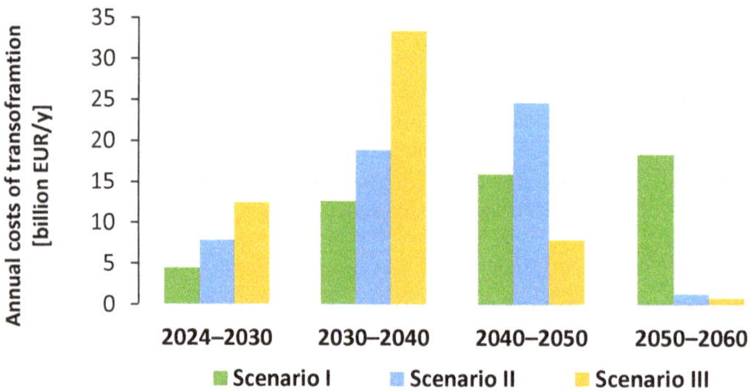

Figure 5. The annual cost of energy system transformation under different scenarios.

Table 5. Results of power system parameter calculation for three scenarios at various intermediate stages in the years 2030, 2040, 2050, and 2060.

Scenario	Year	Wind On-shore	Wind Offshore	PV	Nuclear	Electrolysers	Heat Pumps	Biomethane	Batteries	EV	PSH	Heat Storage (Industry)	Heat Storage (Household)	Transportation	Industry Heat	Household Heat	Electricity
		GW					GW_{el}	mld m^3	GWh					Electrification—Share of Current Demand			
	2024	9	0	17	0	0	0.5	0.0	0	0	0	0	0	0%	0%	0%	100%
I	2030	15	3	29	0	0	1.1	0.6	12	0	0	0	0	0%	0%	0%	103%
	2040	25	13	59	3	5	3.1	2.6	52	46	10	5	10	20%	20%	20%	108%
	2050	44	23	96	6	12	6.1	4.6	92	138	20	15	30	60%	60%	60%	114%
	2060	64	33	136	10	22	9.8	7.8	140	230	30	35	70	100%	100%	100%	120%
II	2030	15	6	41	0	3	1.1	0.6	24	28	0	0	0	12%	0%	0%	103%
	2040	35	19	81	4	8	4.6	2.6	74	120	15	5	10	52%	50%	50%	108%
	2050	60	32	131	10	20	9.1	7.8	134	230	30	25	50	100%	100%	100%	114%
	2060	64	33	136	10	22	9.8	7.8	140	230	30	35	70	100%	100%	100%	120%
III	2030	21	9	44	0	3	1.7	0.6	30	55	6	6	6	24%	12%	12%	103%
	2040	51	29	114	8	21	7.7	6.4	100	207	26	21	41	90%	90%	90%	108%
	2050	61	32	134	10	22	8.8	7.8	135	230	30	31	61	100%	100%	100%	114%
	2060	64	33	136	10	22	9.8	7.8	140	230	30	35	70	100%	100%	100%	120%

Table 6. Changes per year for each power system parameter for three scenarios up to 2060.

Scenario	Years	Wind On-shore	Wind Off-shore	PV	Nuclear	Electrolysers	Heat Pumps	Biomethane	Batteries	EV	PSH	Heat Storage (Indus-try)	Heat Storage (House-hold)	Transportation	Industry Heat	Household Heat	Electricity
		GW/y					GW_{el}/y	mld m^3/y	GWh/y					Annual Growth in Electrification Degree			Relative Annual Growth
I	2024–2030	1.0	0.5	2.0	0.0	0.0	0.10	0.10	2.0	0.00	0.0	0.0	0.0	0.0%	0.0%	0.0%	0.5%
	2030–2040	1.0	1.0	3.0	0.3	0.5	0.20	0.20	4.0	4.60	1.0	0.5	1.0	2.0%	2.0%	2.0%	0.5%
	2040–2050	1.9	1.0	3.7	0.3	0.7	0.30	0.20	4.0	9.20	1.0	1.0	2.0	4.0%	4.0%	4.0%	0.5%
	2050–2060	2.0	1.0	4.0	0.4	1.0	0.37	0.32	4.8	9.20	1.0	2.0	4.0	4.0%	4.0%	4.0%	0.5%
II	2024–2030	1.0	1.0	4.0	0.0	0.5	0.10	0.10	4.0	4.60	0.0	0.0	0.0	2.0%	0.0%	0.0%	0.5%
	2030–2040	2.0	1.3	4.0	0.4	0.5	0.35	0.20	5.0	9.20	1.5	0.5	1.0	4.0%	5.0%	5.0%	0.5%
	2040–2050	2.5	1.3	5.0	0.6	1.2	0.45	0.52	6.0	11.04	1.5	2.0	4.0	4.8%	5.0%	5.0%	0.5%
	2050–2060	0.4	0.1	0.5	0.0	0.2	0.07	0.00	0.6	0.00	0.0	1.0	2.0	0.0%	0.0%	0.0%	0.5%
III	2024–2030	2.0	1.5	4.5	0.0	0.5	0.20	0.10	5.0	9.20	1.0	1.0	1.0	4.0%	2.0%	2.0%	0.5%
	2030–2040	3.0	2.0	7.0	0.8	1.8	0.60	0.58	7.0	15.18	2.0	1.5	3.5	6.6%	7.8%	7.8%	0.5%
	2040–2050	1.0	0.3	2.0	0.2	0.1	0.11	0.14	3.5	2.30	0.4	1.0	2.0	1.0%	1.0%	1.0%	0.5%
	2050–2060	0.3	0.1	0.2	0.0	0.0	0.10	0.00	0.5	0.00	0.0	0.4	0.9	0.0%	0.0%	0.0%	0.5%

On the other hand, considering the depleting carbon budget [56], Scenarios II and III are more desirable compared to Scenario I. In Scenario I, the transport and heating sectors contribute significantly to current GHG emissions, accounting for 80% in 2040 and 40% in 2050, respectively (Table 7). Meanwhile, the industrial sector in Scenarios I and II still emits over 70% of its current GHG emissions by 2040. The simulation results indicate that achieving net zero policy targets is feasible by 2050, and a 90% reduction in GHG emissions across all sectors can be reached as early as 2040. However, this would present a significant challenge for all branches of industry.

Table 7. Comparison of emission reductions under different scenarios and their share of the current value.

Year	Scenario	Electricity	Industry (Hydrogen)	Transportation	Heat
2024	I, II, III	100.0%	100.0%	100.0%	100.0%
2030	I	62.3%	100.0%	100.0%	100.0%
	II	52.1%	97.8%	88.0%	100.0%
	III	50.3%	96.1%	76.0%	88.0%
2040	I	17.5%	81.5%	80.0%	80.0%
	II	26.1%	71.4%	48.0%	50.0%
	III	9.9%	9.5%	10.0%	10.0%
2050	I	9.3%	44.6%	40.0%	40.0%
	II, III	0.0%	0.0%	0.0%	0.0%
2060	I, II, III	0.0%	0.0%	0.0%	0.0%

In the subsequent part of the study, the focus is limited to a detailed presentation of the transformation process for Scenario I, as it is considered to be the most feasible for implementation (Figure 6). For Scenarios II and III, the transformations will follow a similar pattern; however, they will occur faster (Table 5). The simulation results for various time horizons in Scenario I indicate that, currently, utility power plants fuelled by fossil fuels (here referred to as methane) are responsible for 71.1% of the energy supply. According to the adopted assumptions, no significant changes in the electrification of transport and heating will occur by 2030, and hydrogen for industrial use will still be almost entirely produced from non-renewable sources. From 2030, the growth rate of renewable energy capacity will double compared to 2024–2030, and the first units from nuclear power plants will come online. At the same time, the electrification of transport and heating will begin. With the increasing capacity of renewable energy sources (RESs), surplus energy will emerge, a portion of which will be absorbed by electrolysers to produce green hydrogen. By 2040, hydrogen is expected to account for approximately 5% of energy demand, but it will not be until 2060 that its production will be sufficient to meet the demand fully.

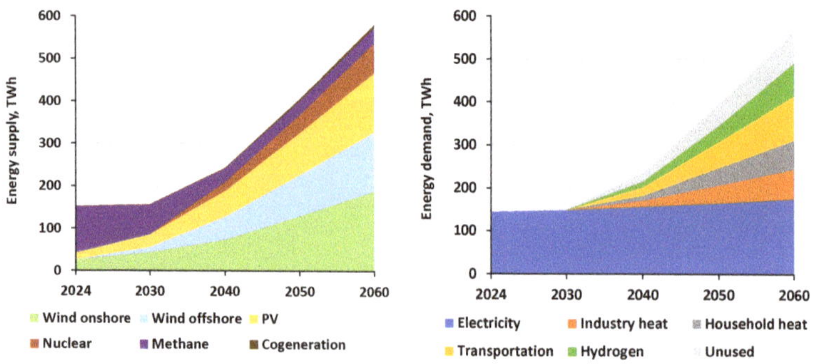

Figure 6. Net zero energy balance in Poland up to 2060—scenario I (weather conditions as of 2023).

To address the energy balance deficits, utility gas power plants fuelled by biomethane are proposed. As mentioned above, biomethane would be sourced exclusively from domestic production and generated continuously at a steady output in the proposed system. The produced biomethane could be used to periodically fuel gas power plants, while any surplus would be stored. Similarly, green hydrogen would be produced during periods of surplus energy availability and continuously consumed by industry, with any surplus also stored. In this case, it is also assumed that the demand for hydrogen remains constant throughout every hour of the year. Based on these assumptions, a long-term balance of biomethane and green hydrogen resources was prepared for weather conditions similar to those from the years 2015–2023, with energy demand and supply levels projected for 2060 (Figure 7). It was assumed that, at the beginning of the analysed period, both storage facilities would be empty. An additional assumption was the inclusion of an element of international trade (imports). The analysis indicates that, during the first three months of the simulation, deficits in both hydrogen and biomethane occur. Biomethane storage shortages continue in the subsequent four years at the beginning of the year (January–March). In contrast, hydrogen shortages are observed primarily during the first two years of the analysis. In later years, there is no longer a need to purchase biomethane or hydrogen, as storage levels remain high. At the end of the 9-year analysed period, the green hydrogen storage level reached 5.4 billion m^3 (21,365 TWh) and the biomethane storage level reached 7.7 billion m^3 (38,188 TWh). Currently, in Poland, the active capacity of underground natural gas storage is approximately 3.3 billion m^3 [57]. Therefore, the simulated volumes of biomethane and hydrogen exceed the capacity of national storage facilities. In such a scenario, several options arise: expanding the capacities of domestic gas storage facilities or exploring favourable conditions for new applications of these gases in the domestic market. Alternatively, the produced gases could be sold to other countries, or, if storage or utilisation is not possible, the production of these gases could be halted. Moreover, despite the adopted assumptions, the levels of biomethane and hydrogen imports were found to be insignificant (Figure 8). During the 9 years, it would be necessary to import a total of 38.2 TWh of biomethane and 20.1 TWh of hydrogen, while simultaneously ending the period with large surpluses of these gases and a significant amount of unutilised energy (581 TWh).

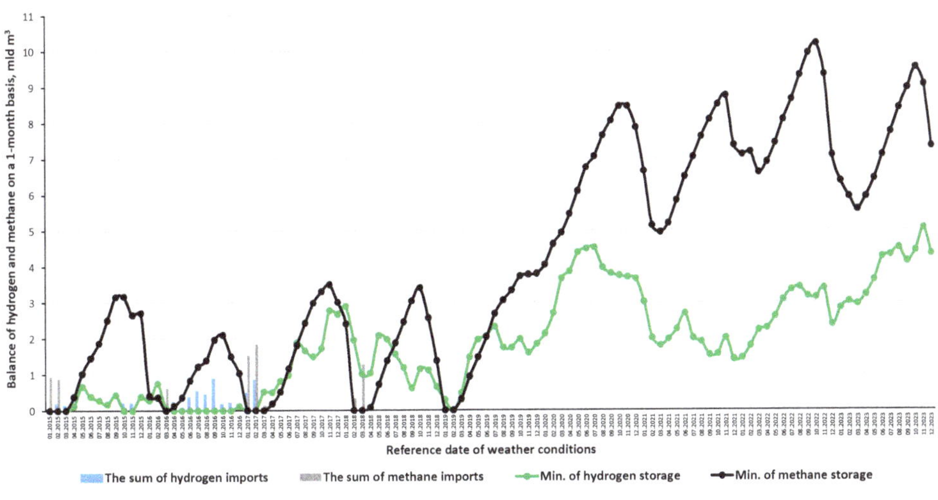

Figure 7. Simulation of the operation of methane and hydrogen storage systems in the years corresponding to weather conditions from 2015 to 2023 (constant energy supply and demand as of 2060).

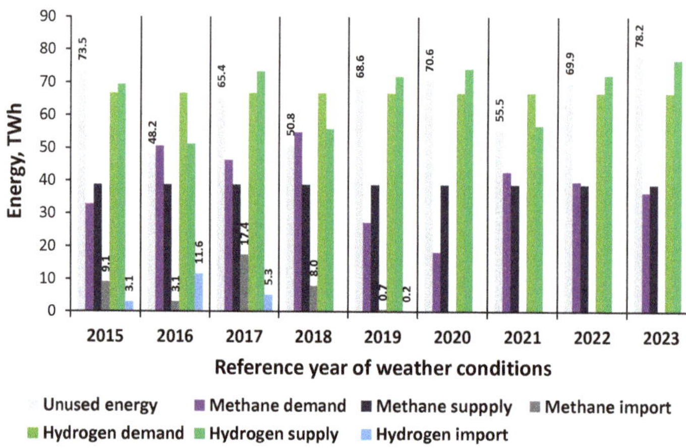

Figure 8. Balance of unused energy, methane, and hydrogen in the years corresponding to weather conditions as in 2015–2023 (import quantities from the simulation of storage system operation and energy supply and demand as of 2060).

A long-term analysis of gas power plants fuelled by biomethane during 2015–2023 revealed that their operation is required for approximately 15.5% of the time on an hourly basis. The peak power output of these plants is necessary to ensure that the balance of the energy system reaches 83.2 GW (Figure 9, 26 February 2018, 19:30). Assuming that this is the total installed capacity of gas power plants, they would operate, on average, at only 5.2% of their capacity, or approximately 4.3 GW on average. For comparison, data on electrolysers with an assumed capacity of 22 GW operating under similar weather conditions during 2015–2023 show that they are operational for 40.4% of the time on an hourly basis and that their capacity is 34.7%, with an average power output of 7.6 GW. This level of utilisation is more acceptable than that of utility gas power plants.

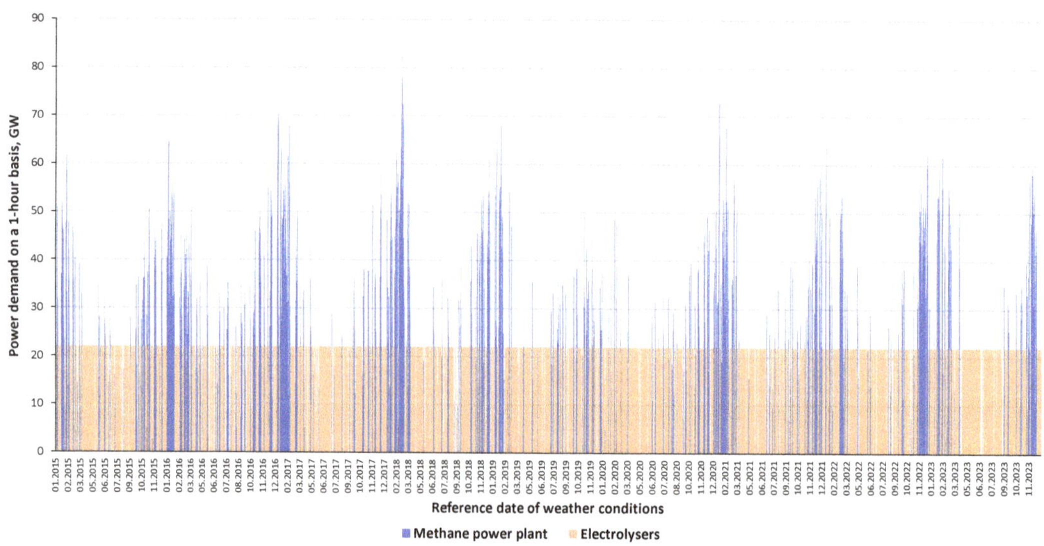

Figure 9. Simulation of the operation of methane power plants, electrolysers, and H_2 storage systems in the years corresponding to weather conditions as in 2015–2023 on a 1 h basis (energy supply and demand as of 2060).

Furthermore, when examining monthly energy demand (Figure 10, with a maximum of 16.8 TWh in January 2017) or annual energy demand (Figure 8, with a maximum of 54.9 TWh in 2018) for biomethane-fuelled energy generation to compensate for renewable energy shortages, and comparing these figures to annual surplus energy (Figure 8, 48–78 TWh), a positive energy balance emerges on an annual basis. This is despite occasional deficits at a variable but not excessively high level. It should be noted that the simulation does not include cross-border trade, as this aspect falls outside the capabilities of the simulator used. In practice, the construction of large capacities (around 80 GW) is not economically justified. The optimal capacity of utility gas power plants, and thus the desired biomethane demand and the required capacities of biomethane facilities, could be the subject of a separate simulation.

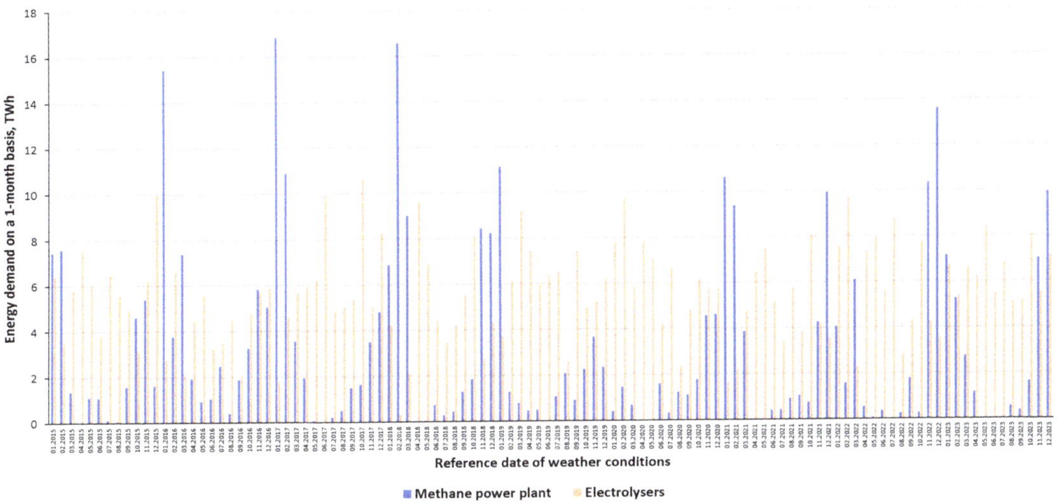

Figure 10. Simulation of the operation of methane power plants, electrolysers, and H_2 storage systems in the years corresponding to weather conditions as in 2015–2023 on a 1-month basis (energy supply and demand as of 2060).

Another way to manage surplus electricity production could be the production of green hydrogen in quantities exceeding the proposed domestic industrial demand (Table 3). The hydrogen obtained could be used to develop existing sectors (such as the fertiliser industry, the polymer industry, etc.), in other areas of the economy (such as energy storage, to power fuel cell vehicles, in the space industry, as a substrate for CO_2 capture processes from the atmosphere, etc.), or it could be sold on the international market. The simulation results (Figure 11) indicate that, in the analysed example, fully utilising the surplus energy production would require as much as 130 GW of electrolyser capacity, which would be utilised only 13.6% of the time. Installation of such a large electrolyser capacity does not appear to be technically or economically justified. Similarly to the case of installed capacity in gas power plants, determining the optimal levels of installed electrolyser capacity also requires multifactor simulations that consider many parameters not included in this study.

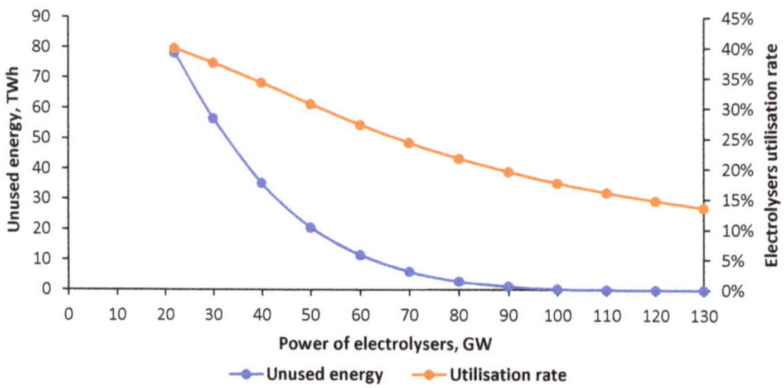

Figure 11. Simulation of the effects of the increase in electrolyser power and energy supply and demand as of 2060.

5. Conclusions

Research has shown that it is possible to achieve a self-sufficient, sustainable, zero-emission, and stable energy system in Poland by 2060, even though the energy sources it relies on are uncontrollable and primarily based on weather-dependent systems. The capacities assumed in the simulation, namely 64 GW of onshore wind, 33 GW of offshore wind, 136 GW of PV, 10 GW of nuclear power, and 22 GW of electrolysers, together with an appropriate system of electricity and heat storage, allow for a net positive energy balance in the zero-emission system under the weather conditions observed between 2015 and 2023. The stability of the system requires the electrolysers to operate in a centralised system, as dispatchable receiving units are controlled centrally.

It has been demonstrated that the energy transition toward a net zero policy requires intensive changes in almost all sectors of the economy, including a complete shift in transportation and heating systems to electricity-based solutions and industrial activities to those based on green hydrogen. The cost of building, maintaining, and operating individual system components has been estimated to be a maximum of EUR 18.4 billion per year for 2050–2060. A challenge appears to be preparing the power grid to handle an annual energy transmission 2.9 times greater than the current level.

The simulation results indicate that sectors of the economy related to green hydrogen and biomethane production will also need to increase their current production capacities significantly. It has been proven that the electrolyser system will play a key role in managing energy production surpluses. As a result of the long-term nine-year simulation, it has been shown that energy self-sufficiency is achievable. Moreover, significant amounts of surplus green hydrogen (5.4 billion m^3 or 21,365 TWh), biomethane (7.7 billion m^3 or 38,188 TWh), and 581 TWh of unused energy are generated. This provides opportunities for the development of all current and future sectors of the economy, and the green hydrogen produced can find numerous applications, meeting both current and new potential demand.

Thanks to the strategy presented here, which combines the management of weather-dependent energy resources with hydrogen production implemented using electrolysers operating as centrally dispatchable elements, the energy system gains the ability to increase the capacity of installations based on renewable energy significantly. This combination enables the efficient utilisation of energy surpluses produced during periods of high output from renewable sources such as solar and wind energy. Such an oversizing of weather-dependent sources relative to current power demand not only increases the share of renewable energy in the energy mix, but also contributes to system stability by storing surplus energy in the form of hydrogen. This is a step towards a more sustainable and flexible energy system that can better address the challenges of ensuring energy security and reducing CO_2 emissions.

Author Contributions: Conceptualization, W.Ż., G.B.-P. and J.P.; methodology, software, and validation, D.B., M.O. and G.B.-P.; formal analysis, investigation, and data curation, W.Ż., D.B. and M.O.; writing—original draft preparation, D.B., J.P., M.O. and G.B.-P.; writing—review and editing, D.B., W.Ż. and G.B.-P. All authors have read and agreed to the published version of the manuscript.

Funding: This research received no external funding.

Data Availability Statement: The raw data supporting the conclusions of this article will be made available by the authors on request.

Conflicts of Interest: The authors declare no conflicts of interest.

References

1. Fetting, C. *The European Green Deal. ESDN Report*; ESDN: Vienna, Austria, 2020.
2. European Environment Agency. EEA Greenhouse Gases—Data Viewer. Available online: https://www.eea.europa.eu/en/analysis/maps-and-charts/greenhouse-gases-viewer-data-viewers (accessed on 5 December 2024).
3. Ministry for Climate and Environment of the Republic of Poland. Poland's National Inventory Report 2024. Available online: https://www.kobize.pl/uploads/materialy/materialy_do_pobrania/krajowa_inwentaryzacja_emisji/NIR_2024_raport_syntetyczny_PL.pdf (accessed on 5 December 2024). (In Polish)
4. Friedlingstein, P.; O'Sullivan, M.; Jones, M.W.; Andrew, R.M.; Bakker, D.C.E.; Hauck, J.; Landschützer, P.; Le Quéré, C.; Luijkx, I.T.; Peters, G.P.; et al. Global Carbon Budget 2023. *Earth Syst. Sci. Data* **2023**, *15*, 5301–5369. [CrossRef]
5. Chen, Z.Y.; Zhao, L.T.; Cheng, L.; Qiu, R.X. How Does China Respond to the Carbon Border Adjustment Mechanism? An Approach of Global Trade Analysis. *Energy Policy* **2025**, *198*, 114486. [CrossRef]
6. Rogelj, J.; Geden, O.; Cowie, A.; Reisinger, A. Net-Zero Emissions Targets Are Vague: Three Ways to Fix. *Nature* **2021**, *591*, 365–368. [CrossRef] [PubMed]
7. Allen, M.R.; Friedlingstein, P.; Girardin, C.A.J.; Jenkins, S.; Malhi, Y.; Mitchell-Larson, E.; Peters, G.P.; Rajamani, L. Net Zero: Science, Origins, and Implications. *Annu. Rev. Environ. Resour.* **2022**, *47*, 849–887. [CrossRef]
8. DeAngelo, J.; Azevedo, I.; Bistline, J.; Clarke, L.; Luderer, G.; Byers, E.; Davis, S.J. Energy Systems in Scenarios at Net-Zero CO_2 Emissions. *Nat. Commun.* **2021**, *12*, 6096. [CrossRef]
9. Osman, A.I.; Hefny, M.; Abdel Maksoud, M.I.A.; Elgarahy, A.M.; Rooney, D.W. *Recent Advances in Carbon Capture Storage and Utilisation Technologies: A Review*; Springer International Publishing: New York, NY, USA, 2021; Volume 19, ISBN 0123456789.
10. Ang, L.; Yongming, L.; Xi, C.; Zhongyi, Z.; Yu, P. Review of CO_2 Sequestration Mechanism in Saline Aquifers. *Nat. Gas Ind. B* **2022**, *9*, 383–393. [CrossRef]
11. Ge, L.; Zhang, B.; Huang, W.; Li, Y.; Hou, L.; Xiao, J.; Mao, Z.; Li, X. A Review of Hydrogen Generation, Storage, and Applications in Power System. *J. Energy Storage* **2024**, *75*, 109307. [CrossRef]
12. Dehghani-Sanij, A.R.; Tharumalingam, E.; Dusseault, M.B.; Fraser, R. Study of Energy Storage Systems and Environmental Challenges of Batteries. *Renew. Sustain. Energy Rev.* **2019**, *104*, 192–208. [CrossRef]
13. Blakers, A.; Stocks, M.; Lu, B.; Cheng, C. A Review of Pumped Hydro Energy Storage. *Prog. Energy* **2021**, *3*, 022003. [CrossRef]
14. Rabi, A.M.; Radulovic, J.; Buick, J.M. Comprehensive Review of Compressed Air Energy Storage (CAES) Technologies. *Thermo* **2023**, *3*, 104–126. [CrossRef]
15. Maka, A.O.M.; Alabid, J.M. Solar Energy Technology and Its Roles in Sustainable Development. *Clean Energy* **2022**, *6*, 476–483. [CrossRef]
16. García Márquez, F.P.; Peinado Gonzalo, A. A Comprehensive Review of Artificial Intelligence and Wind Energy. *Arch. Comput. Methods Eng.* **2022**, *29*, 2935–2958. [CrossRef]
17. Abid, K.; Sharma, A.; Ahmed, S.; Srivastava, S.; Velazco, A.T.; Teodoriu, C. A Review on Geothermal Energy and HPHT Packers for Geothermal Applications. *Energies* **2022**, *15*, 7357. [CrossRef]
18. Polish Transmission System Operator (Polskie Sieci Elektroenergetyczne S.A.). Types of Generation Sources and Definitions, How the National Power System Operates. Available online: https://www.pse.pl/jak-funkcjonuje-krajowy-system-elektroenergetyczny/rodzaje-zrodel-wytworczych-i-definicje (accessed on 5 December 2024). (In Polish).
19. International Energy Agency Projected Costs of Generating Electricity. Available online: https://www.iea.org/reports/projected-costs-of-generating-electricity-2020 (accessed on 5 December 2024).

20. Tchorek, G.; Targowski, F.; Mikusek, P.; Grzybowski, M.; Matraszek, M.; Kupecki, J.H.J.; Jóźwiak, P. Forecast of Demand for Renewable Hydrogen (RFNBO) in Poland by 2030. Available online: https://ien.com.pl/images/konferencje/Prognoza-zapotrzebowania-na-wod%C3%B3r-odnawialny-raport.pdf (accessed on 5 December 2024). (In Polish).
21. Brauers, H.; Oei, P.Y. The Political Economy of Coal in Poland: Drivers and Barriers for a Shift Away from Fossil Fuels. *Energy Policy* **2020**, *144*, 111621. [CrossRef]
22. International Energy Agency (IEA). Energy System of Poland. Available online: https://www.iea.org/countries/poland (accessed on 5 December 2024). (In Polish).
23. Polish Transmission System Operator (Polskie Sieci Elektroenergetyczne S.A.) Quantitative Data Summary on the Operation of the National Power System (KSE) in 2023. Available online: https://www.pse.pl/dane-systemowe/funkcjonowanie-kse/raporty-roczne-z-funkcjonowania-kse-za-rok/raporty-za-rok-2023 (accessed on 5 December 2024). (In Polish).
24. Institute for Renewable Energy Wind Power Plants and Wind Farms in Operation in Poland. Available online: https://sklepieo.pl/funkcjonujace-elektrownie-i-farmy-wiatrowe-w-polsce-listopad-2024.html (accessed on 5 December 2024). (In Polish).
25. Polish Wind Energy Association (Polskie Stowarzyszenie Energetyki Wiatrowej) Rethinking Wind Energy: Necessary Legislative Changes and Onshore Potential Analysis. Available online: https://www.psew.pl/wp-content/uploads/2024/09/Psew_2024_Energia_na_nowo_poprawione.pdf (accessed on 5 December 2024). (In Polish).
26. Polish Transmission System Operator (Polskie Sieci Elektroenergetyczne S.A.) Energy from Offshore Wind Farms Will Flow Thanks to PSE Infrastructure. Available online: https://www.pse.pl/-/energia-z-morskich-farm-wiatrowych-poplynie-dzieki-infrastrukturze-pse (accessed on 5 December 2024). (In Polish).
27. Institute for Renewable Energy Photovoltaic Market in Poland. Available online: https://sklepieo.pl/photovoltaic-market-in-poland-66.html (accessed on 5 December 2024).
28. Polish Wind Energy Association (Polskie Stowarzyszenie Energetyki Wiatrowej). Wind Energy in Poland. Report. Available online: https://www.psew.pl/wp-content/uploads/2023/12/Energetyka-wiatrowa-w-Polsce_2023_internet-1.pdf (accessed on 5 December 2024).
29. International Atomic Energy Agency (IAEA). The Database on Nuclear Power Reactors. Available online: https://pris.iaea.org/PRIS/home.aspx (accessed on 5 December 2024).
30. Macieja, E.A.; Kim, J. Quantitative Assessment of Gaseous Effluents during Routine Operation: A Comparative Study of Planned Nuclear Power Plants at Lubiatowo-Kopalino and Pątnów Sites in Poland. *Sci. Technol. Nucl. Install.* **2024**, *2024*, 8699926. [CrossRef]
31. Weibezahn, J.; Steigerwald, B. Fission for Funds: The Financing of Nuclear Power Plants. *Energy Policy* **2024**, *195*, 114382. [CrossRef]
32. Zarębski, P.; Katarzyński, D. Small Modular Reactors (SMRs) as a Solution for Renewable Energy Gaps: Spatial Analysis for Polish Strategy. *Energies* **2023**, *16*, 6491. [CrossRef]
33. Polish Transmission System Operator (Polskie Sieci Elektroenergetyczne S.A.). The Development Plan for Meeting Current and Future Electricity Demand for the Period 2025–2034. Available online: https://www.cire.pl/files/portal/186/news/347185/045da606d26837065c04be4af2035c743a9e50a1424c01903b52981a4a169a4d.pdf (accessed on 5 December 2024). (In Polish).
34. Ciuła, J.; Wiewiórska, I.; Banaś, M.; Pająk, T.; Szewczyk, P. Balance and Energy Use of Biogas in Poland: Prospects and Directions of Development for the Circular Economy. *Energies* **2023**, *16*, 3910. [CrossRef]
35. Rogala, Z.; Stanclik, M.; Łuszkiewicz, D.; Malecha, Z. Perspectives for the Use of Biogas and Biomethane in the Context of the Green Energy Transformation on the Example of an EU Country. *Energies* **2023**, *16*, 1911. [CrossRef]
36. Bórawski, P.; Bełdycka-Bórawska, A.; Kapsdorferová, Z.; Rokicki, T.; Parzonko, A.; Holden, L. Perspectives of Electricity Production from Biogas in the European Union. *Energies* **2024**, *17*, 1169. [CrossRef]
37. Polish Biomethane Association Biogas and Biomethane in Poland Report. Available online: https://magazynbiomasa.pl/biogas-and-biomethane-in-poland-report-pobierz-za-darmo/ (accessed on 5 December 2024).
38. National Center of Research and Development Polish Power System Simulator. Available online: https://symulatorsystemuenergetycznego.ncbr.gov.pl/en/ (accessed on 5 December 2024).
39. Popkiewicz, M. NCBR's Zero-Emission Energy System Symulator. *Energetyka* **2024**, *5*, 181–198. (In Polish)
40. Photovoltaic Geographical Information System (PVGIS). Available online: https://re.jrc.ec.europa.eu/pvg_tools/en/ (accessed on 5 December 2024).
41. ENTSO-E Transparency Platform. Available online: https://transparency.entsoe.eu/ (accessed on 5 December 2024).
42. Open-Meteo. Available online: https://open-meteo.com/ (accessed on 5 December 2024).
43. Instrat Foundation. Available online: https://energy.instrat.pl/en/electrical-system/generation-capacity-are/ (accessed on 5 December 2024).
44. Ahaotu, M.C.; Ogbogu, C.E.; Thornburg, J.; Akwukwaegbu, I.O. Simulation of PEM Electrolyzer Power Management with Renewable Generation in Owerri, Nigeria. *Energies* **2025**, *18*, 208. [CrossRef]

45. Henkel, V.; Wagner, L.P.; Gehlhoff, F.; Fay, A. Combination of Site-Wide and Real-Time Optimization for the Control of Systems of Electrolyzers. *Energies* **2024**, *17*, 4396. [CrossRef]
46. Lei, J.; Ma, H.; Qin, G.; Guo, Z.; Xia, P.; Hao, C. A Comprehensive Review on the Power Supply System of Hydrogen Production Electrolyzers for Future Integrated Energy Systems. *Energies* **2024**, *17*, 935. [CrossRef]
47. Ricciardi Celsi, M.; Ricciardi Celsi, L. Quantum Computing as a Game Changer on the Path towards a Net-Zero Economy: A Review of the Main Challenges in the Energy Domain. *Energies* **2024**, *17*, 1039. [CrossRef]
48. Nikulins, A.; Sudars, K.; Edelmers, E.; Namatevs, I.; Ozols, K.; Komasilovs, V.; Zacepins, A.; Kviesis, A.; Reinhardt, A. Deep Learning for Wind and Solar Energy Forecasting in Hydrogen Production. *Energies* **2024**, *17*, 1053. [CrossRef]
49. Son, Y.G.; Kim, S.Y.; Bae, I.S. Optimal Coordination of Energy Coupling System Considering Uncertainty of Renewable Energy Sources. *Energies* **2024**, *17*, 931. [CrossRef]
50. The State Council Information Office of the People's Republic of China Responding to Climate Change: China's Policies and Actions. Available online: http://www.ncsc.org.cn/yjcg/cbw/202111/P020211117406748792288.pdf (accessed on 5 December 2024).
51. European Commission. Commission Staff Working Document. Impact Assessment Report. Part 1. SWD(2024) 63 Final, EUR-Lex—52024SC0063—EN—EUR-Lex (Europa.Eu). Available online: https://eur-lex.europa.eu/resource.html?uri=cellar:6c154426-c5a6-11ee-95d9-01aa75ed71a1.0001.02/DOC_1&format=PDF (accessed on 5 December 2024).
52. Czyżak, P.; Mindeková, T. Empowering Farmers in Central Europe: The Case for Agri-PV. Available online: https://ember-energy.org/app/uploads/2024/09/Empowering-farmers-in-Central-Europe-1.pdf (accessed on 5 December 2024).
53. Ministry of the Climate of the Republic of Poland. The Polish Nuclear Power Programme. Annex to Resolution No. 141 of the Cabinet of Ministers of 2 October 2020, Item 946. Available online: https://isap.sejm.gov.pl/isap.nsf/DocDetails.xsp?id=WMP20200000946 (accessed on 5 December 2024). (In Polish)
54. Krasuska, E.; Waliszewska, H.; Krzymiński, M.; Lenart, K.; Popkiewicz, M.; Racięcki, W. The Real Potential for Biomethane Production in Poland. Available online: https://www.gov.pl/attachment/1d27ccd0-14be-46ee-9535-68cef31dd0b0 (accessed on 5 December 2024). (In Polish)
55. Dusiło, M. Energy Transition in Poland. Available online: https://www.forum-energii.eu/en/download/download/transformacja-edycja-2024 (accessed on 5 December 2024).
56. *United Nations Environment Programme Emissions Gap Report 2024: No More Hot Air … Please! with a Massive Gap Between Rhetoric and Reality, Countries Draft New Climate Commitments*; United Nations Environment Programme: Nairobi, Kenya, 2024; Volume 447.
57. Włodek, T.; Szurlej, A. The Role and Importance of Underground Gas Storage (UGS) in the Context of Changes in the Natural Gas Market—Analysis of Selected Cases. Available online: https://se.min-pan.krakow.pl/pelne_teksty36/k36_prezentacje/k36_Wlodek.pdf (accessed on 5 December 2024). (In Polish).

energies

MDPI

Article

A Win-Win Coordinated Scheduling Strategy Between Flexible Load Resource Operators and Smart Grid in 5G Era

Nan Zhang [1,*], Di Liu [1], Tianbao Liu [1], Xueyan Zhang [1], Jing Guo [2], Fusheng Lan [3], Qingyao Li [1], Weiyi Lu [1] and Xiaolong Yang [1]

1 State Grid Information & Telecommunication Group Co., Ltd., Beijing 100761, China
2 State Grid Jiangsu Electric Power Co., Ltd., Nanjing 210024, China
3 State Grid Fujian Electric Power Co., Ltd., Fuzhou 350003, China
* Correspondence: 13911644647@139.com

Abstract: With the rapid expansion of 5G base stations, the increasing energy consumption and fluctuations in power grid loads pose significant challenges to both network operators and grid stability. This paper proposes a coordinated scheduling strategy designed to address these pressing issues by leveraging the flexible load management capabilities of 5G base stations and their potential for inter-regional power demand response within the smart grid framework. This study begins by quantifying the dispatch potential of 5G base stations through a detailed analysis of their load dynamics, particularly under tidal fluctuations, which are critical for understanding the temporal variability of energy consumption. Building on this foundation, dormancy and load transfer strategies are introduced to model the scheduling potential for regional energy storage, enabling more efficient utilization of available resources. To further enhance the optimization of energy distribution, a many-to-many proportional energy-sharing algorithm is developed, which facilitates the aggregation of scheduling capacities across multiple regions. Finally, a comprehensive multi-objective, two-layer collaborative dispatching strategy is proposed, aiming to mitigate grid load volatility and reduce electricity procurement costs for 5G operators. Extensive simulation results demonstrate the effectiveness of this strategy, showing a significant reduction in grid load variance by 37.88% and a notable decrease in operational electricity costs for 5G base stations from CNY 4616.0 to 3024.1. These outcomes highlight the potential of the proposed approach to achieve a win-win scenario, benefiting both base station operators and the smart grid by enhancing energy efficiency and grid stability.

Keywords: 5G base station; adjustment potential; load transfer; energy sharing; scheduling strategy

1. Introduction

In recent years, China has emphasized the acceleration of new infrastructure development. As a key component of this "new infrastructure", the construction of 5G communication base stations is progressing steadily, with the number of base stations expected to reach 8 million by 2025 [1]. Given that the power consumption of 5G base stations is three to four times higher than that of 4G base stations [2,3], and more stations are required to cover the same communication range, the energy consumption issue is particularly prominent [4]. This, in turn, exacerbates the fluctuation of power grid load [5]. In addition, the large-scale integration of renewable energy sources further amplifies grid fluctuations [6]. Therefore, reducing energy consumption and grid volatility while meeting the growing demand for 5G base stations is of significant importance.

To ensure a reliable power supply, 5G base stations are typically equipped with electrochemical energy storage systems as backup power sources. By 2025, the demand for backup batteries in these stations is expected to reach 78.6 GW·h [7]. Given that base station energy storage mostly remains idle and floating charged for long periods, and considering the variation in communication loads across different regions, their dual source-load characteristics and inter-regional collaborative scheduling capabilities can be leveraged as flexible loads for power demand response. Optimizing the base station–grid interactive dispatching strategy can maximize the schedulable potential of 5G base station clusters, improve the utilization of 5G network resources, aid in peak-load shaving and valley-filling for the supply side, and ease the operational pressure on the power grid.

Extensive research exists on the potential of 5G base station energy storage. Reference [8] correlates the lower limit of the state of charge (SOC) of base station energy storage with the station's load state, constructing a schedulable potential model. Reference [9] designs the backup time for base station energy storage, considering mains reliability, and evaluates schedulable capacity based on communication load and backup time. Reference [10] employs a Markov model approach to enhance communication reliability with minimal repair time. However, most of these studies focus solely on the original transmitting power and do not fully utilize the potential of dormant load transfer, limiting the effective release of base station energy storage regulation potential.

Research on dispatching strategies for base station backup energy storage in grid demand response has also been conducted. Reference [11] establishes an economic dispatch model for the 5G base station light-storage system, considering hourly electricity prices and base station load. Reference [12] proposes an energy storage regulation algorithm based on the upper confidence interval of battery state perception to maximize the weighted difference between regulation benefits and load curve variance. Yet, reference [13] explores transferring excess energy storage from 5G base stations to the grid. However, these studies mostly focus on the base station clusters' interaction with the grid, neglecting energy flow and regional cooperativity among the clusters. In contrast, reference [14] connects base stations and the power distribution network, while reference [15] proposes an optimal power purchase algorithm based on one-sided matching and introducing a bidirectional preference list for energy storage regulation in 5G distribution network base stations. However, the majority of current studies explore single or few-to-many energy-sharing models, with limited analysis of many-to-many configurations.

In response, this paper proposes an interactive scheduling strategy between base stations and the power grid to maximize the schedulable potential of 5G base station clusters. By considering base station dormancy and communication load transfer, a potential model for 5G base stations is developed, and a many-to-many proportional energy-sharing strategy is proposed to aggregate multi-region schedulable capacities. A multi-objective, two-layer optimization framework is established for the interaction between 5G base station clusters and the power grid, aiming to minimize grid load volatility and power purchase costs for base stations. The strategy layers objectives through interactions between 5G cluster aggregators and the grid. Finally, the proposed model's effectiveness is verified.

The key innovations of this study include:

- The introduction of load transfer and energy-sharing mechanisms, which significantly expand the schedulable capacity of energy storage systems by enabling flexible resource allocation across multiple regions;

- The development of a many-to-many proportional allocation algorithm, which optimizes energy distribution among 5G base station clusters and enhances the efficiency of multi-region coordination.

2. Overview of Multi-Objective Two-Layer Optimization Architecture of 5G Base Station Clusters

The interactive dispatching scenario diagram of a 5G base station cluster and power distribution network to maximize the schedulable potential established in this paper is shown in Figure 1.

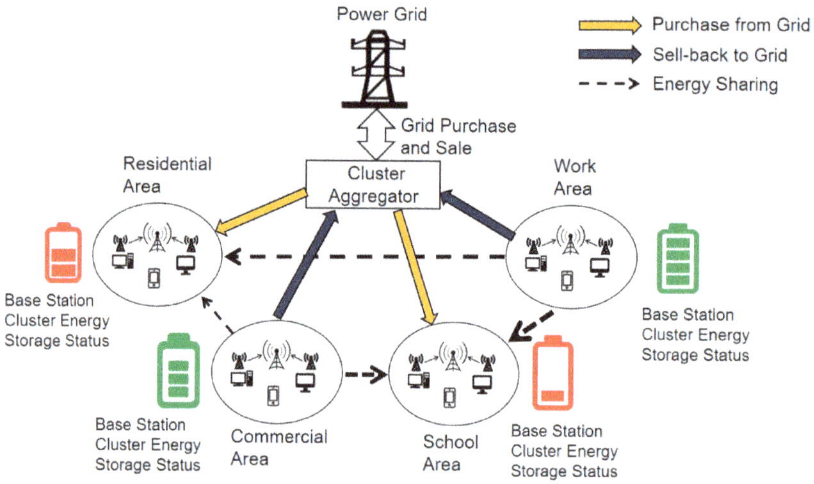

Figure 1. Interactive dispatching diagram of 5G base station clusters and the power distribution network to optimize scheduling potential.

As shown in Figure 1, the study area in this paper is divided into four functional zones: work, commercial, school, and residential. These zones include entities such as 5G micro and macro base stations, the power distribution network, and 5G cluster aggregator. During interactions between the 5G base stations and the power grid, the backup time is first calculated based on the reliability parameters of each regional distribution network. Load transfer is then performed for micro base stations that meet the criteria. Potential models for individual and regional 5G base stations are established using real-time load and backup time data from the transferred macro base stations. Next, a many-to-many proportional allocation algorithm is used to achieve energy sharing between functional zones, transferring excess power from surplus zones to deficit zones, thereby enhancing the overall dispatchable potential of the 5G base station cluster. Finally, the cluster status information is integrated and uploaded to the power grid. The power grid, after considering its own load and other factors, issues electricity price information and an overall charge and discharge plan to the cluster aggregator. The aggregator then formulates real-time control strategies and issues control instructions to each zone. In each zone, energy storage is charged when there is a demand for power absorption from the grid. When the grid load is heavy, energy storage discharges to alleviate grid pressure, ultimately achieving energy interaction between the base stations and the power grid.

3. Analysis and Modeling of Scheduling Energy Storage Potential for 5G Base Stations

3.1. Analysis of Energy Storage Scheduling Potential of 5G Base Station Based on Spatio-Temporal Tidal Characteristics

This paper examines a 5G base station cluster composed of macro and micro base stations. The macro stations provide extensive coverage with stable signals and possess energy storage, while micro stations enhance local signal quality and network capacity. The active antenna unit (AAU) in 5G base station electrical equipment accounts for nearly 90% of total power consumption, making it the primary energy consumer. The AAU's power consumption fluctuates with the communication load of the base station, resulting in the overall power consumption of the 5G base station being highly dependent on real-time communication load [16,17]. As user positions and communication behaviors vary over time, the communication load of a base station exhibits a tidal effect. 5G base stations are typically equipped with energy storage batteries sized based on the peak load to prevent power supply interruptions [18]. Due to the pronounced peak and valley characteristics of communication loads, backup energy storage often includes redundant resources, presenting an optimization opportunity. Consequently, 5G base stations, as flexible loads, are well-suited for participating in power grid regulation.

The dense deployment of 5G base stations creates overlapping coverage areas, allowing for load migration while maintaining communication service quality [19]. Given the linear relationship between communication and power load, micro base stations can transfer communication loads to macro stations when dormancy thresholds are met. This transfer increases the real-time power consumption of the macro station, raises its charging upper limit, and expands its schedulable capacity. Thus, load transfer optimizes cluster power consumption, reduces electricity costs, and enhances the energy storage capacity of macro stations, boosting their scheduling potential.

The load curves of 5G base stations in different functional areas show distinct differences and complementarity [20]. Regional aggregation through real-time energy sharing can further enhance overall scheduling potential. High-load functional areas can prioritize energy use from adjacent surplus areas, reducing reliance on power grid purchases and mitigating peak grid loads. This strategy significantly improves the scheduling potential of 5G base station clusters. The construction process of the 5G base station energy storage scheduling potential model is shown in Figure 2.

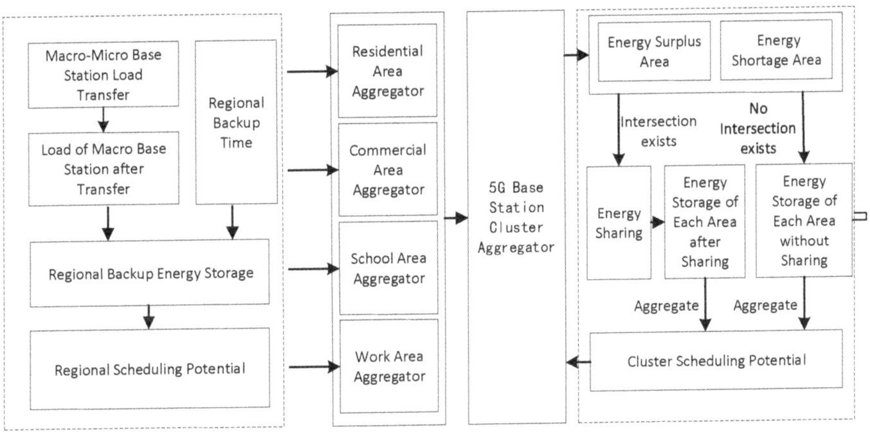

Figure 2. Flow diagram of scheduling potential model of base station cluster.

3.2. Construction of a Scheduling Potential Model for 5G Base Station Units

3.2.1. Energy Storage Backup Time of 5G Base Station

This paper divides the load area into four parts: work, business, school, and residence, which is recorded as a collection $\Pi = \{\text{res}, \text{com}, \text{sch}, \text{work}\}$. The average power outage time and average power outage times of different functional areas were normalized, and the average backup time of residential areas was taken as the benchmark [21], and the minimum backup power time for energy storage of base stations in different functional areas was obtained:

$$T_{\min}^{j,s\in\Pi} = T_{\min}^{j,res} \frac{\eta^j r^j}{\eta^{j,\text{res}} r^{j,\text{res}}} \quad (1)$$

Therefore, the outage duration on the power distribution network side can be obtained.

3.2.2. The Real-Time Power Consumption of the 5G Base Stations

Load transfer refers to the process of redistributing electrical load from one area or system to another to optimize energy usage, balance grid demand, and enhance operational efficiency. For the 5G macro base station j, the power consumption can be expressed as follows:

$$P_{\text{b}}^{\prime j}(t) = P_{\text{b,s}}^{j}(t) + \beta_j P_{\text{b,r}}^{\prime j}(t) \quad (2)$$

where β_j is the efficiency coefficient of macro base station j, taken as the reciprocal of the efficiency of the power amplifier; $P_{\text{b,r}}^{\prime j}(t)$ is the transmission power of base station j, which increases with the number of users accessing it [22].

For 5G micro base station i, there are two states of active and dormant in each time period, and its power consumption can be expressed as follows:

$$P_{\text{b}}^{i}(t) = \begin{cases} P_{\text{sleep}}^{i}, q = 1 \\ P_{\text{b,s}}^{i}(t) + \beta P_{\text{b,r}}^{i}(t), q = 0 \end{cases} \quad (3)$$

where $q = 1$ indicates that the micro base station i is in sleep mode, and $q = 0$ indicates that the micro base station i is in active mode. When the dynamic power consumption of the micro base station is less than the sleep threshold, $q = 1$ is set to 1, otherwise it is set to 0.

When $q = 1$, the distance between the micro base station i and the nearest macro base station k is calculated, and the product of the load and distance of the micro base station i is calculated:

$$W_i = P_{\text{b,r}}^{i}(t) \times s_{i,k} \quad (4)$$

The product of the load transfer to macro base station k is sorted and further classified according to the number of covered Acer stations. The collection of micro-base stations in multi-coverage areas is

$$\Psi' = \{i \| C_i| > 1\} \quad (5)$$

For the micro base station with overlapping coverage area of the multi-macro station, the restart probability is low, which reduces the conversion times of the base station state, thus reducing the power consumption more efficiently [23]. Therefore, when the macro station cannot accommodate the transfer load of all micro base stations, these micro base stations in the overlapping coverage area are preferentially dormant. According to the above sorting rules, the set of micro base stations that can be contained is

$$\Psi = \left\{ i | \sum_k P_{\text{b}}^{i}(t) \leq P_{k,\max} \right\} \quad (6)$$

Therefore, for the macro base station k after load transfer, the power expression is

$$P_{\mathrm{b}}^{k}(t) = P_{\mathrm{b,s}}^{j}(t) + \beta_k P_{\mathrm{b}}^{\prime k}(t) + \sum_{i=1}^{i\in\Psi} \frac{P_{\mathrm{b,r}}^{i}(t)}{\beta_i} \tag{7}$$

3.2.3. G Base Station Scheduling Capacity Model

The main function of 5G base station energy storage is to serve as a backup power supply to ensure the uninterrupted power supply demand of the base station. Therefore, the base station energy storage only participates in the power dispatch during the normal power supply, and the power supply only for the base station itself when the city power is cut off. Considering the upper and lower limits of charge and discharge of energy storage batteries and the requirements of standby energy storage, the rechargeable and discharging dispatching capacity of a single base station j during time t is

$$E_{\mathrm{c,lim}}^{j}(t) = (E_{j,\max} - E_j(t))B_j(t) \tag{8}$$

$$E_{\mathrm{dis,lim}}^{j}(t) = (E_j(t) - E_{j,\min} - E_{\min}(t))B_j(t) \tag{9}$$

$$E_{\min}(t) = \int_{t}^{t+T_{\min}^{j}} P_{\mathrm{b}}^{j}(t)dt \tag{10}$$

where $(E_{j,\max} - E_j(t))$ indicates the upper limit of energy storage, and $(E_j(t) - E_{j,\min} - E_{\min}(t))$ indicates the lower limit of the scheduling capacity of energy storage, which is obtained by subtracting the inherent lower limit of the energy storage itself from the maximum capacity, and then further subtracting the standby energy needed to ensure communication reliability; $\int_{t}^{t+T_{\min}^{j}} P_{\mathrm{b}}^{j}(t)dt$ represents the standby energy, which is calculated based on the standby duration and real-time load. $B_j(m)$ is a 0–1 variable, where $B_j(m) = 1$ indicates that the grid is supplying power normally and the standby energy of the base station can participate in adjustment, while $B_j(m) = 0$ indicates that the grid is failing to supply power and the standby energy of the base station is used to supply power to the base station communication equipment.

According to the rechargeable capacity and discharging scheduling capacity, the maximum charge and discharge power of a single base station j standby energy storage within a scheduling interval is

$$P_{\mathrm{b,c}}^{j}(t) \leq \min\left\{\frac{E_{j,\max} - E_j(t)}{\alpha_c \Delta t}, P_{\mathrm{c,max}}^{j}\right\}B_j(t) \tag{11}$$

$$P_{\mathrm{b,dis}}^{j}(t) \leq \min\left\{\frac{E_j(t) - E_{j,\min} - E_{\min}(t)\alpha_{\mathrm{dis}}}{\Delta t}, P_{\mathrm{dis,max}}^{j}\right\}B_j(t) \tag{12}$$

3.3. Construction of the Scheduling Potential Model of 5G Base Station Cluster Based on Energy Sharing

Energy sharing involves the collaborative distribution of surplus energy among multiple nodes or systems to improve resource utilization and enhance overall efficiency in energy management. In the process of participating in the optimal operation of the power grid, the optional regulation modes can be divided into decentralized regulation and aggregate regulation. Because 5G base station energy storage is characterized by large quantity, wide distribution, and small individual capacity, it will bring excessive computing burden and workload [23,24] to the power grid under the decentralized regulation mode. Therefore, this paper will adopt the aggregate regulation mode and introduce regional aggregators as intermediate agents to construct a distributed regulation architecture. Assuming that n is the total number of base stations involved in the aggregation in a certain functional area, and considering the constraints of base station backup energy storage and SOC, the rechargeable and discharging dispatching capacity in a certain functional area is obtained:

$$E^{m}_{\mathrm{c,lim}}(t) = \sum_{j=1}^{n} (E_{j,\max} - E_j(t))B_j(t) \tag{13}$$

$$E^{m}_{\mathrm{dis,lim}}(t) = \sum_{j=1}^{n} (E^{j}(t) - E^{j}_{\min} - E_{\min}(t))B_j(t) \tag{14}$$

As mentioned in Section 3.1 of this paper, there is a tidal effect in 5G base station load, and there are obvious differences in load peak and trough periods between different regions, which can make full use of this complementary feature for energy sharing. Based on this, this paper puts forward a kind of energy sharing strategy based on multi-to-proportional allocation, namely in each period, through the real-time calculation of surplus power and power shortage, calculate the optimal energy distribution ratio, to ensure the efficient transfer and utilization of energy between different areas, and realize the reasonable distribution and optimization of energy. The specific steps are as follows.

1. Determine the energy state: Based on the load and storage conditions of each region, calculate the total power demand and current storage surplus or shortage. Regions where storage capacity cannot cover upcoming standby demands are energy-scarce, while those exceeding standby demands are energy-rich. Energy-rich regions are denoted as $\theta^{F}(t) = \{F_h, F_{h+1,\ldots}, F_H\}$, and energy-scarce regions are denoted as $\theta^{D}(t) = \{D_g, D_{g+1,\ldots}, D_G\}$.
2. Calculate the total energy surplus and power shortage. Calculate the total surplus energy $\sum_{h=1}^{H} E^{h}_{F}$ and power shortage $\sum_{g=1}^{G} E^{g}_{D}$ in each region, and calculate the proportion of energy surplus of each region $r^{h}_{F} = E^{h}_{F} / \sum_{h=1}^{H} E^{h}_{F}$ and the proportion of energy shortage of each region $r^{g}_{d} = E^{g}_{D} / \sum_{g=1}^{G} E^{g}_{D}$. According to Equation (15), the energy transfer target is obtained as

$$E_{F\to D} = \min\left\{ \sum_{h=1}^{H} E^{h}_{F}, \sum_{g=1}^{G} E^{g}_{D} \right\} \tag{15}$$

3. Construct the energy transmission matrix. To achieve optimal energy allocation, construct the energy transmission matrix and use the maximum matching rule for selective calculation. First, define the energy transmission matrix: $E = \begin{bmatrix} E_{F_1\to D_1} & \cdots & E_{F_1\to D_G} \\ \vdots & \ddots & \vdots \\ E_{F_H\to D_1} & \cdots & E_{F_H\to D_G} \end{bmatrix}$ where $E_{F_h\to D_g}$ represents the transmission energy from the energy-rich region F_h to the energy-scarce region D_g. Next, perform a selective calculation: (1) Identify the region F_{h^*} with the largest current surplus energy and the region D_{g^*} with the largest current energy deficit. (2) Calculate the transferable energy at this moment as $\Delta E = \min\left\{ E^{h^*}_{F}, E^{g^*}_{D} \right\}$. (3) Update the total surplus energy and total deficit energy, and determine whether there are still energy surplus or deficit regions unprocessed. If they exist, return to step (1) and continue with the selective calculation until the energy transfer between all surplus and deficit regions is complete.

4. Complete the inter-regional energy transmission. The final energy transfer matrix is the best energy transfer scheme obtained under the maximum value matching principle. After the energy transfer scheme is implemented in each functional area, the updated power is the equilibrium power realized under the scheme, so as to maximize the scheduling potential of the 5G base station cluster. The shared scheduling potential of the 5G base station cluster is expressed as follows:

$$E_{\text{c,lim}}(t) = \sum_{m=1}^{N_a} (E_{\max}^{m} - E_{\text{c,lim}}^{m}(t))B_k(t) \tag{16}$$

$$E_{\text{dis,lim}}(t) = \sum_{m=1}^{N_a} (E^{m} - E_{\min}^{m}(t) - E_{\min}(t))B_k(t) \tag{17}$$

4. Multi-Objective Two-Layer Optimization Scheduling Strategy for 5G Base Station Cluster

Based on the power reduction demand of the 5G base station and the grid side peak load filling demand, on the basis of the cluster aggregation and schedulable capacity model established in Section 3, a multi-objective two-layer optimization architecture of the 5G base station cluster is proposed in Figure 3.

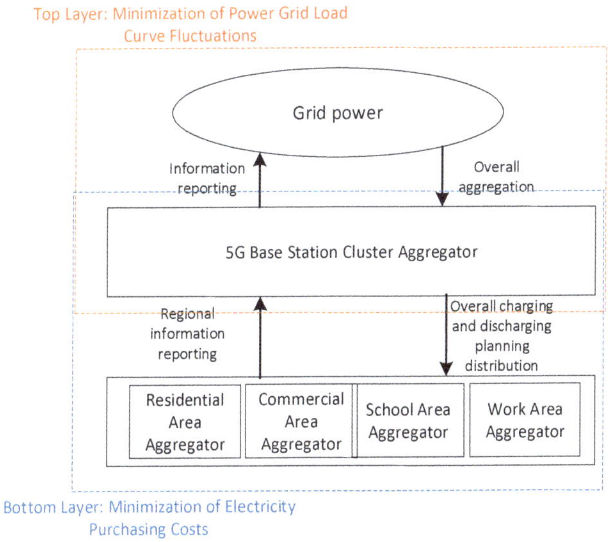

Figure 3. Diagram of multi-objective two-layer optimization architecture of 5G base stations.

In the top layer, the 5G macro-micro base station integrates the cluster load status information and energy storage status information and uploads it to the superior power grid, participates in the power grid dispatching in the form of cluster aggregator, minimizes the variance of the load curve of the power grid, and obtains the overall charge and discharge plan based on the lower optimization, further dispatching between regions. At this point, with the economy of a single region as the goal, the specific scheduling strategies of each region at different times are solved, respectively, so that each region can complete the charge and discharge action together.

4.1. Interaction Model of 5G Base Station Cluster and Power Grid

The top layer of the 5G base station cluster multi-objective two-layer regulation model is the interaction between the 5G base station cluster aggregator and power purchase and sale. Since this paper takes the power grid as the leader and the 5G base station cluster aggregator as the follower, the power grid peak load filling has a higher priority, so the objective function of the top layer model is to minimize the variance of the power grid load curve:

$$\min f_1 = \frac{\sum_{t=1}^{T} [P_{\text{load}}(t) - (P_{\text{B,D}}^{\text{total}}(t) - P_{\text{B,C}}^{\text{total}}(t)) - P_{\text{a}}]^2}{T} \tag{18}$$

$$P_{\text{a}} = \sum_{t=1}^{T} [P(t) - (P_{\text{B,D}}^{\text{total}}(t) - P_{\text{B,C}}^{\text{total}}(t))]/T \tag{19}$$

In the top layer, the overall charging and discharge power of the 5G base station cluster and the upper limit of transmission power of the power grid, is as follows:

1. Cluster Aggregator energy storage charge and discharge power constraint:

$$0 \le P_{\text{B,D}}^{\text{total}}(t) \le \sum_{m=1}^{N_a} P_{\text{B,D,max}}^{m}(t) \tag{20}$$

$$0 \le P_{\text{B,C}}^{\text{total}}(t) \le \sum_{i=1}^{N_a} P_{\text{B,C,max}}^{m}(t) \tag{21}$$

2. Cluster aggregator energy storage energy equation constraint:

$$E(t) = E_0 + \sum_{l=1}^{t-1} [\alpha_c P_{\text{B,C}}^{\text{total}}(l) - \frac{P_{\text{B,D}}^{\text{total}}(l)}{\alpha_{\text{dis}}}]\Delta t \tag{22}$$

3. Cluster aggregator energy storage SOC constraints:

$$\max\left\{\sum_{m=1}^{N_a} E_{\text{min}}^{m}, \sum_{m=1}^{N_a} E_{\text{min}}^{m}(t)\right\} \le E(t) \le \sum_{m=1}^{N_a} E_{\text{max}}^{m} \tag{23}$$

4. Power constraints on the grid:

$$P_{\text{grid}}^{\text{min}} \le P_{\text{B,D}}^{\text{total}}(t) \le P_{\text{grid}}^{\text{max}} \tag{24}$$

$$P_{\text{grid}}^{\text{min}} \le P_{\text{C,D}}^{\text{total}}(t) \le P_{\text{grid}}^{\text{max}} \tag{25}$$

4.2. Energy Allocation Optimization Model for Regional Coordination

At each moment at the end of the energy sharing, a cluster aggregator will be on the upper 5G base station cluster and the power grid interaction overall charge and discharge plan area issued, at the same time as using an electricity purchase/electricity dynamic time-sharing double electricity price mechanism, encouraging the 5G base station area cluster in the case of energy surplus to the power grid, a target function for the 5G base station cluster economic cost minimum:

$$\max f_2 = \sum_{m=1}^{N_a} \sum_{t=1}^{T} (F_1 + F_2 - C_1) \tag{26}$$

$$F_1 = (P_{\text{B,D}_{\text{load}}}^{m}(t) - P_{\text{B,C}}^{m}(t))\varepsilon_1(t) \tag{27}$$

$$F_2 = P^{m}_{\mathrm{B,D_{sell}}}(t)\varepsilon_2(t) \tag{28}$$

$$C_1 = \sum_{j=1}^{n} C_{\mathrm{b}}^{j}(P^{j}_{\mathrm{b,dis}}(t) + P^{j}_{\mathrm{b,c}}(t)) \tag{29}$$

$$P^{m}_{\mathrm{B,D}}(t) = P^{m}_{\mathrm{B,D_{load}}}(t) + P^{m}_{\mathrm{B,D_{sell}}}(t) \tag{30}$$

where F_1 is the storage revenue of the base station; F_2 is the revenue of the sale to grid; C_1 is the charge and discharge loss cost of the base station storage.

The specific constraints are provided as follows:

1. Regional aggregator energy storage charge and discharge power constraint:

$$0 \leq P^{m}_{\mathrm{B,D}}(t) \leq \sum_{j=1}^{n} \min\left\{\frac{E_{j,\max} - E_j(t)}{\eta_{\mathrm{c}}\Delta t}, P^{j}_{\mathrm{c,max}}\right\} \tag{31}$$

$$0 \leq P^{m}_{\mathrm{B,C}}(t) \leq \sum_{j=1}^{n} \min\left\{\frac{E_j(t) - E_{j,\min} - E_{\min}(t)\eta_{\mathrm{dis}}}{\Delta t}, P^{j}_{\mathrm{dis,max}}\right\} \tag{32}$$

2. Regional aggregator energy storage energy equation constraints:

$$E^{m}(t) = E^{m}_{0} + \sum_{l=1}^{t-1} [\alpha_{\mathrm{c}} P^{m}_{\mathrm{B,C}}(l) - \frac{P^{m}_{\mathrm{B,D}}(l)}{\alpha_{\mathrm{dis}}}]\Delta t \tag{33}$$

3. Regional aggregator energy storage SOC constraints:

$$\max\left\{\sum_{j=1}^{n} E^{j}_{\min}, \sum_{j=1}^{n} E^{j}_{\min}(t)\right\} \leq E^{m}(t) \leq E^{m}_{\max} \tag{34}$$

4. Overall charge and discharge power constraint:

$$P^{\mathrm{total}}_{\mathrm{B,D}}(t) = \sum_{m=1}^{N_a} P^{m}_{\mathrm{B,D}}(t) \tag{35}$$

$$P^{\mathrm{total}}_{\mathrm{B,C}}(t) = \sum_{m=1}^{N_a} P^{m}_{\mathrm{B,C}}(t) \tag{36}$$

5. Charge and discharge state constraint:

$$P^{m}_{\mathrm{B,C}}(t) P^{m}_{\mathrm{B,D}}(t) = 0 \tag{37}$$

5. Example and Analysis

5.1. Basic Data

This paper takes the residential, commercial, school, and working areas in a certain area of 5 km × 5 km as an example. Each functional area is evenly distributed with 10 5G macro base stations, while the entire region is evenly distributed with 200 5G micro base stations. The coverage radius of a macro base station is 500 m. The base station load curve characteristics of each functional area and the grid load curve of the area are set by reference [8]. Each macro base station is equipped with two groups of 48 V/500 Ah lithium iron phosphate cascade batteries as backup energy storage. The rated capacity of the battery is 48 kW·h, the charging and discharging efficiency of energy storage is 0.95, the charging and discharging loss coefficient is 0.14, the upper limit of battery SOC is 0.9, and the lower limit is 0.1. The base station parameters, including load characteristics and energy consumption patterns, are derived from actual traffic measurements in a specific urban area. The simulation is based on the MATLAB 2021a software platform, implemented by YALMIP toolbox programming (version R20230622), and solved by the GUROBI solver (version 12.0). The hardware environment for the simulation includes an AMD Ryzen 7 3800X 8-Core processor @ 3.89 GHz and 16.00 GB RAM sourced from Advanced Micro Devices (AMD), headquartered in Santa Clara, CA, USA, ensuring efficient computation and accurate results. Time-sharing purchase and sale price is used in the model and set according to reference [15].

5.2. Analysis of the Results

5.2.1. Analysis of Scheduling Potential of Energy Storage in 5G Base Stations

Based on the backup time and real-time load, the initial backup energy storage curve and the backup energy storage curve after load transfer are obtained, as shown in Figure 4.

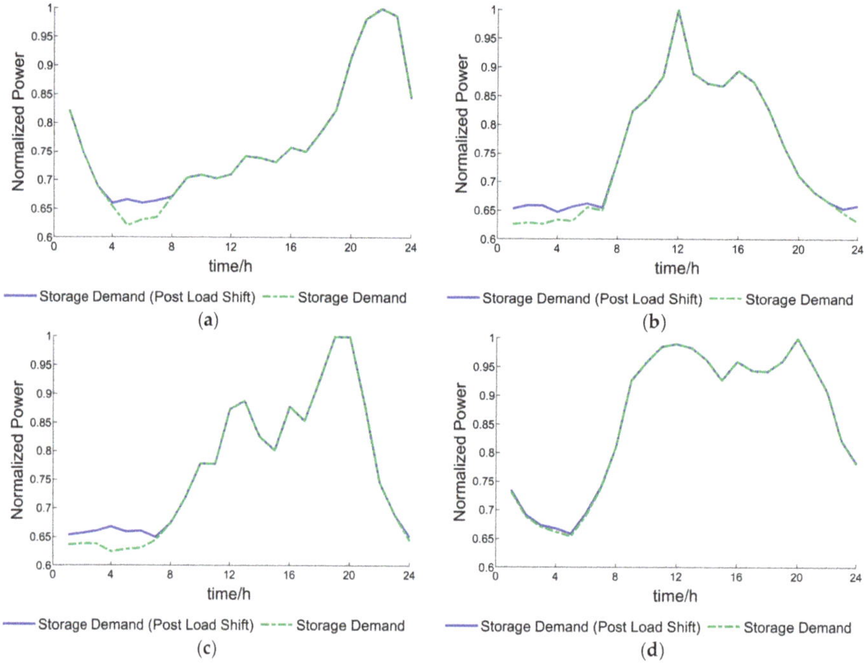

Figure 4. Comparison of the original backup energy storage demand curve after load transfer. (**a**) Residential area, (**b**) commercial area, (**c**) school area, (**d**) work area.

Figure 4 shows a comparison of the energy storage demand (after load transfer) and the original energy storage demand in different time periods in residential, commercial, school, and working areas. As can be seen from Figure 4, the load trough period of each region is distributed at 0:00–7:00 and 22:00–24:00, which coincides with the grid valley period. Therefore, during this period, each region increases the energy storage demand through load transfer, which can broaden the dispatching capacity of each region in this period, which is conducive to the 5G base station cluster to help the grid side to complete the valley filling.

In this paper, the multi-to-multi energy sharing algorithm is adopted to complete the energy flow from the surplus area to the power shortage area. The energy storage shortage pair in each area before and after the algorithm is used, as shown in Figure 5.

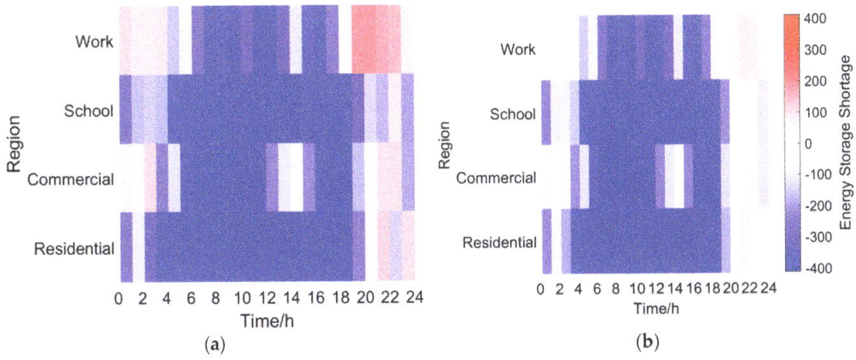

Figure 5. Comparison of the lack of energy storage before and after energy sharing. (**a**) Before energy sharing, (**b**) after energy sharing.

In Figure 5, blue indicates the power surplus and red indicates the power shortage. As can be seen from Figure 5, before the energy sharing of regional energy storage, there is a serious surplus/power shortage in each region. After the energy sharing, the energy storage in each region is balanced and the power shortage is alleviated. This is because under the tidal effect, the load of 5G base stations in different functional areas has staggered peaks. At the same time, due to the complementarity of the schedulable capacity of different regions at the same time, the power shortage areas can be met through energy sharing between regions, and the power purchased from the grid side can be effectively reduced during the peak of the grid load.

5.2.2. Optimization Scheduling Results of 5G Base Station Clusters

The results of upper-layer optimization scheduling in the multi-objective two-layer optimization model of the 5G base station cluster are shown in Figure 6.

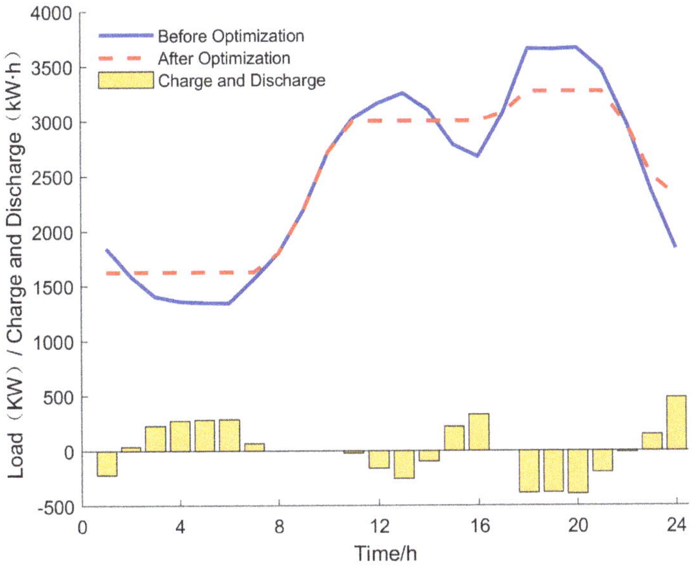

Figure 6. Optimization Scheduling Results of 5G Base Station Clusters and Grid Interaction.

As can be seen from Figure 6, the peak time of the power grid is mainly 12:00–14:00 and 18:00–21:00, and the valley time is mainly 0:00–7:00 and 22:00–24:00. The energy storage of the 5G base station conducts overall discharge at the peak and charge at the valley, which has the effect of peak shifting and valley filling on the power grid curve, leveling the overall load curve, and will not cause new curve fluctuations due to the additional charging and discharge of energy storage.

After the interaction between the 5G base station cluster and the power grid obtains the overall charging and discharge plan, the 5G base station will consider its own economy and distribute the functional areas of the 5G base station. The results of charge and discharge plan allocation for each functional area are shown in Figure 7.

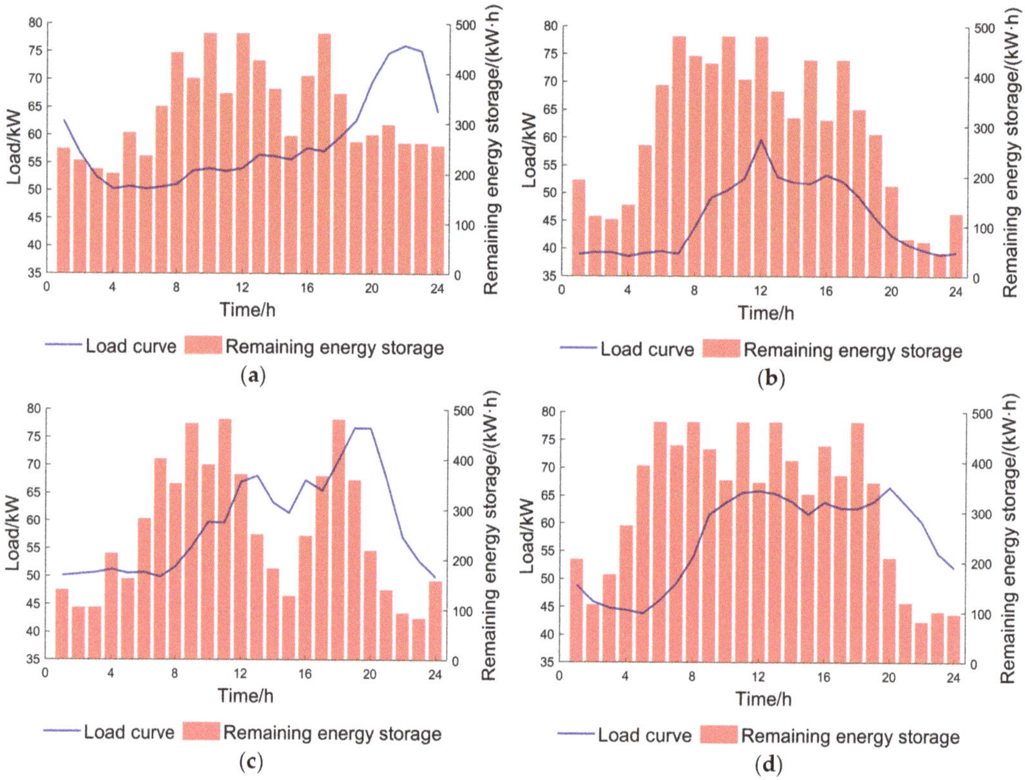

Figure 7. Results of charge and discharge plan allocation for each functional area. (**a**) Residential area, (**b**) commercial area, (**c**) school area, (**d**) work area.

As shown in Figure 7, from 0:00 to 7:00, the optimized overall charge and discharge plan reduces grid load fluctuations, resulting in a rise and then fall in charging. Figure 8 illustrates that, after distributing the overall plan, each region charges according to its own load situation. Between 3:00 and 6:00, the overall charging power peaks, accumulating energy storage to meet grid valley filling demands. During non-peak periods, regions maintain backup energy storage to ensure communication reliability. From 12:00 to 14:00 and 17:00 to 21:00, when the grid peaks, base station energy storage discharges to help with peak load. Significant charge–discharge cooperation is observed due to tidal effects and complementary storage needs. For instance, residential peak load occurs from 19:00 to 24:00, maintaining stable energy storage for backup, while commercial areas discharge to meet peak grid demands. This cooperation ensures balanced dispatching, enhancing flexibility and reliability of the 5G base station system and smoothing the grid load curve.

5.2.3. Comparison Results Under Different Methods

The scheduling results from the multi-objective two-layer optimization model of the 5G base station cluster in this paper are compared with the original power grid load curve, without single-target collaborative scheduling or optimization measures. The results are presented in Figure 8 and Table 1.

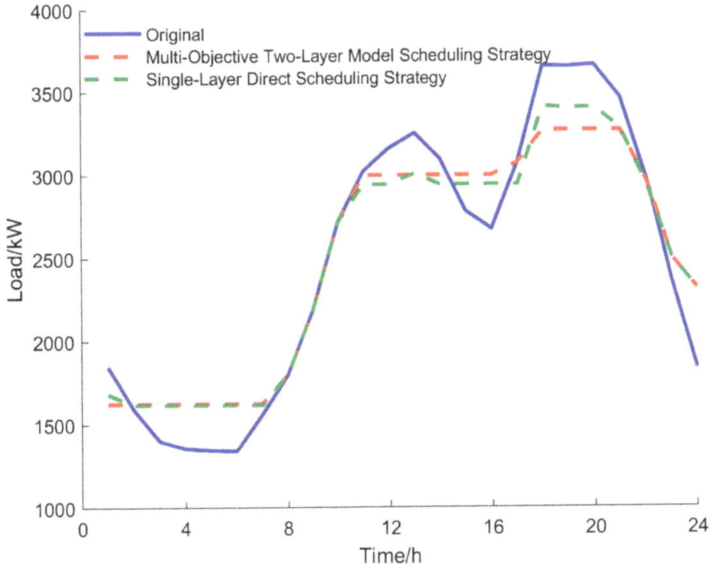

Figure 8. Comparison of power grid load optimization curves under different charge and discharge strategies.

Table 1. Performance of the multi-objective two-level optimization scheduling model.

Scene	Grid Load Curve Variance 105 ($kW^2 \cdot h^2$)	Power Purchase Cost (RMB)
No optimization	6.81	4616.0
Single-Layer Direct Scheduling Strategy	4.66	3466.2
Multi-Objective Two-Layer Model Scheduling Strategy	4.53	3024.1

Figure 8 and Table 1 demonstrate that the multi-objective two-layer dispatching strategy reduces load curve variance by 37.88% compared to the single-target optimization without dispatching, indicating suppressed grid load fluctuations. Power purchase costs decrease from CNY 4616.0 to 3024.1, significantly enhancing the economy of the 5G base station cluster. This improvement is due to regional coordination and energy sharing, where power-deficit areas of 5G base stations prioritize dispatching from surplus areas, reducing grid power purchases during peak times and aiding in grid peak load management while meeting their own needs. Additionally, energy sharing promotes efficient regional energy transmission, reducing grid reliance and minimizing power purchase costs, thus verifying the effectiveness of the 5G base station cluster's multi-objective two-layer optimization model.

6. Conclusions

Aiming at the problem of 5G base station energy consumption and the problem of power grid load fluctuation, this paper constructs the 5G base station energy storage scheduling potential model of the 5G base station that considers the communication load transfer and base station dormancy, and proposes the multi-objective two-layer cooperative scheduling strategy of the 5G base station cluster considering energy sharing and power grid transaction. The following conclusions can be obtained through an example simulation:

1. The proposed method can effectively expand the upper limit of backup energy storage during the trough period. On the one hand, it can better meet the charging demand of the grid side and increase the adjustable range of the Acer station by 35.932 kW; on the other hand, the 5G micro base station enters dormancy when meeting the dormancy threshold, reducing the overall power consumption of the base station side, and the maximum energy-saving efficiency can reach 9.13%;
2. The multi-objective and two-layer optimization scheduling model of the 5G base station cluster designed in this paper can well take into account the demand of peak load shifting and valley filling on the power grid side and the economy of the 5G base station side. Using the spatial and temporal complementarity of each functional region, the mechanism of energy sharing and collaborative scheduling optimization among regions is constructed, so that the 5G cluster shows the characteristics of peak discharge and trough charging, and the variance of the power grid load curve is reduced by 37.88%. At the same time, inter-regional coordination can effectively reduce the overall outsourcing power of 5G base station peak time, and obtain benefits through the low charging and high charging of energy storage, reducing the power purchase cost from CNY 4616.0 to 3024.1, realizing a win-win situation between 5G base station operators and the power grid.

In this paper, there are ideal and single energy storage resources, and mobile energy storage [25] represented by electric vehicles are not considered. Meanwhile, the development of 6G communication technology will bring higher bandwidth, lower delay, and stronger network coverage [26], which can further optimize resource allocation and improve network performance. How to combine with cutting-edge communication technology to build a comprehensive optimal scheduling model considering the participation of multiple subjects is the author's next research focus.

Author Contributions: Conceptualization, N.Z.; methodology, D.L.; software, T.L.; validation, X.Z.; formal analysis, Q.L.; investigation, W.L.; resources, X.Y.; data curation, N.Z.; writing—original draft preparation, W.L.; writing—review and editing, D.L.; visualization, T.L.; supervision, X.Z.; project administration, J.G.; funding acquisition, F.L. All authors have read and agreed to the published version of the manuscript.

Funding: This research was funded by the science and technology project of State Grid Corporation of China (Item No: 5700-202441247A-1-1-ZN).

Data Availability Statement: The original contributions presented in the study are included in the article, further inquiries can be directed to the corresponding author.

Conflicts of Interest: Authors Nan Zhang, Di Liu, Tianbao Liu, Xueyan Zhang, Qingyao Li, Weiyi Lu and Xiaolong Yang were employed by the company State Grid Information & Telecommunication Group Co., Ltd. Author Jing Guo was employed by the company State Grid Jiangsu Electric Power Co., Ltd. Author Fusheng Lan was employed by the company State Grid Fujian Electric Power Co., Ltd.

Nomenclature

Parameter	Description
i, j	Indices for micro and macro station
m	Indices for region
n	Number of base stations in a certain region
N_a	Number of regions
t	Index for time
$T_{min}^{s \in \Pi}$	Minimum standby time(s)
T_{min}^{res}	Minimum standby time in residential areas(s)
η	Average outage frequency
r	Average outage duration
P_b'	Real-time power consumption (KW·h)
$P_{b,s}$	Static power consumption (KW·h)
β	Efficiency coefficient
P_{sleep}	Sleep power consumption of micro base station (KW·h)
$P_{b,r}, P_{b,s}$	Static and dynamic power consumption (KW·h)
$\|C\|$	Number of macro base stations being covered
$P_{k,max}$	Maximum loading capacity of the base station k (KW·h)
P_b^k	Real-time function of the base station k after load transfer (KW·h)
$E_{c,lim}, E_{dis,lim}$	Adjustable charge and discharge capacity of the energy storage (KW)
E_{max}, E_{min}	Upper and minimum capacity limit of the energy storage (KW)
E	Energy state of the storage (KW)
$P_{b,c}, P_{b,dis}$	Charge and discharge power (KW·h)
$P_{c,max}^j, P_{dis,max}^j$	Maximum charge and discharge power limit of the energy storage battery (KW·h)
α	Charge and discharge efficiency of energy storage
P_{load}	Power load of the grid (KW·h)
$P_{B,D}^{total}, P_{B,C}^{total}$	Total discharge and charge power of the 5G base station clusters (KW·h)
$P'_{B,D,max}, P_{B,C,max}$	Maximum discharge and charge power of energy storage in region (KW·h)
E_0	Initial energy storage (KW)
$P_{grid}^{max}, P_{grid}^{min}$	Upper and lower limits of the transmission power allowed by the grid (KW·h)
$P_{B,D}, P_{B,C}$	Charge and discharge power rates of the energy storage participating in grid regulation (KW·h)
$P_{B,D_{load}}$	Discharge power rate that energy storage provides for the base station load (KW·h)
$P_{B,D_{sell}}$	Discharge power rate that energy storage sells to the grid (KW·h)
$\varepsilon_1, \varepsilon_2$	Time-varying purchase and sell electricity price (RMB)
C_b^j	Loss cost coefficient of energy storage

References

1. Yang, C.; Tan, R.; He, D. Flexible Interconnection Coordinated Planning Strategy for Distribution Station Area Containing 5G Base Stations. *Electr. Power Sci. Eng.* **2024**, *40*, 33–41.
2. Rawat, P.; Haddad, M.; Altman, E. Towards Efficient Disaster Management: 5G and Device to Device Communication. In Proceedings of the International Conference on Information & Communication Technologies for Disaster Management, Kyoto, Japan, 16–18 March 2016.
3. Borsatti, D.; Grasselli, C.; Contoli, C.; Micciullo, L.; Spinacci, L.; Settembre, M.; Cerroni, W.; Callegati, F. Mission Critical Communications Support with 5G and Network Slicing. *IEEE Trans. Netw. Serv. Manag.* **2023**, *20*, 595–607. [CrossRef]
4. Spantideas, S.T.; Giannopoulos, A.E.; Trakadas, P. Smart Mission Critical Service Management: Architecture, Deployment Options, and Experimental Results. *IEEE Trans. Netw. Serv. Manag.* 2025. [CrossRef]
5. Li, X.; Hu, C.; Luo, S.; Lu, H.; Piao, Z.; Jing, L. Distributed hybrid-triggered observer-based secondary control of multi-bus DC microgrids over directed networks. *IEEE Trans. Circuits Syst. I Regul. Pap.* **2024**. [CrossRef]

6. Hu, Z.; Su, R.; Veerasamy, V.; Huang, L.; Ma, R. Resilient frequency regulation for microgrids under phasor measurement unit faults and communication intermittency. *IEEE Trans. Ind. Inform.* **2025**, *21*, 1941–1949. [CrossRef]
7. Foresight Net. Analysis of the Market Status and Development Prospect of China's Energy Storage Battery Industry in 2022: The Demand for 5G Base Station Backup Power Storage Will Reach 78.6 GWh. Available online: https://mp.ofweek.com/libattery/a856714352297 (accessed on 25 October 2021).
8. Ma, X.; Meng, X.; Zhu, Q.; Duan, Y.; Wang, Z. Control strategy of 5G base station energy storage considering communication load. *Trans. China Electrotech. Soc.* **2022**, *37*, 2878–2887.
9. Mao, A.; Zhang, L.; Sheng, Q. Research on optimal scheduling of 5G base station energy storage aggregators considering communication reliability. *Trans. China Electrotech. Soc.* **2023**, *38*, 2364–2374.
10. Qiao, R.; Wang, J.; Wei, D.; Cai, C.; Shao, J.; Liu, S. Multi-time-scale optimization method for 5G base station considering backup energy storage and air-conditioning scheduling potential. *Autom. Electr. Power Syst.* **2023**, *47*, 111–120.
11. Han, Z.; Wang, S.; Zhao, Q.; Zheng, Z. A capacity optimization configuration method for photovoltaic and energy storage system of 5g base station considering time-of-use electricity price. *Electr. Power* **2022**, *55*, 8–15.
12. Yin, X.; Wang, Z.; Liu, Y.; Lu, Z.; Lv, G.; Pan, C. Energy storage dispatch of 5G base station energy storage for peak shaving and valley filling of distribution networks. *Power Syst. Clean Energy* **2024**, *40*, 97–102.
13. Yong, P.; Zhang, N.; Hou, Q.; Liu, Y.; Teng, F.; Ci, S.; Kang, C. Evaluating the dispatchable capacity of base station backup batteries in distribution networks. *IEEE Trans. Smart Grid* **2021**, *12*, 3966–3979. [CrossRef]
14. Yin, X.; Lyu, G.; Wang, Z.; Lu, Z.; Liu, Y.; Li, C.; Zhou, Z.; Ma, X. Base station energy sharing method for 5G-integrated distribution network. *Mod. Electr. Power* **2024**, *41*, 310–317.
15. Wang, Y.; Yin, X.; Ou, Q.; Ma, W.; Liu, H.; Du, Z. Energy storage regulation method of base stations in 5G integrated distribution network based on energy sharing and trading coordination. *Electr. Power* **2023**, *56*, 61–70.
16. Yong, P.; Zhang, N.; Ci, S.; Kang, C. 5G communication base stations participating in demand response: Key technologies and prospects. *Proc. CSEE* **2021**, *41*, 5540–5551.
17. Gao, H.; Li, C.; Wang, G.; Jin, T.; Hu, J.; Zhu, J.; Chen, Q.; Kang, C. Dynamic construction and response demonstration of virtual power plant with aggregated large-scale 5G base stations. *Autom. Electr. Power Syst.* **2024**, *48*, 47–55.
18. Lin, G.; Gao, C.; Song, M.; Guo, M.; Lv, R.; Fei, F. Construction and dispatch method of virtual power plant with backup energy storage in communication base stations. *Autom. Electr. Power Syst.* **2022**, *46*, 99–107.
19. Ma, X.; Liu, Z.; Wang, Y.; Feng, S. Research on photovoltaic absorption capacity of 5G base station considering communication load migration and energy storage dynamic backup. *Trans. China Electrotech. Soc.* **2023**, *38*, 5832–5845.
20. Rossoni, P.; Da, R.; Belati, E. Linearized AC load flow applied to analysis in electric power systems. *IEEE Lat. Am. Trans.* **2016**, *14*, 4048–4053. [CrossRef]
21. Li, D. Research on Base Station Sleep Mode Technology Under 5G Dense Heterogeneous Network. Master's Thesis, Beijing University of Posts and Telecommunications, Beijing, China, 2018.
22. Al Haj Hassan, H.; Renga, D.; Meo, M.; Nuaymi, L. A novel energy model for renewable energy-enabled cellular networks providing ancillary services to the smart grid. *IEEE Trans. Green Commun. Netw.* **2019**, *3*, 381–396. [CrossRef]
23. Wang, Z.; Li, Y. Micro-station sleeping algorithm based on coverage area priority in 5G network. *Telecommun. Sci.* **2022**, *38*, 88–95.
24. Jang, T.; Xie, L.; Du, Y.; Ouyang, C. Dispatching strategy of energy storage for 5G base stations based on deep reinforcement learning. *Autom. Electr. Power Syst.* **2023**, *47*, 147–157.
25. Zhang, W.; Zhu, T.; Su, J. Demand response strategy for power-cyber-transportation coupling network considering electric vehicles and 5G base stations. *Autom. Electr. Power Syst.* **2024**, *48*, 116–126.
26. Apostolakis, K.C.; Margetis, G.; Stephanidis, C. Cloud-Native 5G Infrastructure and Network Applications (NetApps) for Public Protection and Disaster Relief: The 5G-EPICENTRE Project. In Proceedings of the 2021 European Conference on Networks and Communications and 6G Summit (EuCNC/6G Summit), Porto, Portugal, 8–11 June 2021.

energies

MDPI

Article

A Deep-Neural-Network-Based Surrogate Model for DC/AC Converter Topology Selection Using Multi-Domain Simulations

Gabriel Avila Saccol, Van-Hai Bui and Wencong Su *

Department of Electrical and Computer Engineering, University of Michigan-Dearborn, Dearborn, MI 48128, USA; gsaccol@umich.edu (G.A.S.); vhbui@umich.edu (V.-H.B.)
* Correspondence: wencong@umich.edu

Abstract: The selection of optimal DC/AC power converter topologies for specific applications is often a time-consuming and complex task, which can lead to suboptimal choices. This paper proposes an AI-assisted methodology to identify the most efficient DC/AC converter based on a set of input design parameters. Separate deep-neural-network-based surrogate models are developed for each considered topology, trained by a large dataset of simulation results obtained from MATLAB/Simulink and PSIM, so that the efficiency of each converter can be determined without performing additional simulations. The proposed methodology allows for quick and accurate efficiency estimation, significantly reducing the analysis time for topology selection. A case study for the two-level converter is also presented, demonstrating that additional parameters, such as the semiconductors junction temperature and output current distortion, can also be predicted using a similar methodology. Results are presented to demonstrate the feasibility of the proposed method.

Keywords: AI-assisted design; DC/AC converter; deep neural network; surrogate model

1. Introduction

DC/AC converters are primarily used for converting direct voltage/current into alternating voltage/current according to electrical energy quality standards [1–3]. This is achieved through the control of the switching states of the semiconductor devices to generate a pattern that reproduces a sinusoidal waveform at the converter output [4]. This feature is essential in renewable energy, uninterruptible power supplies, machine driving applications, and energy storage systems [5–7].

The design process of DC/AC converters is summarized in Figure 1. Firstly, it is necessary to identify the input and output specifications, as well as efficiency requirements and regulatory compliance [8,9]. Next, an appropriate converter topology must be selected, considering application-specific requirements [10,11]. The power stage can then be designed with the appropriate switching devices and gate driver circuits. A suitable modulation and control strategy must be implemented to regulate the output variables. Effective thermal management, including heatsinks and cooling systems, must be designed to ensure component safety [12]. Following these steps, protection mechanisms are required to guard the converter against overcurrent and overvoltage [13]. The design is validated through simulations, followed by the printed circuit board design and initial experimental testing. Finally, the prototype must be subjected to performance tests, thermal behavior, and regulatory compliance.

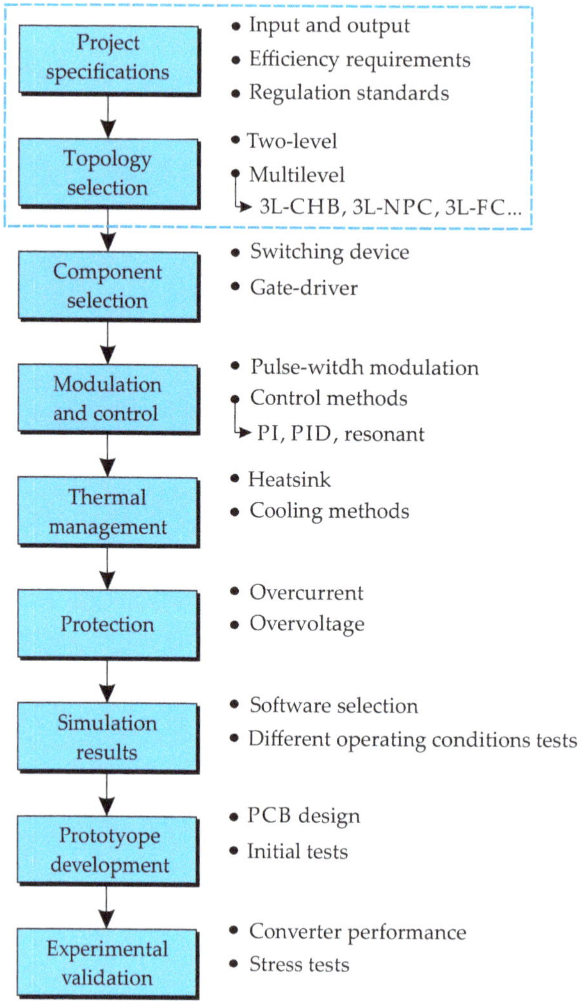

Figure 1. Converter design steps.

Due to a wide range of applications, the DC/AC converter performance must be analyzed specifically and independently for each case [14]. One of the initial steps for this evaluation consists on selecting the topology that best aligns to the input parameters of the converter right after identifying the project specifications, according to Figure 1. However, this is a challenging and time-consuming process that requires the expertise of experienced specialists [15,16]. In some cases, to avoid complex and extensive analysis, superficial evaluations are conducted for the topology selection, which can result in a suboptimal choice.

In parallel, recent advances in machine learning algorithms have enabled this technology to be used in power electronics applications [17,18]. By using a large dataset, it is possible to train a surrogate model that can facilitate the selection of the DC/AC converter topology without requiring any simulations. Surrogate models can approximate complex and resource-intensive simulations, providing significant computational savings and enabling faster design iterations. This data-driven approach enables a quick and automatic identification of the converter that best adapts to a set of predefined input parameters.

Some papers have addressed the use of machine learning algorithms in a converter design. In [19], a deep reinforcement learning (DRL)-based algorithm is proposed to optimize the design parameters of power converters using a deep neural network (DNN)-based surrogate model. However, this work is essentially focused on DC/DC converters, and DC/AC configurations have not been addressed. An automatic method for designing the switching frequency and the passive components while minimizing the converter mass and power losses is presented in [20], but the paper is only focused on the high-gain converter-synchronous rectification low-inductance current secondary boost converter (SLIC-QBC). The potential of an artificial neural network combined with the 3D finite element method for modeling and optimizing magnetic components is explored in [21]. This method is able to simulate inductors with different geometries, windings strandings, core materials, and waveforms; however, it is focused on DC/DC applications only.

A broad range of surrogate models for different power electronics elements is presented in [22], addressing thermal modeling, magnetic component design, and reliability. However, this work lacks focused practical applications and detailed case studies to demonstrate the effectiveness of the proposed methods. An evaluation of the performance and computational costs of four surrogate modeling techniques is performed in [23], where it is concluded that the polynomial surrogate model (PSM) is the most effective for small datasets, while the generalized regression neural network (GRNN), the support vector machines (SVM), and the Kriging demonstrates better performance with larger datasets. Nevertheless, this study is focused on power delivery networks and it does not explore power electronics applications.

A comparative analysis between the recent works on the field is provided in Table 1. One can observe that some existing works have already discussed the use of intelligent algorithms for converter design. Nonetheless, the analyses are mainly focused on DC/DC topologies and DC/AC configurations are not usually explored in this field of study. To address the current research gaps, this paper proposes a deep reinforcement learning method to autonomously obtain the efficiency of multiple DC/AC converter topologies for any given set of input parameters, which can accelerate the process of topology selection by dispensing the use of extensive simulations. A case study is also presented, demonstrating that other parameters such as semiconductors junction temperature and output current distortion can also be determined using a similar approach. Firstly, separate DNN-based surrogate models are trained offline for each of the considered DC/AC configurations. Next, the surrogate models are used to quickly estimate a desired parameter without performing any simulation. These results can be compared to provide an initial concept of the most appropriate topology for a predefined set of input parameters.

In the context of DC/AC converter design, the initial steps highlighted in Figure 1 are critical yet time-consuming and complex. These steps require careful consideration of multiple parameters, which can lead to suboptimal choices if not thoroughly analyzed. The proposed AI-assisted methodology aims to streamline these early design stages by leveraging DNN-based surrogate models. By providing estimations for different converter topologies based on a dataset derived from extensive simulation results, this approach not only reduces the time required for identifying the optimal converter topology, but also enhances the accuracy of the selection process.

This paper is organized as follows. Section 2 describes the development of the DNN-based surrogate model. The dataset collection method used to train the surrogate model of each converter to obtain its respective efficiency is presented in Section 3. A case study to obtain the semiconductors temperature and the output current distortion of a two-level topology is carried out in Section 4. Simulation results and a comparative analysis between the surrogate models are shown in Section 5, to demonstrate the effectiveness of the proposed method.

Table 1. Comparative analysis of recent works.

Work	Methodology	Focus	Scope
[19]	DNN	DC/DC converters	Efficiency optimization
[20]	Artificial neural network	SLIC-QBC	Minimize mass and power loss
[21]	Artificial neural network	DC/DC converters	Magnetic components optimization
[22]	Polynominal chaos expansion and neural network	Broad range of power electronics elements	Heatsink and magnetic component design and reliability evaluation
[23]	PSM, GRNN, SVM, and Kriging	Power delivery networks	Design and optimization

2. Surrogate Model for DC/AC Converter Topologies

The analysis proposed in this work is based on classic topologies, which are those that have already been extensively studied in the literature and are the most commercialized in the industry: the two-level (2L), the three-level cascaded H-bridge (3L-CHB), the three-level neutral point clamped (3L-NPC), and the three-level flying capacitor (3L-FC), as shown in Figure 2 [24,25]. In general, design parameters such as operating power, switching frequency and semiconductors characteristics have different impacts on each topology performance and separately simulating each structure can be particularly time-consuming.

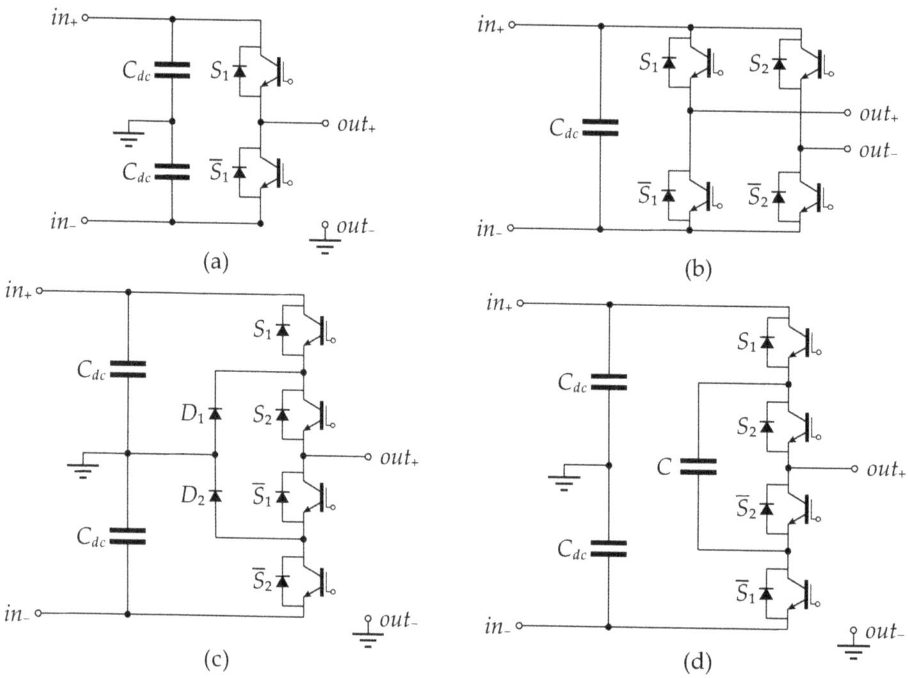

Figure 2. Classic DC/AC converter topologies. (**a**) 2L. (**b**) 3L-CHB. (**c**) 3L-NPC. (**d**) 3L-FC.

Simulation data for each topology were collected using MATLAB/Simulink R2023b software to train the surrogate models that replicate the behavior of their respective converter as close as possible. By using this approach, it is possible to determine the efficiency of each topology without separately simulating each one.

Figure 3 demonstrates the methodology for developing the surrogate models. This technique is implemented separately for each topology presented in Figure 2, resulting in four different surrogate models. The variables that have an impact on the converter efficiency are considered as the simulation input parameters, including the DC-link voltage, output power, output power factor, amplitude modulation index, and switching frequency. Thus, it is possible to generate n different combinations of input parameters, represented by the vectors x_k, $k = 1, 2, \ldots, n$. Each combination is individually simulated to obtain the converter efficiency y_k. This resulting dataset, composed of the x_k and the y_k variables, is then used to train and test a surrogate model that can quickly map the input parameters to the output with high accuracy.

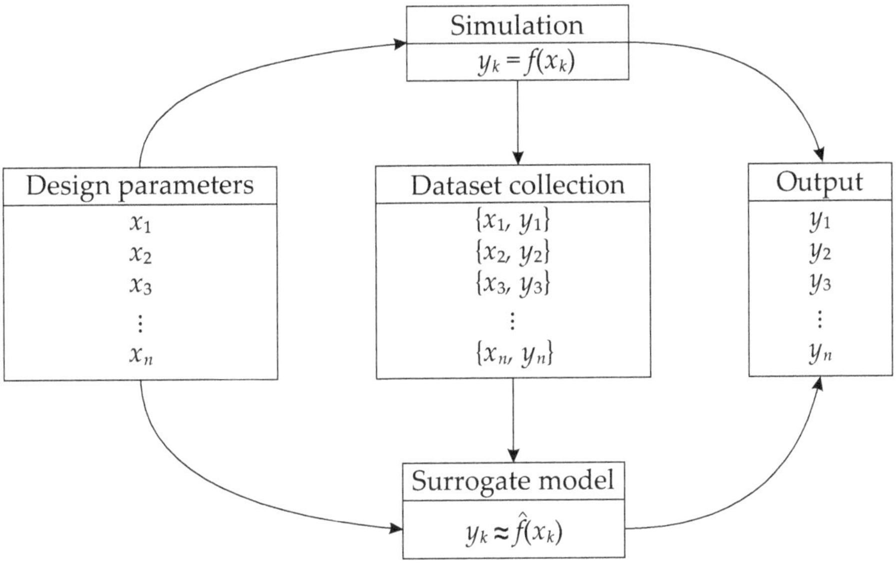

Figure 3. Surrogate model development.

In this work, a separate DNN-based surrogate model is developed for each topology presented in Figure 2. First, a dataset is generated with multiple different combinations of input parameters for one specific converter. In an initial offline training, a portion of this dataset is used to train the surrogate model for this specific converter, while another portion is used to test the results, as shown in Figure 4. Through this process, the weight values of the hidden layer are updated and the efficiency of this converter can be determined from the input parameters provided to the input layer. This process is then repeated for the remaining topologies to obtain their respective surrogate models.

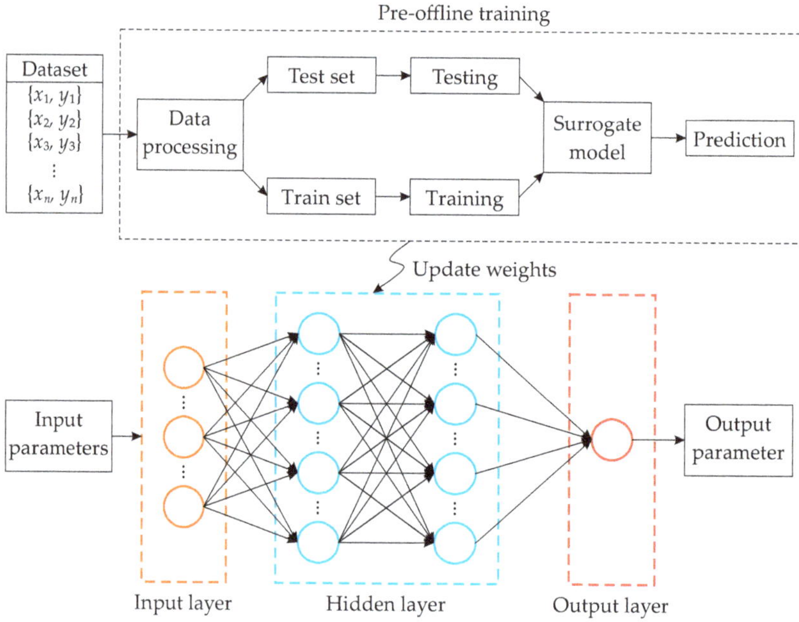

Figure 4. DNN-based surrogate model.

3. Data Acquisition

According to [19], a dataset with at least 30,000 data points is sufficient to ensure a good performance for predicting efficiency values. Therefore, 45,000 simulations were conducted in MATLAB/Simulink for each topology presented in Figure 2, resulting in 180,000 simulations that were used to train and test the surrogate models. The simulations are set up according to Figure 5, where the DC/AC converter can correspond to any of the configurations presented in Figure 2.

Conduction losses are a significant factor that directly affects the overall efficiency of power electronics converters. To effectively model these losses, two key parameters were included into the simulation: the forward voltage drop and the on-state resistance of the IGBTs and their internal diodes. By measuring the voltage and current at the converter terminals, it is possible to determine the input power P_{in} and output power P_{out} to determine each topology efficiency:

$$\eta = \frac{P_{out}}{P_{in}} = \frac{\int_{t_i}^{t_f} v_{in}(t) i_{in}(t)}{\int_{t_i}^{t_f} v_{out}(t) i_{out}(t)}, \quad (1)$$

where $v_{in}(t)$ and $i_{in}(t)$ are the voltage and current measured at the input terminals, respectively, and $v_{out}(t)$ and $i_{out}(t)$ are the voltage and current measured at the output terminals. The difference between the initial time t_i and the final time t_f must be equal to the period of the fundamental component of the output voltage $v_{out}(t)$. This ratio provides a clear measure of each topology performance, with specific emphasis on conduction losses.

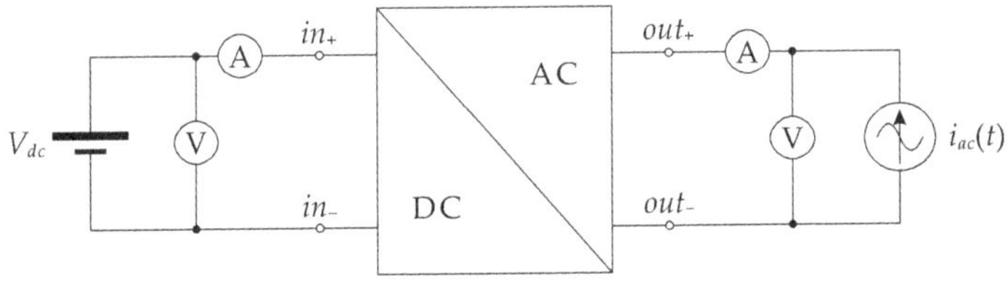

Figure 5. Simulation structure.

To reduce the complexity of the converter response under controlled conditions in steady state, the output load was considered as a controlled current source, whose amplitude I_{ac} and phase displacement φ in relation to the voltage reference are given by:

$$I_{ac} = \frac{2P_{out}}{mV_{dc}\cos(\varphi)} \tag{2}$$

and

$$\varphi = \arccos(\mathrm{PF}), \tag{3}$$

where m and V_{dc} are the DC/AC converter amplitude modulation index and input voltage and PF is the output power factor.

The process of obtaining the dataset is the same for all configurations presented in Figure 2, with variations in the values of the input parameters according to the specific characteristics of each topology. The switching states of the 2L converter are obtained by a conventional pulse width modulation (PWM) strategy with sinusoidal reference, while a phase-disposition (PD) PWM is used to control the switching states of the 3L-NPC converter and a phase-shift (PS) PWM generates the switching signals for the 3L-CHB and the 3L-FC. Furthermore, to simplify the simulations, all circuits were implemented in open-loop, and the output current is defined by the ideal current source presented in Figure 5 and the DC-link capacitors of the converters were replaced by ideal voltage sources. Note that DC-link capacitors have internal resistances and voltage ripples in real applications, which contribute to reduce the converter efficiency. A similar effect occurs by selecting an ideal current source as the converter output, as there are will be no current harmonic components injected into the load. These effects are disregarded in this work and the results focus specifically on the semiconductors power losses.

As previously mentioned, a dataset is required to train and test the surrogate models. The dataset is generated through MATLAB/Simulink simulations with multiple different combinations of input parameters and the efficiency is determined according to Equation (1). Five numerical input parameters for each converter topology are considered, varying within the ranges specified in Table 2. Additionally, three different semiconductor devices were chosen based on the output power P_{out} and DC-link voltage V_{dc} ratings, according to Tables 3 and 4. The voltage and the power values were defined based on the semiconductors limits. Given that the 3L-NPC and the 3L-FC configurations have series connected semiconductors, these topologies can withstand double the DC-link voltage values of the 2L and 3L-CHB converters and, consequently, double the power levels. Note that the semiconductor devices are not treated as input parameters, but were chosen according to the specific power and voltage ratings. The input and output dimensions of the DNN-based surrogate model are organized as follows:

(i) Input layer: five nodes corresponding to five input parameters;
(ii) Output layer: one node corresponding to efficiency.

Table 2. Input parameters for efficiency prediction.

Parameter	Operational Boundary of Input Parameters			
	2L	3L-CHB	3L-NPC	3L-FC
DC-link voltage (kV)	[0.3, 6.5]	[0.3, 6.5]	[0.6, 13]	[0.6, 13]
Output power (kW)	[0.54, 1219]	[0.54, 1219]	[1.08, 2440]	[1.08, 2440]
Amplitude modulation index	[0.8, 1]			
Switching frequency (kHz)	[1, 10]			
Output power factor	[0.9, 1]			

Table 3. Semiconductor devices for 2L and 3L-CHB converters.

Semiconductor Device	Output Power Range (kW)	DC-Link Voltage Range (kV)
RGSX5T565DHR	$0.54 \leq P_{out} \leq 13$	$0.3 \leq V_{dc} \leq 0.65$
FZ600R17KE4	$9.36 \leq P_{out} \leq 212.5$	$0.65 < V_{dc} \leq 1.7$
FZ750R65KE3	$153 \leq P_{out} \leq 1219$	$1.7 < V_{dc} \leq 6.5$

Table 4. Semiconductor devices for 3L-NPC and 3L-FC converters.

Semiconductor Device	Output Power Range (kW)	DC-Link Voltage Range (kV)
RGSX5T565DHR	$1.08 \leq P_{out} \leq 26$	$0.6 \leq V_{dc} \leq 1.3$
FZ600R17KE4	$18.72 \leq P_{out} \leq 425$	$1.3 < V_{dc} \leq 3.4$
FZ750R65KE3	$306 \leq P_{out} \leq 2440$	$3.4 < V_{dc} \leq 13$

The simulations were organized so that the efficiency values were obtained for the converter operating in steady state, with a total simulation time of 0.04 s and models that accurately represents the power electronics converters with detailed switching behavior. The variable-step variable-order numerical solver ode15s, specifically designed to solve ordinary differential equations and differential-algebraic equations, is used to implement the simulations. The implicit nature of the ode15s can manage the complex interactions between the passive and active components of each topology, with a max step size of 10^{-5} s and relative tolerance of 10^{-3}.

The methodology for implementing the simulations is based on a systematic nested loop approach over the maximum and minimum ranges of each topology for a specific semiconductor device. For example, for the 3L-CHB operating with the FZ600R17KE4, the minimum and maximum values of output power are 9.36 kW and 212.5 kW, respectively. The script begins by iterating the DC-link voltage, varying it from its minimum value to its maximum value, divided into five steps. For each DC-link voltage value, the script introduces a nested loop to vary the power, also divided into five steps. Subsequently, within each combination of DC-link voltage and output power, the modulation index is varied from 0.8 to 1 across five steps. This process continues until all variables are included. This methodology is repeated three times for all topologies, for each semiconductor device. This meticulously detailed multi-dimensional parameter sweep allows the evaluation of each converter, with a specific semiconductor device, under a wide range of different operating conditions.

The setup of the DNN-based surrogate model is specified in Table 5. In this work, 45,000 simulations were performed for each topology presented in Figure 2, with multiple different combinations of input parameters, according to the limits presented in Table 2. The obtained efficiency values y_k, shown in Figure 6 for each configuration, were stored in a dataset with their respective simulation input parameters x_k, for $k = 1, 2, \ldots, n$. The frequency distribution of the efficiency values can be visualized in the histogram of Figure 7, where it can be observed that although most simulations present high efficiency, there is still a considerable amount of data with relatively low efficiency, where a detailed analysis needs to be performed to select topologies with better performance.

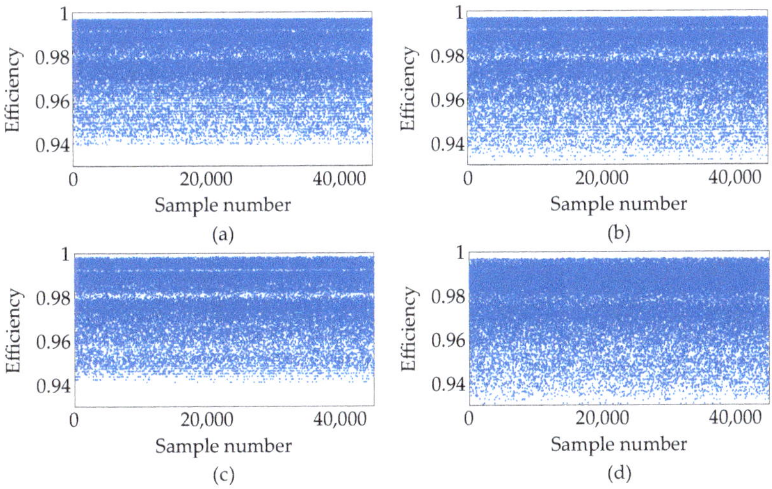

Figure 6. Efficiency dataset for each topology. (**a**) 2L. (**b**) 3L-CHB. (**c**) 3L-NPC. (**d**) 3L-FC.

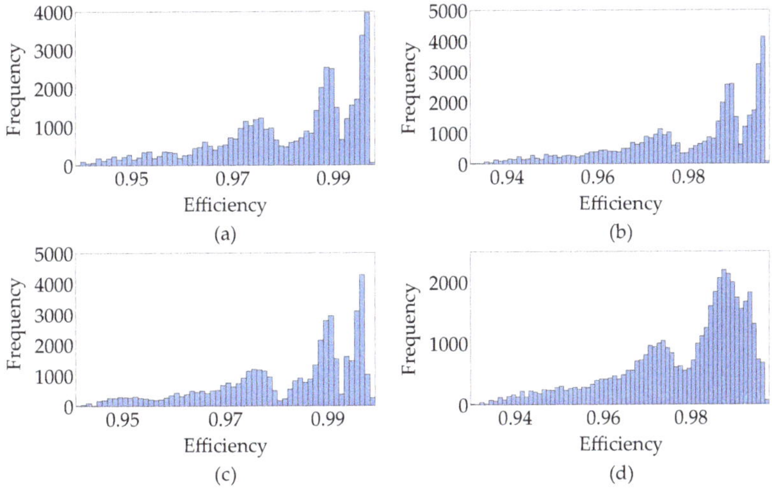

Figure 7. Efficiency values based on frequency of occurrence. (**a**) 2L. (**b**) 3L-CHB. (**c**) 3L-NPC. (**d**) 3L-FC.

To demonstrate the accuracy of the surrogate models, 100 simulations were performed for each topology, with random parameters within the limits defined in Table 2. The efficiency values obtained through the surrogate models exhibit a similar pattern when compared to the simulated ones, according to Figure 8. Based on these results, the absolute errors $e_{\eta,k}$ between the simulated values and the predicted values can be obtained for each topology by:

$$e_{\eta,k} = |\eta_{p,k} - \eta_{s,k}|, k = 1, 2, 3, \ldots, 100, \tag{4}$$

where $\eta_{p,k}$ and $\eta_{s,k}$ are the predicted and simulated efficiency values for the sample number k. This result shows that the surrogate model is capable of predicting efficiency values with a maximum error of 0.006 for this set of data samples, according to Figure 9.

Table 5. Setup of the DNN-based surrogate model.

Parameter	Value
Input size	5
Output size	1
Learning rate	0.001
Hidden layer 1	32
Hidden layer 2	32
Activation function	Relu
Optimizer	Adam
Batch size	64
Loss	MSE

The output data of the surrogate models for this set of simulations can be evaluated through the coefficient of determination R^2_η [26]. This coefficient is a statistical measure that indicates the accuracy of the surrogate models in predicting the correct outcome and it is defined by:

$$R^2_\eta = 1 - \frac{\sum_{k=1}^{100} (\eta_{s,k} - \eta_{p,k})^2}{\sum_{k=1}^{100} (\eta_{s,k} - \overline{\eta}_s)}, \quad (5)$$

where $\overline{\eta}_s$ is the average simulated efficiency value obtained among the 100 considered samples. The closer R^2_η is to unity, the better the predicted data fit the model. Conversely, a null R^2_η value indicates that the predictions of the surrogate model do not correspond to the actual data values. The R^2_η values for the topologies presented in Figure 2 are presented in Table 6, demonstrating high reliability of the surrogate models in predicting the simulated values.

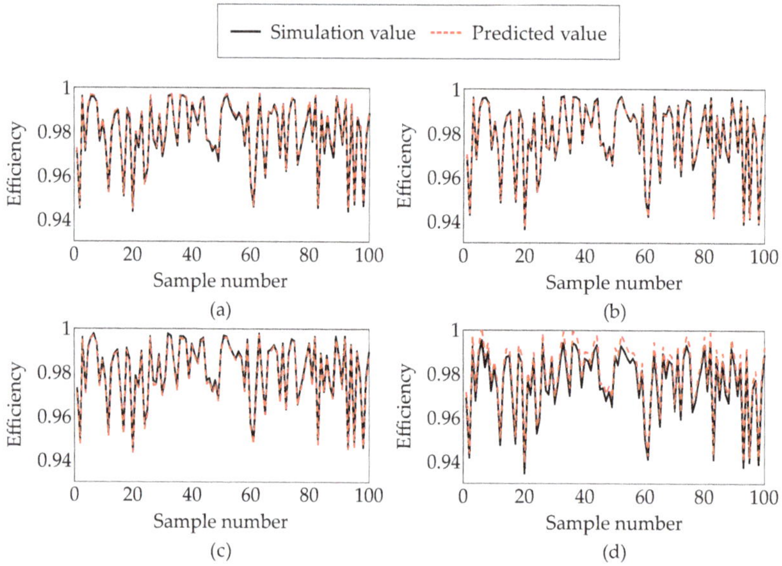

Figure 8. Efficiency simulation and predicted values for 100 data samples. (**a**) 2L. (**b**) 3L-CHB. (**c**) 3L-NPC. (**d**) 3L-FC.

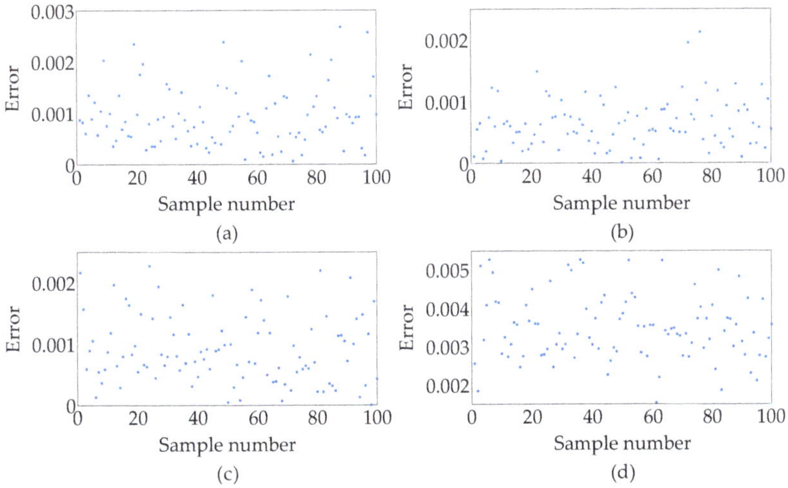

Figure 9. Efficiency prediction error for 100 data samples. (**a**) 2L. (**b**) 3L-CHB. (**c**) 3L-NPC. (**d**) 3L-FC.

Table 6. Coefficient of determination R^2_η for each converter topology.

Converter Topology	R^2_η
2L	0.994
3L-CHB	0.994
3L-NPC	0.937
3L-FC	0.997

4. Case Study: Extending Predictive Capabilities for Additional Performance Indicators for the 2L Converter

The prediction of the efficiency is an important factor to determine the best topology for a given application, but it represents just one element of the converter overall performance. To achieve a better understanding of the best topology for a given application, it is essential to predict additional parameters such as the semiconductors junction temperature and the total harmonic distortion (THD). In this section, the DNN-based surrogate models are extended to cover these additional parameters for the 2L topology.

In this scenario, to determine the semiconductors junction temperature, a larger number of input variables x_k needs to be considered, as stated in Table 7, which increases the simulation complexity. Furthermore, in thermal simulations, power losses are the most critical input [27]. Thus, conduction and switching losses of the IGBTs are considered at this stage and are adjusted according to the junction temperature and collector current of the semiconductors devices. To address these issues, the software PSIM 2021b was integrated with MATLAB/Simulink through the SimCoupler Module [28]. The PSIM Thermal Module [29] can provide feedback to electrical simulation. So, as the temperature of the devices increases, their characteristics adjust accordingly, affecting the behavior of the electrical circuit. In total, 30,000 simulations were performed using the semiconductor FF450R12KE4, available in the PSIM library. The simulations were configured with a total time of 2 s and a time step of 10^{-6} s to determine the final steady-state temperature of the IGBTs. PSIM uses the backward Euler method as its primary solver for circuit simulation [30] to effectively manage the nonlinearities of the switching devices as diodes and transistors.

Table 7. Input parameters for temperature prediction.

Parameter	Operational Boundary of Input Parameters
DC-link voltage (V)	[100, 800]
Output power (kW)	[0.23, 69]
Amplitude modulation index	[0.8, 1]
Switching frequency (kHz)	[1, 4]
Output power factor	[0.8, 1]
Gate resistance (Ω)	[1, 4]
Ambient temperature (°C)	[0, 40]
Thermal resistance heatsink to ambient (°C/W)	[0.1, 0.6]
Thermal resistance case to heatsink (°C/W)	[0.05, 0.3]

However, the obtained dataset presents some peculiarities. The semiconductors temperatures are determined based on their respective power losses. Therefore, considerable power losses values can theoretically result in higher temperature values than the semiconductors can withstand. To prevent these inconsistent values from being used to obtain the converter surrogate model, 8161 simulations in which the junction temperature of the semiconductor devices S_1 or $\overline{S}_1$ was greater than 200 °C were removed from the dataset, so that 21,839 data points were used to obtain the surrogate model, as shown in Figure 10.

The histogram of Figure 11 shows the frequency distribution of the temperature values. According to [31], the probability of semiconductor failure increases with higher operating temperatures. Given the substantial effort required to determine the junction temperature of semiconductor devices, the development of a surrogate model can significantly reduce the time needed to identify the topology with the lowest operating temperatures.

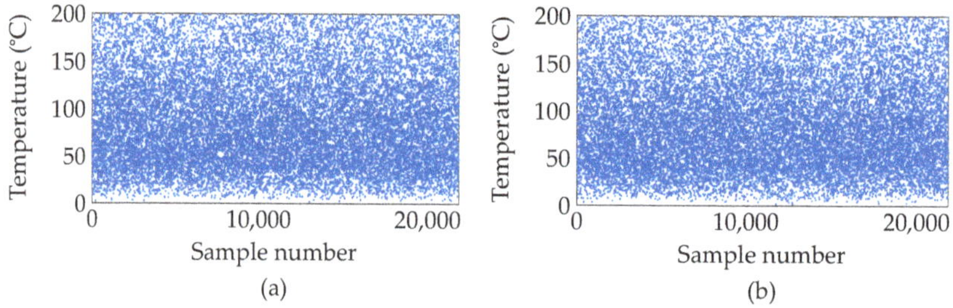

Figure 10. Temperature dataset for the 2L converter. (**a**) Semiconductor S_1. (**b**) Semiconductor $\overline{S}_1$.

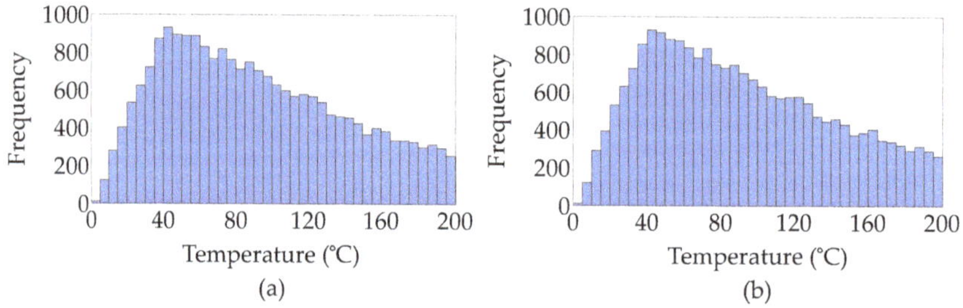

Figure 11. Temperature values based on frequency of occurrence. (**a**) Semiconductor S_1. (**b**) Semiconductor $\overline{S}_1$.

The precision of the surrogate model is shown in Figure 12, where 100 temperature values obtained from simulations were compared with the model results. The absolute errors $e_{T,k}$ between the temperature simulated values and predicted values are determined according to:

$$e_{T,k} = |T_{p,k} - T_{s,k}|, k = 1, 2, 3, \ldots, 100, \quad (6)$$

where $T_{p,k}$ and $T_{s,k}$ are the predicted and simulated temperature values for the sample number k.

These results indicate that even with a reduced dataset of 21,839 samples, the surrogate model can accurately estimate the temperature, as the maximum error for the data samples is less than 5 °C, as shown in Figure 13. The preciseness of the surrogate model for temperature prediction can be evaluated separately for S_1 and $\overline{S}_1$ through the coefficient of determination R_T^2 according to:

$$R_T^2 = 1 - \frac{\sum_{k=1}^{100} (T_{s,k} - T_{p,k})^2}{\sum_{k=1}^{100} (T_{s,k} - \overline{T}_s)}, \tag{7}$$

where $\overline{T}_s$ is the average simulated temperature value obtained among the 100 considered simulations. The surrogate models R_T^2 for both semiconductor devices (S_1 and $\overline{S}_1$) result in the same value of 0.998, indicating high reliability for temperature prediciton.

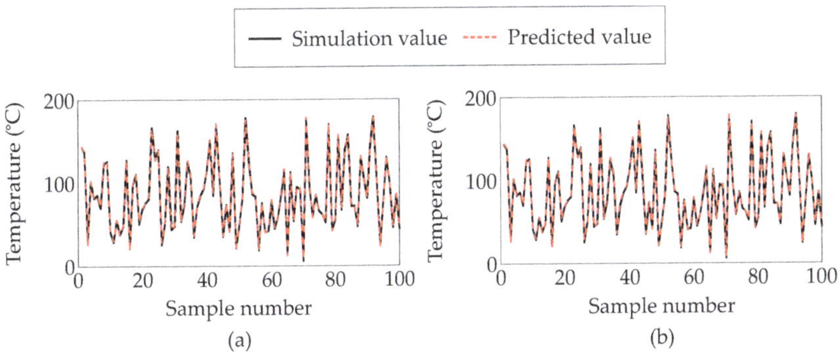

Figure 12. Temperature simulated and predicted values for 100 data samples. (**a**) Semiconductor S_1. (**b**) Semiconductor $\overline{S}_1$.

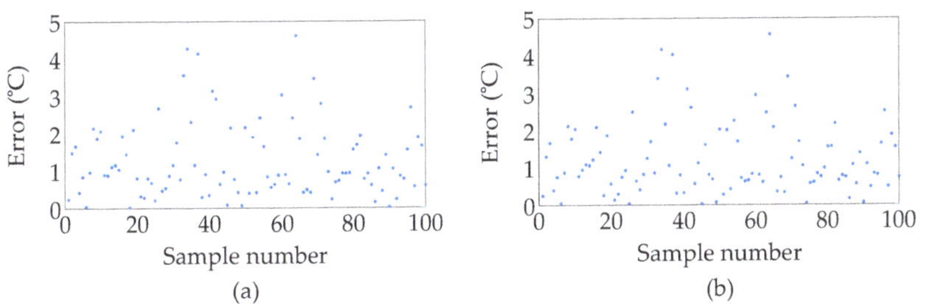

Figure 13. Temperature prediction error for 100 data samples. (**a**) Semiconductor S_1. (**b**) Semiconductor $\overline{S}_1$.

Furthermore, the THD of the output current can be a parameter of interest when a converter is designed. The simulation input parameters to obtain the THD are presented in Table 8. The filter inductor strongly affects the output current THD and this component must be properly designed to meet the IEEE 1547 criteria [32], which regulates the interconnection of distributed resources with the electric power grid. According to this standard, the maximum current THD injected into the grid by any load must not exceed 5%.

Given the distinct patterns of higher and lower THD values, the dataset was divided into two groups: low THD, for THD values less than or equal to 0.2, and high THD, for THD values above 0.2, according to Figure 14. Given the data heterogeneity in these scenarios, this division greatly enhances the accuracy of THD values prediction. The frequency distribution of the dataset is shown in Figure 15. Note that a considerable amount of the dataset exhibits current distortion values above 5%, so implementing a surrogate model for each topology can help to identify the converter with lower THD.

Table 8. Input parameters for THD prediction.

Parameter	Operational Boundary of Input Parameters
DC-link voltage (V)	[100, 800]
Output power (kW)	[0.23, 69]
Amplitude modulation index	[0.8, 1]
Switching frequency (kHz)	[1, 4]
Gate resistance (Ω)	[1, 4]
Ambient temperature (°C)	[0, 40]
Thermal resistance heatsink to ambient (°C/W)	[0.1, 0.6]
Thermal resistance case to heatsink (°C/W)	[0.05, 0.3]
Filter inductor (mH)	[0.003, 26]

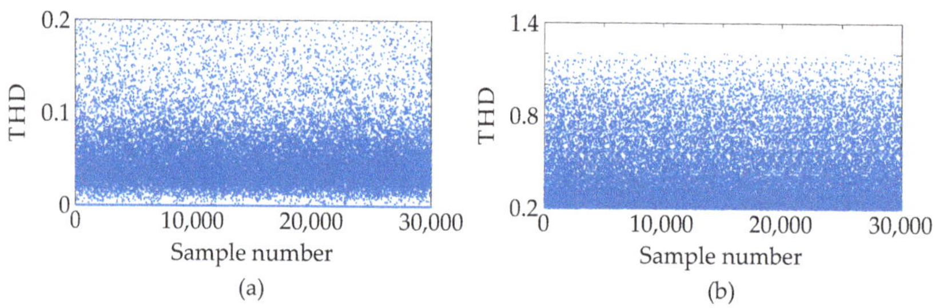

Figure 14. THD dataset for the 2L converter. (**a**) Low THD. (**b**) High THD.

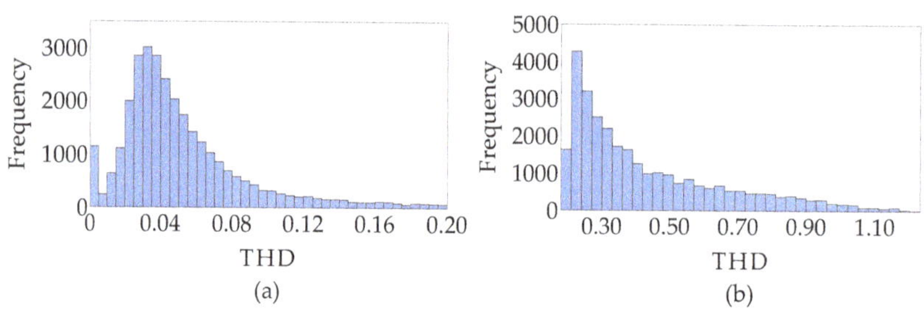

Figure 15. THD values based on frequency of occurrence. (**a**) Low THD. (**b**) High THD.

The accuracy of the surrogate model is illustrated in Figure 16, where 100 THD values obtained from simulations are compared with the surrogate model predictions. The absolute errors $e_{THD,k}$ between simulated and predicted THD values are determined as follows:

$$e_{THD,k} = |THD_{p,k} - THD_{s,k}|, k = 1, 2, 3, \ldots, 100, \quad (8)$$

where $THD_{p,k}$ and $THD_{s,k}$ represent the predicted and simulated THD values, respectively, for the sample number k. The results suggest that the surrogate models can also estimate the THD with a maximum error of 0.04, according to Figure 17.

The coefficient of determination R^2_{THD} is determined separately for low THD results and for high THD results and is obtained through:

$$R^2_{THD} = 1 - \frac{\sum_{k=1}^{100}(THD_{s,k} - THD_{p,k})^2}{\sum_{k=1}^{100}(THD_{s,k} - \overline{THD}_s)}, \quad (9)$$

where $\overline{THD}_s$ is the average THD value for 100 considered samples (of high THD values or low THD values). The coefficients of determination resulted in 0.992 and 0.863 for the high THD and the low THD predictions, respectively. Note that, although the low THD values have less accuracy, the predictions still present acceptable results.

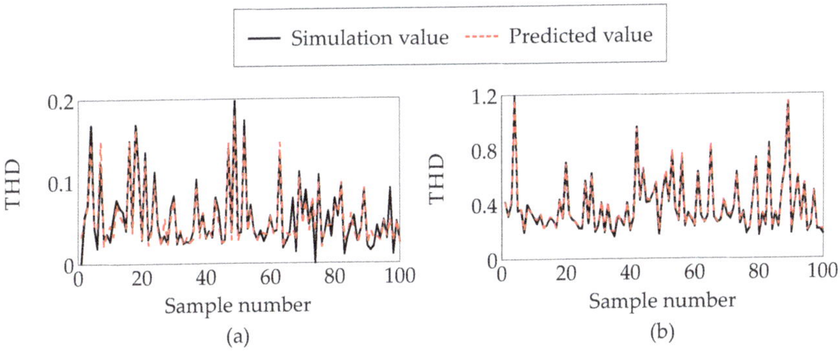

Figure 16. THD simulated and predicted values for 100 data samples. (**a**) Low THD. (**b**) High THD.

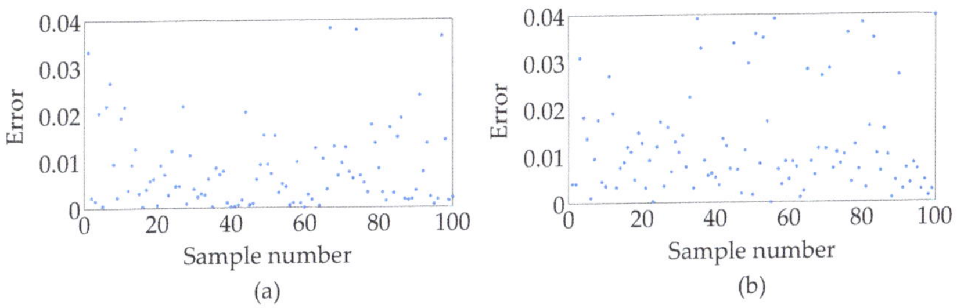

Figure 17. THD prediction error for 100 data samples. (**a**) Low THD. (**b**) High THD.

5. Results

Simulation results were obtained for all topologies presented in Figure 2. The simulations for the efficiency dataset generation were performed using the MATLAB/Simulink software, using the parameters defined in Table 2. The PSIM software was included to obtain the results for the 2L converter junction temperature and output current THD, whose input parameters are defined in Table 7 and Table 8, respectively.

Four surrogate models are generated to determine the efficiency of each topology presented in Figure 2. The performance of these models are illustrated in Figure 18. The performance of these models are illustrated in Figure 18. In this case, a dataset with 7000 simulated values and predicted values for each topology (a total of 28,000 simulated and predicted values for all four topologies) is prepared and the absolute errors are determined for each sample according to Equation (4), for $k = 7000$. Then, these errors are displayed based on the frequency of occurrence within specified ranges. It can be observed that most of the test data have a small prediction error.

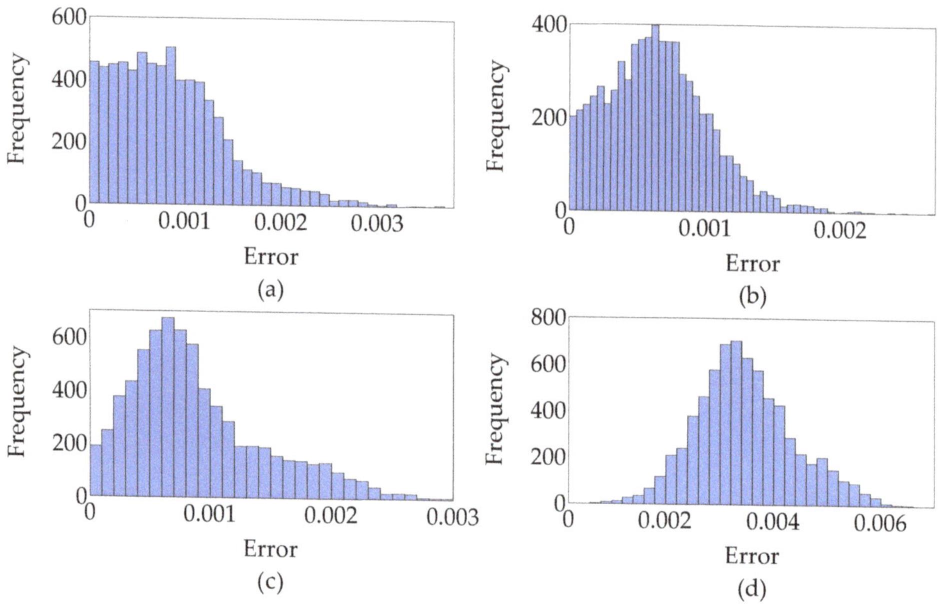

Figure 18. Absolute errors values between efficiency simulated values and predicted values based on frequency of occurrence. (**a**) 2L. (**b**) 3L-CHB. (**c**) 3L-NPC. (**d**) 3L-FC.

The maximum error values between simulated and predicted efficiencies for each converter are summarized in Table 9. Therefore, it is possible to automatically define the highest efficiency converter configuration for a specific set of input parameters without performing any simulation. All proposed surrogate models provide acceptable results, with a maximum error of 0.0068.

Table 9. Maximum predicted value error for each topology.

Converter	Maximum Error
2L	0.0037
3L-CHB	0.0027
3L-NPC	0.0030
3L-FC	0.0068

A semiconductor junction temperature dataset was generated for the 2L converter, by integrating the Matlab/Simulink R2023b software with PSIM. The performance of this model is illustrated in Figure 19, where the absolute error between simulated and predicted values is shown for 4500 data points. It can be observed that, for most cases, the error value does not exceed 5 °C. A similar approach is used for the THD using the same number of data points, as shown in Figure 20. The majority of error data do not exceed 0.04 for both low THD and high THD.

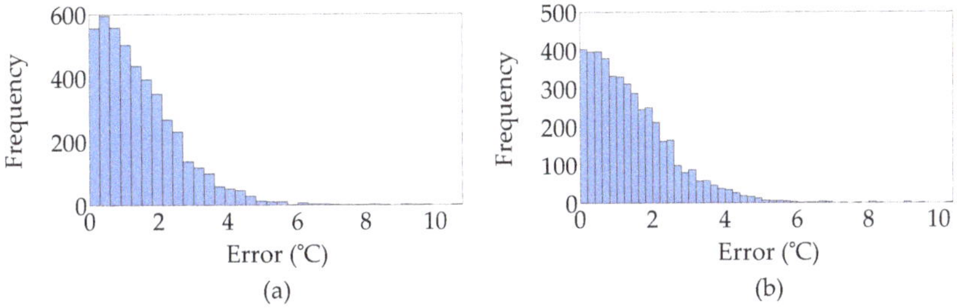

Figure 19. Absolute error values between temperature simulated values and predicted values based on frequency of occurrence for the 2L converter. (**a**) Semiconductor S_1. (**b**) Semiconductor $\overline{S}_1$.

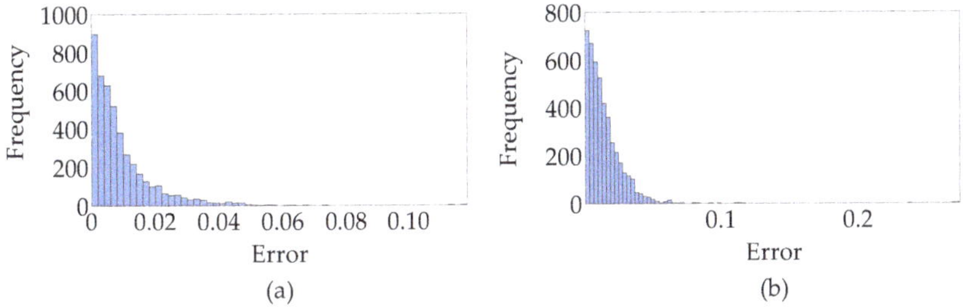

Figure 20. Absolute error values between THD simulated values and predicted values based on frequency of occurrence for the 2L converter. (**a**) Low THD. (**b**) High THD.

6. Conclusions

This paper proposed an AI-assisted methodology to accelerate the DC/AC power converter topology selection using separate DNN-based surrogate models for each of the classic topologies: 2L, 3L-NPC, 3L-FC, and 3L-CHB. Although hardware validation in demonstrating real-world applicability is recognized, this study is focused on a simulation-based analysis, which aligns with the initial steps of the converter topology definition process. Furthermore, the focus on efficiency estimation was intended to address the these critical initial steps of converter selection process, and stability and disturbance analyses are beyond the primary scope of this simulation-based study. Firstly, 45,000 distinct scenarios were simulated for each converter using MATLAB/Simulink, and an efficiency dataset for each converter was generated to train its respective surrogate model. From that, the weights of the training agent were updated and the efficiency of each topology could be estimated without any simulation. A similar approach was conducted for a case study for the 2L converter, to predict its semiconductors junction temperature and output current distortion. The performance of this method was validated through the comparison of simulated and predicted efficiency values, which demonstrated a high degree of reliability. This capability not only reduces the chance of choosing suboptimal topologies but also accelerates the selection of the DC/AC converter configuration.

Author Contributions: Conceptualization, W.S.; methodology, G.A.S. and V.-H.B.; software, G.A.S. and V.-H.B.; validation, V.-H.B.; formal analysis, G.A.S.; investigation, G.A.S. and V.-H.B.; writing—original draft preparation, G.A.S.; writing—review and editing, V.-H.B. and W.S.; supervision, W.S.; project administration, W.S.; funding acquisition, W.S. All authors have read and agreed to the published version of the manuscript.

Funding: This research was funded by the project "Commercializing ML-ACCEPT Software Machine-Learning-Enhanced Automated Circuit Configuration", by the Michigan Translation Research and Commercialization (MTRAC) for Advanced Transportation Innovation Hub, 2023–2024.

Data Availability Statement: The original contributions presented in the study are included in the article, further inquiries can be directed to the corresponding author.

Conflicts of Interest: The authors declare no conflicts of interest.

References

1. Choudhury, S.; Bajaj, M.; Dash, T.; Kamel, S.; Jurado, F. Multilevel Inverter: A Survey on Classical and Advanced Topologies, Control Schemes, Applications to Power System and Future Prospects. *Energies* **2021**, *14*, 5773. [CrossRef]
2. Bughneda, A.; Salem, M.; Richelli, A.; Ishak, D.; Alatai, S. Review of Multilevel Inverters for PV Energy System Applications. *Energies* **2021**, *14*, 1585. [CrossRef]
3. Ali Khan, M.Y.; Liu, H.; Yang, Z.; Yuan, X. A Comprehensive Review on Grid Connected Photovoltaic Inverters, Their Modulation Techniques, and Control Strategies. *Energies* **2020**, *13*, 4185. [CrossRef]
4. Haq, S.; Biswas, S.P.; Hosain, M.K.; Rahman, M.A.; Islam, M.R.; Jahan, S. A Modular Multilevel Converter with an Advanced PWM Control Technique for Grid-Tied Photovoltaic System. *Energies* **2021**, *14*, 331. [CrossRef]
5. Prado, E.O.; Bolsi, P.C.; Sartori, H.C.; Pinheiro, J.R. Design of Uninterruptible Power Supply Inverters for Different Modulation Techniques Using Pareto Front for Cost and Efficiency Optimization. *Energies* **2023**, *16*, 1314. [CrossRef]
6. Saccol, G.A.; Vilerá, K.V.; Carnielutti, F.M.; Rech, C. Power Distribution Method for Single-Phase Cascaded Multilevel Converters Based on Space Vector Modulation. *IEEE J. Emerg. Sel. Top. Power Electron.* **2023**, *11*, 3445–3456. [CrossRef]
7. Leon, J.I.; Vazquez, S.; Franquelo, L.G. Multilevel Converters: Control and Modulation Techniques for Their Operation and Industrial Applications. *Proc. IEEE* **2017**, *11*, 2066–2081. [CrossRef]
8. Texas Instruments Design Guide: TIDM-HV-1PH-DCAC Voltage Source Inverter Reference Design. Available online: https://www.ti.com/lit/ug/tiduay6e/tiduay6e.pdf?ts=1732406676905&ref_url=https%253A%252F%252Fwww.google.com%252F (accessed on 24 November 2024).
9. Hasan, M.A.; Vemula, N.K.; Devarapalli, R.; Knypiński, Ł. Investigation into PV Inverter Topologies from the Standards Compliance Viewpoint. *Energies* **2024**, *17*, 3879. [CrossRef]
10. Zhang, X.; Zhao, T.; Mao, W.; Tan, D.; Chang, L. Multilevel Inverters for Grid-Connected Photovoltaic Applications: Examining Emerging Trends. *IEEE Power Electron. Mag.* **2018**, *5*, 32–41. [CrossRef]
11. Rauf, A.M.; Abdel-Monem, M.; Geury, T.; Hegazy, O. A Review on Multilevel Converters for Efficient Integration of Battery Systems in Stationary Applications. *Energies* **2023**, *16*, 4133. [CrossRef]
12. Wu, T.; Wang, Z.; Ozpineci, B.; Chinthavali, M.; Campbell, S. Automated Heatsink Optimization for Air-Cooled Power Semiconductor Modules. *IEEE Trans. Power Electron.* **2019**, *34*, 5027–5031. [CrossRef]
13. Rahimpour, S.; Husev, O.; Vinnikov, D.; Kurdkandi, N.V.; Tarzamni, H. Fault Management Techniques to Enhance the Reliability of Power Electronics Converters: An Overview. *IEEE Access* **2023**, *11*, 13432–13446. [CrossRef]
14. Saridakis, S.; Koutroulis, E.; Blaabjerg, F. Optimal Design of Modern Transformerless PV Inverter Topologies. *IEEE Trans. Energy Convers.* **2013**, *28*, 394–404. [CrossRef]
15. Abdel-Aziz, A.; Elgenedy, M.A.; Williams, B. A Comparative Review of Three Different Power Inverters for DC–AC Applications. *Energies* **2023**, *16*, 7254. [CrossRef]
16. Khan, M.N.H.; Forouzesh, M.; Siwakoti, Y.P.; Li, L.; Kerekes, T.; Blaabjerg, F. Transformerless Inverter Topologies for Single-Phase Photovoltaic Systems: A Comparative Review. *IEEE J. Emerg. Sel. Top. Power Electron.* **2020**, *8*, 805–835. [CrossRef]
17. Silva, F.L.d.; Glatt, R.; Su, W.; Bui, V.-H.; Chang, F.; Chaturvedi, S.; Wang, M.; Murphey, Y.L. AutoTG: Reinforcement Learning-Based Symbolic Optimization for AI-Assisted Power Converter Design. *IEEE J. Emerg. Sel. Top. Ind. Electron.* **2024**, *5*, 680–689. [CrossRef]
18. An, Y.; Sun, X.; Ren, B.; Zhang, X. Open-Circuit Fault Diagnosis for a Modular Multilevel Converter Based on Hybrid Machine Learning. *IEEE Access* **2024**, *12*, 61529–61541. [CrossRef]
19. Bui, V.-H.; Chang, F.; Su, W.; Wang, M.; Murphey, Y.L.; Da Silva, F.L.; Huang, C.; Xue, L.; Glatt, R. Deep Neural Network-Based Surrogate Model for Optimal Component Sizing of Power Converters Using Deep Reinforcement Learning. *IEEE Access* **2022**, *10*, 78702–78712. [CrossRef]
20. Liu, G.; Kui, L.; Gao, Y.; Cui, W.; Liu, F.; Wang, W. Machine Learning-Based Optimization of Synchronous Rectification Low-Inductance Current Secondary Boost Converter (SLIC-QBC). *Energies* **2023**, *16*, 6690. [CrossRef]
21. Guillod, T.; Papamanolis, P.; Kolar, J.W. Artificial Neural Network (ANN) Based Fast and Accurate Inductor Modelling and Design. *IEEE J. Power Electron.* **2020**, *1*, 284–299. [CrossRef]
22. Zhang, Y.; Xu, Y.; Blaabjerg, F. *Chapter 11—Surrogate Models for Power Electronics Systems Applying Machine Learning Techniques*; Academic Press: Cambridge, MA, USA, 2024; pp. 333–352.
23. Leal-Romo, F.J.; Rayas-Sánchez, J.E.; Chávez-Urtado, J.L. Surrogate-Based Analysis and Design Optimization of Power Delivery Networks. *IEEE Trans. Electromagn. Compat.* **2020**, *62*, 2528–2537. [CrossRef]
24. Salem, M.; Richelli, A.; Yahya, K.; Hamidi, M.N.; Ang, T.-Z.; Alhamrouni, I. A Comprehensive Review on Multilevel Inverters for Grid-Tied System Applications. *Energies* **2022**, *15*, 6315. [CrossRef]
25. Akagi, H. Multilevel Converters: Fundamental Circuits and Systems. *Proc. IEEE* **2017**, *105*, 2048–2065. [CrossRef]
26. Qiu, B.; Zhang, J.; Yang, Y.; Qin, G.; Zhou, Z.; Ying, C. Research on Oil Well Production Prediction Based on GRU-KAN Model Optimized by PSO. *Energies* **2024**, *17*, 5502. [CrossRef]

27. Shahjalal, M.; Shams, T.; Hossain, S.B.; Ahmed, M.R.; Ahsan, M.; Haider, J.; Goswami, R.; Alam, S.B.; Iqbal, A. Thermal analysis of Si-IGBT based power electronic modules in 50kW traction inverter application. *E-Prime Adv. Electr. Eng. Electron. Energy* **2023**, *3*, 2772–6711. [CrossRef]
28. Simcoupler Module. Available online: https://www.powersimtech.com/wp-content/uploads/2020/12/PSIM-SimCoupler-Module.pdf (accessed on 23 September 2024).
29. Thermal Module. Available online: https://powersimtech.com/wp-content/uploads/2020/12/PSIM-Thermal-Module.pdf (accessed on 22 November 2024).
30. PSIM User's Manual. Available online: https://powersimtech.com/wp-content/uploads/2021/01/PSIM-User-Manual.pdf (accessed on 22 November 2024).
31. Nguyen, M.H.; Kwak, S. Enhance Reliability of Semiconductor Devices in Power Converters. *Electronics* **2020**, *9*, 2068. [CrossRef]
32. *IEEE Std 1547-2018*; IEEE Standard for Interconnection and Interoperability of Distributed Energy Resources with Associated Electric Power Systems Interfaces. IEEE: Piscataway, NJ, USA, 2018; Revision of IEEE Std 1547-2003; pp. 1–138.

energies

MDPI

Article

Controller Hardware in the Loop Platform for Evaluating Current-Sharing and Hot-Swap in Microgrids

Juan Martínez-Nolasco [1], Víctor Sámano-Ortega [1,*], Heriberto Rodriguez-Estrada [1], Mauro Santoyo-Mora [1], Elias Rodriguez-Segura [2] and José Zavala-Villalpando [1]

1 Departamento de Ingeniería Mecatrónica, Tecnológico Nacional de México/IT de Celaya, Antonio García Cubas 600, Celaya 38010, Mexico; juan.martinez@itcelaya.edu.mx (J.M.-N.)

2 Departamento de Ingeniería Eléctrica y Electrónica, Tecnológico Nacional de México/IT de Celaya, Antonio García Cubas 600, Celaya 38010, Mexico

* Correspondence: victor.samano@itcelaya.edu.mx

Abstract: Microgrids have increased in popularity thanks to both the integration of renewable energy resources and their energy distribution capability for remote locations. Moreover, the microgrids, mainly using multiple generators connected in parallel, acquire additional advantages by using both Hot-Swap and Current-Sharing techniques. This paper presents the development of a Hardware in the Loop platform to test Current-Sharing algorithms. It is reinforced that the use of a real-time simulation based on Hardware in the Loop is a viable and cost-effective alternative in the validation of controllers. The platform was developed in a graphical programming environment (LabVIEW 2015) and implemented with NI MyRIO 1900 (National Instruments Corp., Austin, TX, USA) development boards for easier reproducibility. The entire code project is openly available and provided in this paper. A system of photovoltaic energy generators was used to evaluate the performance of the HIL platform. As a result, the platform was able to reproduce a similar behavior to the photovoltaic generator, presenting average mean errors of 0.4 V and 0.2 A in its voltage and current, respectively. Additionally, the platform showed its capability to test Current-Sharing algorithms in the occurrence of Hot-Swap events. This work contributes with a validation tool for energy management systems applied to microgrids.

Keywords: current-sharing; energy management system; Hardware in the Loop; hot-swap; microgrid; photovoltaics

1. Introduction

In recent years, DC microgrids have increased in popularity as a plausible alternative to both generate and distribute energy in comparison to conventional energy systems. Among the multiple advantages of a DC microgrid (DC-μG), one that stands out is its energy generation using renewable energy resources (RES). This opens the possibility of reducing the dependency on electrical centrals using fossil fuels, thus diminishing the negative impact on the environment. Between the employed RES in a DC-μG, the use of photovoltaic panels (PV) as generators is very common due to both their financial and technical benefits [1,2]. A microgrid (μG) is structured with five key elements interconnected in a common power bus: distributed generators, energy storage devices, an interconnection system with the AC main grid, loads, and a control unit [3]. The power electronic converters (PECs) play a key role inside a DC-μG. The PECs add control and flexibility to the grid since they act as the main interface between the interconnected elements inside the DC-μG and the distribution bus for the connected loads. Moreover, the DC microgrids (DC-μGs) can deliver successfully a power demand in a controlled and stable manner; therefore, they can be considered as an alternative power generation system for remote locations, data centers, and buildings [4].

The operation of a DC-μG requires a hierarchical control based on two main control levels or layers, i.e., a local control (primary layer) and a global or supervisory control (secondary layer) [5,6]. On the one hand, the local control has the main purpose of regulating the distribution bus voltage; this level of control can be enhanced by including strategies such as Hot-Swap (HS) and Current-Sharing (CS) [7]. On the other hand, the supervisory control could integrate a set of secondary controls to minimize oscillations in the voltage bus, or a set of tertiary controls to handle a connection/disconnection to the AC main grid. Additionally, the supervisory control level can include an Energy Management System (EMS), which is expected to achieve ecological, economical, and calendrical goals [8,9]. As stated by the IEC 61970 norm, an EMS is defined as "a computer system comprising a software platform providing essential support services and a set of applications providing the functionality needed for the effective operation of electrical generation and transmission facilities to assure adequate security of energy supply at minimum cost" [10]. Assuredly, the EMS is responsible for guaranteeing a continuous energy supply in the DC-μG. This becomes possible after managing the available energy in the grid among the generation resources, the storage devices, and the AC main grid. To assure an optimal operation of the DC-μG, the EMS must maximize the usage of RES by exploiting their highest efficiency and reducing the energy losses. Following this optimal operation, the EMS can take account of the emitted greenhouse gases by the grid, too [11]. Based on these approaches for the supervisory level, the DC-μG could have multiple state variants during its operation [12]. Consequently, an EMS must determine the optimal power produced by each generation source while it maintains the stability of the DC-μG [13].

A DC-μG can operate in two different modes: connected mode and islanded mode. In the first case, the DC-μG interchanges energy with the AC main grid through a common coupling point. As this interchange is bidirectional, it is possible to maintain a DC bus regulated in cases such as an energy deficit or an energy surplus. In the islanded mode, the DC-μG must maintain an energy balance between its components by the time that the AC main grid is not accessible. Here, the DC-μG management defines as main goals to preserve a regulated setpoint in the DC voltage bus and to couple the energy demand based on the available energetic resources [14]. The islanded DC-μGs, whose RES include PV and wind turbines, present as the best alternative to satisfy the energy demand in remote locations. However, the unpredictability and variable nature of the RES reflects uncertainty during their operation; therefore, an adequate EMS becomes strictly necessary [15]. Above all, the design of EMS for DC-μGs operating in islanded mode (e.g., the development of electric vehicles and energy supply systems for remote locations) has found new challenges in the creation of new environment-friendly technologies. Consequently, the design of EMS for DC-μGs focuses on a wider addition of RES in the electrical systems in a safe, stable, reliable, robust, and coordinated manner [16].

Due to the number of elements composing a DC-μG, as well as its operational configurations, an EMS must be validated before its implementation on the field. The real-time simulation based on Hardware in the Loop (HIL) appears as an affordable, viable, and commonly used technique to both designing and testing DC-μGs. This technique allows the developer to evaluate a wide variety of conditions during the operation of energy systems, the debugging and optimizing of controllers, and the viability analysis of electrical components connected to the system [17–20]. Thanks to its advantages, the HIL has been recently employed to validate control techniques in the elements forming a μG, as presented by the following authors. First, in [21] is presented the validation of a controller for a μG. Here, the μG is formed by a diesel generator, a PV system, a battery-based storage system, and electric loads. The diesel generator is emulated and connected to the μG with the Power HIL technique; the emulation employs a digital real-time simulator NovaCor from RTDS Technologies. Based on the close correlation between the Power HIL and its real counterpart, the authors conclude that the Power HIL provides flexibility in validation tests for controllers. In [22], a Current-Sharing Controller (CS-C) is proposed to limit the currents in a low-voltage DC-μG that contains a set of PV generators connected in parallel. The controller was validated with a real-time emulator Typhoon HIL 602, which reproduces the behavior of PV arrays, a battery-based storage system, Buck PECs, and the AC main grid. This work concludes that HIL simulations are useful to validate the effectiveness of CS strategies. Similarly, in [23] was employed the real-time emulator Typhoon HIL to simulate an AC-μG with a 5 μs latency. Here, the authors describe a validation of both centralized and decentralized frequency control schemes for an islanded mode operation. Finally, in [24] is described the conversion control of a wind turbine system and its validation with HIL. The wind turbine system is emulated with two Back-To-Back triphasic PECs, where one PEC emulates the grid side, while the other emulates the machine side.

Besides all the inherent advantages in the use of HIL, several commercial platforms are expensive and out of reach for some researchers. To tackle this problem, several proposals have been developed as low-cost alternatives. As an example, in [25] is presented the use of FPGAs to develop HIL simulations, where the analysis of the HIL development included both the numerical data types and the description method. In this work, it is concluded that the development of HIL simulations using a graphical language such as LabVIEW reduces the effort and time for its implementation. With an educative purpose, the authors in [26] present the development of a HIL platform using a dual-core digital signal processor TMS320F28379D (Texas Instruments Incorporated, Dallas, TX, USA) and an Atmega328P microcontroller (Microchip Technology Inc., Chandler, AZ, USA). To demonstrate the helpfulness of this platform, the authors emulate a Buck PEC and a two-container system. Finally, in [27] is described the implementation of a HIL using two DUE Arduino to simulate the control of dynamic systems. One of the Arduinos emulates a dynamic system while the other works as a DAQ and as a controller at the same time. This platform supports the emulation of dynamic systems such as an RLC circuit, a twin rotor MIMO system, and a microgrid.

Usually, a DC-μG must count with a mechanism to monitor the energy demand on the grid while it has more than one generation system. This guarantees a proportional distribution of the energy demand. In fact, this mechanism can be included as part of the EMS and can be designed to cover the energy demand under diverse circumstances. This accomplishment becomes possible as the EMS considers the energy production capabilities by each generation system in the grid. On the one hand, the monitoring mechanism can intercommunicate globally all the generation systems. Consequently, the EMS also performs global actions to accomplish with the energy demand in the grid. On the other hand, the monitoring mechanism can be coupled to each generation system to be locally controlled. In this last approach, the generation systems do not depend on a direct communication with the EMS to establish an optimal power output [28]. Generally, a PV is directly connected to a power converter, whose joint action can produce a series of perturbances after being connected/disconnected to the DC bus in the grid. Therefore, it is recommended to implement an HS technique as part of the energy management in a DC-μG that has at least two photovoltaic generators parallelly interconnected [29]. Additionally, the HS improves the DC-μG operation by adding a redundancy in the PEC, which increments the reliability, adds up fault tolerance, increases the system's flexibility and robustness, and reduces the updating and maintaining times of the grid. Another common strategy for PECs connected in parallel is the CS. This strategy is focused on keeping all the converters contributing with an equitable fraction from the demanded current by the load, while the voltage in the DC bus is always maintained regulated [30]. To maintain equitable output currents from the PECs, the CS-C requires a communication channel between the PECs to monitor their available current [31]. Consequently, the equitable current demand from the PECs avoids a working overload, which results in a longer lifespan of their components, and the system maintains a stability thanks to a well-regulated DC bus. In the literature can be found some works where HS and CS strategies are employed. As an example, in [32] is presented a CS technique based on the injection of an AC signal to the DC bus and a controller. Here, the supplied energy from multiple PECs is regulated based on the induced reactive power by an AC signal. This system was evaluated in the face of HS connections.

Similarly, in [33] is proposed a secondary layer that operates based on a consensus scheme, which is applied for a Current-Sharing management for distributed generators connected in parallel; the scheme was validated in the face of HS connections also. In [34] is presented an adaptive droop control approach for CS and voltage regulation in DC-μG applications. A virtual resistor is used to modify the reference voltage on each parallel converter; thus, the shared current error is reduced. The adaptive control approach was tested using a real-time simulator OPAL-RT OP4510, achieving a significative error reduction for different loading conditions. Similarly, in [35] is proposed a consensus algorithm based on a model predictive control for both CS and voltage regulation of a DC-μG. The functionality of the controller was demonstrated with a DC-μG that contained four distributed generators interconnected; the DC-μG allowed for the configuration of the input constraints for each generator. Finally, a multi-bus DC-μG with a novel control scheme is proposed in [36]. The novel scheme considers simultaneously both voltage regulation and CS for the grid. To evaluate the effectiveness of the novel scheme, the authors considered a grid with six distributed generators with six interconnected buses through transmission lines.

The aim of this work is to describe the development of a HIL platform to implement and validate CS algorithms. Particularly, the platform emulates three photovoltaic energy generator modules (PEGM) connected in parallel to a DC bus. Each PEGM was built with a 250 W PV, a DC/DC boost PEC with a PI controller for voltage control, a CS-C, and a switching device to produce an HS effect. In addition to having the PEGM, the DC-μG includes an energy storage system and the interconnectivity with the AC main grid. The management of a DC-μG requires that each of its elements has the capability to commute between different operating modes. For example, a photovoltaic generation system can switch from a Maximum Power Point Tracking (MPPT) mode to a voltage regulation mode in the DC bus, or vice versa. In this work, the proposed CS-C is analyzed assuming that the PEGM regulate the DC bus in the DC-μG. Following this, the inclusion of both CS and HS techniques presents several advantages. Generally, it avoids either the overcharge or inactivity of any PV. Furthermore, it allows the modules to connect/disconnect to the DC bus without causing any disturbance. Finally, it keeps a regulated voltage in the face of load changes. The main objective of this research is to provide a platform which could be useful to validate a CS-C in the face of an HS and before being implemented in an EMS. Nevertheless, other operating modes of the PEGM, such as the MPPT, are still outside of the scope for this work.

To validate the HIL platform, an experiment was conducted under the following conditions: a variable working load and an HS connection/disconnection of the PEGM. As a result, the platform emulated the output voltage and the inductor current of the PEGM with mean absolute errors of 0.4 V and 0.2 A, respectively. No doubt, it was possible to conclude the usefulness of the platform to validate the CS-C.

Admittedly, a CS-C not only contributes with a satisfactory coupling of RES, but also enhances their performance. This enhanced performance is a consequence of a reliable, flexible, and robust CS controller. An additional effect of this enhancement can be seen with the reduction of perturbances in the DC bus. With a technological tool such as a HIL platform, the researcher will be able to identify and to correct failures of a CS algorithm, even before implementing it on the field. It is important to mention that this platform contributes to the development and evaluation of EMS for DC-μG during an islanded mode, i.e., the voltage in the bus exclusively depends on the RES (e.g., a main AC grid shortage, or the absence of the main AC grid in remote locations). The islanded mode becomes a critical state for a DC-μG as it must be reliable and offer resilience to the grid.

The HIL platform was built using the graphical programming language of LabVIEW and NI MyRIO 1900 (National Instruments Corp., Austin, TX, USA) boards. The selection of the programming language was made by considering two key ideas: first, the easiness to be reproduced, even by not-experienced researchers in embedded systems; second, to offer a more economical alternative in comparison to dedicated HIL platforms. The complete set of codes developed for this project is available for the reader to use.

The present work is ordered as follows: Section 2 presents the complete description of the system implemented under the HIL platform. Next, Section 3 shows the proposed methodology to implement the HIL platform. This is followed by the results of the experimental stage, which are presented in Section 4. Finally, Section 5 contains the conclusions and future work.

2. System Description

The complete HIL platform, with both the CS-C and the HS, contains three PEGM connected in parallel. The proposed system in this paper provides a DC-μG with a power of 0.75 kW through a 190 V DC bus by using photovoltaic generation. The PV are placed in parallel in the DC-μG; a resistive load is also connected to the grid. This configuration of the DC-μG was used to evaluate both the CS and HS techniques, as well as the performance of the grid during an islanded mode operation. As a future work, it is proposed the addition of more elements such as a battery energy storage system, a wind turbine, and the interconnection with the main AC grid. The forthcoming actions related to the future work proposal is explained with detail in Section 5.

The proposed topology has a measuring line called Current-Sharing bus (CSB), where a voltage set point (v_s) was established. This set point is employed by an algorithm to manage the current distribution for the connected loads. All the elements of the system can be seen in Figure 1. To evaluate the CS-C under different conditions during an energy demand, a variable resistive load was connected to the DC grid. With this type of load, a starting point was established in the analysis of a DC-μG using HIL; future analyses are expected to consider both inductive and capacitive loads. Additionally, the e_P switching devices of each PEGM allows a connection/disconnection to the DC grid to produce the desired HS effect.

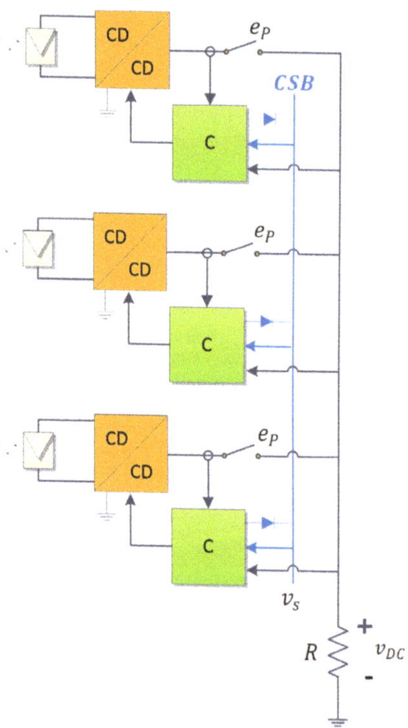

Figure 1. Developed system for CS-C validation.

As can be seen in Figure 1, each PEGM is composed by a PP, a boost DC/DC PEC, and a controller (represented by the green block C). This controller regulates the output voltage from the PEC and already includes a CS-C. The structure of the PEGM and its elements is showed in detail in Figure 2, whose description is made in the following paragraphs.

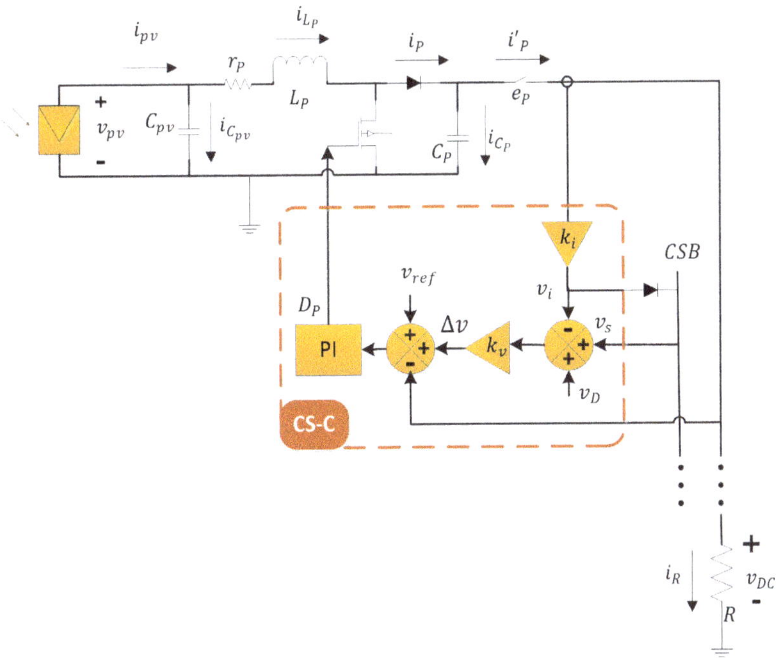

Figure 2. Photovoltaic Energy Generation Module (PEGM).

The PEGM was modeled based on a 250 W VIRTUS II PV (RENESOLA, Stamford, CT, USA). The main characteristics of this PV include a maximum power output (P_{max}) of 250 W, a maximum power voltage (V_{max}) of 30.1 V, a maximum power current (I_{max}) of 8.31 A, an open-circuit voltage (V_{OC}) of 37.4 V, and a short-circuit current (I_{SC}) of 8.83 A.

The boost PEC was designed as not isolated and to obtain the highest gain and efficiency from each PV [37]. Accordingly, the design parameters of the boost converter were defined as follows: an input voltage (V_{in}) of 30 V, an output voltage (V_{out}) of 190 V, a switching frequency (f_{sw}) of 130 kHz, and a power output (P_o) of 250 W. If the PEC receives an input voltage of 30 V to achieve a 190 V output, then it must operate with a gain of 6.33, maintaining a stability under a wide duty cycle condition [38]. Since the DC bus is considered as a large value capacitor with an inherent slow dynamic, the DC bus regulation is viable for this gain. Additionally, the converter was designed to work in a continuous conduction mode, therefore, the converter has a slow response.

The set of PEGM work collaboratively to achieve a voltage regulation in the DC bus. However, each PEGM locally regulates its outputs using a control for voltage regulation and a strategy to obtain a balanced output current. On the one hand, the voltage regulation from the boost converter is made with a PI controller set in a voltage control mode. In similar applications, it is common that the power electronic converters apply an inner loop for a current mode control. Particularly, this type of controller executes the control actions in the PEC on each commutation cycle of the power switching devices. However, the hardware employed for this HIL platform does not tolerate the development of high-switching-frequency models [39]. Then, it is not viable to implement an inner loop with the required latency. Conversely, it is possible to evaluate the proposed CS-C in this work because it has an effect on the outer loop reference in voltage mode to achieve a balanced current distribution, and the voltage dynamics in the DC bus are relatively slower than the current dynamics on each PEC. The present analysis of the CS-C can be extended to similar controllers that have an effect over the outer loop voltage without considering the inner current mode control.

On the other hand, the CS-C monitors the injected current (i'_P) to the power bus, which through the gain k_i is converted to a voltage reference (v_i); v_i is transferred through a diode to CSB. As the diodes on each converter are connected to a common bus, the CSB voltage is defined as the maximum voltage v_i minus the voltage drop offset in the diode (v_D), i.e., $v_s = v_i|_{max} - v_D$. This v_s value acts as feedback for the CS-C to be compared versus the present voltage v_i on each module. Next, a voltage increment (Δv) is computed after multiplying the k_v gain by the voltage difference between the maximum voltage v_i and the output voltage v_i of each module. The value of Δv is then added to the PEC reference voltage (v_{ref}). In this way, if a module is supplying less energy than any other module, a positive disturbance is induced to the PEC ($\Delta v > 0$) and has the effect of demanding more energy from this module. On the contrary, when the module is supplying the grid with more energy than any other module, a negative disturbance ($\Delta v < 0$) is induced to the PEC, which is interpreted into a decreasing of the supplied energy by the module. As a result, a balanced energy demand from all the connected modules to the CSB and the power bus is achieved [40]. The CS algorithm is executed continuously on each CS-C. To ease the comprehension of the algorithm, its sequence is shown in the flow chart in Figure 3. Finally, the HS connection is produced by opening and closing the e_P switch.

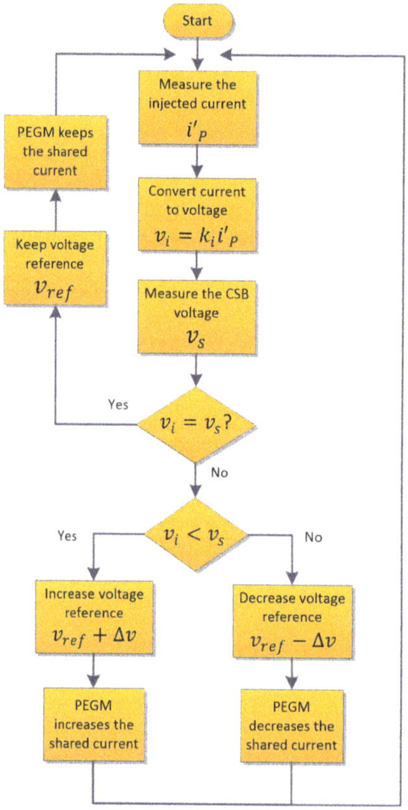

Figure 3. Current-Sharing algorithm flow diagram.

3. HIL Platform Description

Described here is the implemented methodology to both model and embed the elements that integrate the HIL platform described in the previous section.

3.1. System Modeling

A PV model was embedded using a 1024-value look-up table (LUT). These values were obtained from a Keysight E4360A (Keysight Technologies, Inc., Santa Rosa, CA, USA) emulator model; a similar approach is described in [41]. This emulator is widely used in research to substitute real PV [42–45]. The input parameters for this model are the open circuit voltage, the short circuit current, the maximum power point voltage, and the maximum power point current [46]; these values were defined based on the characteristics described in Section 2 for the 250 W VIRTUS II panel.

Similarly, to embed the DC/DC boost PEC, an average model was defined based on the circuit showed in Figure 2 [47] and deriving the discrete Equations (1)–(3). For these equations, T represents the integration period, $v_{pv}(k)$ is the voltage in the input capacitor (C_{pv}), $i_{L_P}(k)$ is the current in the L_P inductor, v_B is the voltage in the output capacitor (C_P), and $i_P(k-1)$ is the current passing through the diode. The HIL platform was implemented in a System on Chip (SoC) that includes a real-time processor (RTP) and an FPGA with a 40 MHz clock system. Particularly, these characteristics allow the device to be considered as a plausible option for developing high-latency applications with low cost; however, it is not viable to develop high-switching-frequency (>20 kHz) models due to the lack of an acceptable resolution [39]. Admittedly, this fact could be a disadvantage in comparison to dedicated HIL platforms. Nevertheless, the low cost of the hardware increases its reproducibility. For this reason, this work makes use of the average model of a converter instead of its commuted model.

$$v_{pv}(k) = v_{pv}(k-1) + \frac{T}{C_{pv}} i_{pv}(k-1) - \frac{T}{C_{pv}} i_{L_P}(k-1) \tag{1}$$

$$i_{L_P}(k) = \left(1 - \frac{r_P T}{L_P}\right) i_{L_P}(k-1) - \frac{(1-L_P)T}{L_P} v_{DC}(k-1) + \frac{T}{L_P} v_{pv}(k-1) \tag{2}$$

$$i_P(k-1) = (1-D_P) i_{L_P}(k-1) \tag{3}$$

Equation (1)–(3) describe the voltage dynamics $v_{pv}(k)$ in the input capacitor, the current $i_{L_P}(k)$ through the inductor, and the current $i_{L_P}(k)$ passing through the diode, respectively. As all the PECs are interconnected in parallel when their switches are closed, Equation (4) describes the voltage $v_{DC}(k)$ for the output capacitor, where e_iC_i is the product of both the output capacitance (C_P) of each PEC and the switching state e_i of the PEC ($e_i = 0$ when the switch is open, $e_i = 1$ otherwise).

$$v_{DC}(k) = \frac{T}{\sum e_i C_i} \sum e_i i_i(k-1) + \left(1 - \frac{T}{R \sum e_i C_i}\right) v_{DC}(k-1) \tag{4}$$

The PI controller regulating the output voltage in the PEC was embedded based on its discrete model. The model is represented by Equation (5), where P is the proportional gain, I refers to the integral gain, $u(k)$ is the control signal, and $e(k)$ is the error signal computed after Equation (6).

$$u(k) = u(k-1) + Pe(k) + (TI - P)e(k-1) \tag{5}$$

$$e(k) = \left(v_{ref} - v_{DC}\right) + \Delta v \tag{6}$$

Finally, the CSB voltage was emulated using Equation (7), where v_s is the shared voltage in the bus, k_i is the gain for shared current algorithm, $max(i_1, i_2, \ldots, i_n)$ returns the current with the highest value, and v_D represents the voltage drop offset in the diode that connects the amplified shared current and the shared bus.

$$v_s = k_i max(i_1, i_2, \ldots, i_n) - v_D \quad (7)$$

The values of the used constants from Equation (1)–(7) are shown in Table 1.

Table 1. Values of the constants included in Equations (1)–(7).

Constant	Value	Units
C_{pv}	330	μF
r_P	0.4	Ω
L_P	110	μH
T	28	μs
C_P	47	μF
P	0.004	--
I	1	--
k_i	6	V/A
k_v	2	--

3.2. Platform Implementation

To execute the evaluation of the proposed system in this work, it was necessary to embed the HIL platform, the set of PEGM, and the voltage dynamics entities of both the power and shared buses. For this purpose, it was required to use two development boards (DB) NI MyRIO 1900 (National Instruments Corp., Austin, TX, USA) These DB have a Xilinx Zynq-7010 SoC (Advanced Micro Devices, Inc., Santa Clara, CA, USA) that includes an RTP and an FPGA with a 40 MHz clock system. Particularly, this FPGA has 28 k logic cells, a 2.1 Mb RAM block, and 80 DSP slices. Under these conditions, each DB is restricted on the number of elements that can be emulated. Thus, the proposed HIL system was emulated with two DB (DB1 and DB2 in Figure 4). The communication between the RTP and the FPGA is through the ARM Advanced Microcontroller Bus Architecture (AMBA) standard. Additionally, the NI MyRIO 1900 (National Instruments Corp., Austin, TX, USA) board has 40 digital input/output (I/O) ports, 8 analog inputs, and 4 analog outputs. The analog inputs use an analog-to-digital converter (ADC) to quantify analog signals. On the contrary, a digital-to-analog converter (DAC) is used to decode digital signals for the analog outputs. Both the ADC and DAC have a 12-bit resolution. The Table 2 contains the main characteristics of the implemented DB for this work.

The connection of the embedded elements in the development boards DB1 and DB2 is shown in Figure 4, as well as the interconnection between the boards. As shown in Figure 4, the PV symbol represents the implemented LUT for a PV, which is connected to a DC/DC block. As it was previously mentioned, the response of the DC/DC block is based on Equations (1)–(3). Similarly, the C block works based on Equation (5) and the CS algorithm described in Section 2. Finally, the v_B and v_s blocks behave as described by Equation (4) and Equation (7), respectively. Additionally, a third development board (identified as DB3 in Figure 4) was required in order to perform signal acquisition during the evaluation tests. The signal acquisition was set to work with a 0.15 ms sampling rate and to maintain a sampling stage of 300 ms. During this stage, the acquired signals were recorded into FIFO registers. At the end of the sampling stage, the recorded data were transmitted to a PC through the RTP to be displayed and analyzed.

In the FPGA from the DB1 was embedded a PEGM (which contains a PP, a PEC with a controller, and the CS-C) and both blocks v_s and v_B. Conversely, the FPGA from the DB2 allocated the remaining PGEM, i.e., two additional PGEM. It is important to mention that the signals i_1, i_2, and i_3 correspond to the currents on each PEGM and are present as i_P in Equation (3). The values of the i_P currents, as well as the v_B and v_s voltages are shared between the boards by using the analog ports. In the case of the switching elements e_i, these are enabled/disabled with a set of digital signals, which are generated in the RTP of DB1. The signal for the inner PEGM of DB1 is sent from the RTP to the FPGA directly, using the AMBA protocol. Correspondingly, the signals for the PEGM 2 and 3 are sent using the wired digital I/O pins between DB1 and DB2. The wiring of the development boards DB1 and DB2 as well as the implemented HIL platform can be seen in Figure 5.

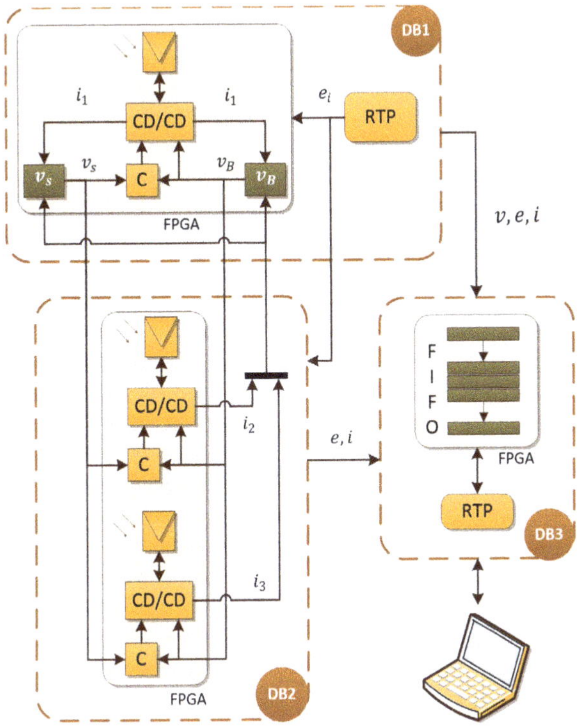

Figure 4. Implemented HIL system.

Table 2. Important parameters of the NI MyRIO DB.

Category	Parameter	Value
Processor	Type Speed Cores	Xilinx Z-7010 (SoC) 667 MHz 2
FPGA	Type Clock frequency Logic cells DSP slices Block RAM	Xilinx Z-7010 (SoC) 40 MHz 28 k 80 2.1 Mb
Analog Inputs	Sample rate Resolution Configuration	500 kS/s 12 bits 8 single-ended + 2 differentials
Analog Outputs	Update rate Resolution	345 kS/s 12 bits
Digital I/O	Configuration Logic Level	Two 16 DIO lines ports + 8 DIO lines port 5 V compatible LVTTL input; 3.3 V LVTTL output

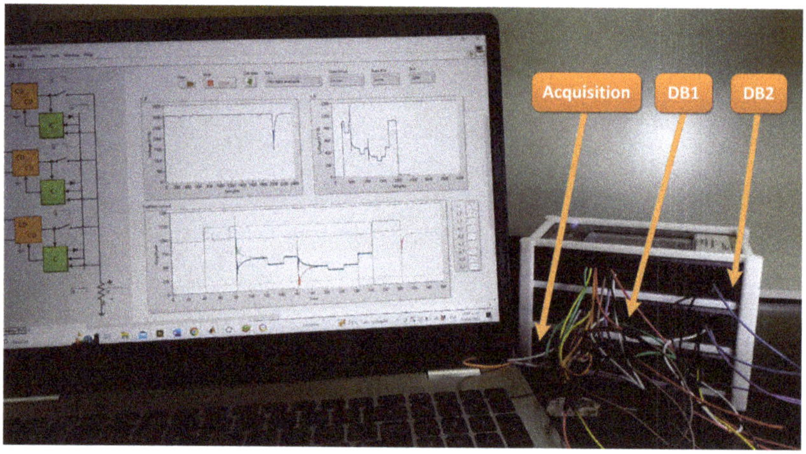

Figure 5. Implemented HIL platform.

To ease the reproduction of the HIL platform presented in this work, it was created with the development environment LabVIEW and high-performance blocks. Figure 6 shows the block diagrams developed to reproduce the PV and PEC models. These diagrams are an example of the algorithm developed for the HIL platform. The complete project and codes for the three DB are openly available in [48].

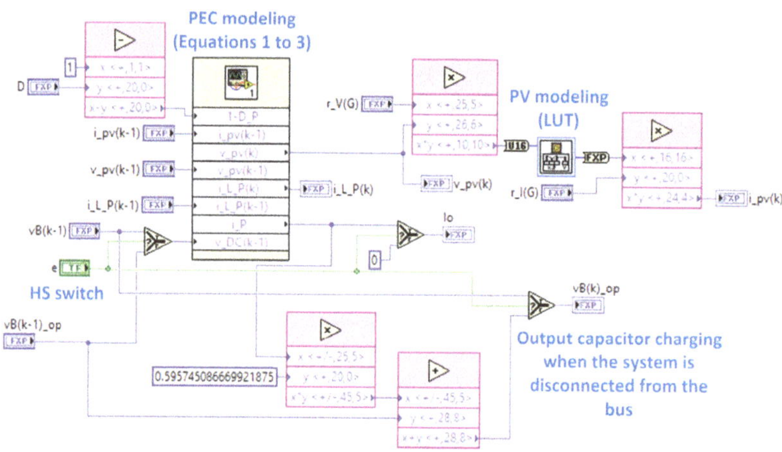

Figure 6. PV and PEC blocks developed in LabVIEW programming.

4. Results

The first step in the validation process for the HIL platform consisted in testing a PEGM individually. This test only took account of the PV, its PEC, and its PI controller without the Current-Sharing algorithm. Using MATLAB and Simulink (R2014a), a software simulation was developed for the continuous model of the PEGM, which was used to compare the accuracy of the HIL platform. Following this, the PEGM was turned on without any load and kept in this condition for 40 ms until the system achieved its stability. Next, a load was connected to demand 10% of the maximum power (25 W) from the PV for another 20 ms. Then, the load was changed to demand from the PV all its maximum power (250 W). For this test, the bus voltage and the inductor current were measured with a 0.15-ms sampling rate. A comparison of the bus voltage response between the

PEGM emulation and the PEGM simulation can be seen in Figure 7. Similarly, Figure 8 shows a comparison for the inductor current response. For both Figures 7 and 8, the letter a shows the response of the voltage bus (see Figure 7a) and the inductor current (see Figure 8a) during the whole test. The turn-on response for the bus voltage and the inductor current can be seen in the graphs labeled with the letter b in both Figures 7 and 8. Additionally, the response of the system under a 10% demand of the maximum power corresponds to letter c, where Figure 7c shows the bus voltage response and Figure 8c shows the inductor current response. Lastly, the demand of the maximum power from the PV (labeled as letter d) reflects the voltage bus response in Figure 7d, meanwhile Figure 8d shows the corresponding response for the inductor current. During the test, there were observed mean absolute errors of 0.4 V for the bus voltage and 0.2 A for the inductor current, whereas the maximum absolute error for these variables were 25.3 V and 19.3 A, respectively. Admittedly, these maximum absolute errors are present as oscillations at the beginning of the experiment while the system starts from its initial conditions. These oscillations can be an effect of the inner initialization algorithm in the simulation software. Since these variations in the results are large only during the beginning of the experiment, it is possible to consider them acceptable in the evaluation of the HIL platform performance.

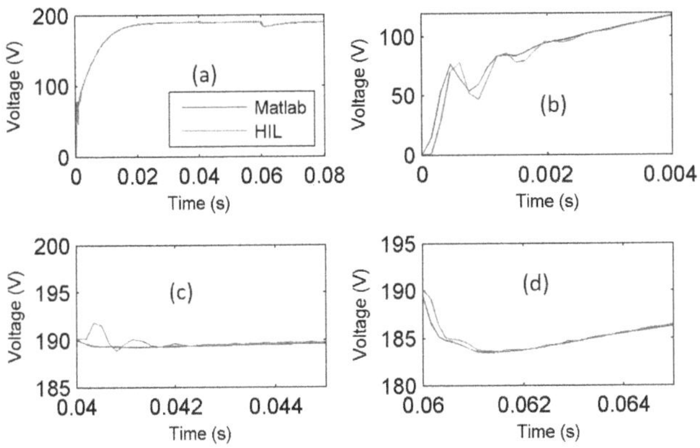

Figure 7. Response of the bus voltage for the individual PGEM test. (**a**) whole test response. (**b**) turn-on response.(**c**) 10% demand response. (**d**) maximum power demand response.

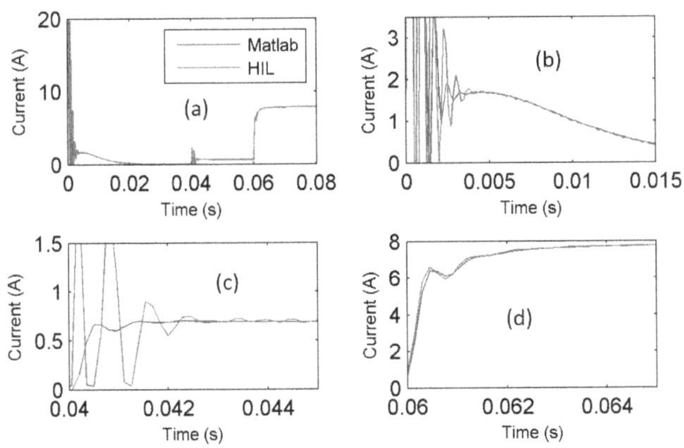

Figure 8. Response of the inductor current for the individual PGEM test. (**a**) whole test response. (**b**) turn-on response. (**c**) 10% demand response. (**d**) maximum power demand response.

Once the HIL platform demonstrated its capability to emulate a single PEGM, the next step was very clear: evaluate the functionality of the platform with the addition of a CS-C. For this test, first a PEGM was turned on (PEGM 1 connection stage); eventually, at t = 40 ms, a load was connected to demand the maximum power from the PV (Load connection stage), i.e., 250 W. Next, PEGM 2 was connected at t = 80 ms (PEGM 2 connection stage); PEGM 3 was connected (PEGM connection stage) after 80 ms, or t = 160 ms. During this test, the variation on the load considered three stages where the PV was required to work with 70% of the maximum power during 20 ms. The load was variated for the instants t = 50 ms (First load variation stage), t = 120 ms (Second load variation stage), and t = 200 ms (Third load variation stage). Each stage of the described test and the responses of both voltage and current are shown in Figures 9 and 10. On the one hand, Figure 9 shows the values of the three i_P currents present in the PEGM, the enabling of the e_p switches, and the bus voltage response scaled by a 0.007 factor. On the other hand, Figure 10 shows a detailed response of the bus voltage enclosed by two lines representing the steady state increased by +2% (green line) and decreased by −2% (blue line).

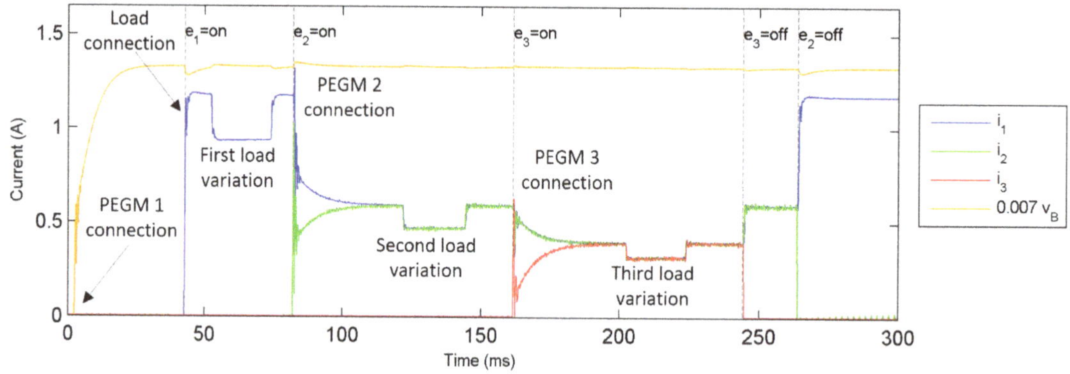

Figure 9. Response of the system during the test of Current-Sharing (CS) and Hot-Swap (HS).

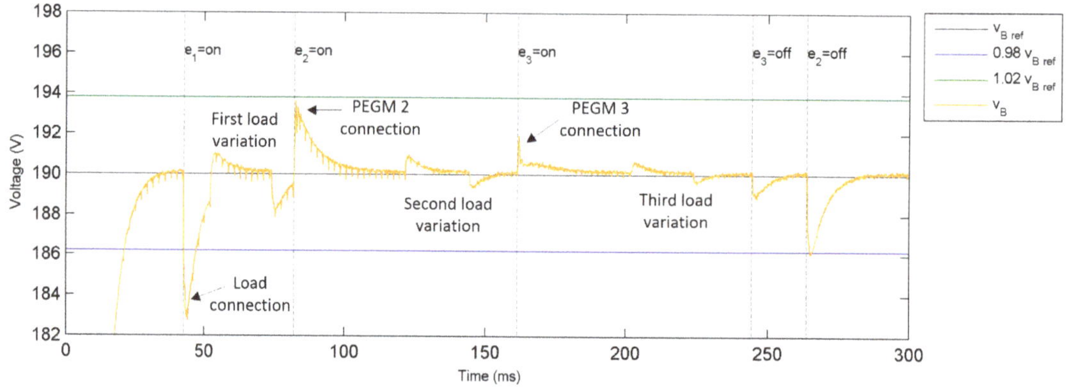

Figure 10. Detailed response of the bus voltage during the test of Current-Sharing (CS) and Hot-Swap (HS).

5. Conclusions

After comparing the response of the HIL platform versus the simulation of the continuous model, it was possible to demonstrate the accuracy of the HIL platform to reproduce the behavior of the selected PEGM for this work, where the mean absolute errors during the evaluation were 0.4 V for the bus voltage and 0.2 A for the inductor current. This accuracy allows the asseveration that the developed HIL platform is reliable for the validation of a Current-Sharing algorithm. Therefore, it could be utilized as a tool in the validation of CS-Cs. In the long run, this type of controllers have a direct effect in the enhancement of the attributes of a DC-µG, including its reliability, flexibility, robustness, and the reduction of disturbances.

Even though the objective of this work was to propose a HIL platform to validate any type of CS-C but not a specific CS-C, it was possible to conclude that the implemented controller was effective. This effectiveness was shown with an even distribution of the demanded current and a stable DC bus voltage. Considering a stability band criteria of 2%, it was appreciated with the aid of Figure 10 that the bus voltage maintained its stability under different conditions, e.g., during HS connections and load variations. After a load connection, a single variation above the 2% tolerance band was observed. Consequently, the DC bus suffered a 7.2 V decay, which represents a 3.8% decrease from its nominal value. In the case of low-voltage systems, the IEEE Std 1547-2018 [49] allows a distributed generator to induce a rapid voltage change when it is below 5% from the nominal value. Unquestionably, as the HIL platform had a 3.8% decrease from the nominal value, it accomplishes the standard and, consequently, it maintains the energy quality during the test. The evaluated CS-C in this work corresponds to a conventional droop control. Particularly, a disadvantage of this type of systems is that as the number of connected PEGM increases, the settling time of the system also increases. This is due to the decrease in the perturbance voltage Δv as the injected current by the PEGM decreases. The state of the art not only introduces some control schemes to solve this problem [32–36], but also offers the addition of functionalities for the CS-C such as the error reduction in the shared current [34], the input constrains setting in the available generation systems [35], and the energy interchange between multiple buses [36]. In a similar way, a low performance in the HIL system was observed due to the use of a single voltage loop for the PI controller in the PEC. However, thanks to the independency of each embedded module in the HIL platform (see Figures 4 and 6), it is possible to modify the implemented controllers separately without affecting any other system (e.g., the PV modules or the PEC). As a future work, it is proposed to employ the developed HIL platform here to evaluate and possibly improve different Current-Sharing algorithms.

The CS-C acts over the voltage reference defined for the local control in the PEC (v_{ref}) by either increasing or decreasing its value. These actions have the purpose of inducing changes in the power that is transferred to the DC bus. To extend the application of the HIL platform to different types of PV modules, or even PECs, it is only necessary to update the characteristics of the modules, or to tune the gains (i.e., k_i, k_v, P, and I) for the local controller. To update the characteristics of a PV module, it will be necessary to take from its datasheet the corresponding values to the open circuit voltage, the short circuit current, the maximum power point voltage, and the maximum power point current. Next, these values are introduced in a panel emulator, such as the Keysight E4360A (Keysight Technologies, Inc., Santa Rosa, CA, USA), to compute the V-I curve of the desired panel. With this V-I curve, a LUT is built and stored into the memory of a DB to emulate the PV module behavior. In essence, after the application of the updates in both the PV module and the CS-C gains, any user can obtain a new HIL system for its evaluation. No doubt, this represents an advantage of the proposed platform.

The HIL platform introduced here stands out in comparison to similar proposals as it offers excellent performance at a low cost. Additionally, as it uses a graphical programming language, the user can find an easiness in its reproducibility, development, and modification. To be sure, the performance of the developed HIL platform showed a respectable latency versus other low-cost alternatives such as Arduino. For example, presented in [27] is the use of an Arduino DUE to embed a dynamic system. However, this proposal has a lower latency because the board executes a software simulation at the end. On the contrary, by embedding a system in an FPGA, such as the NI MyRIO, the system is emulated in hardware, which results in a better latency. Similarly, presented in [26] is a system working with a software simulation but running in a joint system of a DSP board with an Arduino. Although this system achieves high latency values during the software simulation by using parallel processing, the use of this type of processing requires an advanced C language developer to make use of threads, which can be found in mechanisms such as inter-processor communication. Finally, after comparing the emulation of power electronic converters in a NI MyRIO with LabVIEW versus other alternatives [25], the authors conclude that the duo does not present an adequate efficiency in terms of space and timing. However, it stands out as a powerful alternative in the aspects of design effort and design time.

In essence, the presented platform is useful to evaluate a CS algorithm in the face of HS for a set of PEGM connected in parallel. Then, it is viable to correct a controller designed to regulate a DC bus through renewable energy generators. This correction is necessary in the EMS operation for its application in a DC-μG, as there is a great possibility that the grid depends completely on the RES to operate. However, this operation mode represents only one of the modes existing for a PEGM, so there are some future studies recommended to complement the findings of this work:

- First is to integrate the HIL platform in a DC-μG that includes more elements of a distinct nature, e.g., a battery energy storage system, a wind turbine, a main grid interconnection, etc. An aspect contributing to this goal includes the validation of the HIL platform not only with a variable load, but also with a single equation describing both the DC bus dynamics and all the connected elements to it.
- Next is to validate the HIL platform in the face of variations of some other parameters such as the power supplied by the PV, shading of one or various panels, the sudden disconnection of loads from the grid, and the connection of R-L, R-C, and R-L-C loads. In a similar way to the previous goal, a single equation describing the DC bus dynamics and one that is independent from the PEC models will avoid an adjustment in the PEGM after modifying the characteristics and behavior of the connected loads to the bus.
- Lastly, with the inclusion of additional functions for the PEGM local controllers, e.g., an MPPT, it will be possible to evaluate more operative modes of a DC-μG. As the weather variables have a slower time response, the MPPT demand a lower latency during an emulation than other elements such as the PECs contained in the PEGM. Therefore, this particularity allows the use of the RTP in the SoC to implement MPPT controllers without affecting the performance of other embedded systems.

Author Contributions: Conceptualization, E.R.-S.; methodology, J.M.-N. and V.S.-O.; software, V.S.-O.; validation, J.Z.-V. and H.R.-E.; formal analysis, H.R.-E.; investigation, J.Z.-V. and E.R.-S.; writing—original draft preparation, J.M.-N. and M.S.-M.; writing—review and editing, M.S.-M.; supervision, J.M.-N.; project administration, V.S.-O. All authors have read and agreed to the published version of the manuscript.

Funding: This research received no external funding.

Data Availability Statement: The original data presented in the study are openly available in [48].

Conflicts of Interest: The authors declare no conflicts of interest.

References

1. Ullah, S.; Haidar, A.M.A.; Zen, H. Assessment of technical and financial benefits of AC and DC microgrids based on solar photovoltaic. *Electr. Eng.* **2020**, *102*, 1297–1310. [CrossRef]
2. Datta, U.; Kalam, A.; Shi, J. Hybrid PV–wind renewable energy sources for microgrid application: An overview. In *Hybrid-Renewable Energy Systems in Microgrids*; Woodhead Publishing: Sawston, UK, 2018; pp. 1–22. [CrossRef]
3. Sood, V.K.; Abdelgawad, H. Microgrids architectures. In *Distributed Energy Resources in Microgrids*; Academic Press: Cambridge, MA, USA, 2019; pp. 1–31. [CrossRef]
4. Kolluri, R.R.; Mareels, I.; de Hoog, J. Controlling DC microgrids in communities, buildings and data centers. *IET Smart Grid* **2020**, *3*, 376–384. [CrossRef]
5. Al-Tameemi, Z.H.A.; Lie, T.T.; Foo, G.; Blaabjerg, F. Optimal Coordinated Control of DC Microgrid Based on Hybrid PSO–GWO Algorithm. *Electricity* **2022**, *3*, 346–364. [CrossRef]

6. Yang, Y.; Qin, Y.; Tan, S.-C.; Hui, S.Y.R. Efficient Improvement of Photovoltaic-Battery Systems in Standalone DC Microgrids Using a Local Hierarchical Control for the Battery System. *IEEE Trans. Power Electron.* **2019**, *34*, 10796–10807. [CrossRef]
7. Dragičević, T.; Lu, X.; Vasquez, J.C.; Guerrero, J.M. DC Microgrids—Part I: A Review of Control Strategies and Stabilization Techniques. *IEEE Trans. Power Electron.* **2016**, *31*, 4876–4891. [CrossRef]
8. Li, Z.; Ren, Z.; Li, H.; Wang, C.; Wang, C.; Hao, L. An Optimal Operation Method of Hybrid AC/DC Distribution Network Under the Goal of Peaking Carbon Dioxide Emissions and Carbon Neutrality. In Proceedings of the 2023 Panda Forum on Power and Energy (PandaFPE), Chengdu, China, 27–30 April 2023; pp. 997–1004. [CrossRef]
9. Tao, P.; Zhang, Y.; Zhao, J. Distributed Energy Storage Cluster Control Method for DC Microgrid Considering Flexibility. *Math. Probl. Eng.* **2022**, *2022*, 6761595. [CrossRef]
10. Nojavan, S.; Pashaei-Didani, H.; Mohammadi, A.; Ahmadi-Nezamabad, H. Energy management concept of AC, DC, and hybrid AC/DC microgrids. In *Risk-Based Energy Management*; Academic Press: Cambridge, MA, USA, 2020; pp. 1–10. [CrossRef]
11. Mostafa, M.H.; Abdel Aleem, S.H.E.; Ali, S.G.; Abdelaziz, A.Y. Energy-management solutions for microgrids. In *Distributed Energy Resources in Microgrids*; Academic Press: Cambridge, MA, USA, 2019; pp. 483–515. [CrossRef]
12. Merai, M.; Naouar, M.W.; Slama-Belkhodja, I. Coordinated control of multiple Multi-Function Grid connected Converters for power quality improvement in micro-grid applications. In Proceedings of the 2016 7th International Renewable Energy Congress (IREC), Hammamet, Tunisia, 22–24 March 2016; pp. 1–6. [CrossRef]
13. Zahraoui, Y.; Alhamrouni, I.; Mekhilef, S.; Basir Khan, M.R.; Seyedmahmoudian, M.; Stojcevski, A.; Horan, B. Energy management system in microgrids: A comprehensive review. *Sustainability* **2021**, *13*, 10492. [CrossRef]
14. Shahgholian, G. A brief review on microgrids: Operation, applications, modeling, and control. *Int. Trans. Electr. Energy Syst.* **2021**, *31*, e12885. [CrossRef]
15. Ishraque, M.F.; Shezan, S.A.; Ali, M.M.; Rashid, M.M. Optimization of load dispatch strategies for an islanded microgrid connected with renewable energy sources. *Appl. Energy* **2021**, *292*, 116879. [CrossRef]
16. Raya-Armenta, J.M.; Bazmohammadi, N.; Avina-Cervantes, J.G.; Sáez, D.; Vasquez, J.C.; Guerrero, J.M. Energy management system optimization in islanded microgrids: An overview and future trends. *Renew. Sustain. Energy Rev.* **2021**, *149*, 111327. [CrossRef]
17. Salcedo, R.; Corbet, E.; Smith, C.; Limpaecher, E.; Rekha, R.; Nowocin, J.; Lauss, G.; Fonkwe, E.; Almeida, M.; Gartner, P.; et al. Banshee distribution network benchmark and prototyping platform for hardware-in-the-loop integration of microgrid and device controllers. *J. Eng.* **2019**, *2019*, 5365–5373. [CrossRef]
18. AhmadiAhangar, R.; Rosin, A.; Niaki, A.N.; Palu, I.; Korõtko, T. A review on real-time simulation and analysis methods of microgrids. *Int. Trans. Electr. Energy Syst.* **2019**, *29*, e12106. [CrossRef]
19. Lauss, G.; Strunz, K. Accurate and stable hardware-in-the-loop (HIL) real-time simulation of integrated power electronics and power systems. *IEEE Trans. Power Electron.* **2020**, *36*, 10920–10932. [CrossRef]
20. Kumar, G.B.; Palanisamy, K. Energy management of renewable energy-based microgrid system with HESS for various operation modes. *Front. Energy Res.* **2022**, *10*, 995034. [CrossRef]
21. Kikusato, H.; Ustun, T.S.; Suzuki, M.; Sugahara, S.; Hashimoto, J.; Otani, K.; Shirakawa, K.; Yabuki, R.; Watanabe, K.; Shimizu, T. Microgrid Controller Testing Using Power Hardware-in-the-Loop. *Energies* **2020**, *13*, 2044. [CrossRef]
22. Ghanbari, N.; Bhattacharya, S. Adaptive Droop Control Method for Suppressing Circulating Currents in DC Microgrids. *IEEE Open Access J. Power Energy* **2020**, *7*, 100–110. [CrossRef]
23. Nigam, S.; Ajala, O.; Domínguez-García, A.D.; Sauer, P.W. Controller hardware in the loop testing of microgrid secondary frequency control schemes. *Electr. Power Syst. Res.* **2021**, *190*, 106757. [CrossRef]
24. Muñoz-Jadán, Y.; Espinoza-Bolaños, M.; Donoso Merlet, F.; Hidalgo-León, R.; Soriano Idrovo, G.; Jácome-Ruíz, P. Hardware-in-the-Loop for Wind Energy Conversion with Resonant Current Control and Active Damping. *IEEE Lat. Am. Trans.* **2019**, *17*, 1146–1154. [CrossRef]
25. Zamiri, E.; Sanchez, A.; Yushkova, M.; Martínez-García, M.S.; de Castro, A. Comparison of different design alternatives for hardware-in-the-loop of power converters. *Electronics* **2021**, *10*, 926. [CrossRef]
26. Jiang, W.; Sun, L.; Chen, Y.; Ma, H.; Hashimoto, S. A hardware-in-the-loop-on-chip development system for teaching and development of dynamic systems. *Electronics* **2021**, *10*, 801. [CrossRef]
27. Martinez-Armero, Y.; Lopez-Blandon, S.; Giraldo, E. Low-cost Arduino-based Hardware-In-the-Loop Platform for Simulation and Control of Dynamic Systems. *IAENG Int. J. Comput. Sci* **2023**, *50*, 4.
28. Ahmad, S.; Shafiullah, M.; Ahmed, C.B.; Alowaifeer, M. A review of microgrid energy management and control strategies. *IEEE Access* **2023**, *11*, 21729–21757. [CrossRef]
29. Sharma, S.; Iyer, V.M.; Bhattacharya, S.; Kikuchi, J.; Zou, K.; Gupta, M. A Methodology for Seamless Hot-Swap of Converters in DC Microgrids. In Proceedings of the IECON 2020 The 46th Annual Conference of the IEEE Industrial Electronics Society, Singapore, 18–21 October 2020; pp. 3143–3150. [CrossRef]
30. Cucuzzella, M.; Trip, S.; De Persis, C.; Cheng, X.; Ferrara, A.; van der Schaft, A. A Robust Consensus Algorithm for Current Sharing and Voltage Regulation in DC Microgrids. *IEEE Trans. Control Syst. Technol.* **2019**, *27*, 1583–1595. [CrossRef]
31. Shebani, M.M.; Iqbal, M.T.; Quaicoe, J.E. Control Algorithm for Equal Current Sharing between Parallel-Connected Boost Converters in a DC Microgrid. *J. Electr. Comput. Eng.* **2020**, *1*, 1–11. [CrossRef]

32. Zhang, Q.; Zhuang, X.; Liu, Y.; Wang, C.; Guo, H. A Novel Autonomous Current-Sharing Control Strategy for Multiple Paralleled DC–DC Converters in Islanded DC Microgrid. *Energies* **2019**, *12*, 3951. [CrossRef]
33. Tucci, M.; Meng, L.; Guerrero, J.M.; Ferrari-Trecate, G. Plug-and-play control and consensus algorithms for current sharing in DC microgrids. *IFAC-PapersOnLine* **2017**, *50*, 12440–12445. [CrossRef]
34. Mesbah, M.A.; Sayed, K.; Ahmed, A.; Aref, M.; Elbarbary, Z.M.S.; Almuflih, A.S.; Mossa, M.A. Adaptive Control Approach for Accurate Current Sharing and Voltage Regulation in DC Microgrid Applications. *Energies* **2024**, *17*, 284. [CrossRef]
35. Noroozi, N.; Trip, S.; Geiselhart, R. Model predictive control of DC microgrids: Current sharing and voltage regulation. *IFAC-PapersOnLine* **2018**, *51*, 124–129. [CrossRef]
36. Bai, H.; Zhang, H.; Li, C. Voltage regulation and current sharing for multi-bus DC microgrids. *IFAC-PapersOnLine* **2020**, *53*, 13018–13023. [CrossRef]
37. Duan, M.; Duan, J.; Sun, L. Sensorless Current-Sharing Scheme for Multiphase DC-DC Boost Converters. *IEEE Trans. Power Electron.* **2023**, *38*, 1398–1405. [CrossRef]
38. Chen, Z.; Xiang, T.; Wu, Y.; Zhao, L.; Tang, X. Current Sharing Strategy of Three-Phase Series Capacitor Buck Converter for Wide Output Voltage Range Application. *IEEE J. Emerg. Sel. Top. Power Electron.* **2023**, *11*, 836–849. [CrossRef]
39. Estrada, L.; Vázquez, N.; Vaquero, J.; de Castro, Á.; Arau, J. Real-Time Hardware in the Loop Simulation Methodology for Power Converters Using LabVIEW FPGA. *Energies* **2020**, *13*, 373. [CrossRef]
40. Ma, D.; Chen, W.; Ruan, X. A Review of Voltage/Current Sharing Techniques for Series–Parallel-Connected Modular Power Conversion Systems. *IEEE Trans. Power Electron.* **2020**, *35*, 12383–12400. [CrossRef]
41. Samano-Ortega, V.; Rodriguez-Estrada, H.; Rodríguez-Segura, E.; Padilla-Medina, J.; Aguilera-Alvarez, J.; Martinez-Nolasco, J. Power Sharing Control in a Grid-Tied DC Microgrid: Controller Hardware in the Loop Validation. *Appl. Sci.* **2021**, *11*, 9295. [CrossRef]
42. Pan, Y.; Yang, Y.; Blaabjerg, F. Distributed Control of Islanded Series PV-Battery-Hybrid Systems with Low Communication Burden. In Proceedings of the 2020 IEEE 11th International Symposium on Power Electronics for Distributed Generation Systems (PEDG), Dubrovnik, Croatia, 28 September–1 October 2020; pp. 315–321. [CrossRef]
43. Kanathipan, K.; Emamalipour, R.; Lam, J. A Single-Switch High-Gain PV Microconverter with Low-Switch-Voltage-to-High-Voltage-Bus Ratio. *IEEE Trans. Power Electron.* **2020**, *35*, 9530–9540. [CrossRef]
44. Pan, Y.; Sangwongwanich, A.; Yang, Y.; Blaabjerg, F. A Series Interharmonic Filter for Cascaded H-bridge PV Inverters. In Proceedings of the 2020 IEEE Energy Conversion Congress and Exposition (ECCE), Detroit, MI, USA, 11–15 October 2020; pp. 341–346. [CrossRef]
45. Lin, Z.; Xie, W.; Dong, M.; Lan, J. Research of S^3MPR Engineering Application on Space Power Sources. In Proceedings of the 2020 IEEE 4th Information Technology, Networking, Electronic and Automation Control Conference (ITNEC), Chongqing, China, 12–14 June 2020; pp. 817–823.
46. Keysight Technologies, Inc. Keysight Technologies Series E4360 Modular Solar Array Simulator. Available online: http://literature.cdn.keysight.com/litweb/pdf/E4360-90001.pdf (accessed on 16 May 2024).
47. Shen, Y.; Qin, Z.; Wang, H. Modeling and Control of DC-DC Converters. In *En Control of Power Electronic Converters and Systems*; Academic Press: Cambridge, MA, USA, 2018; pp. 69–92. [CrossRef]
48. VictorSamano. HiL_Current-Sharing, GitHub Repository. 2024. Available online: https://github.com/VictorSamano/HiL_Current-Sharing (accessed on 2 July 2024).
49. Photovoltaics, D.G.; Storage, E. *IEEE Standard for Interconnection and Interoperability of Distributed Energy Resources with Associated Electric Power Systems Interfaces*; IEEE Std 1547-2018 (Revision of IEEE Std 1547-2003); IEEE: Piscataway, NJ, USA, 2018; pp. 1–138. [CrossRef]

energies

MDPI

Article

ANN for Temperature and Irradiation Prediction and Maximum Power Point Tracking Using MRP-SMC

Mokhtar Jlidi [1,2], Oscar Barambones [2,*], Faiçal Hamidi [1] and Mohamed Aoun [1]

[1] Laboratory Modélisation, Analyse et Commande des Systèmes, University of Gabes, Gabes LR16ES22, Tunisia; m.moktarjlidi@gmail.com (M.J.); faicalhamidi@yahoo.fr (F.H.); mohamed.aoun@macs.tn (M.A.)

[2] Automatic Control and System Engineering Department, University of the Basque Country, UPV/EHU, Nieves Cano 12, 01006 Vitoria-Gasteiz, Spain

* Correspondence: oscar.barambones@ehu.eus; Tel.: +34-94-501-3235; Fax: +34-94-501-3270

Abstract: Currently, artificial intelligence (AI) is emerging as a dominant force in various technologies, owing to its unparalleled efficiency. Among the plethora of AI techniques available, neural networks (NNs) have garnered significant attention due to their adeptness in addressing diverse challenges, particularly for prediction tasks. This study offers a comprehensive review of predominant AI-based approaches to photovoltaic (PV) energy forecasting, with a particular emphasis on artificial neural networks (ANNs). We introduce a revolutionary methodology that amalgamates the predictive capabilities of ANN with the precision control afforded by the minimum-risk problem and sliding mode control (MRP-SMC), thereby revolutionizing the PV panel performance enhancement. Building upon this methodology, our hybrid approach utilizes the ANN as a proficient weather forecaster, accurately predicting the temperature and solar radiation levels impacting the panels. These forecasts serve as guiding principles for the MRP-SMC algorithm, enabling the proactive determination of the Maximum Power Point (MPP). Unlike conventional methods that grapple with weather unpredictability, the MRP-SMC algorithm transforms stochastic optimization challenges into controllable deterministic risk problems. Our method regulates the boost converter's work cycle dynamically. This dynamic adaptation, guided by environmental predictions from ANNs, unlocks the full potential of PV panels, maximizing energy recovery efficiency. To train the model, we utilized a large dataset comprising 60,538 temperature and solar radiation readings from the Department of Systems Engineering and Automation at the Faculty of Engineering in Vitoria (University of the Basque Country). Our approach demonstrates a high regression coefficient (R = 0.99) and low mean square error (MSE = 0.0044), underscoring its exceptional ability to predict real energy values. In essence, this study proposes a potent fusion of artificial intelligence and control mechanisms that unleash the untapped potential of photovoltaic panels. By utilizing forecasts to guide the converter, we are paving the way for a future where solar energy shines brighter than ever.

Keywords: minimal risk problem; forecasting; ANN; SMC; MPPT; boost

1. Introduction

Renewable energies, with solar power at the forefront, are essential to meeting the challenge of a sustainable energy supply [1,2]. However, the unpredictable nature of sunlight prevents the full potential of solar energy from being exploited [3,4]. To overcome this problem, Maximum Power Point Tracking (MPPT) technology and reliable weather forecasts optimize the PV system efficiency and energy production [5,6]. MPPT ensures that solar panels operate at maximum power under varying sunlight conditions [7,8]. Accurate weather forecasts allow proactive adjustments to MPPT algorithms and system operations [9,10]. The integration of MPPT technology and reliable weather forecasts significantly improves the energy yield and efficiency of solar installations [11,12]. These advances pave the way for a sustainable solar-powered energy future [13,14].

Conventional MPPT algorithms, such as Perturb and Observe (P&O), Hill Climbing (HC), and Incremental Conductance (IC), often struggle to efficiently track the Maximum Power Point (MPP) under stochastic conditions [15,16]. These limitations can lead to significant energy losses and reduced system efficiency [17,18].

Advanced MPPTs such as particle swarm optimization (PSO) and JAYA outperform but are not well-suited to addressing large and complex optimization challenges and are prone to triggering at local optima [19,20]. PSO can converge prematurely to sub-optimal solutions [21], while JAYA's sensitivity to initial conditions and limited exploration of the search space can hamper its effectiveness [22]. The large search space inherent in the genetic algorithm (GA) parameter optimization process can significantly hamper the system speed and increase the complexity [23].

These limitations can lead to significant energy losses in solar energy systems, particularly in dynamic environments with fluctuating irradiance and partial shading [24].

This research addresses the challenge of optimizing operational processes under uncertainty by applying a methodology described in [25]. Stochastic optimization problems are addressed by transforming them into equivalent deterministic formulations, in particular, in scenarios with fully admissible solutions. The approach introduces a deterministic optimization problem presented as a minimum-risk problem (MRP), solved efficiently using non-linear mathematical programming methods. This methodology is then applied to derive optimal decision frameworks for simplified industrial applications, with a specific case study demonstrating the practical application in the optimization of solar energy systems.

In addition, the growing need to predict future trends in various fields has fuelled a strong demand for forecasting techniques. In this context, ANNs have emerged as powerful tools. ANNs have been successfully applied in various fields, including tourism (e.g., forecasting tourist numbers or hotel stays [26]), financial trading [27], and the energy sector (e.g., forecasting renewable energy production and energy consumption [28–30]). Numerous studies consistently demonstrate the superior performance of ANNs, particularly when dealing with high-frequency data, compared to traditional forecasting methods.

In [31], the performance of the ANN model was evaluated against various models documented in the literature, including Quadratic Support Vector Machine (QSVM), Decision Tree [32], Convolution Neural Network–Bi-Direction Long Short-Term Memory (CNN-BILSTM) [33], Deep Learning (DL) [34], Adaptive Neuron Fuzzy Inference System (ANFIS) [35], Group Method of Data Handling (GMDH) [36], and ANFIS-PSO [37], to predict different types of solar radiation. The evaluation used the Mean Squared Error (MSE) and regression(R) criteria to ensure a complete and accurate comparison. The results of this comparative analysis unequivocally demonstrated the superior accuracy of the ANN model compared with the aforementioned models. In particular, the study highlighted the challenges associated with determining hyper-parameters and specific parameters in the other models, a concern mitigated by the relative ease of implementation of the ANN model.

In many studies, researchers have frequently used the proportional integration (PI) controller to regulate the controller duty cycle. Specifically, in research [38], the Fuzzy-PI method was used for control and, in another study [39], a combination of ANN and PI was implemented. Despite the effectiveness of the PI controller, it faces challenges related to its inherent linearity, which complicates its applicability to non-linear systems, as well as the tuning complexities associated with selecting optimal parameters. Alternatively, sliding mode control (SMC) is emerging as a robust and versatile control method recognized for its ability to handle uncertainties and disturbances in dynamic systems.

Many approaches have been explored by researchers to improve the efficiency of photovoltaic systems. In [40], the authors presented a PV-TE hybrid system, combining photovoltaic (PV) and thermal (TE) cells, as a solution to the energy efficiency problems of PV technology. In parallel, ref. [41] used a fuzzy method to find the Maximum Power Point (MPP), while [42] used mathematical models for gradient optimization specifically adapted to photovoltaic panels. Despite their effectiveness, these methods do not address stochastic challenges due to the dynamic nature of the environment. Various prediction techniques, including physical methods, have been proposed, but they have some limitations. For example, as reported in [43], physical and indirect forecasting methods struggle to accurately predict future weather conditions. However, in [41], artificial intelligence emerged as a promising avenue for forecasting.

This paper presents significant advances in the applications of control systems for renewable energy sources. A key aspect of this work is taking advantage of artificial neural networks (ANNs) for temperature and solar radiation forecasting. The adoption of artificial neural networks is justified by their versatility and proven effectiveness in various domains, highlighting their suitability for this particular field. In addition, the paper proposes a new approach using the MRP-SMC method to control DC–DC converters, aligning with the essential principle of Maximum Power Point Tracking (MPPT). This innovative methodology improves converter performance, thus optimizing energy extraction in renewable energy systems.

This paper introduces two pivotal contributions to the field. First, it harnesses the power of artificial neural networks (ANNs) to accurately predict temperature and solar radiation levels, taking advantage of their well-established effectiveness in a multitude of domains. Secondly, it adopts an innovative MRP-SMC-based approach to control DC–DC converters according to the principles of Maximum Power Point Tracking (MPPT). The MRP method skillfully addresses the stochastic challenges arising from unpredictable weather conditions, while sliding mode control (SMC) excels in its ability to managing the uncertainties and disturbances inherent in dynamic systems. By seamlessly integrating these two methodologies, the paper improves the ability to explore the point of maximum power and precisely regulate the flow of energy within renewable energy systems. This holistic approach not only improves efficiency but also enhances the reliability of these systems, paving the way for sustainable energy solutions in the face of changing environmental dynamics.

The paper is structured as follows: Section 2 delves into the methodology, discussing the theoretical framework of the PV model and boost converter, as well as the predictive system and controllers utilized for MPPT. Section 3 outlines the neural-network-based predictive models employed, while Section 4 presents an analysis of the PV characteristics and controller performance. In Section 5, a comparative study is conducted between the proposed forecasting method and methods relying on the time series, alongside a comparison of the proposed MPPT method with the JAYA technique to evaluate the efficiency of the combined strategy. Finally, Section 6 concludes with a summary of the research's main achievements.

2. Methodology

Figure 1 illustrates the configuration of the examined system.

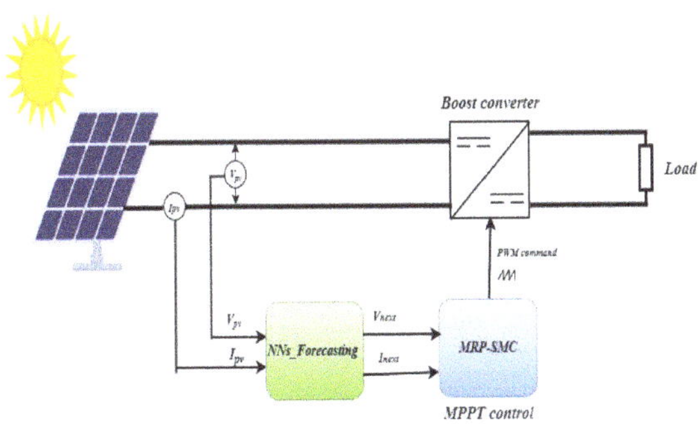

Figure 1. Study system.

2.1. PV Panel Modeling

The study employs the single-diode model due to its balance of accuracy and simplicity [44]. This model represents the photovoltaic (PV) system as a DC source within an electrical circuit, producing a current (Iph) from solar irradiation. It incorporates two types of resistances: Rsh, associated with the diode leakage current at the p-n junction [45], and Rs, which represents the resistance at the PV terminals [46]. Rs negatively affect the system's maximum power output [47]. Figure 2 illustrates the overall model.

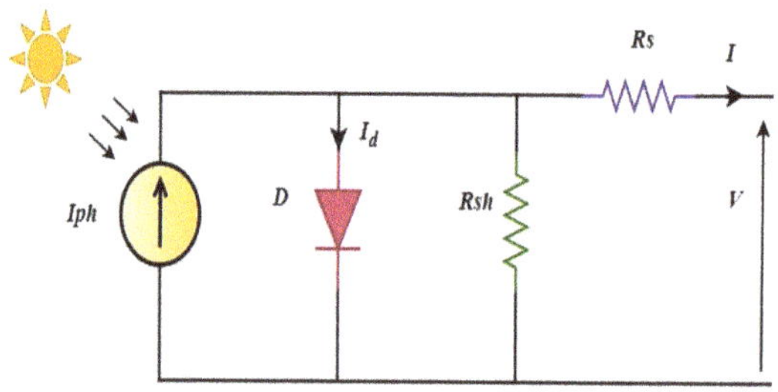

Figure 2. PV panel model.

The mathematical model (1) is derived from the equivalent circuit, representing a photovoltaic (PV) panel with N cells connected in series. While parallel connection of cells is possible, it is rarely seen in real-world scenarios.

$$I = I_{ph} - I_0\left(e^{\frac{q(V+IR_S)}{akTN_S}} - 1\right) - \frac{V + IR_s}{R_{sh}} \tag{1}$$

The initial expression represents the photocurrent, and is shown by Equation (2).

$$I_{ph} = G\frac{I_{sc} + K_i(T - T_{STC})}{G_{STC}} \tag{2}$$

G_{STC} and T_{STC} indicate radiation and temperature when tested under normal conditions. I_{SC} is the current in a short circuit, and K_i is the temperature coefficient.

$$I_0 = \frac{I_{sc} + K_i(T - T_{STC})}{e^{\frac{q(V_{OC}+Kv(T-T_{STC}))}{aKTN_S}} - 1} \tag{3}$$

The equations provided define some important terms: **q** is the constant representing electronic charge, $\mathbf{V_{oc}}$ is the open-circuit voltage, $\mathbf{K_V}$ is the temperature coefficient for open-circuit voltage, **a** is the solar ideal factor, **K** is Boltzmann's constant, and $\mathbf{N_S}$ is the number of series cells. Equation (2) shows that photocurrent is mainly affected by weather conditions like irradiation and temperature. Studies show that irradiation has a bigger effect on **Iph** than temperature, which ultimately influences the total current produced.

2.2. Hybrid MPPT Control

2.2.1. Boost Converter Model

This study used a boost converter to transfer energy from a solar panel to a resistive load. The boost converter increased the voltage from the solar panel to meet the needs of the load. A special boost topology was used for this purpose. Figure 3 shows how the boost converter was connected electrically.

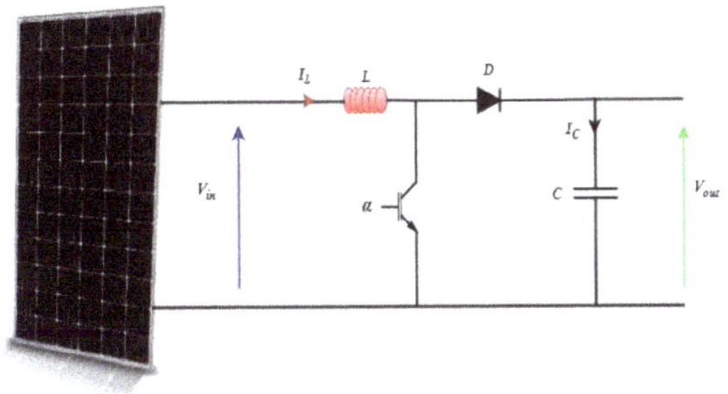

Figure 3. Boost converter's electrical diagram.

The duty cycle (α) determines the relationship between the input and output of the boost converter. This relationship appears in Equation (4). Controlled by a PWM signal generated by a MOSFET, it must be between 0.1 and 0.9 to maintain continuous conduction mode.

$$V_{out} = \frac{1}{1-\alpha} V_{in} \tag{4}$$

The state space representation of the boost converter uses two state variables: x1 for the input current and x2 for the output voltage.

$$\begin{cases} \dot{X} = \begin{bmatrix} 0 & \frac{\alpha-1}{L} \\ \frac{1-\alpha}{C} & \frac{-1}{RC} \end{bmatrix} \begin{bmatrix} x_1 \\ x_2 \end{bmatrix} + \begin{bmatrix} \frac{1}{L} \\ 0 \end{bmatrix} V_{in} \\ Y = \begin{bmatrix} 0 & 1 \end{bmatrix} \begin{bmatrix} x_1 \\ x_2 \end{bmatrix} \end{cases} \tag{5}$$

2.2.2. MRP-SMC Hybrid MPPT Control

1. The MPPT approach

Studying a PV panel trait, denoted as P = f(V), reveals how electrical power relies on the voltage it generates. The solar irradiance, symbolized by I_{rr}, underscores the changing nature of these energy sources. The subsequent expression illustrates a typical representation of a stochastic optimization problem (SOP):

$$\begin{aligned} &\max\left\{P(V) = C^T(I_{rr})f(V)\right\} \\ &\text{a.c.}: \begin{cases} A(I_{rr})V \leq b(I_{rr}) \\ V \geq 0 \end{cases} \end{aligned} \tag{6}$$

The main challenge of this SOP is dealing with uncertainty. We have three variable factors (**A**, **b**, and **C**) for each possible outcome in the events' space **Ω**. To find the optimal decision (**V***), we need to transform the problem into a defined format with acceptable solutions. This allows **V*** to account for all possible variations of **A** and **b**, ensuring that it is the best choice despite the randomness involved.

The initial formulation of the stochastic problem (6) can be replaced by an intermediate problem as shown in Equation (7) to address this complex problem by transforming it into a deterministic risk problem. This approach eliminates uncertainty by examining all possible scenarios and identifying the best decision for every situation. It simplifies technical terms, focusing on managing uncertainty and finding optimal solutions in complex scenarios.

$$\begin{cases} \min\{\alpha\} \\ \text{a.c.} \begin{cases} P\left(C^T f(V) < I_0\right) = \alpha \\ V \in D_{adm} \end{cases} \end{cases} \tag{7}$$

This problem aim to identify the option (V) that minimizes the risk of failing to achieve a target performance level (V_0). This target might be, for instance, the average value of a function (P(V)). Rather than directly minimizing the risk of failure, we focus on minimizing the probability associated with this risk. When a particular assumption about the random factor (c) is valid, we can transform this "minimal risk" problem into a more straightforward fixed problem. The solution to this simplified problem is given by the following equation:

$$\begin{cases} m_i = E(c_i) \\ u_{i,j} = E\left[(c_i - m_i)(c_j - m_j), U = (u_{i,j})\right] \end{cases} \tag{8}$$

A non-linear average optimization problem is considered:

$$\max\left\{P_{max} = m^T f(V)\right\} \tag{9}$$

The problem described in Equation (9) is deterministic, where the vector **m** represents the mean values $\mathbf{m_i}$, resulting in the solution $\mathbf{P^*_{max}}$ The function $\mathbf{P_{max}(V)}$ achieves its maximum value at $\mathbf{P_{max}^* = m^T f(V_{max}^*)}$, and we select $\mathbf{P_0 < P_{max}^*}$. The probability distribution of the random vector is approximated by a parabolic function with parameters $\mathbf{c_0}$, $\mathbf{c_1}$, and $\mathbf{c_2}$, where $\mathbf{P_{max}(V^*) = c_0 V^2 + c_1 V + c_2}$. Consequently, the outcomes are as follows:

$$\alpha = P_{ro}\left(C^T f(V) < P_0\right) = P_{ro}\left(\frac{C^T f(V) - m^T f(V)}{\sqrt{V^T UV}} < \frac{P_0 - m^T f(V)}{\sqrt{V^T UV}}\right) = \Phi\left(\frac{P_0 - m^T f(V)}{\sqrt{V^T UV}}\right) \tag{10}$$

In this context, $\boldsymbol{\alpha}$ represents the minimum risk, $\boldsymbol{\Phi}$ denotes the Laplace function, and $\mathbf{P_{ro}}$ stands for probability. In the earlier equation, **U** is the auto-covariance matrix of the $\mathbf{u_{i,j}}$ values from Equation (8). Because the Laplace function is monotonically increasing and the matrix **U** is positively definite, problem (10) simplifies into a straightforward optimization problem.

$$\min_{V \in D_{adm}} \left(\frac{P_0 - m^T f(V)}{\sqrt{V^T UV}}\right) \tag{11}$$

These problems can be addressed using conventional gradient methods in non-linear programming. This process includes choosing **V*** as the optimal solution for Equation (6). Additionally, Equation (11) matches Equation (12) in the context of maximization.

$$\max\left(\frac{m^T f(V) - P_0}{\sqrt{V^T \cdot U \cdot V}}\right) = \frac{c_0.V^2 + c_1.V + c_2 - P_0}{\sqrt{\begin{bmatrix} V^2 & V & 1 \end{bmatrix} U \begin{bmatrix} V^2 \\ V \\ 1 \end{bmatrix}}} \tag{12}$$

The solution to the problem, which corresponds to the voltage and current linked to the optimal operating point, was calculated using Rosen's Gradient.

2. Sliding Mode Controller

To determine the boost converter's duty cycle, we used SMC, a robust control algorithm that switches between states over time. This method effectively controls the input current without relying on specific system parameters and handles disturbances well. The selected sliding surface is as follows:

$$S = I_{ref} - I_{in} \tag{13}$$

The statement presents $\mathbf{I_{ref}}$ as the optimal current for MPPT control and $\mathbf{I_{in}}$ as the boost converter input current. It focuses on the application of a control law to guide the system on a sliding mode surface in finite time, by describing a specific structure for the control input.

$$U = U_{eq} - U_n \tag{14}$$

In the system, U_{eq}, representing the equivalent control input, governs the behavior on the sliding surface, while U_n, the non-linear switching input, directs and maintains the state on the sliding surface despite uncertainty. To obtain U_{eq} the initial condition is determined by ensuring $\dot{S} = 0$ [48].

$$\dot{S} = \dot{I}_{ref} - \dot{I}_{in} = 0 \tag{15}$$

where $\dot{I}_{in}$ is obtained from Equation (5), the boost converter state space

$$\begin{aligned} &\dot{I}_{ref} - \tfrac{1}{L}((U_{eq} - 1)V_{out} + V_{in}) = 0 \\ &\Rightarrow U_{eq} = 1 - \tfrac{V_{in}}{V_{out}} \end{aligned} \tag{16}$$

By respecting and applying the Lyapunov criterion ($\dot{V} < 0$), U_n was obtained, with:

$$V = \frac{1}{2}S^2 \tag{17}$$

$$\dot{V} = S.\dot{S} < 0 \Leftrightarrow S.(-\frac{1}{L}((U-1)V_{out} + V_{in})) < 0 \tag{18}$$

$$S.(-\frac{1}{L}((U_{eq} - U_n - 1)V_{out} + V_{in})) < 0 \Rightarrow S(\frac{-1}{L}U_n) < 0 \tag{19}$$

$$S.U_n > 0 \text{ where } U_n = k.\text{sign}(S) \tag{20}$$

The positive gain K is crucial: if too small, the controller loses robustness; if too large, it causes oscillations that can lead to chattering, or damage the control unit. Chattering can be reduced by replacing the "sign" function with a hyperbolic tangent function (tanh). Equation (21) presents the new control law.

Equation (21) describes the structure of the command law.

$$U = \frac{1}{2}(1 + \tan h(S)) \tag{21}$$

3. Neural-Network-Based Predictive Models

Artificial intelligence, especially NNs, is becoming more popular for forecasting because they are good at it. NNs work similarly to the brain, with interconnected parts that help them process information. They are being used in many industries, including forecasting, and they are showing great promise in making accurate predictions. The structure of the NN is represented in Figure 4.

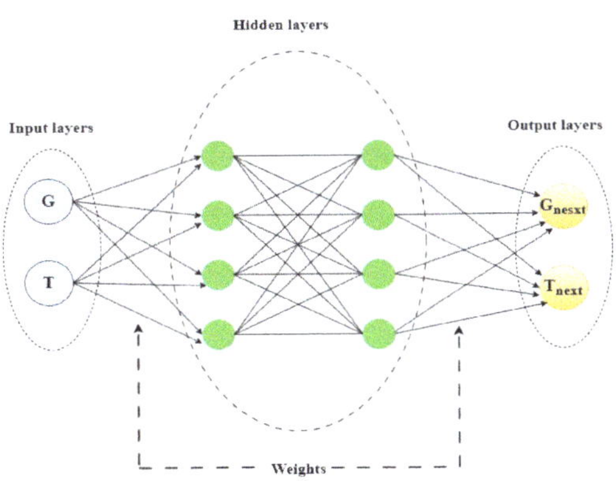

Figure 4. Neural network architecture.

The ANNs process information through a multi-layer architecture. The input layer receives the raw data, which are then transformed, and features are extracted in hidden layers using activation functions. The output layer produces the final prediction. The number of hidden layers, biases, and pooling layers are adjusted based on the complexity of the problem to improve performance. Customizing the network layers and neurons is crucial for each specific problem and dataset. This process is represented by Equation (22).

$$Y = \sigma({W_{i,j}}^{T} * X_j - \beta) = \frac{1}{1 + e^{-({W_{i,j}}^{T} * X_j - \beta)}} \tag{22}$$

$$W = \begin{bmatrix} w_{1,1} & \cdots & w_{1,j} \\ \vdots & \ddots & \vdots \\ w_{i,1} & \cdots & w_{i,j} \end{bmatrix} \tag{23}$$

where

Y: the output layer;
σ: The activation function;
$\mathbf{W_{i,j}}$ represents the weight linking the ith neuron in the preceding layer to the jth neuron in the current layer;
$\mathbf{X_j}$: The output of the ith neuron in the previous layer;
β: Biases.

To predict temperature and solar irradiance, we use an NN with the sigmoid activation function, which introduces non-linearity and ensures output values stay between 0 and 1, making it suitable for these continuous variables.

The process begins with data collection from the Department of Systems Engineering and Automation at Vitoria College of Engineering using a specialized sensor. The data are pre-processed, transformed, and standardized for input to the neural network. A statistical analysis informs the network design, which includes specific input, hidden, and output layers. The network is then trained using the efficient Scaled Conjugate Gradient (SCG) algorithm with adaptive learning rates and second-order optimization. Key metrics such as MSE and R monitor the training progress until the validation error stabilizes. Once trained, the network predicts the temperature and solar radiation (T and G) of new data. Rigorous validation and testing ensure accuracy and reliability, including partitioning data into training, validation, and test sets to adjust weights, prevent over-fitting, and objectively evaluate the performance. This approach ensures accurate forecasts. The following Figures 5 and 6 show the NN performance.

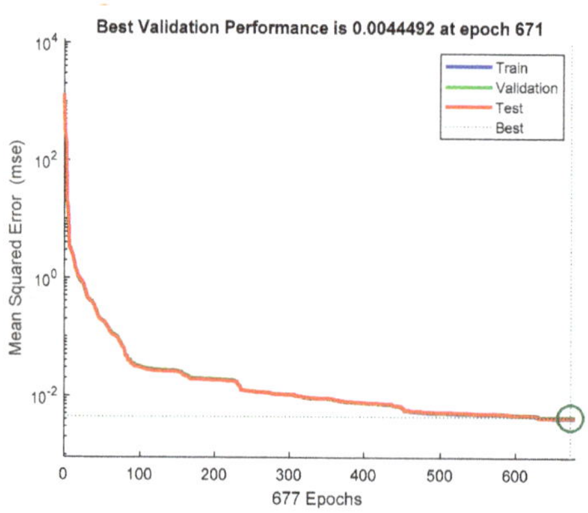

Figure 5. Mean square error.

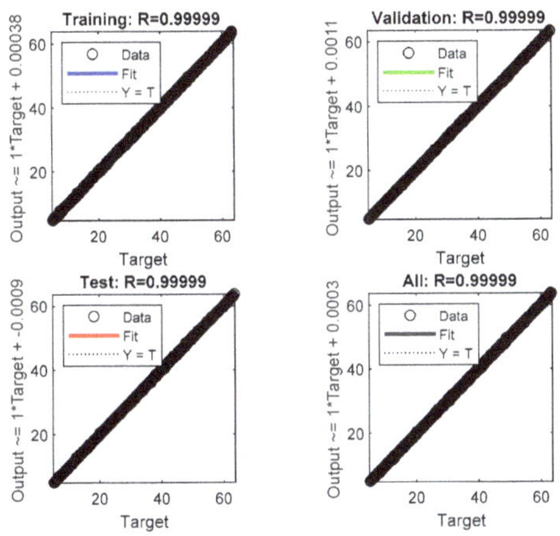

Figure 6. Regression.

During training, errors started repeating after 600 epochs, so the test was stopped at 677 epochs when the gradient was 0.0098. This repetition showed that the data were becoming more important. Epoch 200 was chosen as the baseline, and its weights were used as the final weights. Validation was carried out from Epoch 6 to 677, with errors repeating six times before stopping (see Figure 7).

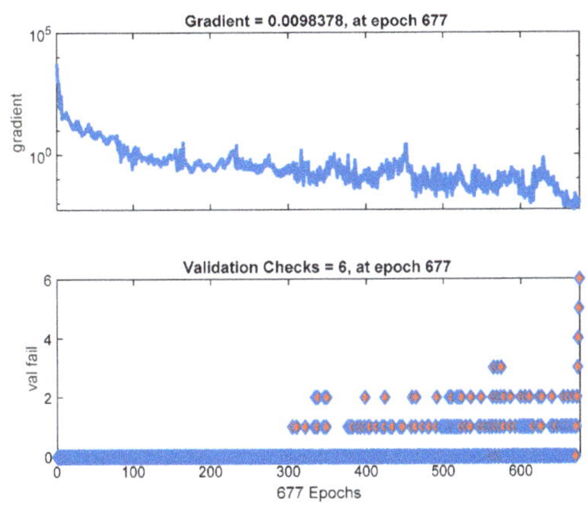

Figure 7. NN training.

From Figure 7, the red points represent the neural network's performance on the validation dataset at each epoch, while the blue line indicates its performance on the training dataset over the epochs.

Figure 8 illustrates the temperature and irradiance prediction outcomes, showcasing the efficacy of the proposed neural-network-based prediction methodology. Rigorously evaluated using real-world data, the technique undergoes a meticulous comparison between the experimental and predicted temperature, as well as irradiance values. The primary goal is to minimize the disparity between the anticipated outcomes and observed data. Encouragingly, a discernible alignment emerges between the predicted values and actual observations, signifying a robust agreement between the predictions and real results.

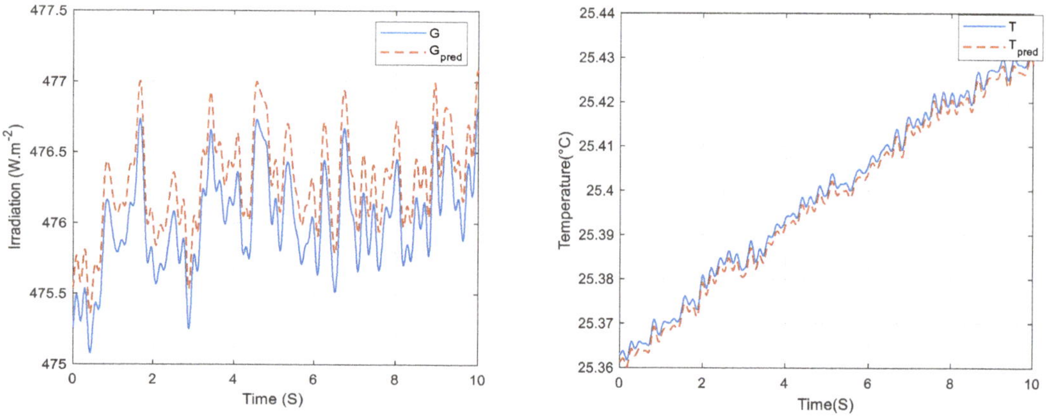

Figure 8. Prediction results.

An analysis of Figure 8 reveals a good agreement between the predicted values and observed results for temperature, with a minimum error of about 0.16 for solar radiation. The results confirm the reliability and robustness of the forecasting process, and confirm its effectiveness in capturing complex dynamics with a high degree of accuracy.

The efficacy of neural networks heavily relies on hyper-parameters. Previous studies, referenced as [48,49], have explored various techniques such as the Bayesian Gaussian substitution process, boosted gradient regression trees, random forest, and heuristic algorithms to identify optimal hyper-parameter configurations. In this current investigation, diverse neural network architectures underwent training and cross-validation. This process involved weight optimization using the back-propagation algorithm, while considering the impact of the number of epochs and hidden-layer NNs' performance.

Through experimentation, the most efficient configuration was determined to be a two-layer feedback network with 15 hidden-layer neurons, trained for 1000 epochs using the SCG algorithm. The dataset comprised 60,538 temperatures and solar radiation values, with 70% allocated to training and 30% to validation and testing sets. The network demonstrated its suitability for regression tasks, evidenced by an MSE of 0.0044 and R of 0.99, as depicted in Figures 5 and 6, respectively. These results indicate a strong correlation between the network's outcomes and the desired objectives, with minimal error.

4. Results

Appendices B and C detail the specifications of the SG340P PV panel and the boost converter, with a resistor used as the load. The initial step was to assess the SG340P panel's performance under various temperature and radiation levels. Figure 9 presents the IV and PV characteristics recorded under these different conditions.

The $\mathbf{P_{PV}}$ curve shows the power generated by the PV panel without tracking, while the $\mathbf{P_{MPP}}$ represents the maximum power achieved with the proposed tracking technique. Figure 11 demonstrates that the PV power curve does not reach its maximum potential. However, the strength of the MPP detection algorithm lies in its precise identification of the Maximum Power Point (MPP), even during sudden fluctuations, a challenge for other methods. This highlights its exceptional ability to adapt to dynamic conditions.

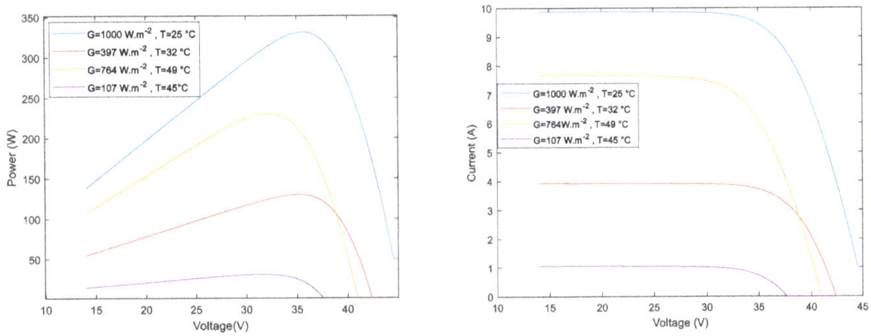

Figure 9. PV panel characteristics.

To accurately evaluate the performance of the Minimum-Risk Problem (MRP) algorithm under different conditions, we conducted a series of tests at different levels of temperature and solar radiation. Figure 10 provides details of the test parameters.

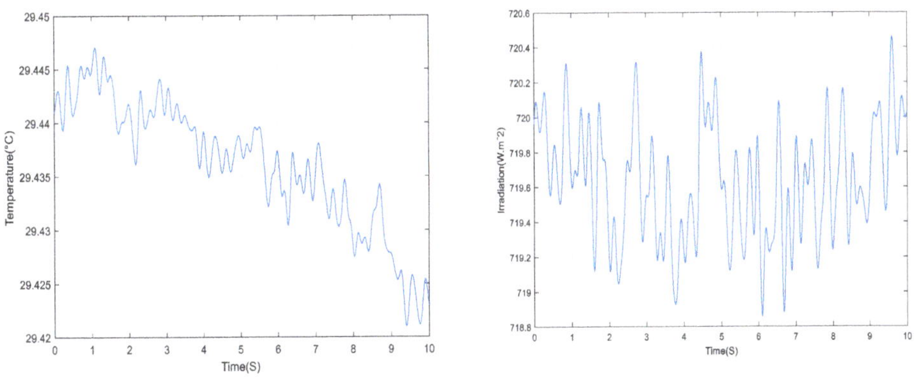

Figure 10. Solar irradiation and temperature profile.

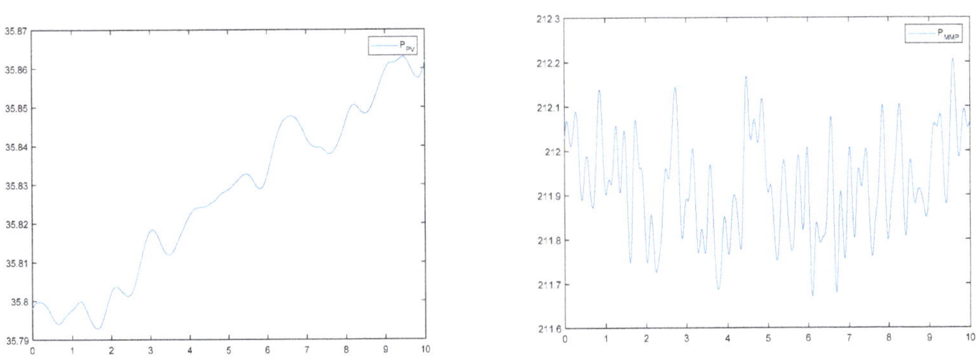

Figure 11. Optimization algorithm performance testing for PV.

Figure 12 illustrates the overall design. The effectiveness of the new neural-network-based prediction technique is evaluated by comparing the actual and predicted temperature and irradiation values, aiming to minimize the difference between the target and the predicted values.

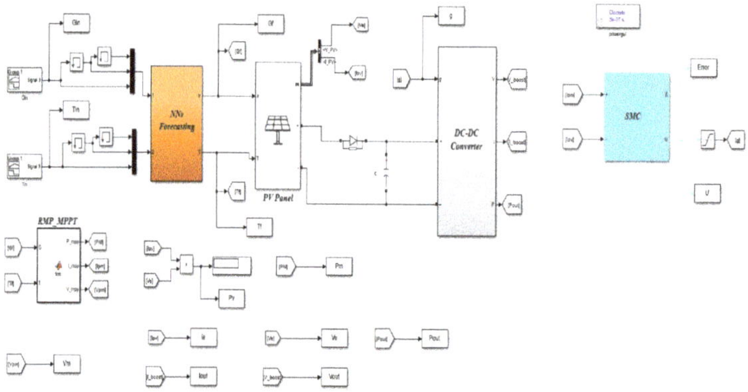

Figure 12. System's overall structure.

We used the predicted temperature and radiation (Tf and Gf) from Figure 13 to simulate the model and forecast the system's behavior under these conditions.

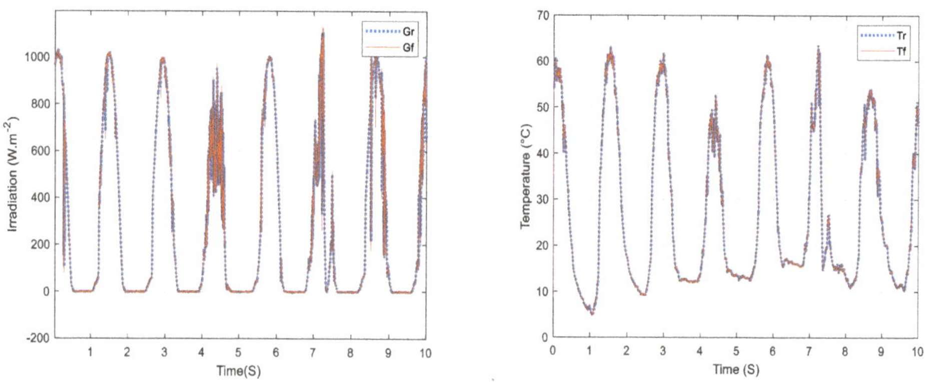

Figure 13. Irradiation and temperature profile.

Figure 14 shows the control signal and the error between the reference current (I_{mpp}) and the boost input current. The duty cycle changes along with the MPPT variations. This demonstrates the controller's ability to adjust to changing system conditions.

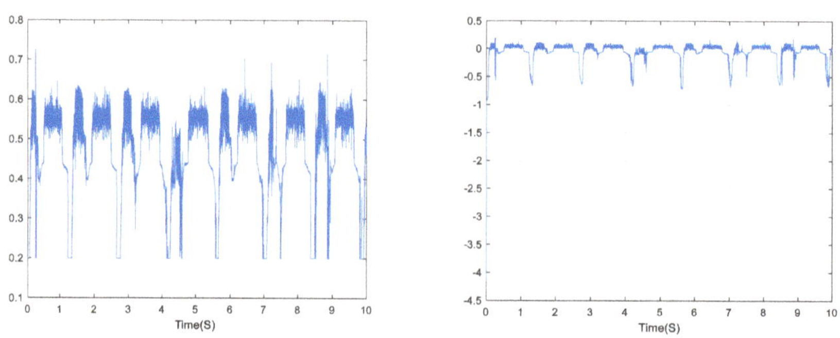

(**a**) (**b**)

Figure 14. Response of SMC: (**a**)controller signal; and (**b**) current error.

The small error shows that the SMC is accurate and effective in controlling the process. Figures 15–17 display, respectively, the voltage, current, and power results for our model, using a load with a resistance of 30 Ω.

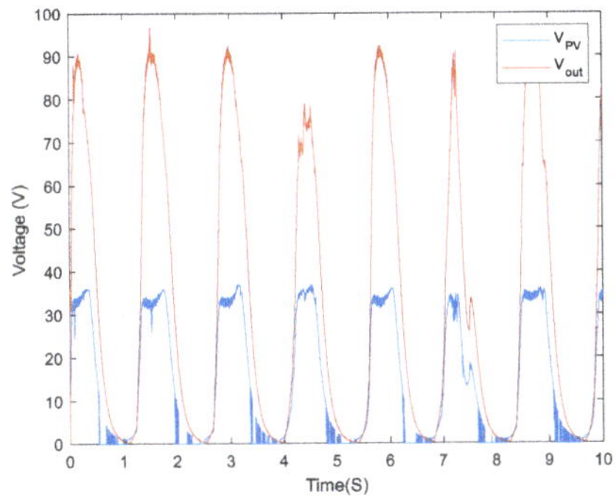

Figure 15. Temporal evolution of the voltage.

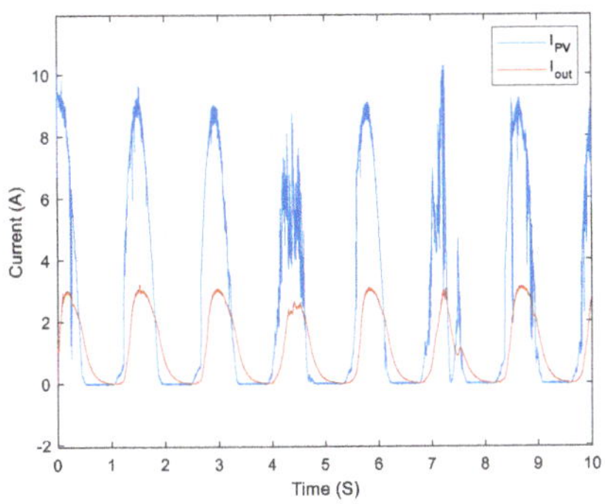

Figure 16. Temporal evolution of the electrical current.

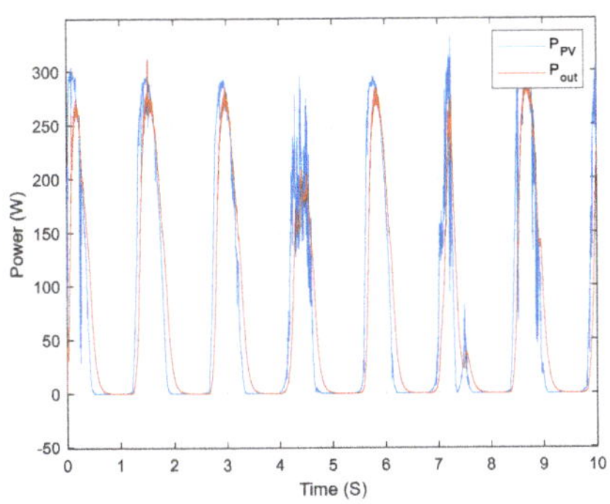

Figure 17. Temporal evolution of the power.

The previous figures demonstrate the optimization method's success in reaching the maximum point. The method adapts well to predicted changes, as seen in Figure 13. The SMC effectively tracks changes from both the optimizer and the PV panel, showing a strong performance during dynamic system transitions.

5. Comparative Study

5.1. Forecasting Method Comparison

To analyze the proposed forecasting solution, a comparative study was performed, and developed with specific time-series methodologies such as Autoregressive Moving Average (ARMA) and Autoregressive Integrated Moving Average (ARIMA). The results are shown in Figure 18 for the predictive temperatures and radiation values generated by each method.

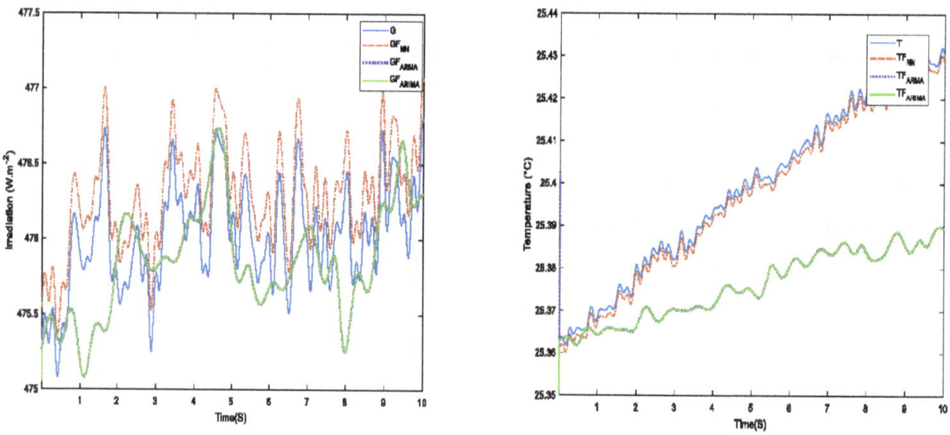

Figure 18. Comparative study of temperature and irradiation forecasting.

To evaluate the forecasting prowess of the ANN model, four metrics are employed, as illustrated in the equation below: R, Mean Absolute Error (MAE), Root Mean Square Error (RMSE), and MSE.

$$MAE = \frac{1}{n}\sum_{i=1}^{n}\left|y_{measured} - y_{predicted}\right| \tag{24}$$

$$MSE = \frac{1}{n}\sum_{i=1}^{n}\left(y_{measured} - y_{predicted}\right)^2 \tag{25}$$

$$RMSE = \sqrt{MSE} \tag{26}$$

$$R^2 = 1 - \frac{\sum\left(y_{measured} - y_{predicted}\right)^2}{\sum\left(y_{measured} - \overline{y}\right)^2} \tag{27}$$

where $\overline{y}$ is the mean value of $y_{measured}$

Table 1 presents analyses comparing our approach against recent studies in forecasting. We assess various aspects such as the system type and prediction model. This underscores the competitiveness of our methodology.

Table 1. Performance metric for prediction.

Model	R^2	MSE	RMSE	MAE	Neural N°	Training Data (%)
ANN	0.98	0.0044	0.066	0.033	15	70
ARIMA	0.99	0.29	0.47	0.088	-	70
ARMA	0.99	0.22	0.47	0.081	-	70

Based on Figure 18 and Table 1, it is clear that the artificial neural network stands out as the best performer among the models evaluated in terms of predictive accuracy. With an impressive R^2 score of 0.98, the proposed ANN model demonstrates a strong relationship between the predicted and actual values. The significantly low MSE of 0.0044, RMSE of 0.066, and MAE of 0.033 also confirm its accuracy and minimal forecast discrepancies. In contrast, while the ARIMA and ARMA models show a strong correlation (R^2: 0.99), they show higher errors compared to ANN. Hence, the proposed ANN model outperforms the time-series-based approach in terms of accuracy and predictive ability.

ANNs outperform the ARIMA and ARMA models due to their ability to capture complex, non-linear patterns without strict assumptions about data distribution. They automatically learn relevant features from raw data, eliminating the need for manual feature engineering. This adaptability allows ANNs to effectively model time-series data with intricate dependencies and irregularities, making them versatile and powerful tools for forecasting. Additionally, ANNs generalize well to unseen data and can perform a wide range of tasks, enhancing their superiority over traditional statistical methods like ARIMA and ARMA.

5.2. MPPT Approach Comparison

To control the input current of the boost converter, we used the SMC, and, to verify the effectiveness of the proposed controller, we will compare it with one of the most famous classical controllers, which is the PI. Through experiments, we have determined the parameters of the PI controller, and the following Figures 19 and 20 show the comparison results of the controllers.

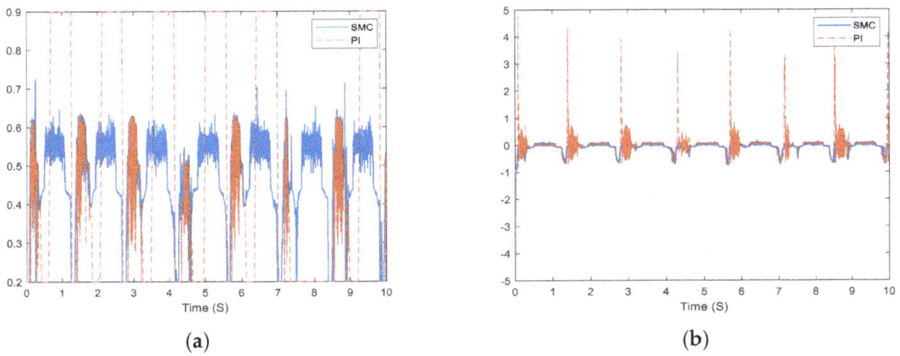

Figure 19. Controller's response: (**a**) controller signal; and (**b**)current error.

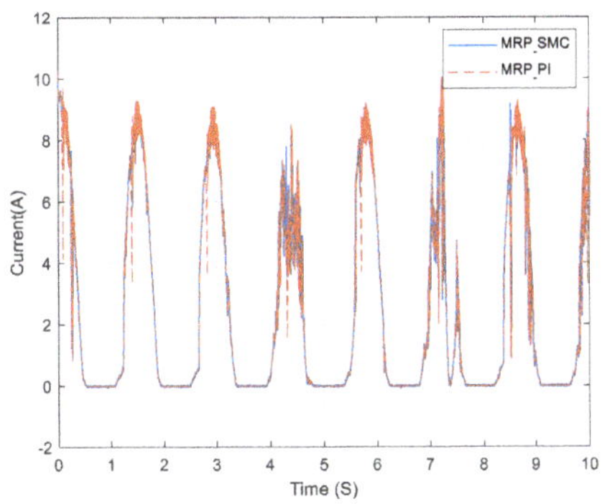

Figure 20. Comparison of the time evolution of PV current.

These figures show that the PI controller has some disadvantages compared to the SMC, especially when sudden changes occur. From Figure 19b, the SMC appears more stable and robust in handling the current error.

To enhance the evaluation of the proposed approach, a comprehensive comparative analysis was performed using the JAYA method. A brief overview of the JAYA algorithm is provided in Appendix D for reference. In order to create a robust framework for evaluating performance, the following three figures illustrate the response of a PV system and serve as a basis for comparison. The Figure 21 shows the comparison between this two methods.

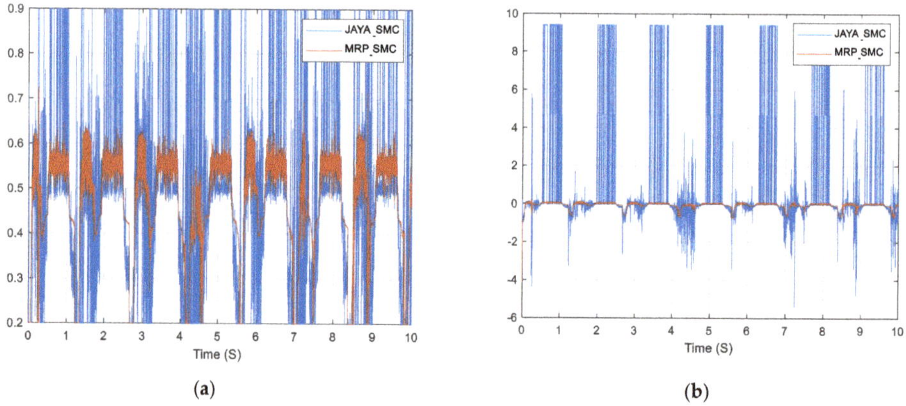

Figure 21. Comparison of SMC response: (**a**) controller signal; and (**b**) current error.

From these two curves, we concluded that the SMC method performs better when used with the MRP method compared to the Jaya method. This is because MRP is faster in the MPP search, making SMC more stable, even with changes in error correction. Figures 22–24 show the time evolution comparison of the output voltage, current, and power, respectively.

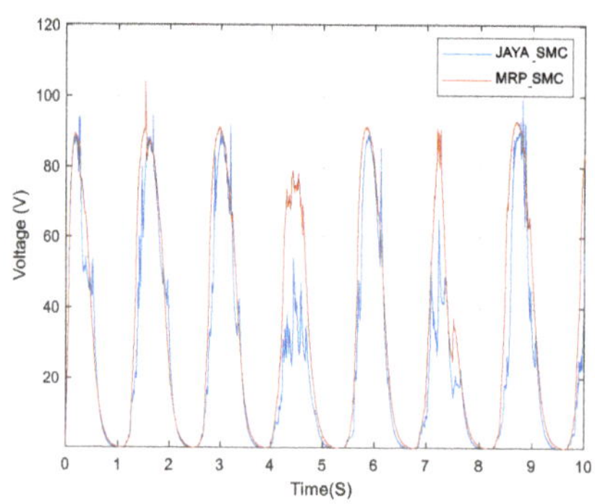

Figure 22. Comparison of the time evolution of output voltage.

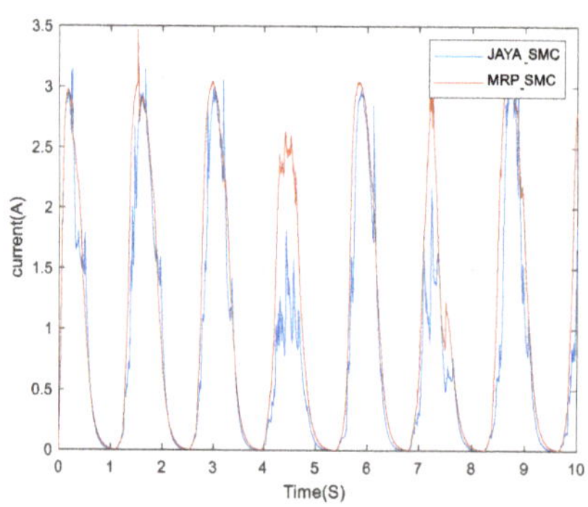

Figure 23. Comparison of the time evolution of output current.

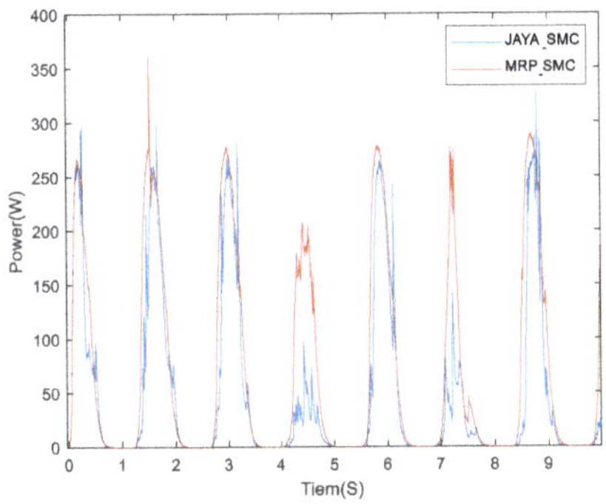

Figure 24. Comparison of the time evolution of output power.

The last three figures show that the MRP_SMC method outperforms JAYA_SMC by effectively tracking maximum points. Utilizing real-time temperature and radiation data in simulations, the MRP_SMC method exhibits superior adaptability to random variations, enabling it to identify maximum points more efficiently than JAYA_SMC. Conversely, JAYA struggles with rapid climate shifts, resulting in output fluctuations that can compromise the system's performance and accuracy, limiting its real-time applicability. In contrast, the MRP approach offers numerous advantages, including resilience to uncertainty, flexibility, support for risk-informed decision-making, integration with control strategies, and applicability across various domains. Additionally, the MRP method facilitates the quantification of uncertainty and provides enhanced decision support, making it invaluable for addressing optimization challenges in complex and uncertain environments.

6. Conclusions

In this study, we have introduced a new way to improve how we predict, control, and optimize the power output from solar panels. We use artificial neurons and a method called Minimal Risk Problem–Sliding Mode Control (MRP-SMC). Our method is efficient because it uses hidden layers to make better predictions and be more accurate. Unlike other methods, it can operate autonomously, even when errors occur or predictions are not perfect. Our method has been tested and shows good results. It has a low mean squared error (MSE) of 0.0044 and a high regression coefficient of 0.99, meaning it is very close to the actual output; the RMSE and MAE values are also low, showing that it predicts with minimal errors. We also use MRP-SMC to find and adjust the maximum power of the solar panels. Our method performs better than other methods like the JAYA method. It excels in adapting to weather-induced changes that impact solar panels. Our approach does not need extra equipment like Open-Circuit Voltage (OCV) or Short-Circuit Current (SCC) techniques. Instead, it controls the input current through the boost converter, using the reference Imp provided by the MRP. Our results show that our predictive model, MPPT, and SMC approaches are more effective than classical controllers. This is demonstrated by the comparison results between SMC and PI.

This research suggests future directions for improving MPPT in solar systems. We plan to develop a robust monitoring system and use intelligent systems to detect and rectify errors in real time.

Author Contributions: O.B.: conceptualization, simulation and data collection, and review. M.J.: simulation and data collection. M.J.: writing—original draft. M.J. and O.B.: writing—review and editing. F.H. and M.A.: supervision. All authors have read and agreed to the published version of the manuscript.

Funding: The authors wish to express their gratitude to the Basque Government, through the project EKOHEGAZ II (ELKARTEK KK-2023/00051); to the Diputación Foral de Álava (DFA), through the project CONAVANTER; to the UPV/EHU, through the project GIU23/002; and to the Mobility Lab Foundation (CONV23/14, CONV23/12) for supporting this work.

Data Availability Statement: The original contributions presented in the study are included in the article, further inquiries can be directed to the corresponding authors.

Acknowledgments: I wish to extend my heartfelt appreciation to the individuals and organizations whose support and contributions have been pivotal in the completion of this article.

Conflicts of Interest: The authors declare no conflicts of interest.

Abbreviations

ANN	Artificial Neural Network
AI	Artificial Intelligence
NN	Neural Network
R	Regression
MSE	Mean Squared Error
MPP	Maximum Power Point
MPPT	Maximum Power Point Tracking
PV	Photovoltaic
P&O	Perturb and Observe
Tanh	Hyperbolic Tangent
OCV	Open-Circuit Voltage
SCC	Short-Circuit Current
SMC	Sliding Mode Control
GA	Genetic Algorithm
PSO	Particle Swarm Optimization
PI	Proportional Integral
DC	Direct Current
IC	Incremental Conductance
QSVM	Quadratic Support Vector Machine
CNN-BILSTM	Convolution Neural Network–Bi-Direction Long Short-Term Memory
ANFIS	Adaptive Neuron Fuzzy Inference System
GMDH	Group Method of Data Handling
ANFIS-PSO	Adaptive Neuron Fuzzy Inference System–Particle Swarm Optimization
SCG	Scaled Conjugate Gradient
SOP	Stochastic Optimization Problem

Appendix A. Sensor Specifications

Properties	Values
Voltage	12–28 V
Irradiance measurement range	Up to 1500 W/m^2
Temperature measurement range	−40 to 90 °C

Appendix B. PV Panel Parameters

Maximum power (Pmax)	340 W
Voltage at Pmax (Vmp)	36.7 V
Current at Pmax (Imp)	9.28 A
Open-circuit voltage (Voc)	45.2 V
Short circuit current (Isc)	9.9 A

Appendix C. The Boost Specifications

Properties	Values
Switching frequency	20 kHz
Maximum input voltage	60 V
Maximum output voltage	250 V
Maximum input current	30 A
Maximum output current	30 A

Appendix D. JAYA Algorithm

Step 1: Define the population size and the maximum number of iterations as N_{pop} and N_{max}, respectively.
Step 2: Identify the best and worst solutions as X_{best} and X_{worst}.
Step 3: While the current generation is less than or equal to N_{max}:

- For each individual in the population:
  - Update the individual's value using the equation $X_n(i,j) = X(i,j) + rand1_{i,j}(X_{best}(j) - |X(i,j)|) - rand2_{i,j}(X_{worst}(j) - |X(i,j)|)$
  - Assess the updated value and replace the previous one if it's more suitable.
  - End the loop.

Step 4: Display the existing solutions X(i) and their corresponding values f(X(i)).

References

1. Song, D.; Liu, Y.; Qin, T.; Gu, H.; Cao, Y.; Shi, H. Overview of the policy instruments for renewable energy development in China. *Energies* **2022**, *15*, 6513. [CrossRef]
2. Asif, M.H.; Zhongfu, T.; Ahmad, B.; Irfan, M.; Razzaq, A.; Ameer, W. Influencing factors of consumers' buying intention of solar energy: A structural equation modeling approach. *Environ. Sci. Pollut. Res.* **2023**, *30*, 30017–30032. [CrossRef] [PubMed]
3. Sarkar, D.; Odyuo, Y. An ab initio issues on renewable energy system integration to grid. *Int. J. Sustain. Energy Plan. Manag.* **2019**, *23*. [CrossRef]
4. Azarpour, A.; Mohammadzadeh, O.; Rezaei, N.; Zendehboudi, S. Current status and future prospects of renewable and sustainable energy in North America: Progress and challenges. *Energy Convers. Manag.* **2022**, *269*, 115945. [CrossRef]
5. Katche, M.L.; Makokha, A.B.; Zachary, S.O.; Adaramola, M.S. A comprehensive review of maximum power point tracking (mppt) techniques used in solar pv systems. *Energies* **2023**, *16*, 2206. [CrossRef]
6. Nkambule, M.S.; Hasan, A.N.; Ali, A.; Hong, J.; Geem, Z.W. Comprehensive evaluation of machine learning MPPT algorithms for a PV system under different weather conditions. *J. Electr. Eng. Technol.* **2021**, *16*, 411–427. [CrossRef]
7. Verma, D.; Nema, S.; Agrawal, R.; Sawle, Y.; Kumar, A. A different approach for maximum power point tracking (MPPT) using impedance matching through non-isolated DC-DC converters in solar photovoltaic systems. *Electronics* **2022**, *11*, 1053. [CrossRef]
8. Khodair, D.; Motahhir, S.; Mostafa, H.H.; Shaker, A.; Munim, H.A.E.; Abouelatta, M.; Saeed, A. Modeling and Simulation of modified MPPT techniques under varying operating climatic conditions. *Energies* **2023**, *16*, 549. [CrossRef]
9. Radia, M.A.; El Nimr, M.K.; Atlam, A.S. IoT-based wireless data acquisition and control system for photovoltaic module performance analysis. *E-Prime-Adv. Electr. Eng. Electron. Energy* **2023**, *6*, 100348. [CrossRef]
10. Rojek, I.; Mikołajewski, D.; Mroziński, A.; Macko, M. Machine Learning-and Artificial Intelligence-Derived Prediction for Home Smart Energy Systems with PV Installation and Battery Energy Storage. *Energies* **2023**, *16*, 6613. [CrossRef]
11. Sharma, A.K.; Pachauri, R.K.; Choudhury, S.; Minai, A.F.; Alotaibi, M.A.; Malik, H.; Márquez, F.P.G. Role of metaheuristic approaches for implementation of integrated MPPT-PV systems: A comprehensive study. *Mathematics* **2023**, *11*, 269. [CrossRef]
12. Mehazzem, F.; André, M.; Calif, R. Efficient Output Photovoltaic Power Prediction Based on MPPT Fuzzy Logic Technique and Solar Spatio-Temporal Forecasting Approach in a Tropical Insular Region. *Energies* **2022**, *15*, 8671. [CrossRef]
13. Koeva, D.; Kutkarska, R.; Zinoviev, V. High Penetration of Renewable Energy Sources and Power Market Formation for Countries in Energy Transition: Assessment via Price Analysis and Energy Forecasting. *Energies* **2023**, *16*, 7788. [CrossRef]
14. Gielen, D.; Boshell, F.; Saygin, D.; Bazilian, M.D.; Wagner, N.; Gorini, R. The role of renewable energy in the global energy transformation. *Energy Strategy Rev.* **2019**, *24*, 38–50. [CrossRef]
15. Sajid, I.; Gautam, A.; Sarwar, A.; Tariq, M.; Liu, H.D.; Ahmad, S.; Sayed, A.E. Optimizing Photovoltaic Power Production in Partial Shading Conditions Using Dandelion Optimizer (DO)-Based MPPT Method. *Processes* **2023**, *11*, 2493. [CrossRef]
16. Akman, D.V.; Malekipirbazari, M.; Yenice, Z.D.; Yeo, A.; Adhikari, N.; Wong, Y.K.; Gumus, A.T. k-best feature selection and ranking via stochastic approximation. *Expert Syst. Appl.* **2023**, *213*, 118864. [CrossRef]

17. Triki, Y.; Bechouche, A.; Seddiki, H.; Abdeslam, D.O. An improved incremental conductance based MPPT algorithm for photovoltaic systems. In Proceedings of the IECON 2021–47th Annual Conference of the IEEE Industrial Electronics Society, Toronto, ON, Canada, 13–16 October 2021; pp. 1–6. [CrossRef]
18. Thirunavukkarasu, G.S.; Seyedmahmoudian, M.; Jamei, E.; Horan, B.; Mekhilef, S.; Stojcevski, A. Role of optimization techniques in microgrid energy management systems—A review. *Energy Strategy Rev.* **2022**, *43*, 100899. [CrossRef]
19. Du, Y.; Xu, F. A hybrid multi-step probability selection particle swarm optimization with dynamic chaotic inertial weight and acceleration coefficients for numerical function optimization. *Symmetry* **2020**, *12*, 922. [CrossRef]
20. Jlidi, M.; Hamidi, F.; Abdelkrim, M.N.; Jerbi, H.; Abbassi, R.; Kchaou, M. Synthesis of an Advanced Maximum Power Point Tracking Method for a Photovoltaic System: A Chaotic Jaya Logistic Approach. In Proceedings of the 2022 4th International Conference on Applied Automation and Industrial Diagnostics (ICAAID), Hail, Saudi Arabia, 29–31 March 2022; pp. 1–6. [CrossRef]
21. Chaitanya, K.; Somayajulu, D.V.; Krishna, P.R. Memory-based approaches for eliminating premature convergence in particle swarm optimization. *Appl. Intell.* **2021**, *51*, 4575–4608. [CrossRef]
22. Alshammari, B.M.; Farah, A.; Alqunun, K.; Guesmi, T. Robust design of dual-input power system stabilizer using chaotic Jaya algorithm. *Energies* **2021**, *14*, 5294. [CrossRef]
23. Yang, J.; Hu, Y.; Zhang, K.; Wu, Y. An improved evolution algorithm using population competition genetic algorithm and self-correction BP neural network based on fitness landscape. *Soft Comput.* **2021**, *25*, 1751–1776. [CrossRef]
24. Zafar, M.H.; Khan, N.M.; Mirza, A.F.; Mansoor, M.; Akhtar, N.; Qadir, M.U.; Moosavi, S.K.R. A novel meta-heuristic optimization algorithm based MPPT control technique for PV systems under complex partial shading condition. *Sustain. Energy Technol. Assess.* **2021**, *47*, 101367. [CrossRef]
25. Colbu, Ş.C.; Popescu, D.; Băncilă, D.M.; Mone, M.A.; Petrescu-Niţă, A. Stochastic Optimization and Risk Problems for Engineering Applications. In Proceedings of the 2022 8th International Conference on Control, Decision and Information Technologies (CoDIT), Istanbul, Turkey, 17–20 May 2022; pp. 775–779. [CrossRef]
26. Ampountolas, A. Modeling and forecasting daily hotel demand: A comparison based on sarimax, neural networks, and garch models. *Forecasting* **2021**, *3*, 580–595. [CrossRef]
27. Kurani, A.; Doshi, P.; Vakharia, A.; Shah, M. A comprehensive comparative study of artificial neural network (ANN) and support vector machines (SVM) on stock forecasting. *Ann. Data Sci.* **2023**, *10*, 183–208. [CrossRef]
28. Gourvenec, S.; Sturt, F.; Reid, E.; Trigos, F. Global assessment of historical, current and forecast ocean energy infrastructure: Implications for marine space planning, sustainable design and end-of-engineered-life management. *Renew. Sustain. Energy Rev.* **2022**, *154*, 111794. [CrossRef]
29. Acharya, S.K.; Wi, Y.M.; Lee, J. Day-ahead forecasting for small-scale photovoltaic power based on similar day detection with selective weather variables. *Electronics* **2020**, *9*, 1117. [CrossRef]
30. Ellahi, M.; Usman, M.R.; Arif, W.; Usman, H.F.; Khan, W.A.; Satrya, G.B.; Shabbir, N. Forecasting of Wind Speed and Power through FFNN and CFNN Using HPSOBA and MHPSO-BAACs Techniques. *Electronics* **2022**, *11*, 4193. [CrossRef]
31. Jlidi, M.; Hamidi, F.; Barambones, O.; Abbassi, R.; Jerbi, H.; Aoun, M.; Karami-Mollaee, A. An Artificial Neural Network for Solar Energy Prediction and Control Using Jaya-SMC. *Electronics* **2023**, *12*, 592. [CrossRef]
32. Kabilan, R.; Chandran, V.; Yogapriya, J.; Karthick, A.; Gandhi, P.P.; Mohanavel, V.; Manoharan, S. Short-term power prediction of building integrated photovoltaic (BIPV) system based on machine learning algorithms. *Int. J. Photoenergy* **2021**, *2021*, 5582418. [CrossRef]
33. Rai, A.; Shrivastava, A.; Jana, K.C. A CNN-BiLSTM based deep learning model for mid-term solar radiation prediction. *Int. Trans. Electr. Energy Syst.* **2021**, *31*, e12664. [CrossRef]
34. Obiora, C.N.; Ali, A.; Hasan, A.N. Estimation of hourly global solar radiation using deep learning algorithms. In Proceedings of the 2020 11th International Renewable Energy Congress (IREC), Hammamet, Tunisia, 29–31 October 2020; pp. 1–6. [CrossRef]
35. Olatomiwa, L.; Mekhilef, S.; Shamshirband, S.; Petković, D. Adaptive neuro-fuzzy approach for solar radiation prediction in Nigeria. *Renew. Sustain. Energy Rev.* **2015**, *51*, 1784–1791. [CrossRef]
36. Khosravi, A.; Nunes, R.O.; Assad, M.E.H.; Machado, L. Comparison of artificial intelligence methods in estimation of daily global solar radiation. *J. Clean. Prod.* **2018**, *194*, 342–358. [CrossRef]
37. Halabi, L.M.; Mekhilef, S.; Hossain, M. Performance evaluation of hybrid adaptive neuro-fuzzy inference system models for predicting monthly global solar radiation. *Appl. Energy* **2018**, *213*, 247–261. [CrossRef]
38. Yilmaz, U.; Kircay, A.; Borekci, S. PV system fuzzy logic MPPT method and PI control as a charge controller. *Renew. Sustain. Energy Rev.* **2018**, *81*, 994–1001. [CrossRef]
39. Idrissi, Y.E.A.; Assalaou, K.; Elmahni, L.; Aitiaz, E. New improved MPPT based on artificial neural network and PI controller for photovoltaic applications. *Int. J. Power Electron. Drive Syst.* **2022**, *13*, 1791. [CrossRef]
40. Tang, J.; Ni, H.; Peng, R.L.; Wang, N.; Zuo, L. A review on energy conversion using hybrid photovoltaic and thermoelectric systems. *J. Power Sources* **2023**, *562*, 232785. [CrossRef]
41. Geethamahalakshmi, G.; Kalaiarasi, N.; Nageswari, D. Fuzzy based MPPT and solar power forecasting using artificial intelligence. *Intell. Autom. Soft Comput.* **2022**, *32*, 1667–1685. [CrossRef]
42. Hamidi, F.; Olteanu, S.C.; Popescu, D.; Jerbi, H.; Dincă, I.; Ben Aoun, S.; Abbassi, R. Model Based Optimisation Algorithm for Maximum Power Point Tracking in Photovoltaic Panels. *Energies* **2020**, *13*, 4798. [CrossRef]

43. Al-Masri, H.M.; Magableh, S.K.; Abuelrub, A.; Saadeh, O.; Ehsani, M. Impact of different photovoltaic models on the design of a combined solar array and pumped hydro storage system. *Appl. Sci.* **2020**, *10*, 3650. [CrossRef]
44. Nguyen-Duc, T.; Nguyen-Duc, H.; Le-Viet, T.; Takano, H. Single-diode models of PV modules: A comparison of conventional approaches and proposal of a novel model. *Energies* **2020**, *13*, 1296. [CrossRef]
45. Rusirawan, D.; Farkas, I. Identification of model parameters of the photovoltaic solar cells. *Energy Procedia* **2014**, *57*, 39–46. [CrossRef]
46. Anani, N.; Ibrahim, H. Adjusting the single-diode model parameters of a photovoltaic module with irradiance and temperature. *Energies* **2020**, *13*, 3226. [CrossRef]
47. Escamilla-García, A.; Soto-Zarazúa, G.M.; Toledano-Ayala, M.; Rivas-Araiza, E.; Gastélum-Barrios, A. Applications of artificial neural networks in greenhouse technology and overview for smart agriculture development. *Appl. Sci.* **2020**, *10*, 3835. [CrossRef]
48. Sajjad, U.; Hussain, I.; Raza, W.; Sultan, M.; Alarifi, I.M.; Wang, C.C. On the critical heat flux assessment of micro-and nanoscale roughened surfaces. *Nanomaterials* **2022**, *12*, 3256. [CrossRef]
49. Gaspar, A.; Oliva, D.; Cuevas, E.; Zaldívar, D.; Pérez, M.; Pajares, G. Hyperparameter optimization in a convolutional neural network using metaheuristic algorithms. In *Metaheuristics in Machine Learning: Theory and Applications*; Springer International Publishing: Cham, Switzerland, 2021; pp. 37–59. [CrossRef]

energies

MDPI

Article

Prosumer Impact on Cellular Power Systems

Jens Maiwald *,† and Tino Schütte †

Faculty of Business Administration and Engineering, Zittau/Goerlitz University of Applied Sciences, Theodor-Koerner-Allee 16, 02763 Zittau, Germany; t.schuette@hszg.de
* Correspondence: jens.maiwald@hszg.de; Tel.: +49-3583-612-4634
† These authors contributed equally to this work.

Abstract: This paper explores the impact of an increasing number of prosumers in electricity supply systems and investigates how market mechanisms can mitigate the negative effects. The Regional Energy Market Model simulates a supply system based on cellular structures, employing agent-based modeling to capture individual behaviors and simulate real market dynamics. This study includes various supply scenarios, such as a solely photovoltaic scenario and a technically diversified scenario with biogas-fueled combined heat and power units. For each scenario, fixed and flexible pricing scenarios are simulated to analyze their effects. The findings reveal that systems heavily reliant on photovoltaics experience negative effects at certain points due to seasonal limitations, while technically diversified supply scenarios demonstrate fewer drawbacks. Flexible pricing systems stimulate demand in a manner beneficial to the system, creating regional added value, and contributing to the balance between generation and consumption, depending on the supply scenario. However, the study underscores that economic incentives alone are insufficient for balancing generation and consumption. The results highlight the importance of exploring opportunities through the interplay of economic incentive mechanisms and technical possibilities.

Keywords: cellular approach; agent-based modeling; regional energy markets; simulating behavior; prosumer; power system modeling

1. Introduction

1.1. Motivation

Once marked by a clear disjuncture between producers and consumers in energy economics, this distinction is now becoming increasingly blurred with the emergence of the phenomenon of prosumers. Prosuming is defined as the (partial) consumption of self-generated electricity, where prosumers embody both producer and consumer roles. Often, prosumers are also the owners of the corresponding generation units. This is particularly evident with low-power generation systems, typically designed to supply single consumers or small consumer groups. Prosumers can either be natural persons or legal entities [1].

Decreasing prices and substantial policy support, such as fixed feed-in tariffs, along with rising electricity market prices, have significantly increased the economic attractiveness of self-supply technologies for many. From a systemic perspective, it is not only about self-supply but also the feed-in of surplus generation into the grid. Thus, the role of prosumers in the energy system has been enlarged while simultaneously facing expectations regarding their contributions to the system [2].

In the overall picture, transitioning from a supply system heavily reliant on conventional energy sources to a more sustainable electricity supply primarily based on renewable energy sources (RESs), prosumers play an important role. Unfortunately, two main problems arise. First, fixed feed-in tariffs are not granted in perpetuity, and prosumer generation units must be profitable at market prices in the long term. Second, an increasing number of prosumers can pose potential challenges to the grid due to the volatility of RESs and the simultaneity of feed-in surpluses, such as voltage and frequency fluctuations, phase mismatches, and supply–demand imbalances [3]. All of these factors could negatively affect the reliability of supply systems.

Numerous papers discuss how technical solutions (e.g., storage systems) or digital aids (e.g., energy management systems) can be used to harmonize volatile generation with rather inflexible demand patterns in different stages of grid levels. There are myriad possibilities, too many to present and discuss in detail within this paper. However, one option has been relatively unexplored so far: economic incentives to flexibilize the demand.

1.2. Theory Development and Hypotheses

Supply systems, especially electricity grids, are currently regulated top-down from the upper to the lower grid levels. However, as the majority of prosumers will operate at the lowest grid level in the future, this control strategy no longer appears suitable given the sheer number of feed-ins. A bottom-up approach needs to be established to enable grid regulation at a lower level. The adoption of so-called microgrids as a solution for the integration of electricity generation from distributed RESs will allow technical problems to be solved in a decentralized fashion, reducing the need for an extremely ramified and complex central coordination [4].

A microgrid can be defined as a group of distributed (renewable) energy sources, energy storage systems, and controllable loads that operate locally as a single controllable entity. They exist in various sizes and can range from simple and small to complex and large networks [3–6]. Due to the energy storage capabilities, a microgrid can operate in grid-connected or islanded modes with a seamless transition between the two modes [6].

Optimally designed, microgrids offer significant benefits, including enhanced security of supply, lower electricity prices, the integration of excess RES generation into the microgrid, and even economic growth in rural areas [7]. However, achieving optimal energy management in microgrids is a challenging task, particularly in terms of efficiently utilizing RESs and energy storage systems while considering the uncertainty associated with load demand and renewable power generation [7].

In the context of this paper's focus, the fundamental contribution of financial incentives to adapting inflexible demand to volatile generation is examined. To sufficiently broaden the potential scope of the investigation, considering a microgrid in the sense of a district solution or smaller does not appear meaningful. The scope should be larger. Therefore, the analysis centers on a regional electricity market within a local supply system. Recently proposed regional market designs, as tested in [8], were found to be financially unattractive for prosumers within the current regulatory framework in the context of the German market. However, as the systematic review in [9] on different time-of-use electricity tariffs shows, there is a definite demand for flexible price tariffs. As demonstrated by [10–12], price signals can trigger reactions and adjustments in electricity demand among different consumer groups, impacting the market. Taking all this into account, a counterfactual market scenario (see Section 2.1) is set up to test the following hypotheses:

Hypothesis 1. *There is a limit to the increase in demand for green electricity, as both the willingness to pay and consumption have finite capacities and cannot perpetually grow.*

It is inherent that consumers' willingness to pay (WTP) and consumption are finite capacities. The WTP varies among individuals based on personal preferences and budget constraints. Likewise, electricity consumption is constrained by actual physical demand and technical limitations, restricting potential increases. Consequently, there is a natural constraint on the growth of market demand for green electricity.

Hypothesis 2. *An increase in the demand for green electricity is hypothesized to result in a reduction in the effort required to achieve an equilibrium between generation and consumption in the market area.*

An increase in demand for green electricity is incentivized by decreasing prices resulting from high feed-in volumes of green electricity. This will reduce surpluses in times of overcapacities. It is also expected that learning effects will occur as a result of successful transactions. Consumers are likely to become more conscious of their energy consumption patterns. This heightened awareness could lead consumers to adjust their electricity usage based on the availability of green electricity. Consequently, the systemic effort required to balance generation and consumption in the market area may decrease.

Hypothesis 3. *As the number of prosumers increases, there will be a positive effect on the supply of green electricity at first. This positive effect turns into a negative effect at a certain point when supply exceeds demand.*

An increasing number of prosumers implies a higher generation of green electricity available during periods of overcapacity in the market, thereby increasing the overall supply. Additionally, as prosumers (partially) consume their self-generated electricity, it reduces their impact on the demand side of the market, consequently increasing the available supply for pure demanders. However, as outlined in Hypothesis 1, the market demand is finite, establishing a saturation volume for supply. With the growing number of suppliers, reaching this saturation point becomes more likely. This effect is further intensified by the earlier mentioned fact: Prosumers do not solely act as consumers; the fewer consumers participating in the market, the smaller the saturation volume.

Hypothesis 4. *A supply system relying solely on solar photovoltaic prosumers will likely experience higher positive effects only during summer but only up to a certain point.*

At first glance, this hypothesis may not appear particularly bold, but it is grounded in several considerations. Photovoltaic (PV) is a highly seasonal and weather-dependent generation technology. During the winter months, the minimal trading volume of green electricity from PV is anticipated in the market. In transitional months, prosumers are likely to consume a substantial part of their self-generated electricity, also resulting in a low trading volume in the market. In summer, midday peaks will lead to significant overcapacity, surpassing the saturation volume of the market. An increasing number of PV prosumers will not counterbalance the negative effects in winter but will amplify the positive effects in summer until they become negative (see Hypotheses 1 and 3). Diversifying generation technologies could partially alleviate these effects by implementing technologies in the system capable of generating green electricity in the cold season and remaining inactive during summer, such as combined heat and power (CHP) units.

2. Materials and Methods

2.1. Structure of the Study

This study's investigations are based on the Regional Energy Market Model (REMM), a model designed to map actor behavior in regional energy markets and investigate the behavioral effects. Only some of the functions implemented in the REMM are required in this paper. Therefore, a complete description of the model will be omitted now and in the following paragraphs and only the part of the required implementations will be discussed. A detailed REMM description can be found in [13].

The REMM was developed in NetLogo (Version 6.2.2, see Appendix A) as a bottom-up approach for integral load management, investigating short-term scenarios in decentralized energy markets. While the REMM can simulate various supply scenarios, this paper focuses on local supply systems. This assumption aligns with [9], where it is proven that there is significant potential for covering energy demand based on renewable energies, especially in smaller communities. Moreover, ref. [9] demonstrates that the transformation of urban energy systems toward the use of local and sustainable energy resources can be the preferred alternative. However, as [9] primarily inspects these issues from a systems perspective using top-down equilibrium models, this paper's model primarily focuses on the consumer groups themselves, utilizing REMM's bottom-up approach.

An exemplary scenario was set up, comparable to the supply system of a small center town in Germany with 26,000 inhabitants. The observed supply system is, therefore, a local distribution grid with its typical entities. This includes private households and small to medium-sized companies from the trade, commerce, and service sectors, as well as the local utility company (LUC) acting as the grid operator. Each individual entity is represented by an agent in the model. It should be noted that the exclusion of companies from the industrial sector in this study was a deliberate choice. This decision is rooted in the fact that large consumers are typically connected to higher grid levels than the distribution grid, which would contradict the microgrid focus of these investigations. Additionally, the inclusion of industrial companies could significantly influence the model results, given their substantial leverage in demand volume compared to individual households or small businesses.

The REMM inherently possesses the capability to represent microgrids. As mentioned earlier, the model adopts a bottom-up approach, known as the cellular approach (CA; for further details on CA, refer to Section 2.2). Given that the REMM is an energy economics model, it incorporates the CA from an economic standpoint. Each model entity is tasked with achieving the equilibrium between generation and consumption independently. In cases where this equilibrium is unattainable, entities can actively seek suitable trading partners within the regional market. This regional market is not treated as an isolated market but as an integrated unit in the wholesale market. Therefore, if no suitable regional trading partner is found, wholesale trading becomes an alternative.

As the focus of this paper is not on optimizing the overall system outcome but rather on simulating emerging phenomena due to certain behaviors of market participants, agent-based modeling (ABM; for further details, refer to Sections 2.3 and 2.4) was the method of choice. Alongside the energy economic model, the REMM includes a behavioral model grounded in ABM. This behavioral model is inspired by the analysis presented in [12], specifically investigating environmentally conscious decisions related to the purchase of green electricity by residential customers, with a focus on customer preferences. This paper adopts and extends this approach to encompass the preference for regionality, applying it not only to one specific consumer group but to all consumer groups within the supply system.

To assess the presented hypotheses, the analyses in the REMM are structured in the following steps:

Step 1: A base scenario is calculated, serving as the starting point for all subsequent analyses. This scenario includes a fixed price system and assumes the approximated current expansion level of 10% PV for households and businesses [14].

Step 2: According to the base scenario, additional scenarios with expansion levels of 20%, 30%, and 50% PV are calculated. The 50% scenario represents an extreme case assumption. This step will deliver the first insights for the assessment of Hypotheses 3 and 4.

Step 3: Scenarios with a PV expansion of 10%, 20%, 30%, and 50% are simulated, each incorporating a flexible pricing system in the REMM. The design of the flexible pricing system is described in Section 3.1.4. The results are compared with those from steps 1 and 2 in order to assess Hypotheses 1 and 2. Moreover, new insights for Hypothesis 3 are gained throughout this step.

Step 4: To conclusively address Hypotheses 3 and 4, fixed pricing system (steps 1 and 2) and flexible pricing system (step 3) scenarios are calculated, incorporating additional CHP expansion from 5%, 10%, 15%, and 25%. It is presumed that the expansion of CHP occurs at a slower rate than that of PV. This assumption is grounded in the more challenging installation process associated with CHP systems compared to PV.

2.2. Cellular Approach

The CA, as a bottom-up approach for energy system planning, focuses on achieving equilibrium between energy generation and consumption at the lowest possible level.

Therefore, energy cells are at the center of the considerations. Each cell has energy-supplying converters, energy-dissipating converters, and one or more local storage units. In simple terms, a cell is defined as being able to generate, consume, and store energy. The goal of every cell is to balance generation and consumption itself. In the event a cell is unable to achieve this equilibrium, it connects with one or more other unbalanced cells. Several cells at one level can be grouped and reflected at the next higher level as a single cell, treated according to the same basic principle (see Figure 1) [15].

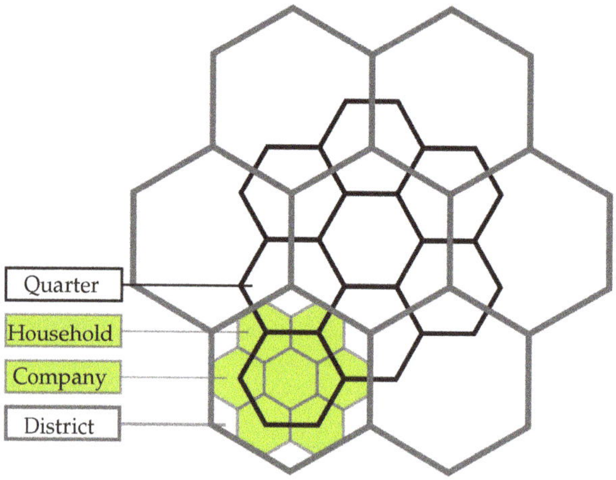

Figure 1. Energy cells (authors' own compilation based on [16]).

Although the basic concept of the CA was not new (see e.g., [17]), the first concrete feasibility study was presented in [15]. However, the adoption of the CA in this paper differs slightly from that study. In its fundamental approach, the CA assumes that cells follow a multi-modal approach, considering various types of energy, such as electricity and heat. As mentioned, this paper's focus is solely on the electricity sector. Additionally, the CA considers energy flows in their physical sense, while this paper's model balances energy quantities, focusing on the economic perspective. Moreover, the 2015 study raises the question of how an energy system would need to be fundamentally structured to implement the cellular principle (Greenfield Approach). In this paper, the CA is applied to the structures of an existing local supply system.

In its key finding, ref. [15] states that the cellular approach (CA) is generally feasible. However, it emphasizes the necessity to expand decentralized structures for future energy supply while simultaneously underlining the importance of interregional energy balancing.

2.3. Agent-Based Modeling

ABM allows for the depiction and investigation of individual interactions between agents, between agents and other factors such as time and space, and the resulting effects on the overall system [18]. Thus, interdependencies between the microscopic and macroscopic levels can be examined. On the microscopic level, agents make decisions based on the variables and rules applicable to them at the current point in time. In turn, on the superordinate, i.e., macroscopic, level, overall systemic phenomena occur without being solely attributable to isolated properties and decisions of individual agents [12,19–21].

2.4. Agent-Based Modeling in Energy Economics

ABM in economics offers a framework for the computational study of economic processes modeled as dynamic systems of interacting agents [22,23]. Events that occur in reality between customers and suppliers during the negotiation and trading processes are often greatly simplified or even neglected in other types of models. Even though the methodological approach is more demanding, ABM allows for the consideration of aspects such as asymmetric information, strategic interaction, expectation formation, transaction costs, external effects, multiple equilibria, and dynamics outside the equilibrium [22]. Within the market, agents communicate with each other and react to market signals by adjusting their individual strategies to different market events. These strategy adjustments (learning effects) are based on decision rules that are implemented for each agent. In such models, a market consists of repeating bidding rounds, with the possibility of strategy adjustments between each round [24,25].

Compared to other markets, special conditions apply to electricity markets, which result primarily from technical restrictions and characteristics of electricity as a commodity. In particular, the limited storability of electricity and its transmission through the grid is based on physical laws rather than economic rules [24,26]. Therefore, the trading of electricity is a balance sheet trade, as the physical fulfillment of the transaction is separate from the actual trade. In addition, the public grid represents a natural monopoly for which usage rights must be acquired and paid for in the form of grid fees.

Models in energy economics therefore usually comprise two levels to adequately depict this complexity: a technical and an economic level. In the case of ABM, sometimes even a socio-economic level is included. While the technical level contains various supply technologies and simulates the operation of the energy system, the market participants (agents) act on the socio-economic level, sometimes changing the structure of the technical system and concluding contracts for energy supplies [27]. Pure economic or financial models, on the other hand, can only poorly explain what happens in the electricity market [26].

3. The Regional Energy Market Model

3.1. Model Theory

3.1.1. Demand Side

Agents representing private households are characterized by the (dynamic) standard load profile H0 and are, therefore, referred to as Residential with Standard Load Profile (RSL) within the REMM. Trade, commerce, and service companies are characterized by the standard load profile G0 and are referred to as Business with Standard Load Profile (BSL). Both profiles were published by the German Electricity Association (see Appendix A). Since they are standardized to an annual consumption of 1000 kWh, each hourly value of these profiles has to be multiplied by a scale factor to make them usable in the model. The scale factor assigned to each agent was randomly chosen from a given domain during the model setup (see Table 1). Constraints ensure that the REMM accurately depicts the overall distribution of households or business sizes in Germany and thus their overall electricity consumption (see [13]). In total, the REMM comprised 15,407 RSL and 1638 BSL agents.

Table 1. Scale factor domains.

	Residential	Business
Scale factor	$\in [1;5]$	$\in [1;12]$

3.1.2. Supply Side

Given the focus on cellular structures and prosumers in this observation, it is assumed that demand-side agents operate all generation units. In line with the steps for assessing hypotheses (see Section 2.1), PV and CHP units are implemented in the model. Generation from solar PV accounts for the variable nature of the availability of RESs. CHP units, fueled with biogas, represent characteristics of controllable renewable generation.

The model randomly selects agents to become prosumers. Each prosumer can operate either only one type of generation unit or both combined. The individual generation capacities of these units are aligned with the annual electricity consumption of the operating agents, based on specific criteria (see [13]). Generation from PV relies heavily on solar radiation. CHP units operate in a heat-controlled mode, and therefore, their performance depends on the temperature. To accurately depict these generation patterns, a database is linked to the model, providing authentic local weather data for wind speed, solar radiation, and temperature. These data were obtained from the Test Reference Years of the German Meteorological Service (see Appendix A).

Each prosumer prefers to consume their self-generated electricity to cover their consumption. During periods of overcapacities, that means generation exceeds consumption, and prosumers sell the surplus electricity in the regional market. Conversely, if generation is insufficient to cover consumption, prosumers will purchase the additional electricity they need from the market.

3.1.3. Local Utility Company

In the model's observations, the focus is on the effects of prosumer and consumer behavior. Therefore, the LUC is modeled to act passively. In this context, 'passive' implies that the LUC is modeled without any consumption, demand, or supply. Moreover, the LUC operates without the intention of making a profit. However, given the existence of a local market and supply system, a system operator is essential. In the REMM, the LUC serves as this enabler, maintaining the overall system, responding to market conditions, and undertaking market clearing. Additionally, the LUC serves as the connector between the regional and interregional wholesale markets, selling or purchasing electricity based on regional over- or undercapacities, respectively.

Figure 2 gives an overview of the REMM's structure and interdependencies while summarizing Sections 3.1.1–3.1.3.

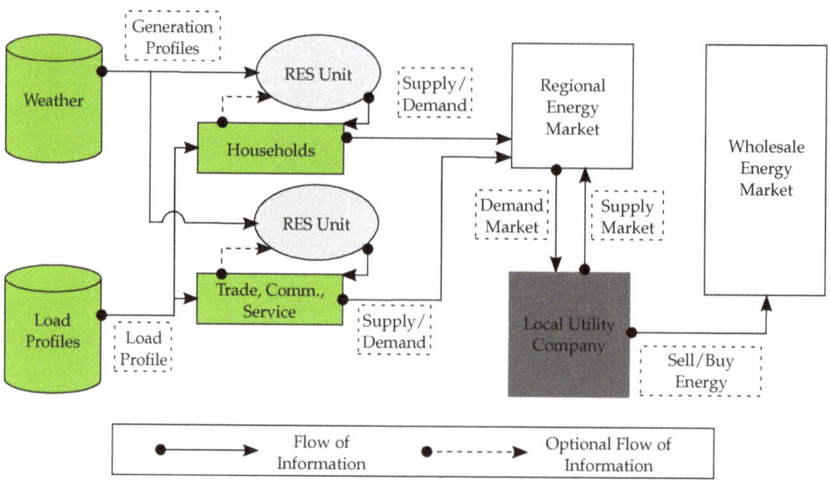

Figure 2. Structure of the REMM.

3.1.4. Market, Products, and Pricing

The REMM was developed to map the effects of preference-based consumer behavior (see Section 3.1.5). Consequently, it only makes sense to structure the product range in such a way that these preferences are reflected. The market was therefore divided into four different products incorporating a modular pricing system. The differentiation is based on the energy source from which the electricity is generated and the location where the electricity is generated. Thus, there are green and grey products, each with regional and interregional background (see Table 2).

It is worth noting that, at first glance, it seems obsolete to include a grey regional product if there is no local conventional production in the system. However, as will become clear later in the paper, this is nevertheless appropriate.

Table 2. Electricity products and price components in REMM.

	Origin	
Source	I—Green & Regional ($p_{base} + p_{green} + p_{regio}$)	II—Green & Interregional ($p_{base} + p_{green} + p_{trans}$)
	III—Grey & Regional ($p_{base} + p_{regio}$)	IV—Grey & Interregional ($p_{base} + p_{trans}$)

3.1.5. Decision Making

Agent behavior in the REMM is influenced by three properties: (I) environmental awareness, (II) regional awareness, and (III) cost sensitivity (see Table 3). Environmental awareness describes the individual preference for green energy sources. Regional awareness represents the preference for electricity generated in a local context. For both properties, a value close to one indicates a high preference, while a value close to zero indicates a low preference. Cost sensitivity describes the individual assessment of higher costs, which is directly dependent on the income for RSL or earnings for BSL. A value close to one indicates a high sensitivity to costs, while a value close to zero indicates a low sensitivity, meaning that these agents would be willing to pay higher prices.

Table 3. Consumer preference domains.

Environmental Awareness	Regional Awareness	Cost Sensitivity
$e \in]0;1]$	$l \in]0;1]$	$c \in]0;1]$

At each time step, all purchasing entities make a new decision regarding their preferred electricity product. This generally involves a two-stage decision process. These agents compare the grey and green products, as well as the regional and interregional products, by calculating their utility values U. Ultimately, the product is chosen based on the one that offers the highest personal value.

3.2. Model Implementation

3.2.1. Economic Model

The REMM characterizes entities by two main attributes, consumption con and demand dem. Consumption describes the total electricity need of an agent i per time step t, while demand represents the electricity purchase from the grid. The demand of a prosumer agent depends on his generation gen (see Equation (1)). Since the generation of a consumer agent is 0, it follows that $dem = con$.

$$dem_{i,t} = con_{i,t} - gen_{i,t} \quad (i = 1 \text{ to } n,\ t = 1 \text{ to } 8760) \tag{1}$$

The sales volume vol that prosumers offer on the market logically depends on their personal difference between consumption and generation. Since consumers are not able to generate electricity, the sales volume is therefore 0.

$$vol_{i,t} = \begin{cases} 0 & : gen_{i,t} \leq con_{i,t} \\ gen_{i,t} - con_{i,t} & : gen_{i,t} > con_{i,t} \end{cases} \tag{2}$$

The electricity price p_{el} results as the sum of three components: (I) source of electricity; (II) grid fee; and (III) taxes, levies, and apportionments. Since the third component is incurred for each product, it is considered as 0 and, thus, not taken into account in further analysis.

$$p_{el} = p_{source} + p_{grid} + p_{tax} \tag{3}$$

Buyers purchasing grey electricity have to pay a base price p_{base}. For those choosing green electricity, an additional price premium p_{green} is applied. This premium serves as an incentive for supporting renewable energy sources, resembling past initiatives such as the levy under the German Renewable Energies Act.

$$p_{source} = \begin{cases} p_{base} & : \textit{Grey product} \\ p_{base} + p_{green} & : \textit{Green product} \end{cases} \quad (4)$$

All prosumer entities selling self-generated electricity are considered producers offering regional electricity. Conversely, electricity generated outside the observed area is classified as interregional. In the REMM, regional purchases can incur a different grid fee than interregional purchases.

$$p_{grid} = \begin{cases} p_{reg} & : \textit{Regional Purchase} \\ p_{trans} & : \textit{Interregional Purchase} \end{cases} \quad (5)$$

The regional market is highly volatile. Consequently, situations can arise where parts of the preference-driven demand cannot be met. In REMM, the LUC is responsible for performing the market clearing. Situations characterized by a regional oversupply are not crucial for the simulation, given the assumption that leftover electricity could be sold at the wholesale market at any time. It is also not critical if there is no sales volume for green electricity in the market. However, situations with undersupply, or in this case, a shortage of sales volume compared to the demand for green regional electricity, can become challenging. A decision has to be made regarding which agents get served and who has to switch to another product, along with the criteria for making this decision.

In this special case, the decision on which agent to serve is based on their WTP, derived from utility functions described in Section 3.2.2. The green regional market section is cleared first. All applying agents are listed based on their WTP for green electricity, with those having a higher WTP served before those with a lower WTP. Agents who cannot be served are then forced to switch to an alternative product. The second step involves clearing the grey regional market section. Since there is no regional conventional generation discussed in this paper, and, consequently, sales volume is 0, it is mandatory to take this step due to later model procedures. All agents not served in the first two clearing steps, as well as those who initially decided to purchase interregional electricity, are served in the third step. For more information on determining the WTP and switching to alternative products, refer to [13].

3.2.2. Behavioral Model

Similar to the approach in [12], purchasing agents calculate the utility value U by comparing an intrinsic value u_{intr} with the negative value of (higher) costs u_{cost} in a two-stage decision process. The first stage is the decision on whether to choose a green or grey electricity product. The second stage involves deciding between purchasing a regional or interregional product. It should be noted that even if there is no consideration of regional conventional energy generation in this paper, it is still essential to allow all agents to decide on this product. This is necessary due to the interdependencies with alternative products required to calculate the utility values. Moreover, it is necessary within the market clearing process of the REMM.

Decision I: Calculating the intrinsic value for green electricity, it is assumed that one unit of the additional price premium p_{green} for electricity from RESs can be converted into exactly one unit of an abstract personal good. This abstract personal good can be interpreted as well-being or moral satisfaction. The intrinsic value is derived from the agent's environmental awareness e combined with its price sensitivity c and the amount of p_{green}. Consequently, the intrinsic value for the purchase of grey electricity is 0.

$$u_{intr,i,t} = \begin{cases} 0 & : Grey \\ e_i \cdot \sqrt{p_{green,t}} \cdot c_i & : Green \end{cases} \tag{6}$$

This intrinsic value is compared with the (negative) effect of higher costs caused by the extra mark-up for green electricity. The effect is determined by the agent's price sensitivity c and the amount of the price components p.

$$u_{cost,i,t} = \begin{cases} c_i^2 \cdot p_{base,t} & : Grey \\ c_i^2 \cdot (p_{base,t} + p_{green,t}) & : Green \end{cases} \tag{7}$$

Hence, the overall utility functions for the purchase of green and grey electricity are formed as follows:

$$U_{grey,i,t} = 0 - c_i^2 \cdot p_{base,t} \tag{8}$$

$$U_{green,i,t} = e_i \cdot \sqrt{p_{green,t}} \cdot c_i - c_i^2 \cdot (p_{base,t} + p_{green,t}) \tag{9}$$

Decision II: This stage's logic follows the approach from stage I. Here, it is assumed that an intrinsic value exists for the purchase of electricity generated in a regional context. The intrinsic value for interregional purchase is consequently 0.

$$u_{intr,i,t} = \begin{cases} l_i^2 \cdot \sqrt{p_{reg,t}} \cdot c_i & : Regional\ purchase \\ 0 & : Interregional\ purchase \end{cases} \tag{10}$$

The utility value for costs is determined analogously to the approach in stage I, taking into account the different price components for regional and interregional purchases.

$$u_{cost,i,t} = \begin{cases} c_i^2 \cdot p_{reg,t} & : Regional\ purchase \\ c_i^2 \cdot p_{trans,t} & : Interregional\ purchase \end{cases} \tag{11}$$

Combining both, the overall utility functions for regional and interregional purchases appear as follows:

$$U_{reg,i,t} = l_i^2 \cdot \sqrt{p_{reg,t}} \cdot c_i - c_i^2 \cdot p_{reg,t} \tag{12}$$

$$U_{inter,i,t} = 0 - c_i^2 \cdot p_{trans,t} \tag{13}$$

4. Results

4.1. Scenario Framework

The following scenario framework was created to implement the steps outlined in Section 2.1. Table 4 provides a summary of the different expansion stages to be simulated. Initially, a supply system equipped solely with PV units was simulated. After that, this PV system underwent further expansion to include CHP.

Table 4. Expansion stages for RESs.

Path	(a)	(b)	(c)	(d)
PV	10%	20%	30%	50%
PV & CHP	10%; 5%	20%; 10%	30%; 15%	50%; 25%

In addition, different price scenarios were simulated for these RES expansion stages. Table 5 provides an overview of the details. The price assumptions were derived from the average electricity prices for households in Germany in 2015, which also served as the reference year for the weather data and distribution variables for RSL and BSL agents [28]. The main difference between them was that the price component for green electricity in the flexible price scenario changed based on the regional utilization of the RES units; the higher the regional generation, the more affordable the price component, and vice versa. Finally, it was examined whether an additional reduction in the price component of the regional grid fee compared to the grid fee for interregional purchases had an additional effect.

Table 5. Price components for different scenarios.

	Fix	Flex	Flex & Regio
p_{base}	7.5 ct/kWh	7.5 ct/kWh	7.5 ct/kWh
p_{green}	6.5 ct/kWh	[2;6.5] ct/kWh	[2;6.5] ct/kWh
p_{regio}	7.5 ct/kWh	7.5 ct/kWh	6 ct/kWh
p_{trans}	7.5 ct/kWh	7.5 ct/kWh	7.5 ct/kWh

In order to account for the randomness inherent in the REMM, arising from distributions of agent properties, five model runs were calculated for each scenario and summarized at the end as an arithmetic mean.

4.2. Evaluation Framework

Essentially, this investigation focuses on two key issues and their outcomes: (I) the influence of RES expansion on generation and market sales volume; and (II) shifts in demand arising from the flexibility of electricity prices. Therefore, all scenarios, including PV (a)–(d) and PV and CHP (a)–(d), underwent simulation within both the fixed price and flex price systems, followed by subsequent analysis.

Since the total consumption in the system remains rather constant due to the fixed number of agents, the focus of the analyses is on demand. However, it is important to note that a change in demand does not equate to an adjustment in consumption. Instead, it signifies a shift from an alternative product to a green regional product, while overall consumption remains unchanged.

Two cases were of particular interest for the analyses: (I) the scenario where the sales volume for green regional electricity exceeded the demand for this product, implying that regional value creation could be realized with an increase in demand, and (II) the situation where the sales volume surpassed the total regional consumer consumption, indicating that an increase in consumption could result in regional value creation.

Therefore, the initial focus is on examining the ordered residual load curve of the supply system for each scenario (a)–(d). This involves subtracting the amount of generated RES electricity from the amount of consumption in each hour, enabling to draw conclusions regarding the duration of under- or overcapacities within the system.

The focus then shifts to the actual demand, specifically examining the changes that occur due to flexible prices and how the demand aligns with the sales volume during the period under review. For the sake of visualization, the analysis cannot be conducted individually for each hour. Therefore, a quantitative evaluation is performed in relation to the months of the year under review to present an overview of flexibilization effects and seasonal influences.

Finally, an analysis variable specifically developed for the REMM is introduced: the residual sales volume of the market. This variable is based on the logic of the residual load of the supply system, which primarily focuses on load (consumption or demand). However, for market evaluation in the REMM, the focus is on regional supply, specifically the sales volume. Consequently, the logic is applied in the opposite direction, where the actual demand is subtracted from the sales volume for green electricity. This approach allows to draw conclusions about when the market is oversupplied or undersupplied and to assess the effects of flexibilization.

4.3. Scenario Results: Solely PV

Figure 3 displays the ordered residual load curve for the period under review across the different scenarios (a)–(d) in relation to the entire supply system. It is noticeable that scenario (a) experiences no regional oversupply at any time. Hours with overcapacity begin to occur from scenario (b). Scenarios (c) and (d), with around 1000 and 1700 h, respectively, exhibit a large number of hours with sometimes immensely high overcapacities. In combination with an undercapacity of around 4000 h, these scenarios are not recommended from a systemic perspective.

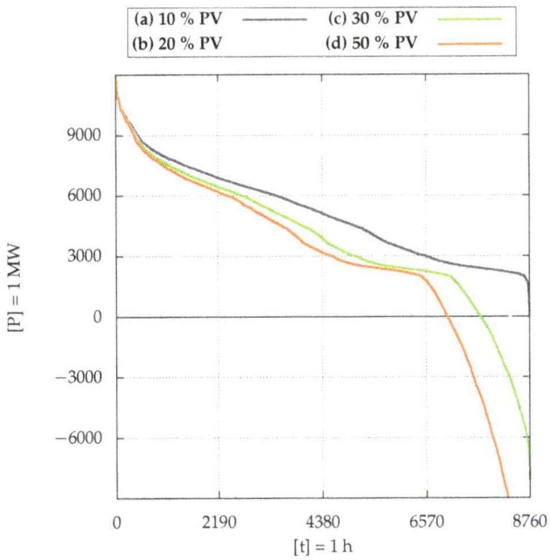

Figure 3. Ordered residual load profile (region) of the supply system over one year for different PV expansion stages (a)–(d) .

Figures 4 and 5 illustrate the variations in demand for green regional electricity in the regional market in the fixed price and flex price scenario, aggregated over one month. Figure 4 depicts the PV expansion scenarios (a) and (b), and Figure 5 shows scenarios (c) and (d). Two notable trends emerge: (I) Across all scenarios, demand under flexible pricing consistently surpasses that under fixed prices; and (II) with the growth of RES expansion in the system, the overall market demand decreases. This is primarily driven by the increasing numbers of prosumers, who participate less frequently as 'demanders' in the market compared to pure consumers.

Figures 6 and 7 extend the findings from Figures 4 and 5, including the available sales volume for green regional electricity in the regional market. Figure 6 depicts PV expansion scenarios (a) and (b), and Figure 7 shows scenarios (c) and (d). It should be noted that these figures only show aggregated quantities and do not display hourly analyses. In scenario (a), supply and fixed price demand balance relatively well from May to August. At this point, the increase in demand due to flexibilization has seemingly no effect, as there is not enough left-over sales volume to meet the increased demand. However, during months with limited solar radiation, the generation capacity falls short of covering demand. Scenario (b) extends the period for meeting demand to April to September but results in significant oversupply during summer months, in some cases by a factor of up to 2, even in the flex price scenario. Scenarios (c) and (d) show minimal improvements in meeting demand and, more critically, lead to a significant oversupply during summer months by a factor of 3 in (c) and a factor of 5 in (d).

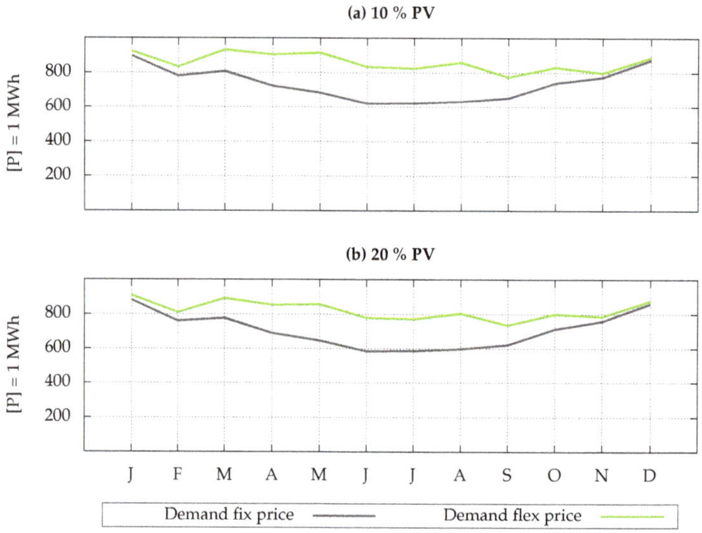

Figure 4. Total demand per month for regional green electricity in fixed and flex price systems over one year for PV expansion stages (a) 10% PV and (b) 20% PV.

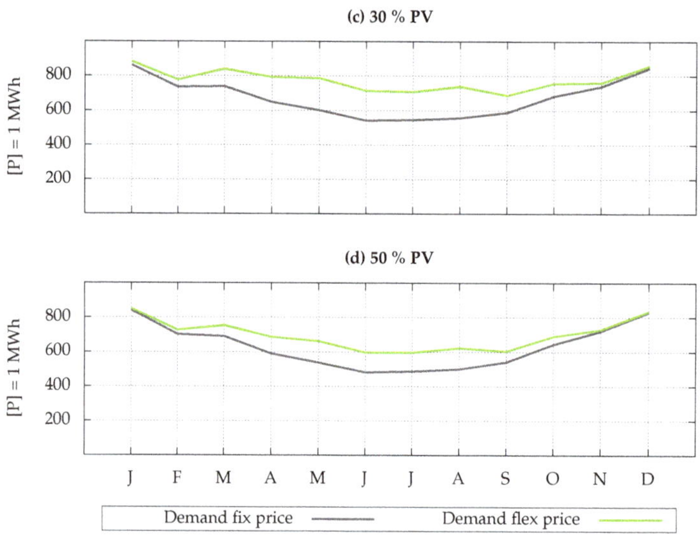

Figure 5. Total demand per month for regional green electricity in fixed and flex price systems over one year for PV expansion stages (c) 30% PV and (d) 50% PV.

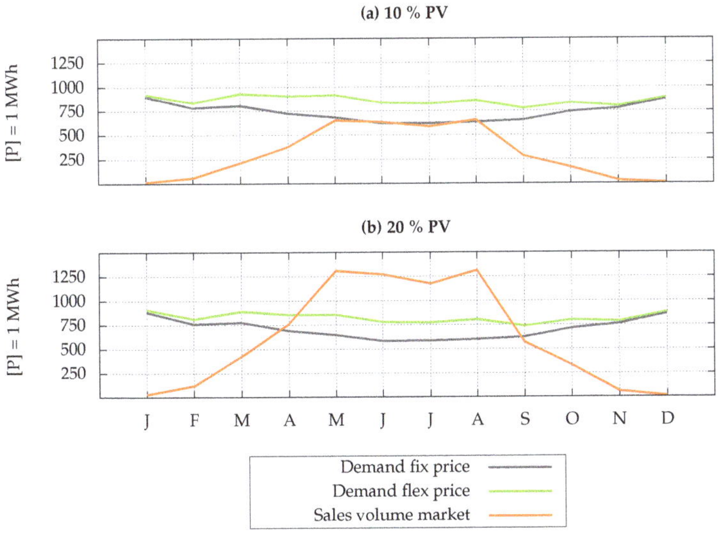

Figure 6. Relation between total demand in fixed and flex price scenarios and total sales volume per month for regional green electricity over one year for PV expansion stages (a) 10% PV and (b) 20% PV.

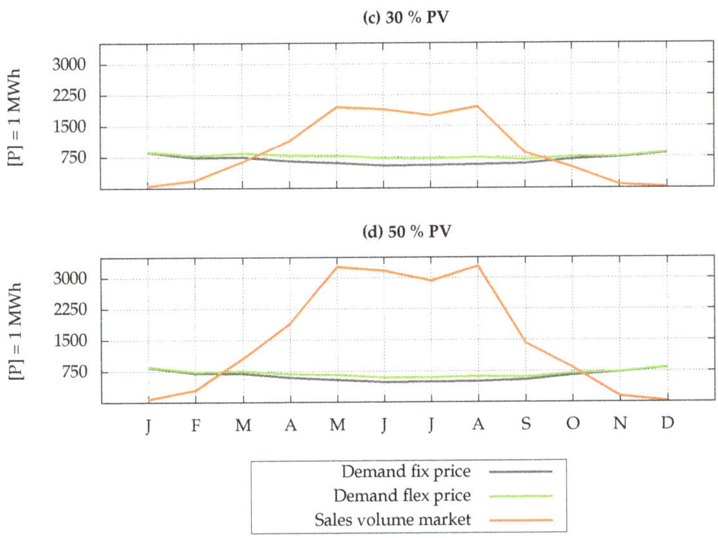

Figure 7. Relation between total demand in fixed and flex price scenarios and total sales volume per month for regional green electricity over one year for PV expansion stages (c) 30% PV and (d) 50% PV.

Figures 8 and 9 display the residual sales volumes for the regional market under both the fixed price and flex price systems. Figure 8 displays the PV expansion scenarios (a) and (b), and Figure 9 shows scenarios (c) and (d). While the quantitative analysis in Figure 6 initially suggests that no effects can be harnessed through flexibilization under scenario (a), this specific analysis reveals that these effects do exist. It indicates that regionally generated electricity could be marketed within the region. Similar effects are observed for scenario (b), although the impact appears limited due to the constraint that demand cannot be increased beyond an upper limit and the growing number of prosumers. In scenarios (c) and (d), these constraints seem to completely block the positive effects of flexibilization.

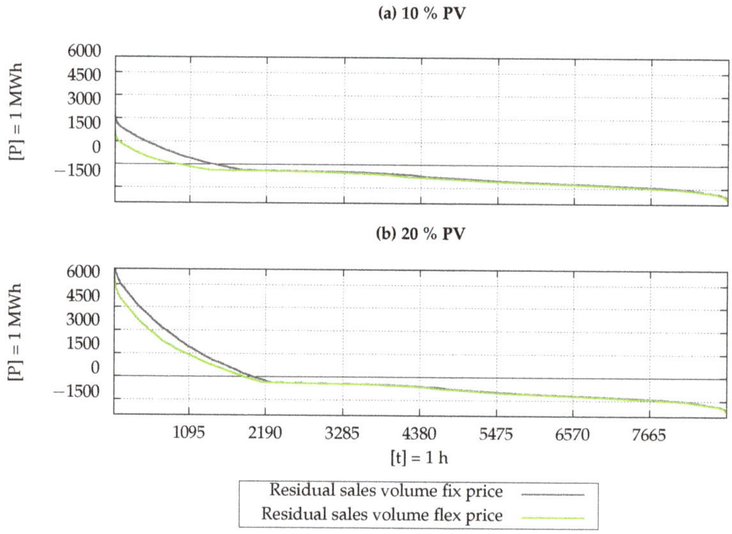

Figure 8. Ordered residual sales volume in fixed and flex price scenarios over one year for PV expansion stages (a) 10% PV and (b) 20% PV.

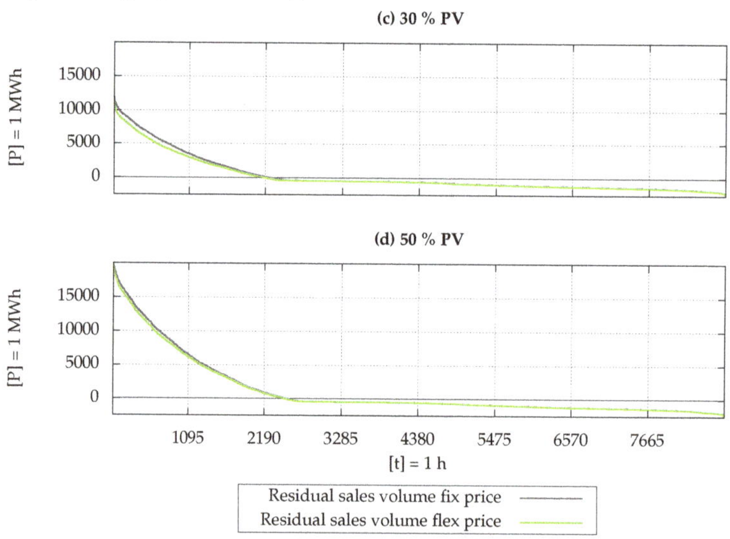

Figure 9. Ordered residual sales volume in fixed and flex price scenarios over one year for PV expansion stages (c) 30% PV and (d) 50% PV.

4.4. Scenario Results: PV and CHP

Figure 10 illustrates the ordered residual load curve for the observed period across scenarios (a)–(d) in relation to the entire supply system. Similar to Figure 3, scenario (a) experiences no regional oversupply at any time. Also, hours with overcapacity begin to occur in scenario (b). Scenarios (c) and (d) exhibit significantly higher numbers of hours with overcapacities, with 1300 and 3400 h, respectively, compared to the solely PV scenario.

In the solely PV expansion scenario, none of the scenarios (a)–(d) has undercapacities by less than 6000 MW in the first 2000 h of the graph. In the PV and CHP expansion scenarios, only scenario (a) exhibits this characteristic. This indicates that in this expansion scenario, generation matches consumption more closely than in the solely PV scenario.

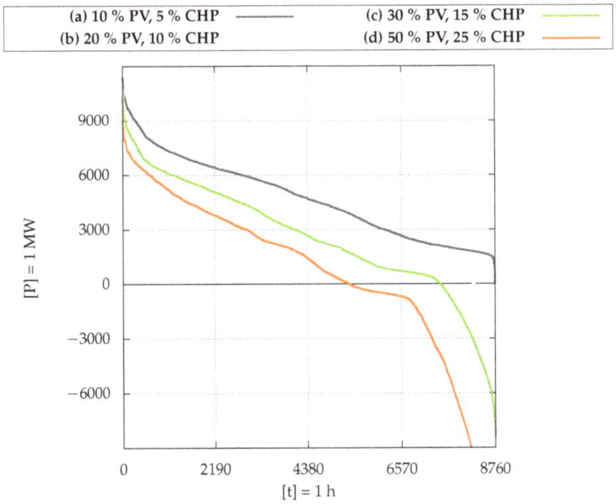

Figure 10. Ordered residual load profile (region) of the supply system over one year for different PV and CHP expansion stages (a)–(d).

Figures 11 and 12 illustrate the variations in demand for green regional electricity in the regional market in the fixed price and flex price scenario, aggregated over one month. Figure 11 depicts the expansion scenarios (a) and (b), and Figure 12 shows scenarios (c) and (d). The two notable trends observed in the solely PV scenario (see Figures 4 and 5) can be observed here as well: (I) Demand under flexible pricing consistently surpasses that under fixed prices in all scenarios, and (II) with the growth of RES expansion in the system, the overall market demand decreases, here, of course, in a more drastic manner than in the solely PV scenario due to the higher number of prosumers. However, a third phenomenon can also be observed. The increased generation due to CHP units within the system in the cold season results in price and, thus, demand effects in the market in the winter and transition months. As a result, the graphs are flatter than in the solely PV scenario.

Figures 13 and 14 extend the findings from Figures 11 and 12, including the available sales volume for green regional electricity in the regional market. Figure 13 depicts the expansion scenarios (a) and (b), and Figure 14 shows scenarios (c) and (d). It is worth noting that, once again, this representation focuses on aggregated quantities rather than providing hourly analyses. What hits the eye first is that the sales volume is higher, and, as inferred from Figures 11 and 12, the demand is generally lower than in the solely PV scenario. This can be categorized as a positive effect on the overall system in scenario (a), even though this figure is a market analysis. Starting from scenario (b), it becomes apparent that the sales volume appears to be disproportionately large for the market's demand situation. In scenarios (c) and (d), the sales volume surpasses the quantity demanded, with (d) even experiencing this oversupply in the winter months.

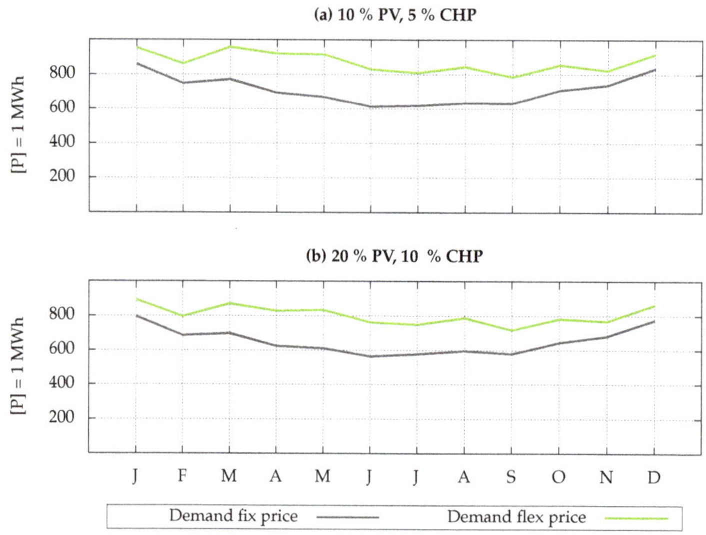

Figure 11. Total demand per month for regional green electricity in fixed and flex price systems over one year for PV and CHP expansion stages (a) 10% PV, 5% CHP and (b) 20% PV, 10% CHP.

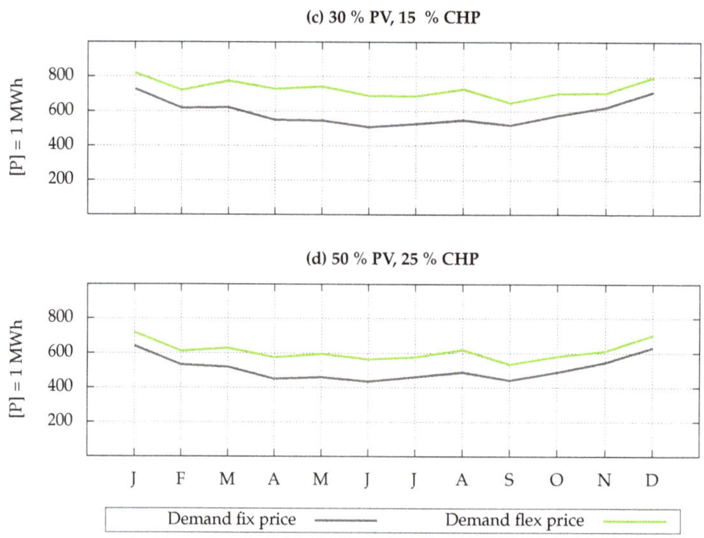

Figure 12. Total demand per month for regional green electricity in fixed and flex price systems over one year for PV and CHP expansion stages (c) 30% PV, 15% CHP and (d) 50% PV, 25% CHP.

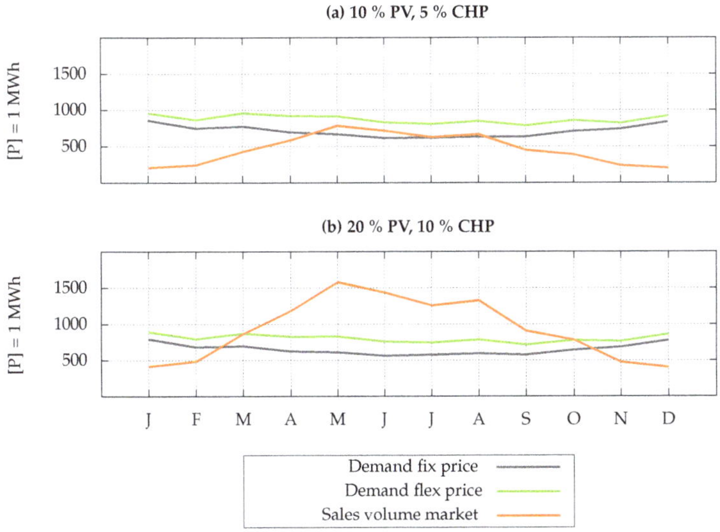

Figure 13. Relation between total demand in fixed and flex price scenarios and total sales volume per month for regional green electricity over one year for PV and CHP expansion stages (a) 10% PV, 5% CHP and (b) 20% PV, 10% CHP.

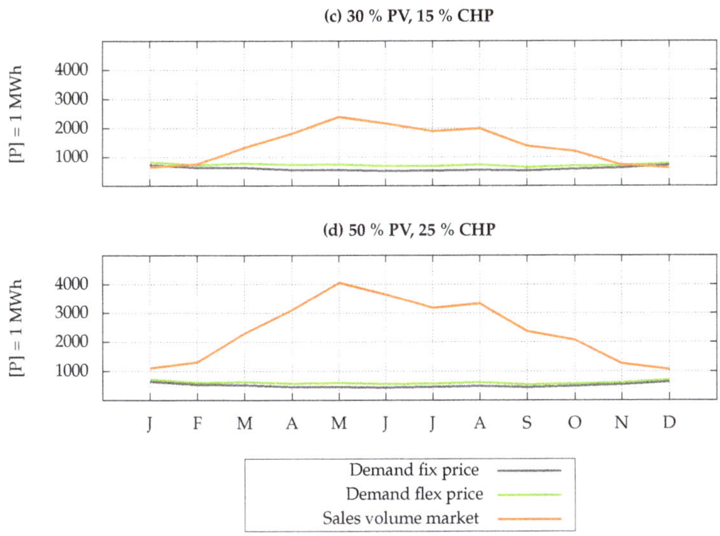

Figure 14. Relation between total demand in fixed and flex price scenarios and total sales volume per month for regional green electricity over one year for PV and CHP expansion stages (c) 30% PV, 15% CHP and (d) 50% PV, 25% CHP.

Figures 15 and 16 display the residual sales volumes for the regional market under both fixed and flex price systems. Figure 15 displays the expansion scenarios (a) and (b), and Figure 16 shows scenarios (c) and (d). Compared to the representation in the solely PV scenario (see Figures 8 and 9), it is noticeable that all graphs (a)–(d) later fall below 0, indicating a generally higher sales volume. However, in case (a), the graph remains at 0 for a while, suggesting a balanced level between demand and supply in the market. Scenarios (a) and (b) also show the positive effect of the flexibilization of the price component for green electricity. All scenarios (a)–(d) share the same maximum values as in the solely PV scenario, suggesting that these values are based on PV feed-in peaks. However, scenarios (c) and (d) show a significantly larger supply surplus than in the solely PV scenario.

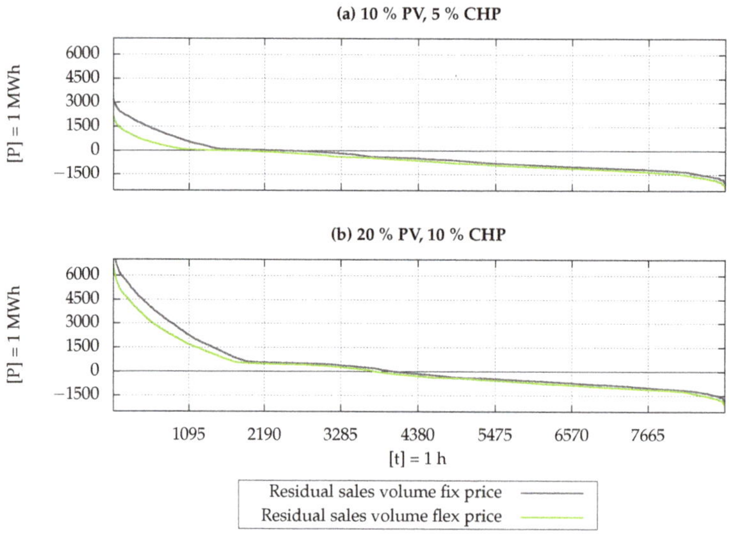

Figure 15. Ordered residual sales volume in fixed and flex price scenarios over one year for PV and CHP expansion stages (a) 10% PV, 5% CHP and (b) 20% PV, 10% CHP.

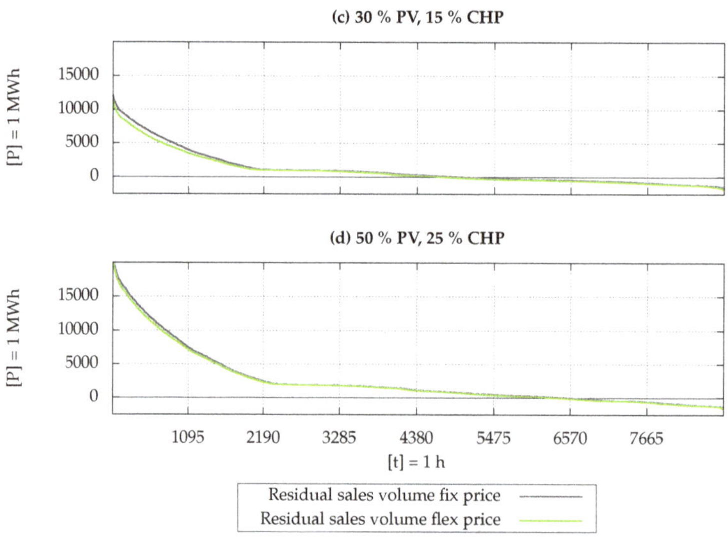

Figure 16. Ordered residual sales volume in fixed and flex price scenarios over one year for PV and CHP expansion stages (c) 30% PV, 15% CHP and (d) 50% PV, 25% CHP.

4.5. Scenario Results: Flex and Regio

Table 6 illustrates changes in demand for green regional electricity in scenarios PV (a)–(d). In these scenarios, the price component for the regional grid fee is discounted, in addition to the flexible price component for green electricity. It can be stated that there is a slight positive effect on the increase in demand resulting from the reduction in the regional grid fee. However, the increase in demand can be categorized as comparatively minor. It is particularly striking that there is no increase in scenario (b); instead, demand stagnates and even tends to decrease minimally.

Table 6. Difference in monthly demand [MWh] between flex and flex and regio price systems in PV scenario (a)–(d).

Month	(a) 10% PV	(b) 20% PV	(c) 30% PV	(d) 50% PV
Jan	3.88	−1.39	1.97	−0.57
Feb	2.43	−0.68	2.72	1.52
Mar	2.17	−0.41	4.43	3.04
Apr	2.26	−0.11	5.46	3.78
May	2.49	−0.10	6.02	4.44
Jun	2.10	−0.15	5.60	4.74
Jul	1.86	−0.33	5.61	4.03
Aug	2.70	−0.39	5.71	3.35
Sep	1.08	−0.45	4.47	3.96
Oct	1.85	−0.83	4.09	2.21
Nov	2.66	−1.22	2.22	0.29
Dec	4.00	−1.78	1.73	−1.27

Table 7 presents the results for the PV and CHP (a)–(d) scenarios, taking into account the discount on the regional grid fee. A positive effect on the increase in demand resulting from the reduction in the grid fee is observed in scenarios (a) and (c), particularly noticeable in winter months when compared to the results in Table 6. However, it is essential to note that the increase in demand can still be categorized as comparatively minor when considering the overall demand. Notably, scenarios (b) and (d) exhibit hardly any or even slightly negative changes in demand.

Table 7. Difference in monthly demand [MWh] between flex and flex and regio price systems in PV & CHP scenario (a)–(d).

Month	(a) 10% PV, 5% CHP	(b) 20% PV, 10% CHP	(c) 30% PV, 15% CHP	(d) 50% PV, 25% CHP
Jan	21.86	2.60	18.30	−1.45
Feb	18.48	2.73	15.77	−1.08
Mar	19.11	3.32	16.10	−0.75
Apr	17.20	2.32	14.65	−0.58
May	15.04	1.59	13.52	−0.45
Jun	13.28	1.95	12.25	−1.31
Jul	13.09	3.05	12.74	−1.56
Aug	12.95	2.41	11.93	−1.40
Sep	15.54	3.03	13.40	−0.70
Oct	18.89	2.80	15.27	−0.80
Nov	20.48	2.16	16.29	−1.21
Dec	22.78	2.18	18.16	−1.77

5. Discussion

Figures 4, 5, 11, and 12, as well as Figures 8, 9, 15, and 16, paint a clear picture. Despite different supply structures, there is an increase and adjustment in demand, but the level of demand volumes remains roughly the same across scenarios (a) to (d) for the respective expansion scenarios and stages. Even with the adjustment of the regional grid fee as an additional discount to the overall electricity price, there was no significant increase in demand. A further discount in the price components, therefore, could potentially tap into more demand, but it is unlikely to surpass the demonstrated level. Thus, Hypothesis 1 can be stated as proven. Empirical evidence of this finding is shown in different econometric analyses about the distribution and preferences concerning the WTP for green electricity. For example, refer to [29,30].

Theorem 1. *The increase in demand for green electricity is constrained by finite capacities in both WTP and consumption. This limitation arises from the fact that these capacities cannot perpetually grow and are subject to inherent constraints.*

As demonstrated, consumer prices linked to the utilization of generation result in an adjustment of demand in a manner that benefits the grid. This phenomenon is particularly evident in Figures 8, 9, 15, and 16, specifically in the residual sales volume. However, in the modeling studies, this adjustment in demand was essentially a shift in available products rather than an actual modification in consumption. Achieving a genuine adjustment in consumption requires the use of technical aids such as storage technologies and controllable consumer appliances. The ongoing technical advances and increasing sales volumes of battery storage systems show the trend of balancing supply and demand already on the lowest possible level of cellular power systems. Additionally, market-available energy management software is widely used in order to mitigate the influence of fluctuating generation. For evidence of statistics, refer to [31,32] and similar.

Hypothesis 2 could therefore only partially be confirmed. The theorem is as follows:

Theorem 2. *With the necessary technical aids in place, an increase in the demand for green electricity is postulated to lead to a corresponding reduction in the effort required to establish an equilibrium between generation and consumption within the market area.*

Figures 3 and 10, as well as Figures 6, 7, 13, and 14, confirm Hypothesis 3 so far. In scenarios (a) and (b) of the respective expansion scenarios, positive effects are observed as a result of the increase in the number of prosumers, namely the total demand on the market decreases and the existing trading volume increases to a manageable level. Scenarios (c) and (d) show a partially exorbitant oversupply in all expansion stages. Figures 8, 9, 15, and 16 reinforce this picture. According to these figures, it must even be stated that the first negative effects can occur in the (b) scenarios without technical aids. The future development of the electricity market will show where this turning point may occur. There is no empirical evidence available up to now. Thus, according to Hypothesis 3, the theorem has to be formulated as follows:

Theorem 3. *In a supply system with a low number of prosumers initially, there is an initial positive impact on the supply of green electricity as the number of prosumers increases. However, this positive effect transitions into a negative impact at some point. Where this point lies cannot be stated generally. It varies case by case and is highly dependent on the specific situation of the supply system itself.*

The response to Hypothesis 4 highlights the pronounced seasonality of PV. Additionally, it has been illustrated that all feed-in peaks are attributed to PV. In Figures 6, 7, 13, and 14, positive effects quickly occur with PV, especially during months with likely high PV feed-in. However, to extend these effects to colder months, an excessively high number of PV installations (scenarios (c) and (d)) is required, resulting in the negative consequence of extreme oversupply in summer. The implementation of CHP mitigates the overall situation within the scenarios. All curves shown for scenarios with CHP are much flatter, indicating overall less volatile generation patterns. Scenario (a) in Figure 15 even exhibits hours with residual sales volume = 0, indicating an absolute system-supporting condition.

Of course, this development will probably be compensated to some extent by other technologies such as wind, water, and biomass power plants in a regional energy market. In order to illustrate these mutual influences, the current REMM needs to be extended.

Theorem 4. *A supply system exclusively dependent on solar PV prosumers is anticipated to witness heightened positive effects primarily during summer, reaching a peak up to a certain threshold. Beyond this point, the positive effects may diminish or even transition into negative consequences, necessitating further considerations for system resilience and adaptability.*

The analyses show which impact prosumers can have on a cellular system and which positive effect flexible market prices can have on the overall system. It provides a good insight into the opportunities and boundaries of regional energy markets based on cellular systems. Nevertheless, these analyses are not claimed to be exhaustive and can and will be further expanded upon.

The next starting point could be the adjustment of the flexible price mechanism, which is currently still based on the ratio of installed capacity to actual local generation and, therefore, only has a positive effect when generation is heading toward the maximum. PV-heavy systems only benefit in summer, and due to their inherent characteristics, there are no positive effects in winter.

Furthermore, no actual possibilities for increasing or shifting the load of consumers were taken into account. If not only demand but also consumption can be increased, this would possibly reinforce the effects. Additional permanent consumption technologies such as heat pumps could also have an equalizing effect.

Furthermore, no larger companies were considered that could be additional consumers of regional green electricity and thus have a smoothing effect on the residual load of the system.

Author Contributions: Conceptualization, J.M. and T.S.; methodology, J.M. and T.S.; software, J.M.; validation, J.M.; formal analysis, J.M. and T.S.; investigation, J.M.; resources, J.M. and T.S.; data curation, J.M. and T.S.; writing—original draft preparation, J.M.; writing—review and editing, T.S. and J.M.; visualization, J.M.; supervision, T.S.; project administration, T.S. All authors have read and agreed to the published version of the manuscript.

Funding: This article is co-funded by the Open Access Publication Fund of Hochschule Zittau/Goerlitz University of Applied Sciences.

Data Availability Statement: All used data are open source. Please see Appendix A for further information.

Conflicts of Interest: The authors declare no conflicts of interest.

Abbreviations

The following abbreviations are used in this manuscript:

ABM	Agent-based modeling
BSL	Business with Standard Load Profile
CA	Cellular approach
CHP	Combined heat and power
LUC	Local utility company
PV	Solar photovoltaic
REMM	Regional energy market model
RES	Renewable energy source
RSL	Residential with Standard Load Profile
WTP	Willingness to pay

Appendix A. Data and Download

For additional information and to download the data, please consult the following:

NetLogo https://ccl.northwestern.edu/netlogo/ (last accessed on 26 January 2024)

Standard Load Profiles https://www.bdew.de/energie/standardlastprofile-strom/ (last accessed on 26 January 2024)

Weather Data https://www.dwd.de/DE/leistungen/testreferenzjahre (last accessed on 26 January 2024)

References

1. Kegel, J.; Lenk, C.; Ouanes, N.; Wiesenthal, J.; Weiß, J. *Prosumerverhalten und Energiewende—Wie wirken sich Verhaltensaenderungen von Prosumerhaushalten auf das Energiesystem aus?* Working Report of the 'EE-Rebound' Research Project; Institut fuer Oekologische Wirtschaftsforschung: Berlin, Germany, 2022. Available online: https://www.ioew.de/fileadmin/user_upload/BILDER_und_Downloaddateien/Publikationen/2022/Kegel_et_al_2022_Prosumerverhalten_und_Energiewende_Energiesystemebene.pdf (accessed on 26 January 2024).
2. Duetschke, E.; Galvin, R.; Brunzema, I. Rebound and Spillovers: Prosumers in Transition. *Front. Psychol.* **2021**, *15*, 636109. [CrossRef]. [CrossRef] [PubMed]
3. Mumbere, K.; Sasaki, Y.; Yorino, N.; Zoka, Y.; Tanioka, Y.; Bedawy, A. A Resilient Prosumer Model for Microgrid Communities with High PV Penetration. *Energies* **2023**, *16*, 621. [CrossRef]. [CrossRef]
4. Olivares, D.; Mehrizi-Sani, A.; Etemadi, A.; Canizares, C.; Iravani, R.; Kazerani, M.; Hajimiragha, A.; Gomis-Bellmunt, O.; Saeedifard, M.; Palma-Behnke, R.; et al. Trends in Microgrid Control. *IEEE Trans. Smart Grid* **2014**, *5*, 1905–1919. [CrossRef]. [CrossRef]
5. Farrokhabadi, M.; Canizares, C.; Simpson-Porco, J.; Nasr, E.; Fan, L.; Mendoza-Araya, P.; Tonkoski, R.; Tamrakar, U.; Hatziargyriou, N.; Lagos, D.; et al. Microgrid Stability Definitions, Analysis, and Examples. *IEEE Trans. Power Syst.* **2020**, *35*, 13–29. [CrossRef]. [CrossRef]
6. Merabet, A.; Al-Durra, A.; El-Saadany, E. Energy management system for optimal cost and storage utilization of renewable hybrid energy microgrid. *Energy Convers. Manag.* **2022**, *252*, 115116. [CrossRef]. [CrossRef]
7. Murty, V.; Kumar, A. Optimal Energy Management and Techno-economic Analysis in Microgrid with Hybrid Renewable Energy Sources. *J. Mod. Power Syst. Clean Energy* **2020**, *8*, 929–940. [CrossRef]. [CrossRef]
8. Lueth, A.; Weibezahn, J.; Zepter, J. On Distributional Effects in Local Electricity Markets Designs-Evidence from a German Case Study. *Energies* **2020**, *13*, 1993. [CrossRef]. [CrossRef]
9. Mainzer, K. Analyse und Optimierung Urbaner Energiesysteme. Ph.D. Thesis, Karlsruhe Institute of Technology (KIT), Karlsruhe, Germany, 2018.
10. Jordehi, A. Optimisation of demand response in electric power systems, a review. *Renew. Sustain. Energy Rev.* **2019**, *103*, 308–319. [CrossRef]. [CrossRef]
11. Zhou, Z.; Zhao, F.; Wang, J. Agent-Based Electricity Market Simulation with Demand Response from Commercial Buildings. In Proceedings of the 2012 IEEE Power and Energy Society General Meeting, San Diego, CA, USA, 22–26 July 2012. [CrossRef].
12. Pokropp, M. Agentenbasierte Netzwerkgetriebene Simulation Umweltbewusster Konsumentenentscheidungen: Eine Modellgestuetzte Analyse Sozialer Einfluesse mit Einer Anwendung auf den Oekostrombezug von Privathaushalten. Ph.D. Thesis, Technical University of Dresden, Dresden, Germany, 2011.
13. Maiwald, J.; Schuette, T. Decentralised Electricity Markets and Proactive Customer Behaviour. *Energies* **2021**, *14*, 781. [CrossRef]. [CrossRef]
14. PV Magazin. KfW: Zwoelf Prozent aller Deutschen Haushalte Nutzen Photovoltaik. Available online: https://www.pv-magazine.de/2023/09/06/kfw-zwoelf-prozent-aller-deutschen-haushalte-nutzen-photovoltaik/ (accessed on 23 February 2024).
15. Verband der Elektrotechnik, Elektronik und Informationstechnik e. V. (VDE). *Der Zellulare Ansatz: Grundlage einer erfolgreichen, regionenuebergreifenden Energiewende*; Energietechnische Gesellschaft (ETG) im VDE: Frankfurt am Main, Germany, 2015.
16. Kiessling, A. Available online: https://energieorganismus.de/energiesystem/ (accessed on 22 February 2024).
17. Reiner Lemoine Institut gGmbH. Vergleich und Optimierung von Zentral und Dezentral Orientierten Ausbaupfaden zu einer Stromversorgung aus Erneuerbaren Energien in Deutschland: Stand 21.10.2013; Haleakala-Stiftung, 100 Prozent Erneuerbar Stiftung, BVMW Bundesverband Mittelstaendische Wirtschaft; Berlin, 2013. Available online: https://100-prozent-erneuerbar.de/wp-content/uploads/2013_Studie_Vergleich_und_Optimierung_von_zentral_und_dezentral.pdf (accessed on 4 January 2024).
18. Hamill, L.; Gilbert, N. *Agent-Based Modelling in Economics*, 1st ed.; John Wiley & Sons: Chichester, UK, 2016; ISBN 978-1-118-45607-1.
19. Kruse, S. Komponentenbasierte Modellierung und Simulation Lernfaehiger Agenten. Ph.D. Thesis, University of Hamburg, Hamburg, Germany, 2014.
20. Railsback, S.; Grimm, V. *Agent-Based and Individual-Based Modeling: A Practical Introduction*, 1st ed.; Princeton University Press: Princeton, NJ , USA, 2012; ISBN 978-0-691-13673-8.
21. Richiardi, M. *Agent-Based Computational Economics—A Short Introduction*; Working Paper No. 69; Laboratorio R. Revelli, Collegio Carlo Alberto: Moncalieri, Italy, 2007. Available online: http://www.laboratoriorevelli.it/_pdf/wp69.pdf (accessed on 22 December 2023).
22. Guerci, E.; Rastegar, M.; Cincotti, S. Agent-based Modeling and Simulation of Competitive Wholesale Electricity Markets. In *Handbook of Power Systems II*; Rebennack, S., Pardalos, P., Pereira, M., Iliadis, N., Eds.; Springer-Verlag GmbH: Berlin/Heidelberg, Germany, 2010. ISBN 978-3-642-12686-4.
23. Tesfatsion, L. A constructive Approach to econmoic Theory. In *Agent-Based Computational Economics*; Tesfatsion, L., Judd, K., Eds.; North-Holland Publishing Co.: Amsterdam, The Netherlands, 2006. ISBN 978-0-444-51253-6.
24. Massimo, G. *Energiewirtschaftliche Analysen des Deutschen Strommarkts mit Agentenbasierter Simulation*, 1st ed.; Nomos Verlagsgesellschaft mbH & Co. KG: Baden-Baden, Germany, 2010. ISBN 978-3-845-22744-3.
25. Moest, D.; Fichtner, W. Einfuehrung zur Energiesystemanalyse. In Proceedings of the Workshop Energiesystemanalyse, Karlsruhe, Germany, 27 November 2008; pp. 11–31. ISBN 978-3-86644-389-1.

26. Ventosa, M.; Baillo, A.; Ramos, A.; Rivier, M. Electricity market modeling trends. *Energy Policy* **2005**, *33*, 897–913. [CrossRef] [CrossRef]
27. Wittmann, T.; Bruckner, T. Agentenbasierte Modellierung urbaner Energiesysteme. *Wirtschaftsinformatik* **2007**, *49*, 352–360. [CrossRef] [CrossRef]
28. Statista. Strompreise für Haushaltskunden in Deutschland in den Jahren 2013 bis 2023 (in Euro-Cent pro Kilowattstunde). Available online: https://de.statista.com/statistik/daten/studie/154908/umfrage/strompreise-fuer-haushaltskunden-seit-2006/ (accessed on 28 February 2024).
29. Sundt, S.; Rehdanz, K. Consumers' willingness to pay for green electricity: A meta-analysis of the literature. *Energy Econ.* **2015**, *51*, 1–8. [CrossRef] [CrossRef]
30. Schmuecker, G. Verbraucher Wuerden mehr für Oekostrom Zahlen. Available online: https://idw-online.de/de/news643669 (accessed on 15 April 2024).
31. Bundesverband Solarwirtschaft. Anzahl Insgesamt Installierter Photovoltaik-Stromspeicher in Deutschland in den Jahren 2013 bis 2023. Available online: https://de.statista.com/statistik/daten/studie/1078876/umfrage/anzahl-installierter-solarstromspeichern-in-deutschland/ (accessed on 15 April 2024).
32. Statistisches Bundesamt. Umsatz in der Deutschen Batteriebranche* in den Jahren 2005 bis 2022 (in Millionen Euro). Available online: https://de.statista.com/statistik/daten/studie/373212/umfrage/umsatz-der-deutschen-batteriebranche/ (accessed on 15 April 2024).

energies

MDPI

Article

Growing Importance of Micro-Meteorology in the New Power System: Review, Analysis and Case Study

Huijun Zhang [1,†], Mingjie Zhang [2,†], Ran Yi [1], Yaxin Liu [1], Qiuzi Han Wen [1,*] and Xin Meng [2]

[1] China Huaneng Clean Energy Research Institute, Beijing 102209, China; hj_zhang@qny.chng.com.cn (H.Z.); r_yi@qny.chng.com.cn (R.Y.); yx_liu@qny.chng.com.cn (Y.L.)
[2] China Huaneng Jilin Branch, Jilin 130012, China
[*] Correspondence: qz_wenhan@qny.chng.com.cn
[†] These authors contributed equally to this work.

Abstract: With the increasing penetration of renewable energy resources, their variable, intermittent and unpredictable characteristics bring new challenges to the power system. These challenges require micro-meteorological data and techniques to provide more support for the power systems, including planning, dispatching, operation, and so on. This paper aims to provide readers with insights into the effects of micro-meteorology on power systems, as well as the actual improvement brought by micro-meteorology in some power system scenarios. This paper provides a review including the relevant micro-meteorological techniques such as observation, assimilation and numerical techniques, as well as artificial intelligence, presenting a relatively complete overview of the most recent and relevant micro-meteorology-related literature associated with power systems. The impact of micro-meteorology on power systems is analyzed in six different forms of power generation and three typical scenarios of different stages in the power system, as well as integrated energy systems and disaster prevention and reduction. Finally, a case study in China is provided. This case takes wind power prediction as an example in a power system to compare the performance when applying micro-meteorological data or not. The experimental results demonstrated that using the micro-meteorological reanalysis dataset with high spatial–temporal resolution for wind power prediction performed better, verifying the improvement of micro-meteorology to the power system to some extent.

Keywords: micro-meteorology; new power system; renewable energy; data analysis

1. Introduction

The current global demand for energy is still growing [1]. Fossil fuels, as the traditional main energy source, cause environmental and health problems [2,3]. The combustion of fossil fuels emits large amounts of greenhouse gases, leading to serious problems such as climate change [4], and posing a threat to humanity [5,6]. Therefore, governments have revisited energy strategies and policies to reduce negative impacts [1], and at the same time, a range of methods have been implemented to effectively curtail greenhouse gas emissions and alleviate related challenges, e.g., promoting clean and renewable energy resources [7,8], developing new devices with minimal environmental impact [9,10] and enhancing current technologies' efficiency [11,12]. Among these suggested methods, transitioning to renewable energy resources is one of the most promising methods to reduce greenhouse gases and secure energy supplies [1].

As renewable energy gradually replaces fossil fuels, the proportion of renewable energy in power generation continues to increase. However, as the penetration rate of renewable energy continues to increase, the variable, intermittent and unpredictable characteristics of renewable energy resources such as wind and solar energy bring new challenges (e.g., flexibility and stability) to the power system [13].

One of the main reasons for these challenges is that the output of various renewable energy resources in the power system is determined by underlying meteorological factors [14,15]. Some researchers have focused on the effect of weather information (especially with high temporal–spatial resolutions) on power systems [16–18]. Weather models are capable of delivering relevant predictions in the planning and operation of power systems [17]. Therefore, meteorology-related data and technologies can provide support for dispatching, distribution, operation and maintenance in power systems. Meteorology plays a vital role throughout the complete lifecycle of power operation (including energy project planning and site selection, design selection, construction, operation, maintenance management, and so on) and among various different forms of power generation. Therefore, meteorological data and technologies are inevitably a cornerstone in power systems.

Focusing on the requirement of higher temporal–spatial resolutions in planning, optimization and other models in power systems [16], this paper targets micro-meteorology. This paper is devoted to presenting the association between micro-meteorology and new power systems, and analyzing the role of micro-meteorology. Micro-meteorology studies the interaction of the lower atmosphere with the Earth's surface [19]. Compared with general meteorology, micro-meteorology focuses on the lower atmosphere, which is closer to the surface, and finer scales. Micro-meteorology is able to provide support for dispatching, distribution, operation, maintenance, etc. within the power system. For example, as large-scale wind power and photovoltaic power are integrated into the power system, the uncertainty of wind power and photovoltaic power pose severe challenges to the balance and dispatch of the power system. Accurate high-resolution gridding observation and micro-meteorology forecasting can establish a solid foundation for accurately predicting power output and load, providing excellent support for power spot market transactions. Furthermore, renewable energy stations are often dispersed over a wide area, with weak resilience in withstanding meteorological disasters. Therefore, high-quality micro-meteorological early warnings can provide great help in preventing losses, casualties, etc., caused by meteorological disasters.

Therefore, this study focuses on examining and analyzing the influence of micro-meteorology on new power systems, and provides real cases for evaluation and support. There are three parts of this work, i.e., (1) a brief review of power-system-related micro-meteorology technologies, (2) a detailed analysis of the impact of micro-meteorology in new power systems, and (3) a case study about a micro-meteorology reanalysis dataset from desert, Gobi and desertification land, and the evaluation of wind power prediction in this micro-meteorological dataset. The overall framework of this study is shown in Figure 1.

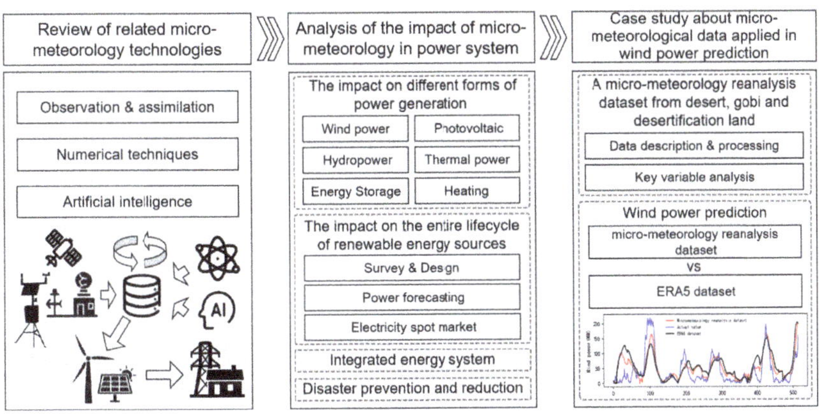

Figure 1. Overall framework of this study.

(1) Review of power-system-related micro-meteorology technologies. This paper surveys several micro-meteorological fields related to the power system (i.e., observation and assimilation, numerical technologies and artificial intelligence), providing a brief introduction of the recent and common-used micro-meteorological technologies applied in power systems. Specifically, this paper reviews the micro-meteorological technologies from three aspects, i.e., observation and assimilation, numerical technologies and artificial intelligence. Observation, assimilation and numerical technologies are the most commonly used micro-meteorological technologies in power systems, involving micro-meteorological data acquisition, micro-meteorological data refinement processing, micro-meteorological feature forecasting, etc. Moreover, with the development of artificial intelligence, artificial-intelligence-based methods present great competitiveness in various micro-meteorological fields, e.g., weather forecasting and downscaling. Therefore, this paper specifically reviews the literature related to micro-meteorology based on artificial intelligence in recent years, which has utilized or has huge application potential in power systems.

(2) Analysis of the impact of micro-meteorology in new power systems. A detailed analysis of the impact of micro-meteorology in new power systems is presented in this study from four aspects, i.e., the impact on different power generation forms, the impact on the entire lifecycle of renewable energy power operations, the integrated energy system, and disaster prevention and reduction. In order to analyze the impact on different power generation forms, this paper discusses how micro-meteorology affects power operation and output of different power generations, as well as the differences in impacts caused by different micro-meteorological factors. For the analysis of the impact on the entire lifecycle of renewable energy power operations, three typical scenarios are selected, corresponding to the early, middle and late stages of the entire lifecycle, respectively, (i.e., survey and design, power prediction, and electricity spot market trading). Regarding the latter two aspects, this paper analyzes the role of micro-meteorology in the integrated energy system and introduces the support for disaster prevention and reduction.

(3) Case study about a micro-meteorological dataset and its evaluation of wind power prediction. This paper presents a case in Jiuquan, China, introducing a micro-meteorological reanalysis dataset constructed in desert, Gobi and desertification land, as well as taking wind power prediction as a representative application for evaluation. Four models are utilized to predict wind power, namely, eXtreme Gradient Boosting (XGBoost), Light Gradient Boosting Machine (LightGBM), long short-term memory (LSTM) and gated recurrent unit (GRU). This paper compares the performance of this micro-meteorological reanalysis dataset with $0.01° \times 0.01°$ resolution and an open source reanalysis dataset (ERA5 [20]) with $0.25° \times 0.25°$ resolution. Two metrics are used for evaluation, i.e., mean absolute error (MAE) and root mean square error (RMSE). The results demonstrate that using the micro-meteorological dataset achieved about 1–3 improvements in MAE and about 1–2 improvements in RMSE, verifying the effectiveness of the micro-meteorological dataset.

The main contributions are as follows:

- This paper reviews the micro-meteorology technology applied and developed in the new power system, including related observations, assimilation, numerical prediction, dynamical downscaling and the artificial intelligence literature.
- This paper analyzes the importance of micro-meteorology in the new power system, including the impact of micro-meteorology on different forms of power generation, the implementation of micro-meteorology in the entire lifecycle of renewable energy power operations, integrated energy system, and disaster prevention and reduction.
- This paper provides a case study in Jiuquan, China to illustrate the importance of micro-meteorology in real applications. Wind power prediction is taken as a representative application to evaluate the performance with a micro-meteorology dataset. The results show that using a micro-meteorological dataset achieved better wind power prediction performance utilized by four common models, presenting the effectiveness of micro-meteorological data.

2. Related Works

This paper chooses Google Scholar (https://scholar.google.com/, accessed on 19 February 2024) as the literature search engine, and uses "power", "energy", "meteorology", "climate", "weather", "power | meteorology", "power | climate", "power | weather", "energy | meteorology", "energy | climate", "energy | weather", "meteorological observation", "assimilation", "wrf", "weather | neural" and "weather | learning" as key words during the search process. Google Scholar provides a service to broadly search for the scholarly literature. Researchers can search in Google Scholar across many disciplines and sources: articles, theses, books, abstracts and court opinions, from academic publishers, professional societies, online repositories, universities and other web sites. Google Scholar contains most of the academic literature across various databases. Compared with citation databases such as Web of Science and Scope, Google Scholar provides more literature from various databases, while the quality of the literature varies. Therefore, for the search results corresponding to each key word, the retrieved literature were sorted according to the key word relevance, and the top 500 were manually checked based on publication sources, relevance and year of publication. Especially, the most relevant literature to the production and application was filtered during the checking process, and, ultimately, 89 were cited in this section.

Two statistical analyses of the 89 cited studies are conducted in this section. As shown in Figure 2, the number of studies carried out by different countries or regions was counted. Researchers from America contributed the most literature in this paper, followed by researchers from China. The proportion of the literature from European countries (e.g., the UK, Portugal, Norway and Germany) is also relatively high. Figure 3 illustrates the amount of research published during different periods in this paper. Most of the literature cited in this paper has been published since 2010. The amount of research published from 2015 to 2019 was the highest among all the divided periods. The literature relevant to artificial intelligent emerged after 2015.

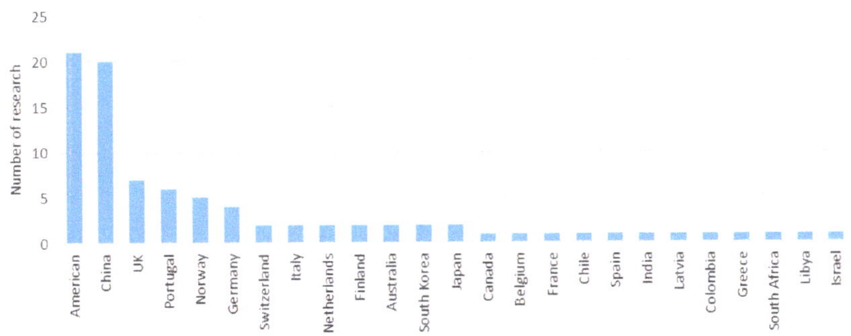

Figure 2. The amount of research carried out by different countries or regions.

2.1. Observation and Assimilation

Meteorological observation is a scientific activity that conducts systematic, comprehensive and quantitative observation and recording of the atmospheric environment. It is the basis of modern meteorological science and an important means of weather forecasting and climate research. It is important for meteorological service agencies, scientific research institutions and related industries. Development is also of great significance. The meteorological observation system involves land, sea, air and space, and can be classified into nine categories: radar observation, high-altitude meteorological observation, ground meteorological observation, ground-based remote sensing vertical observation, etc. [21].

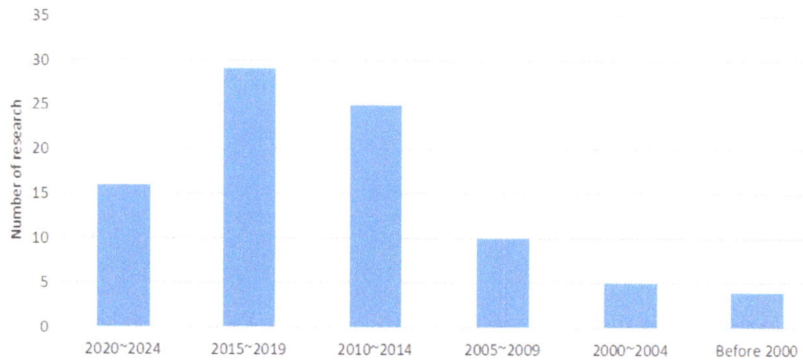

Figure 3. The amount of research published during different periods.

With the development of society and the advancement of science and technology, meteorological observation technology is also constantly improving and innovating. Comprehensive meteorological observation technology is one of the most advanced meteorological observation technologies at present. It includes a variety of meteorological observation methods and technologies and can measure the physics of the five major layers of the earth's climate system: the atmosphere, hydrosphere, cryosphere, lithosphere, and biosphere [22]. It is able to carry out long-term, continuous and systematic observations of chemical and biological characteristics and their changing processes and interactions, thereby providing more accurate and comprehensive meteorological data.

Countries around the world have established meteorological observatories to monitor the physical, chemical, biological, and hydrological aspects [23]. For example, the Blue Hill Meteorological Observatory, situated at the 194 m peak of Great Blue Hill in Milton, MA, USA, was established in 1884/1885 to conduct continuous observations of weather and climate conditions, along with extensive atmospheric science education and outreach activities [24,25]. The Lindenberg Meteorological Observatory, situated roughly 50 km southeast of Berlin's outskirts in Germany, dates back to 1905 and was created for conducting field measurements of the atmosphere is vertical structure and conducting research [26–29]. The Sphinx Observatory, located above the Jungfraujoch peak in Switzerland at an altitude of 3571 m, was constructed in 1936/1937 [30]. The Nepal Climate Observatory–Pyramid is situated at an altitude of 5079 m above sea level in the eastern Nepal Himalaya and was installed in March 2006 to observe trace gases, aerosols, and meteorological parameters [31,32]. The China Meteorological Administration (CMA) selected five national climate observatories (including Dali in Yunnan, Zhangye in Gansu, Xilinhot in Neimenggu, Shouxian in Anhui, and Dianbai in Guangzhou), and launched 24 national climate observatories to carry out long-term, continuous, stereoscopic, and integrated observations [23].

In order to carry out power prediction and operation planning more accurately, it is necessary to monitor the micro-meteorological factors, to provide more accurate and higher temporal–spatial resolution micro-meteorological information [15]. Wang et al. designed micro-meteorological disaster monitoring and pre-warning system in the power grid [33]. Gu et al. constructed an integrated micro-meteorological monitoring system composed of meteorological monitoring stations and clusters [15]. Li et al. proposed a system of micro-meteorological data collection and application (SMDCA), including micro-meteorology collection, data transmission, and data application [18].

In addition, the performance of meteorological satellites has been continuously upgraded. Currently, the spatial and temporal resolution of Fengyun meteorological satellite products can reach 250 m [34]. With the continuous development and improvement of smart meteorology, strengthening smart collaborative observation systems and improving high-precision atmospheric live monitoring products are important development directions for micro-meteorological observation technology.

Although observations are critical for estimating the state of the Earth system, they suffer from two key limitations [35]. The first limitation is that observations contain errors, i.e., bias, randomness, and representativeness [36,37]. The second limitation lies in spatio-temporal gaps [38]. It is necessary to fill in the gaps in the observations, and during this process, a model is needed, e.g., linear interpolation or geostatistical approaches based on the spatial and temporal auto-correlation of the observations [38]. For instance, the model could be a chemistry-transport model (CTM) [39], a general circulation model (GCM) [40], a land surface model (LSM) [41], etc., combining chemical equations and heterogeneous chemistry, incorporating the discretized Navier–Stokes equations, and exploiting the transfer of energy between the land surface and the atmosphere, respectively. Therefore, the observations can be extended by the model and the observational gaps are filled. It is desirable to implement automatic or intelligent interpolation methods, that use models to fill observation information gaps. The method that allows such interpolation is data assimilation [38,42].

Data assimilation is an objective approach of combining observational and model information to provide an estimate of the most likely state of the Earth system and its uncertainty [35]. Data assimilation mainly encompasses two broad categories of numerical algorithms, i.e., variational algorithms (e.g., 4-D variational method) and sequential algorithms (e.g., Kalman filter) [43].

The 4-D variational (4D-Var) method is developed based on the 3-D variational (3D-Var) method via incorporating temporal dimension [43]. Minimization is performed within a time window usually of 6 or 12 h, e.g., for chemical data assimilation [39,44]. To make variational methods more efficient, incremental methods are often used, replacing the non-linear assimilation problem with a series of approximate linear least squares problems [45]. The applicability of 4D-Var in weather forecasting has been demonstrated [46].

The Kalman filter (KF) method [47] is a typical method of sequential assimilation algorithms, in which a recursive sequential algorithm is applied to evolve a forecast. A variant of the KF is the Physical-space Statistical Analysis Scheme (PSAS) [48], which is very effective for systems where the number of observations is much smaller than the model state space dimensions. KF can be generalized to non-linear operators, and the resulting equations is called the Extended Kalman filter (EKF) [43]. Eskes et al. developed a KF method for ozone analysis and prediction [49]. They use observation minus prediction statistics [50] to estimate horizontal error correlations, observation errors, and prediction errors, aiming to reduce the required computer resources. A Variational Kalman filter (VKF) was developed to overcome the limitations of KF and EKF for high-dimensional systems [51], demonstrating comparable results to standard KF and EKF. Following this, Solonen et al. proposed an extension of the VKF to ensemble filtering [52].

Meteorological observation and data assimilation technology are important foundations for micro-meteorological research and the key to application fields such as weather forecasting and climate research. In practice, the spatial and temporal resolution and quality requirements for meteorological observation data are increasing, and micro-meteorological observation and data assimilation technology will become more and more important.

2.2. Numerical Techniques

Traditional numerical techniques used for high-resolution micro-meteorological forecasting and downscaling are generally dynamic forecasting and downscaling techniques. Among micro-meteorological dynamic forecasting and downscaling technologies, the Weather Research and Forecasting (WRF) model is the most common and widely used micro-meteorological forecasting and downscaling technology. WRF is a widely used numerical weather prediction and atmospheric research tool developed by the National Center for Atmospheric Research, the National Oceanic and Atmospheric Administration, and their partners. It is used to model short-term weather forecasts, atmospheric processes, and long-term climate simulations. WRF has higher accuracy than traditional numerical weather prediction models and has higher spatial and temporal resolution, making it an important tool for micro-meteorological research.

In terms of micro-meteorological research and application, WRF has had many successful practices. Heikkila et al. used ERA-40 reanalysis data to drive the WRF model for double-nested downscaling simulations [53]. The results showed that the simulation showed great advantages in the spatial distribution of precipitation, extreme precipitation values, and precipitation frequency. Soares et al. proposed a dynamically downscaled climatology of Portugal, produced by a high resolution (9 km) WRF simulation, forced by 20 years of ERA-Interim reanalysis (1989–2008), nested in an intermediate domain with 27 km of resolution [54]. WRF results at 9 km compare favorably with published results supporting its use as a high-resolution regional climate model. This higher resolution allows for a better representation of extreme events, which is critical for developing mitigation/adaptation strategies by policy makers and downstream users of regional climate models in applications such as flash floods or heatwaves. Evans et al. used NNRP reanalysis data as the lateral boundary conditions of the WRF model and conducted a 24-year long-term simulation of Australia [55]. The resolution of the main region was 50 km and the resolution of the sub-region was 10km. The results showed that the sub-region can be simulated well. It can obtain the statistical characteristics of precipitation from daily to interannual time scales and can capture the spatial distribution characteristics of precipitation anomalies related to ENSO anomaly years. Zhang et al. used the WRF model to conduct a triple nested simulation of the United States [56], and the results proved that increasing the resolution can significantly improve the precipitation simulation effect. In East Asia, relevant studies also showed that the WRF model can simulate the overall characteristics of the East Asian monsoon system and the characteristics of the regional precipitation diurnal cycle, and is a good downscaling tool [57–59]. Liu et al. developed a multi-scale four-dimensional weather data assimilation and large eddy simulation system WRF-RTFDDA-LES [60]. They used the Newton relaxation method to assimilate observation data on kilometer-level mesoscale model grids for large eddy simulation and realize weather forecasts. Weather forecasts on nested grids from large scale (~2000 km) to small scale (~100 m) can capture the wind field characteristics and micro-scale airflow inside many wind farms. Cardoso et al. presented and evaluated precipitation from a high-resolution WRF climate simulation against daily gridded observations in the Iberian Peninsula [61]. The simulation with higher resolution displays a remarkable improvement in representing Iberian precipitation patterns across all time scales, emphasizing the representation of variations and extreme weather statistics. KUSAKA et al. adopted the WRF model at a 3 km horizontal grid increment to achieve dynamical downscaling and predicting heat stress in August 2070 for the Tokyo, Osaka, and Nagoya metropolises [62]. Yang et al. employed the WRF model, coupled with a single-layer Urban Canopy Model, to simulate urban climate in Nanjing, located in eastern China. Drawing from the results of simulations spanning ten summers between 2000 and 2009, they found that the WRF model has the capability to capture high-resolution features of urban climate in the Nanjing area [63]. Marta-Almeida et al. studied a high resolution atmospheric modelling for a 20-year recent historical period [64]. The dynamic downscaling approach adopted used the Max Planck Institute Earth System Model (MPI-ESM) to drive the WRF running in climate mode. Three online nested domains were used covering part of the North Atlantic and Europe, with a resolution 81 km, reaching 9 km in the innermost domain which covers the Iberian Peninsula. To sum up, the WRF model has good capabilities in dynamic forecasting and downscaling and is currently widely used in actual operations.

WRF is also widely adopted in energy-related meteorological and resource fields. Carvalho et al. evaluated the performance of the WRF model in wind simulation under different numerical and physical options for an area of Portugal located in complex terrain and characterized by its significant wind energy resource and compared the results of six different WRF simulations on the wind data collected at thirteen wind measuring stations located in Portugal in areas of high wind energy potential [65,66]. Deppe et al. used the WRF model with 10 km horizontal grid spacing to explore improvements in wind speed forecasts at a typical wind turbine hub height (80 m) [67]. They evaluated version 3.1.1 of the WRF model to accurately reproduce 80 m wind speeds and ramp events by comparing WRF simulations using six different planetary boundary layer (PBL) schemes to observations of 80 m wind speed gathered at the Pomeroy, Iowa wind farm site. Aiming at estimating offshore wind power potential for the central coast of Chile, Mattar et al. adopted the WRF model to simulate wind fields at the regional level on the wind speed data from in situ stations and ERA-Interim reanalysis [68]. Wind field simulations were performed at different heights (20, 30, 40 and 140 m.a.s.l.) and a spatial resolution of 3 × 3 km for the period from 1 February 2006 to 31 January 2007, which comprised the entire series of in situ data available. Tuy et al. used the WRF and Sentinel-1 level 2 ocean (L2 OCN) products to assess the offshore wind power potential in Cambodia, aiming to evaluate potential sites and to estimate the annual energy production (AEP) and equivalent CO_2 reduction [69]. The model was initially calibrated and validated against observed onshore winds at four weather stations before its main simulation for the two-year study period of 2018–2019.

Zhao et al. proposed a new multi-step forecasting method for operational wind forecast, using 96 steps of the next day, termed the CS-FS-WRF-E model, which is based on a WRF ensemble forecast, a novel Fuzzy System, and a Cuckoo Search (CS) algorithm [70]. This novelty model outperformed other approaches in different wind farms and was used for operational wind forecasting within acceptable computations. Next, they proposed to divide multi-step forecast into waves based on their synoptic backgrounds and developed a new multi-step forecasting method termed CSFC-Apriori-WRF, providing a one-day-ahead wind speed and power forecast consisting of 96 steps [71]. For the stability conditions evaluation of the marine atmospheric boundary layer (MABL) in an offshore wind farm development, the WRF model has been applied over the North Sea during March 2005 [72]. The performance of the WRF model was analyzed for sensitivity to different horizontal resolutions, input datasets, PBL parameterizations, and nesting options. When comparing the model results with other modeling studies and high-quality observations recorded at offshore measurement platform FINO1, it was found that the combination of ERA-Interim reanalysis data and the 2.5-level MYNN PBL scheme successfully simulated the MABL over the North Sea. Salvação et al. presented a ten-year wind hindcast for the Iberian Peninsula coast and conducted simulations with the WRF model at a spatial resolution of 9 and 3 km with a 6-hourly output [73]. Prósper et al. performed high-resolution (333 m) WRF forecasting for a farm in complex terrain and conducted detailed validation with observations at the nacelle of each turbine for one full year [74]. The results showed that the forecasts were very skillful with wind speed MAE = 1.87 m/s and wind power NMAE = 14.75%.

2.3. Artificial Intelligence

In recent years, innovative applications of fundamental theories such as mathematics, probability, and statistics, coupled with the widespread adoption of high-performance computing capabilities and the increasing availability of massive data, have propelled the rapid development of artificial intelligence (AI) technology. This has fostered the emergence of a new scientific research paradigm known as "AI for Science", providing researchers with novel methods and tools to enhance the accuracy of weather prediction and forecasting [75,76]. Grover et al. explored forecasting weather as a data-intensive challenge [77]. They proposed a hybrid approach and utilized a deep neural network to model the joint features of meteorological variables. The experiments on real-world meteorological data highlighted the promise of this approach. Volkovs et al. proposed a novel deep learning method for weather forecasting [78]. They primed the model states with atmospheric data, which achieved better results.

AI has already been applied to various finer-divided areas such as temperature forecasting, short-term precipitation forecasting, ENSO prediction, numerical forecasting model interpretation, remote sensing, extreme weather events identification and prediction, and so on [79–92]. For example, on the issue of short-term precipitation forecasting, researchers have developed numerous short-term precipitation forecasting methods based on Doppler radar echo data and machine learning, and have been well verified in operational applications [83,93]. Agrawal et al. proposed the MetNet nowcast precipitation forecast model based on satellite cloud images (GOES) and radar data (MRMS) [94]. They adopted convolutional downsampling, ConvLSTM structure and attention mechanism, achieving an 8-h precipitation forecast with a resolution of 1 km in the United States. The results demonstrated that the methods performed better than the US NOAA integrated forecast system within the 2-h nowcast time limit. Kaparakis et al. utilized UNet as the core model and proposed the Weather Fusion UNet (WF-UNet) model to investigate its performance for precipitation nowcasting [95]. The WF-UNet model integrates precipitation and wind speed variables as input. In the evaluation in Western Europe, the WF-UNet model performed better compared to other UNet-based methods using only precipitation radar input data.

AI technology has also been widely applied in addressing the resolution challenge [96]. For dynamic modeling, researchers trained dynamic convolutional layers, filtering whose weights are updated dynamically based on inputs during the forward pass, in short-term weather forecasting [97]. Vandal et al. proposed a generalized stacked superresolution CNN framework for statistical downscaling of climate variables [98]. Superresolution refers to the ability of a network to produce an output image with a higher resolution than the input. It was suggested that a single trained model can effectively downscale spatial heterogeneous regions, and the deep learning methods have demonstrated advantages over other techniques. Gentine et al. utilized MLP to learn intricate dynamics like convective heating, moistening, and cloud-radiative transfer feedback from high-resolution simulations, with the aim of replacing multiple existing parameterization methods [99]. Furthermore, the climate modeling community is assembling large datasets to facilitate large-scale big data deep learning [100]. Albert et al. utilized GAN to model urban expansion, which is closely linked to water consumption [101]. Trained on images of 30,000 cities, GAN was able to accurately replicate the realistic patterns and distribution of urban masses without any externally imposed restrictions, such as rules prohibiting the construction of cities on bodies of water. Instead, GAN learned these rules through the training process.

Most recently, large-scale weather forecast models based on deep learning have received great attention and development, becoming a strong competitor of traditional numerical forecast models [102–107].

Researchers proposed a data-driven deep learning earth system simulator FourCastNet [102,103]. FourCastNet used the Adaptive Fourier Neural Operator (AFNO) to solve continuous and high-resolution inputs and can generate real-time weather forecasts one week in advance. It is faster than the existing NWP in predicting global weather and generating mesoscale forecasts. It has reached five orders of magnitude and the achieved state-of-the-art accuracy.

Subsequently, Bi et al. proposed a powerful AI-based medium-range global weather forecasting system Pangu-Weather [104]. Pangu-Weather adopted the Swin Transformer as the backbone and designed a three-dimensional Earth-Specific Transformer (3DEST). Earth-specific absolute positional bias related to latitude, longitude and height was introduced to learn the irregular components of each spatial operation. In addition, the Pangu-Weather proposed a hierarchical temporal aggregation strategy and trained four models with different forecast intervals, namely 1-h intervals, 3-h intervals, 6-h intervals, and 24-h intervals. These models were called according to the greedy algorithm. Minimize the number of iterations required to predict meteorological conditions at a specific time. The results on the fifth generation of the European Centre for Medium-Range Weather Forecasts (ECMWF) reanalysis (ERA5) data showed that Pangu-Weather performed well at deterministic forecast and extreme weather forecast, while achieved 10,000 times faster than the Operational IFS.

Next, GraphCast, a state-of-the-art AI model capable of medium-range weather forecasting with unprecedented accuracy, was proposed [105]. GraphCast used graph neural network (GNN) as the basic architecture and adopted an encoding-processing-decoding structural configuration. The encoder mapped the data from the latitude and longitude grid to multi-grid graph nodes to learn features. The processor used a 16-layer deep GNN to learn the information transfer process and update multiple Grid nodes, the final decoder used directed GNN to map the learned multi-grid features back to the latitude and longitude grid and combine them with the input information to form the output. GraphCast was able to predict weather conditions up to 10 days in advance with greater accuracy and speed than the industry gold-standard weather simulation system, the High-Resolution Forecast (HRES), produced by the European Centre for Medium-Range Weather Forecasts (ECMWF).

Following, researchers from the Shanghai Artificial Intelligence Laboratory, University of Science and Technology of China, Shanghai Jiao Tong University, Nanjing University of Information Science and Technology, the Institute of Atmospheric Physics of the Chinese Academy of Sciences, and the Shanghai Central Meteorological Observatory, released the large-scale global medium-range weather forecast model called Fengwu [106]. The model was built based on multi-modal and multi-task deep learning methods. It utilized multi-modal neural networks and multi-task automatic equalization weights to solve the problem of representation and mutual influence of multiple atmospheric variables. Fengwu only took 30 s to generate high-precision global forecast results for the next 10 days.

Chen et al. proposed a cascaded deep learning weather forecast system (called Fuxi) that provides a 15-day global weather forecast [107]. Fuxi was an autoregressive model, consisting of three modules, i.e., Cube Embedding, U-Transformer and a fully connected layer. It used the weather parameters of the first two time steps as input to predict the weather parameters of the next time step (one time step is 6 h), and used an autoregressive multi-step loss to reduce the cumulative error over long lead times. Experimental results showed that Fuxi achieved 15-day forecasts.

Details of the current large-scale global weather forecast models are compared in Table 1.

Table 1. A summary of the current large-scale global weather forecast models.

Methods	Training Data	Forecasting Variables	Parameters	Training Settings	Forecasting Speed
FourCastNet	ERA5 data	20	-	67.4 min with 3072 A100 GPUs	7 s to compute a 100 member, 24 h forecast
Pangu-Weather	ERA5 data	69	0.256 B	16 days with 192 V100 GPUs	1.4 s to achieve global forecast
GraphCast	ERA5 data	227	-	3 weeks with 32 TPUv4	60 s to compute 10 days forecast
Fengwu	ERA5 data	189	-	17 days with 32 A100 GPUs	30 s to compute 10 days forecast
Fuxi	ERA5 data	70	4.5 B	30 h with 8 A100 GPU Clusters	-

3. Analysis

3.1. The Impact of Micro-Meteorology on Power Generation

Micro-meteorology plays a crucial role in the collection and storage processes of various energy sources such as wind power, photovoltaic power, hydropower, and thermal power [108]. Monitoring meteorological changes at a micro-scale in real-time allows for timely adjustments to the operational status of power generation devices, maximizing the capture of renewable energy and optimizing energy storage methods.

3.1.1. Wind Power

The micro-meteorological factors have three main effects on wind power. Firstly, micro-meteorological factors can affect the variations in wind speed and direction. Factors such as terrain, buildings, and vegetation can cause the flow of wind to generate vortices and turbulence, resulting in changes in wind speed and direction in space and time. These factors must be considered in wind farm layout and turbine design to maximize the utilization of available wind energy.

Secondly, micro-meteorological factors can influence turbulence and buffeting effects. Turbulence refers to irregular variations in wind speed and direction, playing a significant role in micro-meteorology. Turbulence can lead to vibration and stress concentration on wind turbine blades, thereby reducing the lifespan and performance of the turbines. Therefore, by measuring and predicting micro-meteorological conditions, it is possible to avoid areas with strong turbulence, reducing vibration and damage to the turbines.

Thirdly, micro-meteorological factors also affect temperature and humidity variations. An increase in temperature can lead to a decrease in air density, which then affects the transmission and conversion efficiency of wind energy. Changes in humidity can also impact air density and wind speed. Therefore, understanding the influence of micro-meteorological factors on temperature and humidity can optimize the operation of wind farms and the performance of turbines.

In practical applications, the collection and processing of micro-meteorological data are critical steps in wind farm construction and operational management. Typically, multiple micro-meteorological observation sites are established at various heights around the wind farm, and the data is transmitted to a central control room for real-time monitoring and analysis through wireless or wired networks. These observation sites measure micro-meteorological parameters such as wind speed, wind direction, temperature, humidity, and other relevant meteorological elements. Through the collection and analysis of these data, the meteorological conditions inside and around the wind farm can be understood, providing effective references and guidance for the operation of wind turbines.

At the same time, researchers also conduct relevant studies and simulations on different micro-meteorological factors [109]. For example, simulations are conducted to evaluate the impact of terrain and buildings on the wind field and to understand the stress distribution on wind turbine blades under different turbulence conditions. Simulations are also carried out to investigate the effects of temperature and humidity variations on the transmission and conversion of wind energy. Through these studies and simulations, a better understanding of the mechanisms by which micro-meteorological factors affect wind power can be achieved, providing a scientific basis for the optimization design and operational management of wind farms.

In conclusion, micro-meteorological factors play a crucial role in wind power, directly influencing the efficiency of wind energy utilization and the reliability of wind turbines. Through in-depth research and real-time monitoring of micro-meteorology, the layout of wind farms and the design of turbines can be optimized, resulting in improved utilization of wind energy and extended lifespan of turbines, thus promoting the development of sustainable energy.

3.1.2. Photovoltaic

Meteorology plays a crucial role in photovoltaic power generation, as it directly influences the performance and operation of photovoltaic systems.

Solar radiation analysis is a key aspect of meteorological observations in photovoltaic power generation. Solar radiation encompasses the process of solar energy propagation in the atmosphere, including total radiation, direct radiation, and scattered radiation, among other components. Solar radiation is the basic energy source for photovoltaic power plants, so accurate measurement and analysis of solar radiation are crucial for assessing the power generation potential of photovoltaic power plants. Meteorological observations allow for the detection of variations in solar radiation across different time periods and seasons, offering a foundation for site selection and power generation forecasting for photovoltaic power plants. In addition, the spatial distribution of solar radiation also affects the layout and installation of photovoltaic power plants. Through the analysis of meteorological data, the optimal tilt angle and orientation of the components can be determined to optimize the utilization of solar radiation resources.

Temperature is another important meteorological parameter in photovoltaic power generation. Temperature directly affects the efficiency and performance of photovoltaic cells. Excessive operating temperature of photovoltaic cells can result in decreased efficiency and diminished power generation. Therefore, comprehending the impact of temperature fluctuations on photovoltaic power generation is crucial. By utilizing meteorological observations and mathematical models, the temperature characteristics of photovoltaic components can be assessed, enabling adjustments in component layout and heat dissipation design to enhance power generation efficiency. Moreover, high temperatures can impact the longevity of photovoltaic components, underscoring the significance of regulating component operating temperatures as a key method to prolong their lifespan.

Wind speed and wind direction are two other meteorological parameters that have a significant impact on the operation of photovoltaic power systems. Strong winds can affect the stability and safety of photovoltaic components, leading to potential damage. Observing the variations in wind speed and wind direction can help take corresponding structural design and protective measures to ensure the normal operation and safety of photovoltaic components. Furthermore, wind can also affect the heat dissipation of photovoltaic components. Appropriate wind speed can help with heat dissipation, lowering component temperature and improving power generation efficiency. Therefore, monitoring and analyzing wind speed and wind direction are necessary in the design and operation of photovoltaic power systems.

Humidity and precipitation are also factors to consider in meteorological observations. High humidity and heavy precipitation can have negative effects on the performance and lifespan of photovoltaic components. Humidity can contribute to the accumulation of dust and dirt on the surface of photovoltaic components, reducing their light absorption capacity. Precipitation can cause water pooling on component surfaces, increasing optical losses and potentially causing electrical short circuits. Through meteorological observations, variations in humidity and precipitation are obtained; thus, appropriate protective measures can be taken to ensure the normal operation and lifespan of photovoltaic components.

In addition to the parameters mentioned above, meteorological observations can also provide a scientific basis for site selection, component layout, tilt angle selection, and power generation prediction in photovoltaic projects. By utilizing meteorological data, the potential and risks of photovoltaic power plants can be assessed, and rational planning and decision-making can be made to improve the performance and economic benefits of photovoltaic power systems. For example, through the analysis of parameters such as solar radiation, temperature, and wind speed, the optimal site for a photovoltaic power plant can be determined to achieve higher power generation. By collecting and analyzing meteorological data, the annual and seasonal power generation of a photovoltaic power plant can be predicted, providing a decision-making basis for its operation and management.

Through meteorological observation and analysis, the impact of factors such as solar radiation, temperature, wind speed, wind direction, humidity, and precipitation on photovoltaic power generation can be assessed. Decisions can then be made regarding component layout, heat dissipation design, protective measures, and power generation prediction to improve the efficiency, reliability, and economic performance of photovoltaic power systems. The study and application of meteorology will provide important support for the development and promotion of the photovoltaic power generation industry.

3.1.3. Hydropower

Micro-meteorology plays an equally important role in the hydropower industry, as it directly affects the operation and efficiency of hydroelectric power plants. In the planning, design, construction, and operation of hydroelectric power plants, the collection and analysis of micro-meteorological data are crucial for ensuring the safety, stability, and efficient operation of the plants. The following will elaborate on the role and impact of micro-meteorology in the hydropower industry.

Firstly, rainfall observation and prediction are among the important applications of micro-meteorology in the hydropower industry. Rainfall is a significant factor in determining the inflow volume and available water resources for hydroelectric power plants. By monitoring rainfall in real time using micro-meteorological observation data and integrating them with meteorological models, precise rainfall predictions can be generated. Precise rainfall forecasts help hydroelectric power plants make reasonable reservoir scheduling to maximize the utilization of water resources and reduce the risks of disasters caused by excessive or insufficient rainfall. Furthermore, rainfall monitoring offers timely alerts and warnings to pertinent personnel at the power plants, empowering them to implement suitable emergency measures for ensuring the safe operation of the plants.

Secondly, temperature observation and its impact are also key applications of micro-meteorology in the hydropower industry. Temperature has a direct influence on the operation and efficiency of hydroelectric power plants. By observing micro-meteorological data, the changes in ambient temperature can be understood, thereby assessing the cooling conditions of the power plants. High temperatures can lead to decreased efficiency and power generation capacity of hydroelectric units, as well as increased heat dissipation loads on the turbines and generating equipment. Through temperature observation and analysis, corresponding measures can be taken, such as increasing cooling equipment and optimizing turbine designs, to improve the efficiency and reliability of the hydroelectric power plants.

Additionally, humidity observation and its impact are also significant applications of micro-meteorology in the hydropower industry. Humidity affects the power generation efficiency and equipment lifespan of hydroelectric power plants. High humidity increases the heat dissipation load on hydroelectric units and may cause surface condensation and corrosion of equipment. By monitoring humidity changes through micro-meteorological observation data, timely adjustments can be made, such as strengthening ventilation and implementing moisture-proof treatments, to ensure the normal operation and extended lifespan of the power plant equipment.

Moreover, wind speed observation and its impact are important applications of micro-meteorology in the hydropower industry. Wind speed is a crucial factor in wind power generation and also has an influence on the operation of hydroelectric power plants. Strong winds may affect the fluctuation of the water surface in reservoirs, thereby impacting the power generation efficiency and safety of hydroelectric units. Real-time monitoring of wind speed through micro-meteorological data, combined with the characteristics of wind power generation, enables wind energy forecasting and adjustments to the power generation plan, maximizing the utilization of wind energy resources.

Furthermore, solar radiation observation and its impact are significant applications of micro-meteorology in the hydropower industry. Solar radiation is the fundamental energy source for photovoltaic power generation and also affects the operation of hydroelectric power plants. The intensity and variations of solar radiation affect factors such as evaporation rates from the water surface and water temperature, which in turn impact the power generation efficiency and reservoir water management of hydroelectric units. Through real-time monitoring of solar radiation using micro-meteorological data, optimization of the power generation plan and scheduling of the hydroelectric power plants can be achieved based on the characteristics of solar energy generation.

By collecting and analyzing micro-meteorological observation data, optimization and improvements can be made in aspects such as rainfall prediction, temperature control, humidity regulation, wind energy utilization, and solar energy generation. These measures contribute to enhancing the power generation efficiency, reliability, and economic viability of hydroelectric power plants, as well as promoting the utilization and conservation of sustainable energy sources. The application of micro-meteorology provides technical support and a decision-making basis for the development of the hydropower industry, making significant contributions to the development of sustainable energy.

3.1.4. Thermal Power

Micro-meteorology also plays an important role in the field of thermal power generated by coal combustion, which is converted into kinetic energy to produce electrical energy, and it has a direct impact on the operation and power generation efficiency of thermal power plants.

Firstly, temperature observation and its impact is one of the key applications of micro-meteorology in the field of thermal power. The combustion process in a thermal power plant necessitates specific temperature conditions to achieve complete fuel combustion and adequate heat supply. Monitoring variations in environmental temperature using micro-meteorological observation data enables the adjustment of boiler operating parameters in thermal power plants to sustain the required combustion temperature. Simultaneously, micro-meteorological data can aid in forecasting the evolving trends in atmospheric temperature, facilitating the development of a sound heating plan and adjustments in thermal power operations to ensure the stable and efficient power generation of the thermal power plant.

Secondly, humidity observation and its impact are also one of the important applications of micro-meteorology in the field of thermal power. Humidity has a direct impact on the combustion process and environmental pollutant emissions of thermal power plants. High humidity may cause incomplete combustion of fuel and increase the water content in the flue gas, thereby reducing the combustion efficiency and environmental emission quality of thermal power plants. By monitoring the changes in humidity in real-time through micro-meteorological observation data, and combining the impact of humidity on combustion, the operating parameters and control equipment of thermal power plants can be adjusted to ensure complete combustion of fuel and reduce the emission of environmental pollutants.

In addition, wind speed observation and its impact are also one of the important applications of micro-meteorology in the field of thermal power. The wind speed directly affects the combustion process and flue gas emissions of thermal power plants. Severe winds can disrupt the combustion stability of the thermal power plant's boiler and alter the direction of flue gas emissions, thereby impacting power generation efficiency and environmental emission quality. Real-time monitoring of wind speed variations using micro-meteorological observation data, and integrating the effects of wind speed on combustion and emissions, allows the adjustment of operating parameters and emission equipment of thermal power plants to ensure combustion stability and reasonable emission of environmental pollutants.

Furthermore, pressure observation and its impact are also one of the important applications of micro-meteorology in the field of thermal power. Pressure has a certain impact on the combustion process of thermal power plant boilers and flue gas emissions. Changes in pressure will affect the stability of combustion air volume, fuel supply, and flue gas emissions. By monitoring the changes in pressure in real-time through micro-meteorological observation data, and combining the impact of pressure on combustion and emissions, the operating parameters and control equipment of thermal power plants can be adjusted to ensure the stability of combustion and reasonable emission of environmental pollutants.

Through the collection and analysis of micro-meteorological observation data, temperature, humidity, wind speed, and pressure parameters can be monitored and adjusted to optimize the combustion process and environmental emissions of thermal power plants and improve power generation efficiency and environmental protection levels. The application of micro-meteorology can provide technical support and a decision-making basis for the development of the thermal power industry and make an important contribution to the sustainable development of the energy industry.

3.1.5. Energy Storage

Micro-meteorology also has a direct impact on the operation and efficiency of energy storage systems. Firstly, solar energy storage is a prevalent method of energy storage closely linked to micro-meteorology. Micro-meteorological observation data can yield critical information such as solar radiation, air temperature, and humidity, essential for the design and operation of solar energy storage systems. Analysis of micro-meteorological data enables the prediction of evolving trends in solar energy resources, facilitating adjustments to the charging and discharging strategies of energy storage systems to maximize the utilization of solar energy resources and enhance energy storage system efficiency.

Secondly, wind energy storage is also a common method of energy storage, and the application of micro-meteorology to wind energy storage is equally important. Micro-meteorological observation data can provide information such as wind speed, wind direction, and air temperature, which are of significant importance for the design and operation of wind energy storage systems. Through the analysis of micro-meteorological data, the changing trends of wind energy resources can be predicted, helping to adjust the charging and discharging strategies of energy storage systems to maximize the use of wind energy resources and improve the efficiency of energy storage systems. Additionally, micro-meteorological data can also be used to evaluate the feasibility and economic viability of wind energy storage systems, providing a basis for project decision-making.

Furthermore, temperature observation and its influence are also one of the important applications of micro-meteorology in the field of energy storage. Temperature has a direct impact on the battery performance and energy efficiency of energy storage systems. High temperatures can accelerate the aging and loss of batteries, reducing the lifespan and efficiency of energy storage systems. Through micro-meteorological observation data, the changing environmental temperature can be monitored in real time, and appropriate measures can be taken to control the temperature, protect the batteries, and improve the efficiency and reliability of energy storage systems.

Through the collection and analysis of micro-meteorological observation data, effective utilization of renewable energy resources such as solar and wind energy can be achieved, optimizing the charging and discharging strategies of energy storage systems, and improving the efficiency and reliability of energy storage.

3.1.6. Heating

Micro-meteorology also plays an important role in the field of heating and has a direct impact on the operation and efficiency of heating systems. Firstly, temperature observation and its influence are crucial applications of micro-meteorology in the heating field. Micro-meteorological observation data can provide real-time information on environmental temperature changes. Based on this data, heating systems can adjust the operating parameters of heating equipment to regulate the output of heat power according to external temperature changes, meeting the heating demands in different seasons and weather conditions. For example, during cold winters when the outdoor temperature is low, the heating system can increase the supply of hot water or steam to ensure stable and comfortable indoor temperatures.

Secondly, humidity observation and its influence are also important applications of micro-meteorology in the heating field. Humidity has a certain impact on the regulation of heating systems and indoor comfort. In dry environments, heating systems can increase the humidity of indoor air to improve comfort. Through micro-meteorological observation data, humidity changes can be monitored in real time, and in combination with the control strategies of heating systems, the control devices for indoor humidity can be adjusted to provide a comfortable heating environment indoors.

Furthermore, wind speed observation and its influence are also important applications of micro-meteorology in the heating field. Wind speed directly affects heat loss and energy consumption in heating systems. Strong winds can increase heat dissipation from building walls and windows, thereby increasing the load on heating systems and energy consumption. Through micro-meteorological observation data, wind speed changes can be monitored in real time, and in combination with the control strategies of heating systems, insulation measures for building walls and windows can be adjusted to reduce heat loss and energy consumption.

In addition, air pressure observation and its influence are also important applications of micro-meteorology in the heating field. Changes in air pressure can affect the delivery and distribution of hot water or steam in heating systems. Through micro-meteorological observation data, air pressure changes can be monitored in real time, and in combination with the control strategies of heating systems, the operating parameters of heating pipelines can be adjusted to ensure the normal operation and heating quality of the systems.

Through the collection and analysis of micro-meteorological observation data, monitoring and regulation of parameters such as temperature, humidity, wind speed, and air pressure can be achieved, optimizing the operation and energy consumption of heating systems, and improving heating efficiency and indoor comfort.

The impact of micro-meteorology on power generation is summarized in Table 2.

Table 2. A summary of the impact of micro-meteorology on different forms of power generation.

Forms of Power Generation	**Associated Micro-Meteorological Elements**	**Influence**
Wind Power	wind speed and direction, temperature, humidity, etc.	(1) cause the flow of wind to generate vortices and turbulence; (2) cause vibration and stress concentration on wind turbine blades and thus reduce the lifespan and performance of the turbines; (3) affect the transmission and conversion efficiency of wind energy.
ine Photovoltaic	solar radiation, temperature, wind speed and wind direction, humidity and precipitation, etc.	(1) basic energy source; (2) affect the efficiency and performance of photovoltaic cells; (3) affect the stability and safety of photovoltaic components, leading to potential damage; (4) negatively effect on the performance and lifespan of photovoltaic components.
ine Hydropower	rainfall, temperature, humidity, wind speed, solar radiation, etc.	(1) determine the inflow volume and available water resources; (2) lead to decreased efficiency and power generation capacity of hydroelectric units; (3) affect the power generation efficiency and equipment lifespan; (4) affect the fluctuation of the water surface in reservoirs; (5) affect factors such as evaporation rates from the water surface and water temperature.
ine Thermal Power	temperature, humidity, wind speed, pressure, etc.	(1) help predict the changing trend of atmospheric temperature, thereby implementing a reasonable heating plan and thermal power adjustment; (2) reduce the combustion efficiency and environmental emission quality of thermal power plants; (3) affect the combustion stability of the boiler and the direction of flue gas emissions; (4) affect the stability of combustion air volume, fuel supply, and flue gas emissions.
ine Energy Storage	solar radiation, air temperature, humidity, wind speed, wind direction, etc.	(1) help adjust the charging and discharging strategies of energy storage systems; (2) help adjust the charging and discharging strategies of energy storage systems; (3) take appropriate measures to control the temperature, protect the batteries, and improve the efficiency and reliability of energy storage systems.
ine Heating	temperature, humidity, wind speed, pressure, etc.	(1) regulate the output of heat power according to external temperature changes, meeting the heating demands in different seasons and weather conditions; (2) adjust the control devices for indoor humidity to provide a comfortable heating environment indoors; (3) increase heat dissipation from building walls and windows, increasing the load on heating systems and energy consumption; (4) affect the delivery and distribution of hot water or steam in heating systems.

3.2. The Impact of Micro-Meteorology on the Entire Lifecycle of New Energy Sources

New energy sources represented by wind and solar power are subject to the constraints and challenges of changeable weather. Micro-meteorological technology plays an important role throughout the entire lifecycle of new energy sources, from site selection in planning, grid connection and operation, to post-evaluation and trading recovery (as shown in Figure 4.

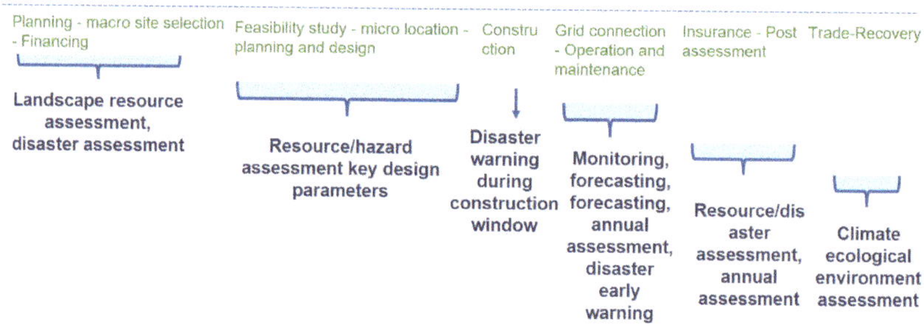

Figure 4. The impact of micro-meteorology on the entire lifecycle of new energy sources.

3.2.1. Survey and Design

In recent years, with the gradual increase in environmental awareness and the continuous growth in energy demand, new energy generation projects have received widespread attention and vigorous promotion. In the detailed survey and parameter design of these projects, micro-meteorological data plays a crucial role.

Firstly, micro-meteorological data is indeed crucial for the site selection and construction of wind power projects. In the preliminary investigation of wind power projects, the collection and analysis of historical meteorological data can identify regions abundant in wind energy resources, laying a foundation for the construction of wind power projects. During actual construction, micro-meteorological data can provide information such as wind speed, wind direction, and wind energy density, helping to determine the layout of wind turbines and the construction of wind farms. High-resolution meteorological data can also provide a more detailed analysis of meteorological conditions, helping wind power projects predict wind energy output, adjust wind turbine operation status, and improve wind power generation efficiency.

Secondly, micro-meteorological data also plays an important role in the site selection and design of photovoltaic power generation projects. Photovoltaic power generation relies on factors such as solar radiation intensity, duration of sunlight, and atmospheric transparency. High-resolution meteorological data can provide more detailed meteorological information. By comparing and analyzing micro-meteorological data from different regions, areas abundant in solar energy resources can be identified, reducing errors in the placement and tilt angle of photovoltaic panels, and improving photovoltaic power generation efficiency.

In addition to its impact on project site selection and design, micro-meteorological data are also crucial for project operation and maintenance. For example, in the actual operation of wind power projects, micro-meteorological data can provide information such as wind speed and direction for wind turbine adjustment and maintenance. In the operation of photovoltaic power generation projects, micro-meteorological data can provide weather forecasts and temperature information for system optimization and maintenance. Furthermore, micro-meteorological data are also important for other power generation projects such as hydroelectric power. For example, in the design of hydroelectric power plants, micro-meteorological data can provide information on rainfall and runoff, helping to design reservoir capacity and select hydroelectric generators.

In addition, accurate short-term weather forecasting is needed during engineering construction and operation to take appropriate construction arrangements and emergency measures, ensuring safe and smooth project construction and operation. High-resolution meteorological data can provide more accurate data support for these predictions, helping to avoid potential meteorological risks and safeguard the smooth progress of engineering projects.

The scientific application of micro-meteorological data is a necessary condition for building more reliable, efficient, and environmentally friendly new energy generation projects. With the continuous advancement of technology and the improvement of data acquisition techniques, the acquisition and analysis of micro-meteorological data will become more precise and comprehensive. In the future, with the continuous expansion and popularization of renewable energy generation projects, the application of micro-meteorological data will become more widespread and in-depth. With the help of micro-meteorological data, more and more renewable energy generation projects will become an important pillar for people's survival and development.

3.2.2. Power Forecasting

Although new energy generation has become relatively mature and widely applied, compared to traditional thermal power plants, it still exhibits strong randomness and volatility. When large-scale integration into the power grid occurs, it will have a significant impact on grid stability. Ensuring the stability of the grid requires the ability to predict power and generation in advance and adjust generation and grid load accordingly.

Compared to traditional power generation methods, renewable energy, mainly wind power and photovoltaic power, is more susceptible to environmental factors such as wind speed, wind direction, sunshine duration, temperature, and atmospheric pressure during the generation process. It has higher randomness, volatility, and uncontrollability. The large-scale integration of renewable energy into the grid poses significant challenges to the stable operation and power dispatch of the system [110].

Therefore, accurate prediction of new energy power is one of the effective methods to solve the aforementioned difficulties. Meteorological data play a vital role in new energy power prediction. Taking wind power prediction as an example, one commonly used method is to establish a wind power prediction model based on historical meteorological and power output data of wind farms, and use numerical weather forecast results as input to predict the power output of wind farms. The precision of weather forecasts significantly impacts the accuracy of power predictions.

The core of new energy power prediction technology lies in exploring the relationship between numerical weather forecasts representing weather changes and the power variations in new energy power stations. If the weather forecasts for the location of the power station are inaccurate, it will result in significant errors in new energy power predictions.

The commonly available meteorological forecasts in the market mainly provide city-level meteorological data at the ground level. The data do not match the spatial distribution of new energy power stations, and the vertical distribution does not match wind power generation. In addition, the current general meteorological forecasts have low resolution and cannot provide high-resolution and more detailed forecasts at the location of the power station [111]. Therefore, conducting power predictions usually requires purchasing commercial numerical weather forecast data, mainly in two types: grid-based data with a spatial resolution of about 3km*3km and specified latitude and longitude points, often located at wind measurement towers or specific points within wind farms.

Therefore, the importance of constructing a micro-meteorological forecast system for new energy generation is becoming increasingly evident. High-resolution micro-meteorological forecast data can support the fine modeling of new energy power prediction at the single-machine level and equipment level, improving the accuracy of power predictions at new energy power stations.

3.2.3. China Electricity Spot Market Trading

In China's spot market environment, accurate micro-meteorological data forecasting is of great significance in new energy power forecasting, network-wide load forecasting, and inter-provincial transactions. It can ultimately effectively guide electricity sales companies and wholesale users to formulate reasonable trading strategies.

Meteorological analysis and prediction are the basis for various load predictions and new energy predictions in the electricity market, with the main goal of preparing for spot market price prediction through the prediction of various loads as well as photovoltaic and wind power output.

Meteorological fluctuations and randomness are key factors influencing the power output variability of new energy stations in the context of predicting new energy power. Wind power generation is strongly related to meteorological factors such as wind speed and air density, while photovoltaic power generation is strongly related to meteorological factors such as solar radiation intensity (solar incidence angle, cloud cover, etc.) and temperature. Accurate meteorological information prediction is crucial for new energy power prediction. The power of new energy stations is often negatively correlated with spot prices. The real-time balance of supply and demand in the power system. With the construction of a new type of power system dominated by new energy sources, the impact of new energy station power on spot prices becomes more prominent. If micro-meteorological forecasting can be achieved one week or even longer in advance and converted into new energy power prediction, market entities can formulate trading strategies proactively and flexibly.

In terms of load, the overall load of the power grid is affected by meteorological factors. Temperature is the most significant meteorological factor affecting the overall load of the power grid, and the load level and temperature are positively correlated in summer and negatively correlated in winter. Under the distributed photovoltaic "self-use, surplus to the grid" mechanism, when the temperature is constant, the distributed photovoltaic greatly reduces the user's demand for grid power supply, and even the distributed photovoltaic sends power back to the grid, leading to a decrease in the grid's coordinated load level. The grid load level directly affects the power generation capacity supply and demand of thermal power enterprises, thereby affecting the spot market price level. High-precision micro-meteorological forecasting is crucial for predicting load levels and spot market price levels.

During the period of inter-provincial market opening, larger-scale high-precision micro-meteorological forecasting is critical for predicting spot market prices in the province. When the temperature in the receiving province is higher in summer or lower in winter, the inter-provincial price has an advantage, and the power generation enterprises sending power to the province raise the provincial spot market quotation level, pulling up the clearing price of the provincial spot market through quotation, obtain settlement income from the provincial spot market, and participate in the inter-provincial market space with maximum power generation capacity to obtain high inter-provincial market income. However, when there is no need for external power to meet the balance of provincial electricity consumption, the inter-provincial market price lacks advantages, and power generation enterprises sending power out of the province often lower the provincial spot market quotation to fully participate in the provincial market.

With the continuous implementation of the further deepening of the electricity trading marketization reform, the impact of meteorological factors on the market supply and demand ends and even spot prices will become more apparent, becoming the core data that affect the quality of trading decisions. Therefore, strengthening the production, optimization, collection, and application of micro-meteorological forecasting data will also become the necessary path for all kinds of market entities to build core trading capabilities.

3.3. Integrated Energy System

Zeng et al. [112] argued that building a comprehensive energy system promotes the integrated development of power generation, transmission, distribution, storage, and multiple energy complementary development. Micro-meteorology provides basic data support for the integrated operation and optimization of comprehensive energy systems. The basic connotation of a comprehensive energy system can be summarized as "multiple energy complementary and coordinated optimization", covering multiple energy resources such as electricity, coal, oil, natural gas, and renewable energy. Micro-meteorology data supports the centralized and distributed energy types to achieve a full complementary supply of multiple energy sources. Based on the smart grid, it can achieve interconnection with various types of networks such as heating networks, natural gas networks, and transportation networks. With micro-meteorology data as one of the main bases, it can play the regulatory capacity of flexible storage resources such as energy storage, which can effectively mobilize the response potential of demand-side resources. It can realize the horizontal multiple energy complementarity and vertical coordinated development between multiple energy systems, thereby constructing a diversified energy supply system and promoting an energy supply revolution.

Micro-meteorology data will provide effective help for the dispatching and optimization of integrated energy systems. Through scientifically dispatching multiple energy sources within the system, a comprehensive energy system can achieve efficient consumption of renewable energy, efficient utilization of comprehensive energy, and safe and economic energy use by users. At the same time, the organic coordination of multiple energy systems also helps to eliminate the supply bottleneck of different energy systems, delay the construction of energy supply systems, and improve the utilization efficiency of various energy equipment. In addition, in the application of virtual power plants, micro-meteorology data aggregate and optimize distributed energy resources (including distributed power generation, energy storage systems, controllable loads, etc.), providing basic data support for power markets and grid operations, and providing management and auxiliary services for distribution and transmission networks.

Therefore, micro-meteorology has a profound impact on the integrated scheduling and multi-energy complementary of comprehensive energy systems. For example, it can be specifically reflected in the following aspects:

(1) Wind and solar power output fluctuations: With the increase in the proportion of new energy, wind and solar power are important components of a comprehensive energy system, and their output fluctuations have an important impact on the integrated scheduling and operation of the power system. Therefore, improving the accuracy of wind and solar power output prediction based on micro-meteorology data is one of the key tasks of the integrated scheduling of comprehensive energy systems.

(2) Extreme weather events: Extreme weather events such as typhoons, heavy rain, and snow may have a serious impact on the operation of the power system, including equipment damage and power interruption. Therefore, based on micro-meteorology observations, risk assessment of extreme weather events and the formulation of corresponding emergency plans are important measures to ensure that the energy system can respond and recover quickly in the face of meteorological risks.

(3) Flexible scheduling of power generation resources: Under the influence of meteorological changes, the energy system needs to flexibly schedule various types of power generation resources. For example, when the output fluctuations of wind and solar power are large, the input of adjustable power generation resources such as thermal power and hydropower can be increased to balance the system load. In addition, by establishing regional power markets, cross-regional power scheduling can be achieved, which can alleviate the pressure caused by meteorological effects to a certain extent. In these solutions, accurate decision-making and scheduling based on micro-meteorology forecasts are crucial.

(4) Application of energy storage technology: The comprehensive energy system needs to balance the impact of meteorological conditions on the stability of energy output, and energy storage technology plays an important role in this process. By applying energy storage technologies such as batteries and pumped storage, it is possible to smooth the load curve of the system during wind and solar power output fluctuations and improve system stability. At the same time, energy storage technology can also provide additional support during peak periods of electricity demand and reduce the pressure on the power

system. Micro-meteorology data have important significance for optimizing and allocating energy storage systems and achieving peak shaving and valley filling.

Therefore, in the integrated scheduling and multi-energy complementary of comprehensive energy systems, it is necessary to fully consider the impact of micro-meteorology factors, take corresponding measures, and improve the ability and stability of comprehensive energy systems.

In summary, the core characteristics of a comprehensive energy system are reflected in interconnection, comprehensive utilization, and optimized sharing. Micro-meteorology plays a very important role in internal optimization and scheduling, power generation planning, event warning, emergency response, and other important links to comprehensive energy systems. It is also among the basic data for the refined planning, allocation, and scheduling of comprehensive energy systems. With the increase in the proportion of new energy, micro-meteorology will become increasingly important in comprehensive energy systems.

3.4. Disaster Prevention and Reduction

In the context of global warming, the frequency and intensity of extreme weather events are continuously increasing [113]. The Intergovernmental Panel on Climate Change (IPCC) Sixth Assessment Report by the Working Group I states unequivocally that human influence has undoubtedly caused the warming of the atmosphere, ocean, and land, leading to widespread and rapid changes in the atmosphere, ocean, cryosphere, and biosphere. The scale and multi-faceted nature of the overall climate system changes are unprecedented over centuries or even thousands of years. Human-induced climate change has already affected the weather and extreme climates in every region worldwide.

Extreme weather and climate events often involve abnormal changes in meteorological elements [114]. For instance, during heatwaves and static stability in the atmosphere, widespread calm wind conditions can render wind turbines virtually ineffective for periods of time. Similarly, in extremely cold and overcast conditions, solar photovoltaic components may not receive sufficient solar radiation for conversion. In times of drought, river flow may decrease, leading to reduced power generation at hydroelectric plants; insufficient reservoir water levels could result in their inability to function normally. During floods, riverbed erosion and dam damage may occur, impacting water turbine generators and power facilities, potentially causing reservoir overflow and disrupting power generation at hydroelectric plants. As renewable energy becomes a dominant component of the new energy structure, heightened attention must be given to the new risks posed by climate change. A micro-meteorological observation network will provide essential foundational data for predicting extreme weather conditions.

The response to extreme weather events, particularly extreme temperature events such as heatwaves and extreme cold, often requires substantial energy reserves and distribution support. Climate change-induced extreme weather events will present an unusually complex situation for future energy supply needs. The extreme snowstorm in Texas in early 2021 [115] resulted in widespread power and water system failures, leaving residents without water and electricity for several days, and some even perished from freezing and starvation at home [116]. Texas faced a shortage of reliable electricity during the cold snap. Report [117] indicated that half of the state's wind turbines froze, causing the proportion of wind power in the electricity mix to drop from 42 percent to 8 percent. In such circumstances, relying on more precise micro-meteorological forecasts to issue early warnings for extreme weather and prevent natural disasters such as geological hazards could greatly enhance disaster prevention and reduction efforts, effectively serving safety, operations, power generation, and market trading, playing a crucial role in energy preparedness, dispatch, and stable grid operation.

4. Case Study

This section provides a case study to illustrate how data and technique of micro-meteorolgy can help to improve the efficiency of the new power system.

China is planning to optimize its energy mix by building massive wind and solar power facilities in the country's Gobi and other desert areas to support the objective of reaching more than 1200 gigawatts of installed solar and wind capacity by 2030. Meteorological data and services are becoming more crucial for renewable energy development and operation. This is clearly reflected in the memorandum of understanding between the International Renewable Energy Agency (IRENA) and the World Meteorological Organization (WMO) on working together to support the transition to renewable energy through tailored weather, water and climate services that was recently signed at the COP27. While China's Gobi desert area has enormous potential for harvesting solar and wind energy, this is also an area where in situ observations and forecasting skills of numerical weather predictions are more limited. In addition, high-influence weather phenomena such as gust, strong wind and sandstorms are more frequent. Thus, power prediction skills for wind and photovoltaic generation are very challenging in this area.

This section shows how high-resolution data that captures the local micro-meteorology conditions of a wind farm well in the Gobi desert area can help improve the accuracy of power prediction.

4.1. Data Description

Firstly, the mesoscale numerical model WRF-ARW4.5 is used to conduct numerical simulation experiments, so as to obtain a micro-meteorology reanalysis dataset in the Gobi desert area. During the experiment, the Grid Nudging is used for data assimilation, and the open source reanalysis dataset-ERA5 is used as the background field. The ERA5 reanalysis data come from a global meteorological dataset provided by ECMWF. The time resolution of the dataset is 1 h and the spatial resolution is 0.25° × 0.25°.

The micro-meteorology reanalysis dataset was collected from March 2022 to February 2023 with a time resolution of 15 min and a spatial resolution of 0.01° × 0.01°. The spatial range covered by the dataset is 40.30° N–41.25° N, 95.60° E–96.62° E. The dataset includes upper-air variables and surface variables. In this section, the temperature, relative humidity, pressure, wind direction, wind speed at different heights and precipitation of grid points adjacent to a wind farm in Jiuquan, China are selected for data analysis and wind power prediction. The details of meteorological variables are shown in Table 3. The first 70% of the data points are used as training samples, and the remaining data points are used as test data.

Table 3. The details of meteorological variables.

Variable	Unit	Heights (m)
Temperature	°C	2/30/50/70/90/110/130/150
Relative humidity	%	2/30/50/70/90/110/130/150
Pressure	kPa	surface/30/50/70/90/110/130/150
Wind direction	°	10/30/50/70/90/110/130/150
Wind speed	m/s	10/30/50/70/90/110/130/150
Precipitation	m	-

4.2. Key Variable Analysis

First, the 90 m wind speed of 144 grid points covering the wind farm at different times is shown in Figure 5. The spatial range covered by 144 grid points is 40.80° N–40.97° N, 96.37° E–6.44° E, so the spatial span of 144 grid points is 0.17° × 0.07°, which is significantly smaller than t9he spatial resolution of ERA5 dataset. It can be seen from Figure 5 that the 90 m wind speed at each grid point changes significantly at the same time. For example, at 6:00 on 1 March 2022, the maximum wind speed in all grid points is 8.68 m/s, and the minimum wind speed is 5.13 m/s. The difference between them reaches 3.55 m/s.

This shows that the wind farm located in the Gobi desert area has local meteorological characteristics, which can be displayed by the micro-meteorology reanalysis dataset, but can not be reflected by the data with low spatial–temporal resolution, such as ERA5 dataset.

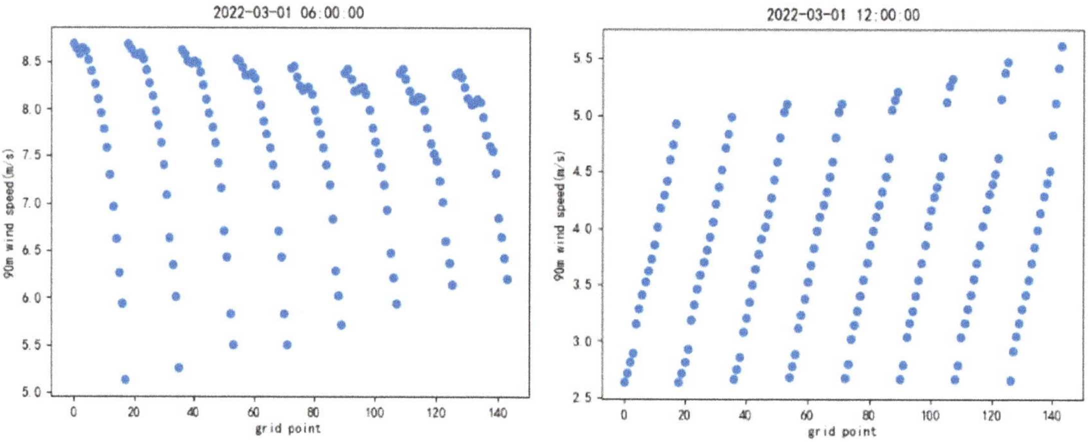

Figure 5. The 90 m wind speed of grid points at different times.

Next, the correlation between 90 m wind speed at different grid points and wind power is analyzed, and the results are shown in Figure 6. It can be seen that the correlation between wind power and 90 m wind speed at most grid points is strong, which is not less than 0.7. In addition, the spatial span of grid points covering the wind farm is also 0.17° × 0.07°, which is significantly smaller than the spatial resolution of the ERA5 dataset. The correlation coefficient between wind power and 90 m wind speed at each grid point is different, which indicates that the wind speed at 90 m height is variable within the spatial span and has local micro-meteorological characteristics, which can not be captured by the ERA5 dataset. It is proved that the micro-meteorology reanalysis dataset can capture the local micro-meteorology conditions of a wind farm in the Gobi desert area well.

4.3. *Wind Power Prediction*

In this section, the wind power prediction based on the micro-meteorology reanalysis dataset and the ERA5 reanalysis data is compared.

4.3.1. Experimental Setting

The ERA5 dataset also includes pressure, temperature, humidity, wind speed, wind direction and other variables in the upper air and surface. But the temporal resolution and spatial resolution of the dataset are lower than those of the micro-meteorology reanalysis dataset. In this section, the ERA5 dataset was also collected from March 2022 to February 2023. The temperature, pressure, wind direction and wind speed at different heights and precipitation of grid points adjacent to the wind farm in Jiuquan were selected for wind power prediction. In addition, the data used for wind power prediction are shown in Table 4. It can be seen that the time resolution of ERA5 does not match the wind power data, so linear interpolation was performed on the ERA5 dataset to make its time resolution 15 min. The training set and the test set were also divided according to the ratio of 7:3.

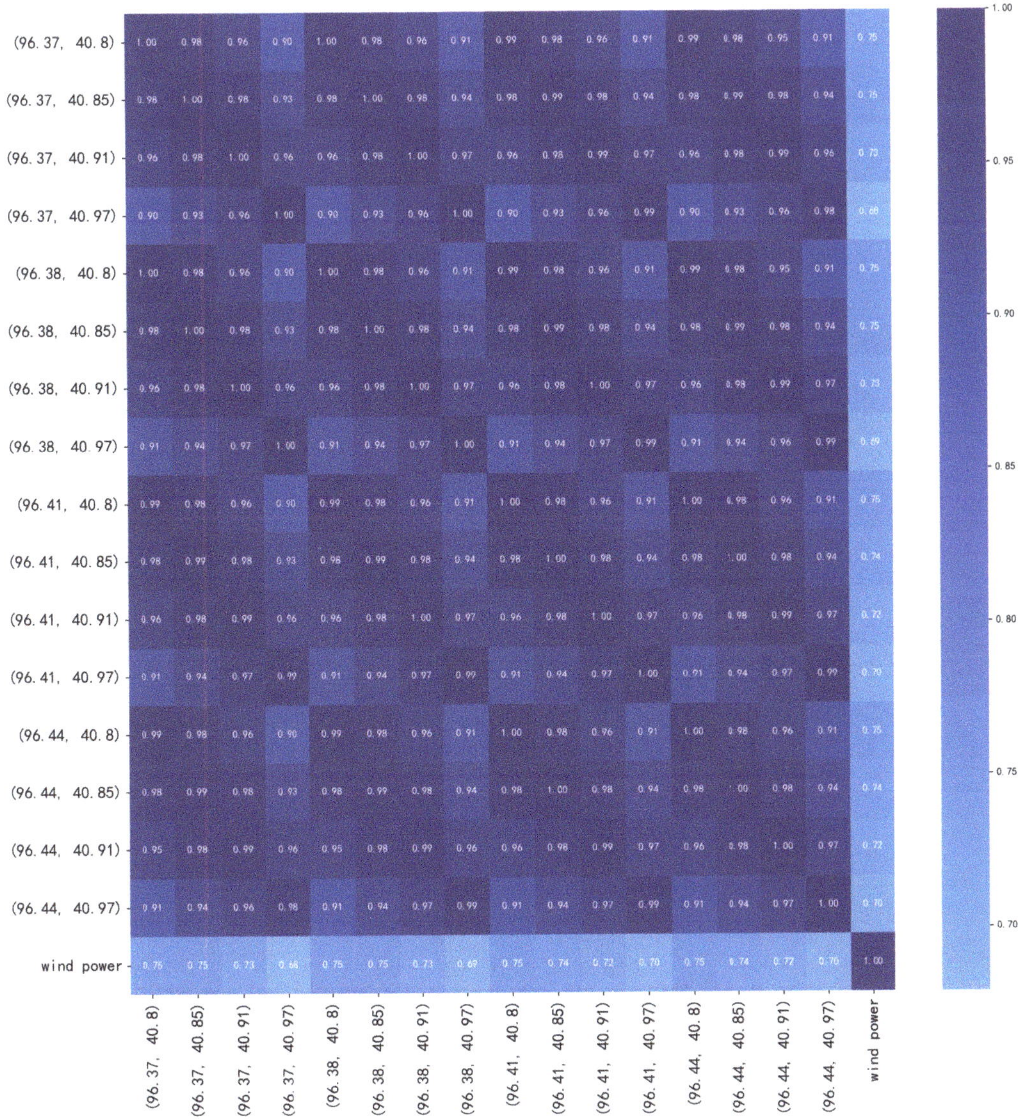

Figure 6. The correlation matrix between 90 m wind speed at different grid points and wind power.

Table 4. Data used for wind power prediction.

Data	Time Frame	Temporal Resolution	Spatial Resolution
Micro-meteorology reanalysis dataset	2022.3–2023.2	15 min	0.01° × 0.01°
ERA5	2022.3–2023.2	1 h	0.25° × 0.25°
Wind power	2022.3–2023.2	15 min	a wind farm

The models uesd in this section are eXtreme Gradient Boosting (XGBoost), Light Gradient Boosting Machine (LightGBM), long short-term memory (LSTM) and gated recurrent unit (GRU). XGBoost and LightGBM are traditional machine learning models and tree-based models, which have been applied to tackle the problem of wind power prediction (WPP). XGBoost is a supervised ensemble learning algorithm, which can be used to solve regression problems. XGBoost uses the decision tree as the base learner to construct multiple weak learners. In the iterative learning process, it continuously trains along the direction of reducing the gradient, and uses the second-order Taylor series to expand the loss function. In order to find the global optimal solution, regularization is added to the objective function to control the accuracy and complexity of the model. XGBoost can process in parallel, prevent overfitting, and has fast speed. LightGBM is an ensemble learning algorithm based on a gradient lifting tree. By adding gradient-based one-side sampling and exclusive feature bundling, LightGBM solves the time-consuming problem caused by the traditional gradient lifting tree which needs to traverse all data at the feature splitting point to calculate the information gain. In summary, traditional machine learning models can adaptively learn the characteristics of data and predict nonlinear WP data accurately. Compared with physical models and traditional statistical models, they have a better prediction performance. However, these models have limitations in expressing complex data and fully reflecting the characteristics of data [118].

Over the past few years, deep learning models have become the most popularly used machine learning models, owing to their intricate structures and potent nonlinear mapping capabilities. They are increasingly being employed in a variety of applications, including WPP. In deep learning models, RNN-based models are widely used in WPP; due to their structure, they are prone to the problem of gradient disappearance or explosion. LSTM and GRU are commonly used RNN-based models, and LSTM introduces a gating mechanism on the basis of RNN, replacing common neuron modules with special memory neuron modules, so it can well solve the long-term dependence problem of sequences [119], and various studies [120] have shown that compared to other prediction methods, LSTM has a higher accuracy in WPP. GRU is a variant of LSTM, which also can solve the problem of gradient disappearance and explosion. Compared with LSTM, GRU does not have a dedicated memory cell, so GRU has a simpler structure, making it be trained faster than LSTM. But LSTM has a better performance for the dataset with longer sequences. In this section, the LSTM and GRU were implemented in PyTorch with a "mean square error (MSE)" loss function and an "Adaptive Moment Estimation (Adam)" optimizer.

In order to measure the prediction performance of the different datasets, three prediction error indicators were used: mean absolute error (MAE), root mean square error (RMSE) and C_R. The equations for these indicators are given in Equations (1)–(4).

$$\mathrm{MAE} = \frac{1}{n}\Sigma_{i=1}^{n}|y_i - \hat{y}_i| \tag{1}$$

where n is the number of test samples, and y_i and $\hat{y}_i$ are the ith actual value and the corresponding estimated value, respectively.

$$\mathrm{RMSE} = \sqrt{\frac{1}{n}\Sigma_{i=1}^{n}(y_i - \hat{y}_i)^2} \tag{2}$$

$$E_{\mathrm{RMSE}} = \sqrt{\frac{1}{n}\Sigma_{i=1}^{n}(\frac{y_i - \hat{y}_i}{C_i})^2} \tag{3}$$

$$C_R = 1 - E_{\mathrm{RMSE}} \tag{4}$$

where C_i is the ith startup capacity of the wind farm.

4.3.2. Results

The wind power prediction results of different models applied to the two datasets are presented in Table 5. In the table, Dataset 1 refers to the micro-meteorology reanalysis dataset, and Dataset 2 refers to the ERA5 dataset. It can be seen that compared with the prediction results using the ERA5 dataset, the wind power prediction results of all models based on the micro-meteorology reanalysis dataset have high accuracy. Moreover, compared to shallow machine learning models LightGBM and XGBoost, deep learning models LSTM and GRU have better prediction performance. In addition, according to RMSE and C_R, LSTM has the highest accuracy among all the models. Therefore, LSTM is chosen as the best prediction model, and the wind power prediction results of LSTM based on the two data sets and the absolute value of the error between the prediction results and the actual values over a period of time are shown in Figure 7. It can be seen that at most times, the absolute value of the error is smaller and the prediction accuracy is higher when using the micro-meteorology reanalysis dataset.

The analysis shows that the meteorological observation stations in the desert, Gobi and desertification land are sparse, and the basic meteorological data resources in these areas are seriously insufficient. In addition, due to the influence of the natural geographical environment, the obvious local micro-meteorology is prone to occur in the wind farm area. Therefore, the micro-meteorology reanalysis dataset can capture the local micro-meteorology conditions of the wind farm in the Gobi desert area well, and using the micro-meteorology reanalysis dataset with high spatial–temporal resolution for wind power prediction can effectively improve the prediction accuracy.

Table 5. Prediction results of different models for two datasets.

Method	MAE (MW)		RMSE (MW)		C_R	
	Dataset 1	Dataset 2	Dataset 1	Dataset 2	Dataset 1	Dataset 2
LSTM	38.0658	40.4811	53.2572	54.5717	86.6857%	86.3570%
GRU	38.4487	40.1762	54.0846	54.8024	86.4789%	86.2994%
LightGBM	41.1604	42.0596	54.0357	54.6014	86.4911%	86.3497%
XGBoost	41.0750	44.0245	54.4686	56.7056	86.3829%	85.8237%

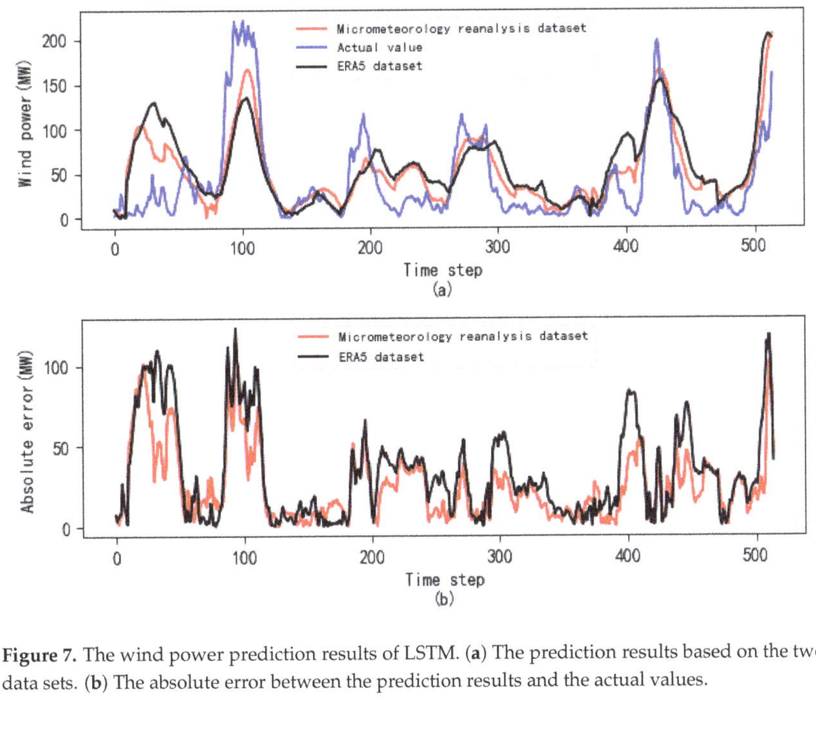

Figure 7. The wind power prediction results of LSTM. (**a**) The prediction results based on the two data sets. (**b**) The absolute error between the prediction results and the actual values.

5. Discussion

Despite its clear benefits analyzed before, the integration of micro-meteorology into the power system faces challenges. This section discusses several challenges from observations, data, techniques, uncertainty, and cooperation.

5.1. Observation System

Due to the spatial heterogeneity of micro-meteorological phenomena, an extensive and dense observation network is required to capture meteorological changes at various spatial locations. This requires reasonable planning of the layout of the observation network to ensure that it covers a sufficient spatial range. It also requires timely collection, transmission and processing of observation data to ensure the accuracy and reliability of the data. However, because new energy development is often located in remote areas, meteorological observation equipment is seriously insufficient. At the same time, because these devices are usually high-precision, high-sensitivity instruments, they require regular calibration and maintenance to maintain their performance and accuracy.

In addition, the establishment and maintenance of a micro-meteorological observation system requires a large amount of financial and technical support. Due to insufficient financial and technical support, the micro-meteorological observation system may have problems such as aging equipment and backward technology, resulting in a decline in the quality of observation data or an inability to meet demand. Therefore, it is necessary to strengthen investment in financial and technical support to ensure the normal operation and development of the micro-meteorological observation system.

5.2. Data

In order for micro-meteorology to play a better role in the new power system, the requirements for real-time data are relatively high, which brings great challenges to data collection and transmission. Micro-meteorological data usually contain a large amount of information, such as wind speed, wind direction, temperature, humidity, air pressure, etc. These data require complex processing and analysis to extract useful information and transform it into understandable prediction results. In this process, professional technical personnel and advanced data analysis tools are needed.

In addition, data quality assurance is also an important challenge. The accuracy and reliability of micro-meteorological data have an important impact on subsequent meteorological prediction and decision-making. Therefore, data calibration and quality assurance are an indispensable part. This requires establishing strict data quality standards and calibration processes to ensure data accuracy and reliability.

Furthermore, because micro-meteorological data have the characteristics of high dimensionality, high frequency, and high spatial correlation, efficient data storage and processing technologies need to be developed to meet real-time and accuracy requirements.

5.3. Reliable Predictive Models

The prediction model based on micro-meteorological data is one of the key technologies to realize the incorporation of micro-meteorology into new power systems. However, due to the high complexity and uncertainty of micro-meteorological phenomena, establishing accurate and reliable prediction models is a challenging task. Currently, commonly used methods include statistical models, physical models, neural networks, etc. These methods each have their own advantages and disadvantages when dealing with specific problems, so they need to be selected and optimized based on actual application scenarios. In addition, the interpretability and generalization ability of the model also needs to be considered to ensure the reliability and stability of the prediction results.

5.4. Uncertainty

Micro-meteorological phenomena are affected by many factors, such as climate change, topography, atmospheric composition, etc., and these factors may bring uncertainty. Therefore, it is necessary to consider these uncertain factors when establishing a forecast model and use probability theory and statistical methods to evaluate and revise the forecast results. An early warning mechanism needs to be established to provide timely warnings of possible meteorological disasters and take effective response measures to reduce their impact on the power system.

5.5. Cooperation and Support

In order to better play the role of micro-meteorology in new power systems, more comprehensive exchanges and cooperation and integration of data, technology and personnel between the two fields are needed. In addition, it is also necessary to actively promote policy and financial support, strengthen the development of interdisciplinary technologies, and promote the building of researchers and teams.

However, these challenges also bring opportunities for innovation and growth. As the energy sector transitions towards more distributed and flexible systems, micro-meteorology can play a more pivotal role. Furthermore, advances in big data and artificial intelligence technology will provide more possibilities for the application of micro-meteorology in power systems.

6. Conclusions

Meteorological conditions affect all aspects of new power systems. The production and operation of the power system are closely related to the atmospheric environment. Especially as the proportion of renewable energy in power systems continues to rise, micro-meteorology is becoming one of the core technologies to ensure the safety and stability of power systems.

In the entire power generation process of the regional power system, including survey and design, wind power, solar power, hydrothermal power generation, energy storage, heating, power prediction, electricity market transactions, etc., micro-meteorology plays an important role. Micro-meteorology provides basic data and technical support systems for the power system, making contributions to dispatch optimization, disaster prevention and early warning, management and control, transaction operation scenarios, etc.

In the future, targeted micro-meteorological technological innovation work will be considered and carried out in various aspects, such as improving the utilization rate of meteorological data resources, enriching meteorological data development and utilization methods and tools, and promoting the transformation of meteorological data resources into data elements. These works will be conducive to the establishment and application of micro-meteorological services for more efficient management and overall optimization of multiple energy forms, improving overall energy efficiency, reducing costs, and achieving accurate predictions of power market trends through the development and utilization of multi-scale meteorological data.

Author Contributions: Conceptualization, M.Z. and Q.H.W.; investigation, H.Z., R.Y. and Y.L.; survey, H.Z.; analysis, R.Y.; case study, Y.L.; validation, Y.L.; writing—original draft preparation, H.Z., R.Y. and Y.L.; writing—review and editing, M.Z., H.Z. and Q.H.W.; visualization, Y.L., H.Z. and R.Y.; supervision, M.Z. and Q.H.W.; project administration, M.Z., X.M. and Q.H.W.; funding acquisition, M.Z., X.M. and Q.H.W. All authors have read and agreed to the published version of the manuscript.

Funding: This research was funded by the Research Fund of China Huaneng Clean Energy Research Institute QNYJJ22-22 and the Soft Science Research Project 2023-2 of China Huaneng Group.

Data Availability Statement: The data presented in this study are available on request from the corresponding author. The data are not publicly available due to commercial restrictions.

Conflicts of Interest: The authors declare that the research was conducted in the absence of any commercial or financial relationships that could be construed as a potential conflict of interest.

References

1. Olabi, A.; Abdelkareem, M.A. Renewable energy and climate change. *Renew. Sustain. Energy Rev.* **2022**, *158*, 112111. [CrossRef]
2. Yang, X.; Pang, J.; Teng, F.; Gong, R.; Springer, C. The environmental co-benefit and economic impact of China's low-carbon pathways: Evidence from linking bottom-up and top-down models. *Renew. Sustain. Energy Rev.* **2021**, *136*, 110438. [CrossRef]
3. Curtin, J.; McInerney, C.; Gallachóir, B.Ó.; Hickey, C.; Deane, P.; Deeney, P. Quantifying stranding risk for fossil fuel assets and implications for renewable energy investment: A review of the literature. *Renew. Sustain. Energy Rev.* **2019**, *116*, 109402. [CrossRef]
4. Olabi, A.; Obaideen, K.; Elsaid, K.; Wilberforce, T.; Sayed, E.T.; Maghrabie, H.M.; Abdelkareem, M.A. Assessment of the pre-combustion carbon capture contribution into sustainable development goals SDGs using novel indicators. *Renew. Sustain. Energy Rev.* **2022**, *153*, 111710. [CrossRef]
5. Gustavsson, L.; Nguyen, T.; Sathre, R.; Tettey, U.Y.A. Climate effects of forestry and substitution of concrete buildings and fossil energy. *Renew. Sustain. Energy Rev.* **2021**, *136*, 110435. [CrossRef]
6. He, X.; Wang, F.; Wallington, T.; Shen, W.; Melaina, M.; Kim, H.; De Kleine, R.; Lin, T.; Zhang, S.; Keoleian, G.; et al. Well-to-wheels emissions, costs, and feedstock potentials for light-duty hydrogen fuel cell vehicles in China in 2017 and 2030. *Renew. Sustain. Energy Rev.* **2021**, *137*, 110477. [CrossRef]
7. Hussain, A.; Arif, S.M.; Aslam, M. Emerging renewable and sustainable energy technologies: State of the art. *Renew. Sustain. Energy Rev.* **2017**, *71*, 12–28. [CrossRef]
8. Madi, E.; Pope, K.; Huang, W.; Iqbal, T. A review of integrating ice detection and mitigation for wind turbine blades. *Renew. Sustain. Energy Rev.* **2019**, *103*, 269–281. [CrossRef]
9. Olabi, A.G.; Abdelkareem, M.A.; Wilberforce, T.; Sayed, E.T. Application of graphene in energy storage device–A review. *Renew. Sustain. Energy Rev.* **2021**, *135*, 110026. [CrossRef]
10. Sayed, E.T.; Abdelkareem, M.A.; Bahaa, A.; Eisa, T.; Alawadhi, H.; Al-Asheh, S.; Chae, K.J.; Olabi, A. Synthesis and performance evaluation of various metal chalcogenides as active anodes for direct urea fuel cells. *Renew. Sustain. Energy Rev.* **2021**, *150*, 111470. [CrossRef]
11. Olabi, A.; Elsaid, K.; Rabaia, M.K.H.; Askalany, A.A.; Abdelkareem, M.A. Waste heat-driven desalination systems: Perspective. *Energy* **2020**, *209*, 118373. [CrossRef]
12. Pili, R.; Martínez, L.G.; Wieland, C.; Spliethoff, H. Techno-economic potential of waste heat recovery from German energy-intensive industry with Organic Rankine Cycle technology. *Renew. Sustain. Energy Rev.* **2020**, *134*, 110324. [CrossRef]
13. Impram, S.; Nese, S.V.; Oral, B. Challenges of renewable energy penetration on power system flexibility: A survey. *Energy Strategy Rev.* **2020**, *31*, 100539. [CrossRef]
14. Li, C.; Shi, H.; Cao, Y.; Wang, J.; Kuang, Y.; Tan, Y.; Wei, J. Comprehensive review of renewable energy curtailment and avoidance: A specific example in China. *Renew. Sustain. Energy Rev.* **2015**, *41*, 1067–1079. [CrossRef]
15. Gu, Y.; Zhang, S.; Dai, X.; Lu, Y.; Wang, Y. Research on Regional Micro-meteorological Measurement System for the Operation Control of Distribution Network List the author names here. In Proceedings of the 2020 International Conference on Advanced Electrical and Energy Systems, Osaka, Japan, 18–21 August 2020; Volume 582, p. 012005.
16. Deng, X.; Lv, T. Power system planning with increasing variable renewable energy: A review of optimization models. *J. Clean. Prod.* **2020**, *246*, 118962. [CrossRef]
17. Papaefthymiou, G.; Dragoon, K. Towards 100% renewable energy systems: Uncapping power system flexibility. *Energy Policy* **2016**, *92*, 69–82. [CrossRef]
18. Li, L.; Li, Z.; Zhang, X.; Li, C. Micrometeorological Data Collection and Application in Internet of Things for Power Systems. *IFAC-PapersOnLine* **2020**, *53*, 431–435. [CrossRef]
19. Moncrieff, J.; Valentini, R.; Greco, S.; Guenther, S.; Ciccioli, P. Trace gas exchange over terrestrial ecosystems: Methods and perspectives in micrometeorology. *J. Exp. Bot.* **1997**, *48*, 1133–1142. [CrossRef]
20. Hersbach, H.; Bell, B.; Berrisford, P.; Hirahara, S.; Horányi, A.; Muñoz-Sabater, J.; Nicolas, J.; Peubey, C.; Radu, R.; Schepers, D.; et al. The ERA5 global reanalysis. *Q. J. R. Meteorol. Soc.* **2020**, *146*, 1999–2049. [CrossRef]
21. Lv, S.; Huang, J.; Zhou, K. Research on effective operation mode of Meteorological Observation Quality Management System. *J. Meteorol. Res. Appl.* **2023**, *44*, 103–108.
22. Change, I.P.O.C. Climate change 2007: The physical science basis. *Agenda* **2007**, *6*, 333.
23. Xu, A.; Li, J. An overview of the integrated meteorological observations in complex terrain region at Dali National Climate Observatory, China. *Atmosphere* **2020**, *11*, 279. [CrossRef]
24. Conover, J.H. *The Blue Hill Meteorological Observatory: The First 100 Years, 1885–1985*; Springer: Berlin/Heidelberg, Germany, 2018.
25. Magee, N.B.; Melaas, E.; Finocchio, P.M.; Jardel, M.; Noonan, A.; Iacono, M.J. Blue Hill Observatory sunshine: Assessment of climate signals in the longest continuous meteorological record in North America. *Bull. Am. Meteorol. Soc.* **2014**, *95*, 1741–1751. [CrossRef]
26. Neisser, J.; Adam, W.; Beyrich, F.; Leiterer, U.; Steinhagen, H. Atmospheric boundary layer monitoring at the Meteorological Observatory Lindenberg as a part of the "Lindenberg Column": Facilities and selected results. *Meteorol. Z.* **2002**, *11*, 241 –253. 10.1127/0941-2948/2002/0011-0241. [CrossRef]
27. Adam, W.; Dier, H.; Leiterer, U. 100 years aerology in Lindenberg and first long-time observations in the free atmosphere. *Meteorol. Z.* **2005**, *14*, 597–607. [CrossRef]

28. Berger, F.H.; Hantel, M. Meteorological Observatory Lindenberg 1905–2005. *Meteorol. Z.* **2005**, *14*, 596–596. 10.1127/0941-2948/2005/0072. [CrossRef]
29. Weller, M.; Gericke, K. Long-term observations of aerosol optical depths at the Meteorological Observatory Lindenberg. *Meteorol. Z.* **2005**, *14*, 651–662. 10.1127/0941-2948/2005/0070. [CrossRef]
30. Overton, A.K. Jungfraujoch high altitude research station. *Weather* **2008**, *63*, 76–79. [CrossRef]
31. Bonasoni, P.; Laj, P.; Angelini, F.; Arduini, J.; Bonafe, U.; Calzolari, F.; Cristofanelli, P.; Decesari, S.; Facchini, M.; Fuzzi, S.; et al. The ABC-Pyramid Atmospheric Research Observatory in Himalaya for aerosol, ozone and halocarbon measurements. *Sci. Total Environ.* **2008**, *391*, 252–261. [CrossRef]
32. Bonasoni, P.; Laj, P.; Marinoni, A.; Sprenger, M.; Angelini, F.; Arduini, J.; Bonafè, U.; Calzolari, F.; Colombo, T.; Decesari, S.; et al. Atmospheric Brown Clouds in the Himalayas: First two years of continuous observations at the Nepal Climate Observatory-Pyramid (5079 m). *Atmos. Chem. Phys.* **2010**, *10*, 7515–7531. [CrossRef]
33. Zhili, W.; Xiaofeng, Z.; Guoliang, W. Design and Realization of Micro-meteorological Disaster Morni ng and Pre-warning System in Power Grid. *Power D Energy* **2014**, *35*, 712–716.
34. Yang, J.; Zhang, P.; Lu, N.; Yang, Z.; Shi, J.; Dong, C. Improvements on global meteorological observations from the current Fengyun 3 satellites and beyond. *Int. J. Digit. Earth* **2012**, *5*, 251–265. [CrossRef]
35. Lahoz, W.A.; Schneider, P. Data assimilation: Making sense of Earth Observation. *Front. Environ. Sci.* **2014**, *2*, 16. [CrossRef]
36. Lahoz, W. Research satellites. In *Data Assimilation: Making Sense of Observations*; Springer: Berlin/Heidelberg, Germany, 2010; pp. 301–321.
37. Ménard, R. Bias estimation. In *Data Assimilation: Making Sense of Observations*; Springer: Berlin/Heidelberg, Germany, 2010; pp. 113–135.
38. Lahoz, W.; Khattatov, B.; Ménard, R. Data assimilation and information. In *Data Assimilation: Making Sense of Observations*; Springer: Berlin/Heidelberg, Germany, 2010; pp. 3–12.
39. Errera, Q.; Daerden, F.; Chabrillat, S.; Lambert, J.; Lahoz, W.; Viscardy, S.; Bonjean, S.; Fonteyn, D. 4D-Var assimilation of MIPAS chemical observations: Ozone and nitrogen dioxide analyses. *Atmos. Chem. Phys.* **2008**, *8*, 6169–6187. [CrossRef]
40. Salby, M.L. *Fundamentals of Atmospheric Physics*; Elsevier: Amsterdam, The Netherlands, 1996.
41. Lahoz, W.A.; De Lannoy, G.J. Closing the gaps in our knowledge of the hydrological cycle over land: Conceptual problems. *Surv. Geophys.* **2014**, *35*, 623–660. [CrossRef]
42. Kalnay, E. *Atmospheric Modeling, Data Assimilation and Predictability*; Cambridge University Press: Cambridge, UK, 2003.
43. Bouttier, F.; Courtier, P. *Data Assimilation Concepts and Methods*; European Centre for Medium-Range Weather Forecasts (ECMWF): Reading, UK, 2002.
44. Lahoz, W.; Errera, Q.; Swinbank, R.; Fonteyn, D. Data assimilation of stratospheric constituents: A review. *Atmos. Chem. Phys.* **2007**, *7*, 5745–5773. [CrossRef]
45. Courtier, P.; Thépaut, J.N.; Hollingsworth, A. A strategy for operational implementation of 4D-Var, using an incremental approach. *Q. J. R. Meteorol. Soc.* **1994**, *120*, 1367–1387.
46. Simmons, A.J.; Hollingsworth, A. Some aspects of the improvement in skill of numerical weather prediction. *Q. J. R. Meteorol. Soc. A J. Atmos. Sci. Appl. Meteorol. Phys. Oceanogr.* **2002**, *128*, 647–677. [CrossRef]
47. Kalman, R.E. A new approach to linear filtering and prediction problems. *J. Basic Eng.* **1960**, *82*, 35–45. [CrossRef]
48. Cohn, S.E.; Da Silva, A.; Guo, J.; Sienkiewicz, M.; Lamich, D. Assessing the effects of data selection with the DAO physical-space statistical analysis system. *Mon. Weather Rev.* **1998**, *126*, 2913–2926. [CrossRef]
49. Eskes, H.; Velthoven, P.V.; Valks, P.; Kelder, H. Assimilation of GOME total-ozone satellite observations in a three-dimensional tracer-transport model. *Q. J. R. Meteorol. Soc. A J. Atmos. Sci. Appl. Meteorol. Phys. Oceanogr.* **2003**, *129*, 1663–1681. [CrossRef]
50. Lahoz, W.; Errera, Q. Constituent assimilation. In *Data Assimilation: Making Sense of Observations*; Springer: Berlin/Heidelberg, Germany, 2010; pp. 449–490.
51. Auvinen, H.; Bardsley, J.M.; Haario, H.; Kauranne, T. The variational Kalman filter and an efficient implementation using limited memory BFGS. *Int. J. Numer. Methods Fluids* **2010**, *64*, 314–335. [CrossRef]
52. Solonen, A.; Haario, H.; Hakkarainen, J.; Auvinen, H.; Amour, I.; Kauranne, T. Variational ensemble Kalman filtering using limited memory BFGS. *Electron. Trans. Numer. Anal.* **2012**, *39*, 271–285.
53. Heikkilä, U.; Sandvik, A.; Sorteberg, A. Dynamical downscaling of ERA-40 in complex terrain using the WRF regional climate model. *Clim. Dyn.* **2011**, *37*, 1551–1564. [CrossRef]
54. Soares, P.M.; Cardoso, R.M.; Miranda, P.M.; de Medeiros, J.; Belo-Pereira, M.; Espirito-Santo, F. WRF high resolution dynamical downscaling of ERA-Interim for Portugal. *Clim. Dyn.* **2012**, *39*, 2497–2522. [CrossRef]
55. Evans, J.; McCabe, M. Regional climate simulation over Australia's Murray-Darling basin: A multitemporal assessment. *J. Geophys. Res. Atmos.* **2010**, *115* . [CrossRef]
56. Zhang, Y.; Dulière, V.; Mote, P.W.; Salathé, E.P. Evaluation of WRF and HadRM mesoscale climate simulations over the US Pacific Northwest. *J. Clim.* **2009**, *22*, 5511–5526. [CrossRef]
57. Koo, M.S.; Hong, S.Y. Diurnal variations of simulated precipitation over East Asia in two regional climate models. *J. Geophys. Res. Atmos.* **2010**, *115*. [CrossRef]
58. Lo, J.C.F.; Yang, Z.L.; Pielke Sr, R.A. Assessment of three dynamical climate downscaling methods using the Weather Research and Forecasting (WRF) model. *J. Geophys. Res. Atmos.* **2008**, *113*. [CrossRef]

59. Zhang, J.; Wu, L.; Dong, W. Land-atmosphere coupling and summer climate variability over East Asia. *J. Geophys. Res. Atmos.* **2011**, *116*. [CrossRef]
60. Liu, Y.; Warner, T.; Liu, Y.; Vincent, C.; Wu, W.; Mahoney, B.; Swerdlin, S.; Parks, K.; Boehnert, J. Simultaneous nested modeling from the synoptic scale to the LES scale for wind energy applications. *J. Wind Eng. Ind. Aerodyn.* **2011**, *99*, 308–319. [CrossRef]
61. Cardoso, R.; Soares, P.; Miranda, P.; Belo-Pereira, M. WRF high resolution simulation of Iberian mean and extreme precipitation climate. *Int. J. Climatol.* **2013**, *33*, 2591–2608. [CrossRef]
62. Kusaka, H.; Hara, M.; Takane, Y. Urban climate projection by the WRF model at 3-km horizontal grid increment: Dynamical downscaling and predicting heat stress in the 2070's August for Tokyo, Osaka, and Nagoya metropolises. *Meteorol. J.* **2012**, *90*, 47–63. [CrossRef]
63. Yang, B.; Zhang, Y.; Qian, Y. Simulation of urban climate with high-resolution WRF model: A case study in Nanjing, China. *Asia-Pac. J. Atmos. Sci.* **2012**, *48*, 227–241. [CrossRef]
64. Marta-Almeida, M.; Teixeira, J.C.; Carvalho, M.J.; Melo-Gonçalves, P.; Rocha, A.M. High resolution WRF climatic simulations for the Iberian Peninsula: Model validation. *Phys. Chem. Earth, Parts A/B/C* **2016**, *94*, 94–105. [CrossRef]
65. Carvalho, D.; Rocha, A.; Gómez-Gesteira, M.; Santos, C. A sensitivity study of the WRF model in wind simulation for an area of high wind energy. *Environ. Model. Softw.* **2012**, *33*, 23–34. [CrossRef]
66. Carvalho, D.; Rocha, A.; Gómez-Gesteira, M.; Santos, C.S. WRF wind simulation and wind energy production estimates forced by different reanalyses: Comparison with observed data for Portugal. *Appl. Energy* **2014**, *117*, 116–126. [CrossRef]
67. Deppe, A.J.; Gallus, W.A.; Takle, E.S. A WRF ensemble for improved wind speed forecasts at turbine height. *Weather Forecast.* **2013**, *28*, 212–228. [CrossRef]
68. Mattar, C.; Borvarán, D. Offshore wind power simulation by using WRF in the central coast of Chile. *Renew. Energy* **2016**, *94*, 22–31. [CrossRef]
69. Tuy, S.; Lee, H.S.; Chreng, K. Integrated assessment of offshore wind power potential using Weather Research and Forecast (WRF) downscaling with Sentinel-1 satellite imagery, optimal sites, annual energy production and equivalent CO_2 reduction. *Renew. Sustain. Energy Rev.* **2022**, *163*, 112501. [CrossRef]
70. Zhao, J.; Guo, Z.H.; Su, Z.Y.; Zhao, Z.Y.; Xiao, X.; Liu, F. An improved multi-step forecasting model based on WRF ensembles and creative fuzzy systems for wind speed. *Appl. Energy* **2016**, *162*, 808–826. [CrossRef]
71. Zhao, J.; Guo, Y.; Xiao, X.; Wang, J.; Chi, D.; Guo, Z. Multi-step wind speed and power forecasts based on a WRF simulation and an optimized association method. *Appl. Energy* **2017**, *197*, 183–202. [CrossRef]
72. Giannakopoulou, E.M.; Nhili, R.; et al. WRF model methodology for offshore wind energy applications. *Adv. Meteorol.* **2014**, *2014*, 319819. [CrossRef]
73. Salvação, N.; Soares, C.G. Wind resource assessment offshore the Atlantic Iberian coast with the WRF model. *Energy* **2018**, *145*, 276–287. [CrossRef]
74. Prósper, M.A.; Otero-Casal, C.; Fernández, F.C.; Miguez-Macho, G. Wind power forecasting for a real onshore wind farm on complex terrain using WRF high resolution simulations. *Renew. Energy* **2019**, *135*, 674–686. [CrossRef]
75. Shen, X.; Meng, K.; Guo, Z.; Zhang, L. Convolutional Long-Short Term Memory Network for convective weather prediction based on 3D Doppler radar data. In Proceedings of the 2020 International Conference on Cyberspace Innovation of Advanced Technologies, Guangzhou, China, 4–6 December 2020; pp. 459–464. [CrossRef]
76. Wang, B.; Lu, J.; Yan, Z.; Luo, H.; Li, T.; Zheng, Y.; Zhang, G. Deep Uncertainty Quantification: A Machine Learning Approach for Weather Forecasting. In Proceedings of the 25th ACM SIGKDD International Conference on Knowledge Discovery & Data Mining, Anchorage, AK, USA, 4–8 August 2019; pp. 2087–2095. [CrossRef]
77. Grover, A.; Kapoor, A.; Horvitz, E. A Deep Hybrid Model for Weather Forecasting. In Proceedings of the 21th ACM SIGKDD International Conference on Knowledge Discovery and Data Mining, Sydney, NSW, Australia, 10–13 August 2015; pp. 379–386. [CrossRef]
78. Volkovs, K.; Urtans, E.; Caune, V. Primed UNet-LSTM for Weather Forecasting. In Proceedings of the 2023 7th International Conference on Advances in Artificial Intelligence, Istanbul, Turkiye, 13–15 October 2023; pp. 13–17. [CrossRef]
79. Zhang, L.; Zhang, L.; Du, B. Deep learning for remote sensing data: A technical tutorial on the state of the art. *IEEE Geosci. Remote Sens. Mag.* **2016**, *4*, 22–40. [CrossRef]
80. Makantasis, K.; Karantzalos, K.; Doulamis, A.; Doulamis, N. Deep supervised learning for hyperspectral data classification through convolutional neural networks. In Proceedings of the 2015 IEEE International Geoscience and Remote Sensing Symposium (IGARSS), Milan, Italy, 26–31 July 2015; pp. 4959–4962.
81. Liu, Y.; Racah, E.; Correa, J.; Khosrowshahi, A.; Lavers, D.; Kunkel, K.; Wehner, M.; Collins, W.; et al. Application of deep convolutional neural networks for detecting extreme weather in climate datasets. *arXiv* **2016**, arXiv:1605.01156.
82. Hernández, E.; Sanchez-Anguix, V.; Julian, V.; Palanca, J.; Duque, N. Rainfall prediction: A deep learning approach. In Proceedings of the 11th International Conference, HAIS 2016, Seville, Spain, 18–20 April 2016; pp. 151–162.
83. Shi, X.; Gao, Z.; Lausen, L.; Wang, H.; Yeung, D.Y.; Wong, W.K.; Woo, W.C. Deep learning for precipitation nowcasting: A benchmark and a new model. *Adv. Neural Inf. Process. Syst.* **2017**, *30*, 5622–5632.

84. You, J.; Li, X.; Low, M.; Lobell, D.; Ermon, S. Deep gaussian process for crop yield prediction based on remote sensing data. In Proceedings of the The Thirty-First AAAI Conference on Artificial Intelligence; The Twenty-Ninth Innovative Applications of Artificial Intelligence Conferencel; The Seventh Symposium on Educational Advances in Artificial Intelligence, San Francisco, CA, USA, 4–9 February 2017; Volume 31.
85. Papernot, N.; McDaniel, P.; Swami, A.; Harang, R. Crafting adversarial input sequences for recurrent neural networks. In Proceedings of the MILCOM 2016—2016 IEEE Military Communications Conference, Baltimore, MD, USA, 1–3 November 2016; pp. 49–54.
86. Lee, W.; Kim, S.; Lee, Y.T.; Lee, H.W.; Choi, M. Deep neural networks for wild fire detection with unmanned aerial vehicle. In Proceedings of the 2017 IEEE International Conference on Consumer Electronics (ICCE), Las Vegas, NV, USA, 8–10 January 2017; pp. 252–253.
87. Sharma, J.; Granmo, O.C.; Goodwin, M.; Fidje, J.T. Deep convolutional neural networks for fire detection in images. In Proceedings of the Engineering Applications of Neural Networks: 18th International Conference, EANN 2017, Athens, Greece, 25–27 August 2017; pp. 183–193.
88. Zhang, Q.; Xu, J.; Xu, L.; Guo, H. Deep convolutional neural networks for forest fire detection. In Proceedings of the 2016 International Forum on Management, Education and Information Technology Application, Guangzhou, China, 30–31 January 2016; pp. 568–575.
89. Chen, I.C.; Hu, S.C. Realizing Specific Weather Forecast through Machine Learning Enabled Prediction Model. In Proceedings of the 2019 3rd High Performance Computing and Cluster Technologies Conference, Guangzhou, China, 22–24 June 2019; pp. 71–74. [CrossRef]
90. Egbueze, P.U.; Wang, Z. Weather Recognition Based on Still Images Using Deep Learning Neural Network with Resnet-15. In Proceedings of the 2022 6th International Conference on Deep Learning Technologies, Xi'an, China, 26–28 July 2022; pp. 8–13. [CrossRef]
91. Li, X.; Wang, Z.; Lu, X. A Multi-Task Framework for Weather Recognition. In Proceedings of the 25th ACM International Conference on Multimedia, Mountain View, CA, USA, 23–27 October 2017; pp. 1318–1326. [CrossRef]
92. Awami, S.H.; Shakmak, Y.H.; Mohammed, R.A. A Novel Approach for Forecasting Average Temperature Using Artificial Neural Networks (Applied to Benghazi City's Weather). In Proceedings of the 6th International Conference on Engineering & MIS 2020, Almaty, Kazakhstan, 14–16 September 2020. [CrossRef]
93. Shi, E.; LI, Q.; GU, D.; Zhao, Z. Weather radar echo extrapolation method based on convolutional neural networks. *J. Comput. Appl.* **2018**, *38*, 661.
94. Agrawal, S.; Barrington, L.; Bromberg, C.; Burge, J.; Gazen, C.; Hickey, J. Machine learning for precipitation nowcasting from radar images. *arXiv* **2019**, arXiv:1912.12132.
95. Kaparakis, C.; Mehrkanoon, S. WF-UNet: Weather Data Fusion using 3D-UNet for Precipitation Nowcasting. *Procedia Comput. Sci.* **2023**, *222*, 223–232. [CrossRef]
96. Shen, C. A transdisciplinary review of deep learning research and its relevance for water resources scientists. *Water Resour. Res.* **2018**, *54*, 8558–8593. [CrossRef]
97. Klein, B.; Wolf, L.; Afek, Y. A dynamic convolutional layer for short range weather prediction. In Proceedings of the IEEE Conference on Computer Vision and Pattern Recognition, Boston, MA, USA, 7–12 June 2015; pp. 4840–4848.
98. Vandal, T.; Kodra, E.; Ganguly, S.; Michaelis, A.; Nemani, R.; Ganguly, A.R. Deepsd: Generating high resolution climate change projections through single image super-resolution. In Proceedings of the 23rd ACM SIGKDD International Conference on Knowledge Discovery and Data Mining, Halifax, NS, Canada, 13–17 August 2017; pp. 1663–1672.
99. Gentine, P.; Pritchard, M.; Rasp, S.; Reinaudi, G.; Yacalis, G. Could machine learning break the convection parameterization deadlock? *Geophys. Res. Lett.* **2018**, *45*, 5742–5751. [CrossRef]
100. Racah, E.; Beckham, C.; Maharaj, T.; Ebrahimi Kahou, S.; Prabhat, M.; Pal, C. Extremeweather: A large-scale climate dataset for semi-supervised detection, localization, and understanding of extreme weather events. *Adv. Neural Inf. Process. Syst.* **2017**, *30*, 3405–3416.
101. Albert, A.; Strano, E.; Kaur, J.; González, M. Modeling urbanization patterns with generative adversarial networks. In Proceedings of the IGARSS 2018—2018 IEEE International Geoscience and Remote Sensing Symposium, Valencia, Spain, 22–27 July 2018; pp. 2095–2098.
102. Pathak, J.; Subramanian, S.; Harrington, P.; Raja, S.; Chattopadhyay, A.; Mardani, M.; Kurth, T.; Hall, D.; Li, Z.; Azizzadenesheli, K.; et al. Fourcastnet: A global data-driven high-resolution weather model using adaptive fourier neural operators. *arXiv* **2022**, arXiv:2202.11214.
103. Kurth, T.; Subramanian, S.; Harrington, P.; Pathak, J.; Mardani, M.; Hall, D.; Miele, A.; Kashinath, K.; Anandkumar, A. Fourcastnet: Accelerating global high-resolution weather forecasting using adaptive fourier neural operators. In Proceedings of the PASC '23: Platform for Advanced Scientific Computing Conference, Davos, Switzerland, 26–28 June 2023; pp. 1–11.
104. Bi, K.; Xie, L.; Zhang, H.; Chen, X.; Gu, X.; Tian, Q. Accurate medium-range global weather forecasting with 3D neural networks. *Nature* **2023**, *619*, 533–538. [CrossRef]
105. Lam, R.; Sanchez-Gonzalez, A.; Willson, M.; Wirnsberger, P.; Fortunato, M.; Alet, F.; Ravuri, S.; Ewalds, T.; Eaton-Rosen, Z.; Hu, W.; et al. Learning skillful medium-range global weather forecasting. *Science* **2023**, *382*, 1416–1421. [CrossRef]

106. Chen, K.; Han, T.; Gong, J.; Bai, L.; Ling, F.; Luo, J.J.; Chen, X.; Ma, L.; Zhang, T.; Su, R.; et al. FengWu: Pushing the Skillful Global Medium-range Weather Forecast beyond 10 Days Lead. *arXiv* **2023**, arXiv:2304.02948.
107. Chen, L.; Zhong, X.; Zhang, F.; Cheng, Y.; Xu, Y.; Qi, Y.; Li, H. FuXi: A cascade machine learning forecasting system for 15-day global weather forecast. *arXiv* **2023**, arXiv:2306.12873.
108. Paraschiv, L.S.; Paraschiv, S. Contribution of renewable energy (hydro, wind, solar and biomass) to decarbonization and transformation of the electricity generation sector for sustainable development. *Energy Rep.* **2023**, *9*, 535–544. [CrossRef]
109. Moonen, P.; Defraeye, T.; Dorer, V.; Blocken, B.; Carmeliet, J. Urban Physics: Effect of the micro-climate on comfort, health and energy demand. *Front. Archit. Res.* **2012**, *1*, 197–228. [CrossRef]
110. Ma, L.; Jiang, J. Challenges and countermeasures for the development of China's wind power industry under the low-carbon energy transition. *China Energy* **2021**, *43*, 8.
111. Zhu, Q.; Li, J.; Qiao, J.; Shi, M.; Wang, C. Application and prospect of artificial intelligence technology in renewable energy forecasting. *Proc. CSEE* **2023**, *43*, 3027–3047. [CrossRef]
112. Zeng, M.; Xu, Y. Essentials of an integrated energy system: Integration of source, grid, load and storage + multi-energy complementation. *China Energy News*, 12 April 2021.
113. Majedul Islam, M. Threats to Humanity from Climate Change. In *Climate Change: The Social and Scientific Construct*; Springer: Cham, Switzerland, 2022; pp. 21–36.
114. McBean, G. Climate Change and Extreme Weather: A Basis for Action. *Nat. Hazards* **2004**, *31*, 177–190. [CrossRef]
115. James, D.G.; David, J.F.; Upmanu, L.; Vijay, M. How unprecedented was the February 2021 Texas cold snap? *Environ. Res. Lett.* **2021**, *16*, 064056.
116. Smead, R.G. Ercot—the eyes of texas (and the world) are upon you: What can be done to avoid a february 2021 repeat. *Clim. Energy* **2021**, *37*, 14–18. [CrossRef]
117. Everhart, K.; Gergely, M. *Severe Power Cuts in Texas Highlight Energy Security Risks Related to Extreme Weather Events*; Technical Report; International Energy Agency: Paris, France, 2021.
118. Wei, H.; Wang, W.S.; Kao, X.X. A novel approach to ultra-short-term wind power prediction based on feature engineering and informer. *Energy Rep.* **2023**, *9*, 1236–1250. [CrossRef]
119. Yu, R.; Gao, J.; Yu, M.; Lu, W.; Xu, T.; Zhao, M.; Zhang, J.; Zhang, R.; Zhang, Z. LSTM-EFG for wind power forecasting based on sequential correlation features. *Future Gener. Comput. Syst.* **2019**, *93*, 33–42. [CrossRef]
120. Yuan, X.; Chen, C.; Jiang, M.; Yuan, Y. Prediction interval of wind power using parameter optimized Beta distribution based LSTM model. *Appl. Soft Comput.* **2019**, *82*, 105550. [CrossRef]

energies

MDPI

Article

Research on Edge-Computing-Based High Concurrency and Availability "Cloud, Edge, and End Collaboration" Substation Operation Support System and Applications

Yun Long [1], Yude Bao [1] and Linjun Zeng [2,*]

[1] Guangzhou Power Supply Bureau, Guangdong Power Grid Co., Ltd., Guangzhou 510260, China; wyl@guangzhou.csg.cn (Y.L.); baoyd@guangzhou.csg.cn (Y.B.)
[2] School of Energy and Power Engineering, Changsha University of Science and Technology, Changsha 410018, China
* Correspondence: zenglj23@csust.edu.cn

Abstract: With the continuous promotion of digital transformation in the field of power transformation, the diversification of application scenarios, and the scale of pilot construction, the real-time, concurrency, and security requirements for data fusion and application support of the power monitoring system, management information system, and other business platforms are getting higher and higher, and this paper puts forward a high concurrency and availability "cloud-side-end collaboration" based on edge computing. This paper proposes a high concurrency and availability "cloud, edge and end collaboration" architecture based on edge computing for substation operation support systems. First, this paper summarizes the development status of domestic substation operation support systems and analyzes the advantages and disadvantages of various technical architectures. Then, a "cloud-side-end cooperative" substation operation support system architecture with "high real-time, high concurrency, high security and high stability" is proposed, which focuses on remote inspection, remote operation, and remote safety control of substation businesses from the perspective of engineering applications. It realizes transparent monitoring of equipment operation, unified management of operation data, and integration of production command and decision-making; solves the problems of dispersed coexistence of multiple systems for dispatching, monitoring, analysis, management, and other businesses, switching between multiple systems, and insufficient real-time and stability of the system; and controls the risks of the grid, reduces the potential safety hazards, and solves the contradiction between the continuous growth of the grid equipment and the shortage of production personnel. The results of engineering application examples show that the proposed architecture compared with the existing system architecture has greater advantages and can meet the requirements of large-scale access to the substation, with feasible popularization and application.

Keywords: power monitoring system; cloud–edge–device collaboration; large-scale digital substation; edge computing; intelligent patrol; intelligent operation; decision fusion

1. Introduction

With the deep integration of substation field end control technology (OT) and information technology (IT), the digital transformation in the field of substations continues to advance, and the level of sensible, knowable, and controllable equipment continues to be enhanced; different scales of pilot applications of substation monitoring systems have appeared in different parts of power grids, and a variety of technological routes and system architectures have also been derived. But due to the concurrency, security, and stability of the system architecture, the digital transformation in the field of substations is still far away from scaled-up promotion [1].

China's preliminary research on power monitoring systems for the substation domain and intelligent substation transformation can be roughly divided into three stages. Stage 1 involves intelligent substation technology utilizing RTU technology to collect station signals through secondary cables, thus realizing the "four-remote" function in a single substation. However, it is generally used in the substation terminal automation system (SAS) which does not require the site, and the old RTU system is often unable to meet the requirements of long-term stable operation of the substation, reconstruction, and expansion of the access to new signals and substation automation monitoring system applications such as the need to carry out renovation [2,3]. Stage 2 comprises the application of centralized distributed control technology, which is often used in the control of active distribution networks due to its high accuracy, effectively improves the performance of the distribution monitoring system (DMS) in the region. However, the centralized control of the corresponding time is slow, not conducive to real-time scheduling, and has high dependence on the centralized control, which is vulnerable to a single point of failure and has poor reliability [4]. In stage 3, with the popularization and application of the IEC 61850 specification, the power monitoring system is enabled to realize hierarchical distributed control [5]. Hierarchical distributed control can overcome the shortcomings of the traditional centralized control mentioned above, and distributed control can realize the cooperative control of each distributed substation with a good control effect, which is the hot spot of research in the field of power transformation at present [6,7].

Although substation monitoring systems (SMSs) have gone through different stages of transformation, current substation monitoring and control management usually requires human beings to carry out equipment inspections, monitoring, and maintenance [8]. The enhancement of the automation level of the relevant equipment in the existing power system requires different operation and management methods for different equipment, which puts forward higher requirements for the basic level of the relevant management personnel and the effectiveness of management [9,10]. However, this approach has obvious defects. Firstly, labor costs are high, as substation equipment needs to operate for long periods of time, requiring extensive manpower for regular inspection and maintenance. Additionally, the challenging and risky environments, such as working at great heights, further increase the difficulty and risks for maintenance personnel. Secondly, operation and maintenance is difficult [11]. To solve these problems, through the construction of advanced, reliable, integrated, low-carbon, and environmentally friendly intelligent equipment in substations, with the goal of improving the overall operation of the system on a network-wide basis, the rendering intelligent of monitoring equipment and the comprehensive analysis of fault information are realized [12].

Although the intelligent transformation of substations has been implemented, there are still some challenges in the operation and maintenance of the power grid substations in China:

(1) The system is complex. There are a large number of devices in the substation operation and maintenance system, and different devices have significant differences in functionality and performance, increasing the difficulty of maintenance management and the occurrence rate of faults.
(2) Maintenance is difficult. Substation equipment is distributed widely, making centralized management difficult.

Therefore, it is urgent to optimize the traditional substation monitoring mode and promote the application of artificial intelligence terminals in safety monitoring at the operation and maintenance site, improve monitoring efficiency, and solve the problem of requiring a large amount of manpower and resources for safety monitoring at the operation and maintenance site [13].

Based on this, power monitoring and management systems based on cloud computing architecture have emerged [14]. Such systems can upload real-time data from power system operation and maintenance sites to the cloud platform for centralized processing, and then provide timely feedback on the processing results to management personnel through the platform. In the field of power system substations, the combination of the Grid Information Model (GIM) and Geographic Information System (GIS) is widely used for data computation, with cloud computing being the predominant approach. Cloud computing refers to the formation of a highly capable system through computer networks, which can store and aggregate relevant resources and be configured on-demand to provide personalized services to users. With the continuous development of cloud computing technology, it is gradually being applied in power monitoring systems [15]. Reference [16] proposed a power video surveillance platform based on cloud computing technology, providing a detailed analysis of its structural design and the application of its functionalities in the system. It verified the accuracy and efficiency of the platform compared to traditional surveillance systems. Ref. [17] builds upon digital image-processing methods for power video and addresses the challenges of a large number of devices and wide geographical coverage in power video surveillance systems. It proposes a new technology architecture for intelligent fault and status detection in power video based on cloud computing, effectively utilizing existing hardware resources to achieve network-wide video device fault diagnosis and status querying. This architecture provides reliable support for power grid operation and maintenance management. Ref. [18] presents a new web-enabled near-real-time power quality (PQ) monitoring system, I-Grid/spl trade, which was developed in the U.S. Its use of an internal modem to send event data over the internet to the I-Grid database server provides more reliable data for monitoring and management of its power system. Ref. [19] presents a system that can be used to monitor power consumption and, if connected over the internet to dedicated cloud services, can display in real-time the monitored quantities in the system. The monitoring system was tested with a programmable logic controllers (PLC)-based system and with one based on field-programmable gate arrays (FPGAs). Monitoring systems can disconnect parts of the monitored system's circuits in order to identify power-hungry components or for studying the implementation of different power-saving modes. Ref. [20] proposes a transformer power monitoring system in India. It measures the current, voltage, and temperature of the transformer and uploads the recorded data to the cloud. This data helps to detect any anomalies in the transformer and alerts the authorities to take action. Ref. [21] proposes an IoT power monitor to fulfill the needs of three-phase power monitoring, which securely transmits real-time data to a cloud server, making the data available anytime, anywhere. Table 1 presents a comparison of related research both domestically and internationally. Based on the comprehensive results in Table 1, it can be seen that cloud computing has not yet formed a widely applicable solution in the field of power substations. Moreover, there are numerous challenges in terms of real-time performance, concurrency, stability, and security when it comes to the unified and centralized migration of power monitoring systems to the cloud.

Table 1. Detailed comparison of the proposed method with published papers.

Ref	Field	Method	Scales	Topicality	Accuracy	Concurrency	Safety	Stability
[1]	SAS	RTU	small	✘	✓	✘	✘	✘
[2]	SAS	RTU + RMON	medium	✘	✓	✓	✘	✓
[4]	DMS	Centralized distributed	medium	✘	✓	✘	✘	✘
[7]	DMS	Hierarchical distributed	large	✘	✓	✓	✘	✘
[14]	DMS	Cloud computing	large	✘	✓	✓	✘	✓
[15]	SAS	Cloud computing	large	✘	✓	✓	✘	✓
[16]	SMS	Cloud computing	large	✘	✓	✓	✘	✓
[17]	DMS	Cloud–edge collaboration	large	✘	✓	✓	✓	✓
This paper	SMS	"Cloud-edge-end" collaboration	large	✓	✓	✓	✓	✓

In response to the aforementioned background, this paper enumerates the existing architectures of power monitoring systems and compares them in terms of real-time performance, concurrency, stability, and security. It proposes a novel architecture for a high-concurrency and highly available "cloud, edge and end collaboration" substation operation support system based on edge computing. The test results demonstrate that compared to existing power monitoring system architectures, the proposed architecture has significant advantages in terms of large-scale substation integration and reasonable distribution of computing resources, meeting the requirements for scalable deployment [22].

2. Current Mainstream Power Monitoring System Architecture

According to the spirit of the document "Notice on Promoting the Optimization of Substation Operation and Maintenance Models and Pilot Construction of Centralized Control Stations" (Equipment Substation [2020] No. 57) issued by the State Grid Corporation of China, the pilot construction of centralized control stations is a key annual task of the State Grid Corporation. It aims to integrate and improve existing resources, optimize operation and maintenance management models, ensure high-quality implementation of pilot projects, and formulate reasonable construction plans for the pilot stations to steadily advance the pilot work. The 1000 kV UHV Dongwu Station and Taizhou Station of State Grid Jiangsu Electric Power Co., Ltd. (Nanjing, China) have been accepted by the State Grid Corporation of China (Beijing, China), making them the first batch of pilot projects for digital substations (converter stations). The achievements of these projects will be replicated and promoted within the State Grid system. Intelligent inspection is one of the key aspects of digital substation construction. A total of 198 visible light and infrared cameras have been deployed within the UHV Taizhou Station, covering all primary and secondary equipment. The construction of digital converter stations not only significantly reduces the workload of frontline personnel but also improves the accuracy of equipment status perception. In addition to the "smart eyes" distributed throughout the entire station, digital substations also possess the "brain" to diagnose equipment faults.

Not only the State Grid Corporation, but also the Southern Power Grid Guangxi Guigang Power Supply Bureau has successfully completed the transition handover period of the two county-level power supply enterprises in the area under its jurisdiction for the intensification of transmission and substation business of 35 kV and above, and it has completed the target one month ahead of time to enter the pilot application stage of the intensification work. Transmission and substation business intensification is an important means for Guangxi Power Grid Company to adapt to the needs of reform and development and promote high-quality development. Through the intensification work, the mature management practices of the prefecture and city power supply bureaus are extended to the whole network, which greatly improves the equipment level, resource allocation, and operation and maintenance quality of county-level power grids. By summarizing the construction of the domestic substation intensive pilot, it can be found that the mainstream power monitoring of domestic substations can be mainly divided into three modes: direct connection to the main station, hierarchical grading, and side-to-side collaboration.

2.1. Mode One: Master Direct Connection Mode

The master station direct connection mode refers to the mode in which edge devices in the substation are directly connected to the central master station. In this mode, edge devices communicate and exchange data with the central master station through communication networks such as ethernet and wireless communication. Therefore, the communication protocols and methods between edge devices and the central master station need to be compatible and consistent to ensure the correct transmission and parsing of data. The master station direct connection mode is a simple and efficient substation operation system architecture mode, suitable for smaller-scale substations and simpler application scenarios. It can provide real-time data transmission and response, facilitating system management and maintenance, but it has relatively weak scalability [23]. As shown in Figure 1, the station-side substation intelligent gateway (without AI functionality) only collects and forwards data, while all task scheduling and intelligent analysis functions are performed by the regional platform.

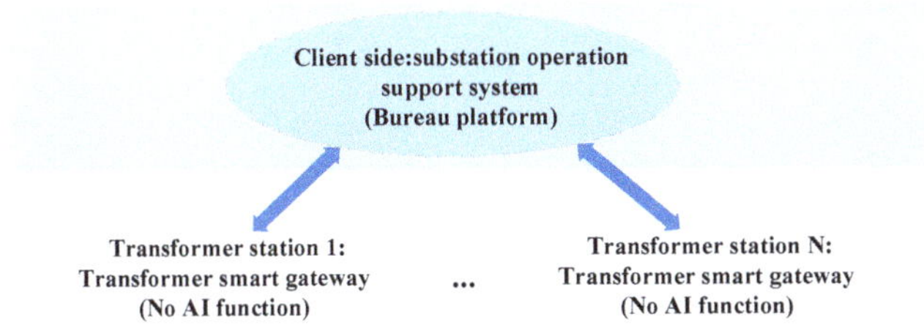

Figure 1. Master direct connection mode diagram.

This mode does not rely on the capabilities of intelligent gateways and achieves functionality and resource consolidation through the regional platform. However, the underlying framework of the system is complex and places high demands on hardware resources such as communication networks, master station computing power, and storage.

2.2. Model Two: Tiered and Graded Model

The layered and hierarchical mode is a pattern that divides and organizes system functions into different levels. By categorizing system functions into a hardware layer, data acquisition and transmission layer, data processing and storage layer, control and decision-making layer, and management and application layer, the substation operation support system can achieve functional isolation and decoupling between different levels, making the system more stable and reliable. At the same time, this mode also facilitates system scalability and upgrades, allowing flexible adjustments and optimizations of different levels according to actual needs [24]. As shown in Figure 2, the station-side substation intelligent gateway (without AI functionality) only collects and forwards data, while task scheduling and intelligent operations are implemented by the regional platform. Several capability support modules are deployed at the regional level to achieve intelligent inspection and intelligent safety functions.

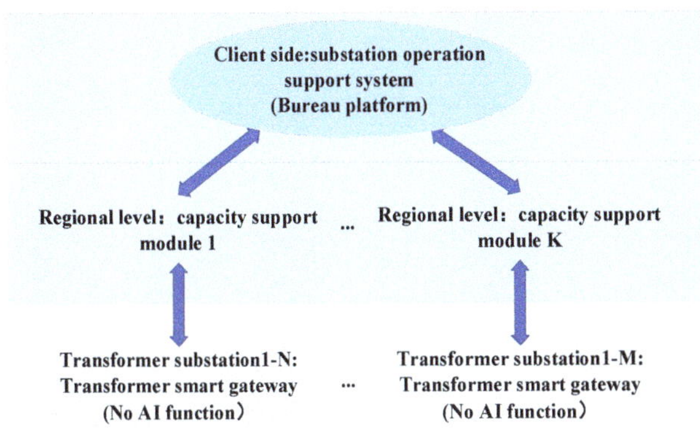

Figure 2. Layered hierarchical model diagram.

This model eases the pressure on bandwidth, arithmetic, and many other aspects through multi-tier deployment. However, it increases the operation and maintenance cost of each regional layer.

2.3. Model Three: Side–End Synergy Model

The edge–end collaborative mode refers to the mode in which data interaction and collaborative work are achieved between edge devices in the substation and the central master station. The central master station, as the core of the entire system, is responsible for receiving and processing data transmitted by edge devices. At the same time, edge devices also have certain edge computing capabilities, allowing them to perform local data processing and decision-making. Through edge computing, edge devices can perform real-time analysis and fault diagnosis on the collected data, enabling quick response to changes in the substation's operational status [25]. As shown in Figure 3, the station-side substation intelligent gateway (with AI functionality) not only has the functions of data acquisition and forwarding but also has task execution and intelligent analysis capabilities. The regional platform only handles task scheduling and operational functions.

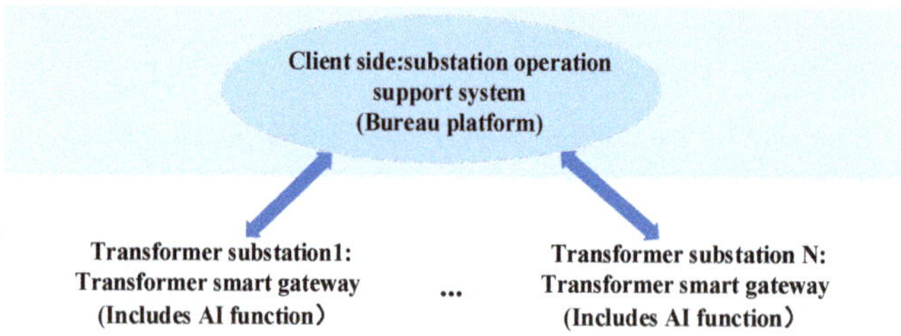

Figure 3. Side–end collaboration model diagram.

This model greatly relieves the pressure of communication and arithmetic scheduling on the side of the regional platform by realizing most of the functions at the station side. However, it relies heavily on the currently immature intelligent gateway technology capability, and the system architecture at the end is still focused on small-scale applications such as inspection and maintenance centers, failing to achieve a large scale.

In summary, each mode has its own advantages and disadvantages:

(1) Mode 1: This mode is simple and easy to implement and maintain. The central master station is responsible for data processing and decision-making. However, it may face challenges such as high pressure on the master station, a single point of failure, high latency, and an inability to perform task execution. This mode is mainly suitable for simple data collection and transmission scenarios.
(2) Mode 2: This mode uses distributed processing, which reduces the pressure on the central master station and enables task execution. It requires hardware resource configuration at different levels and needs to consider data transmission and synchronization issues. This mode is mainly suitable for large-scale data processing and decision-making scenarios.
(3) Mode 3: The edge devices have certain computing capabilities and can perform data processing and decision-making locally, collaborating with the central master station. This improves real-time performance and enables task execution. However, it may face limitations in edge device processing capabilities and challenges in coordination and synchronization between edge devices. This mode is mainly suitable for edge computing and scenarios with high real-time requirements, especially when handling a large amount of data that is difficult or slow to process.

It can be found that between different modes, for Mode 1 and Mode 2 the station-side substation intelligent gateway only collects and forwards data, and cannot realize the task execution function. Compared to Mode 3, they are more simple, but less practical. Mode 3, on the other hand, relies on intelligent gateway technology, which makes it difficult to process or slower to process when a large amount of data enters, and cannot meet the demand for real-time processing of monitoring systems in the field of substation in the future.

3. Comparison of Architecture Patterns

3.1. Business Requirements

The business requirements of domestic mega-city power grids (400–600 substations) are selected as a comparative prototype. The business requirements for the system architecture are formulated in terms of the dimensions of the core business in the production area, such as inspection, operation, security control, and storage requirements [26].

In the remote inspection scenario, with about 2000 inspection points per substation, it takes 1.5 h to complete the inspection task requirements for 600 substations;

In the remote operation scenario, 15 operation tasks are carried out at the same time (3 operation tickets at the same time, each operation ticket involves 5 substations), which needs to satisfy the demand of simultaneously completing the recognition of 15 operation equipment's position within 1 min.

In the remote security control scenario, the fall maintenance peak period is at the same time for the operation of about 150 substations, with an average of each substation having 2 work controls, each job calling for 2 cameras, to meet the demand for 4 h of continuous 600-channel video work security identification.

In terms of storage, it needs to meet the requirements of 10 T of structured data and defective equipment videos and pictures to be saved permanently, and 50 T unstructured data to be stored for more than 3 months.

3.2. Comparison and Analysis

According to the business requirements of mega-city power grids, the advantages and disadvantages of the three construction modes are analyzed and compared from the dimensions of network requirements, hardware requirements, system performance, system reliability, construction cycle, economic costs, and operation and maintenance difficulties, as shown in Table 2.

Table 2. Characterization of different models.

Dimension	Model 1	Model 2	Model 3
Network requirements (bandwidth requirements)	High	Higher	Low
Hardware requirements	34 servers, AI power $\geq$ 21,760 T	3 servers, AI power $\geq$ 1920 T	3 servers, AI power $\geq$ 1920 T
Systems performance	Less	Higher	High
Construction cycle	Long	Longer	Shorter
Specificities	High technical difficulty in realization	Reduced technical difficulty of the platform	Decentralized and independent
Economic cost (overall construction)	Low	Higher	High

In terms of network demand, Mode 1 occupies the platform-side to station-side network in the whole process, with high bandwidth demand; Mode 2 occupies the regional-side to station-side network in the whole process, with partly high bandwidth demand; and Mode 3 occupies it only in the uploading of inspection results, with low bandwidth demand.

In terms of hardware requirements, Mode 1 is expected to require 34 20-core CPU streaming servers on the platform side, with AI power $\geq$ 21,760 T; Mode 2 is expected to require 3 20-core CPU streaming servers on the platform side, with AI power $\geq$ 1920 T, and 32 20-core CPU streaming servers on the regional layer, with AI power $\geq$ 21,500 T; Mode 3 is expected to require 3 20-core CPU streaming servers on the platform side, with AI power $\geq$ 1920 T, and a station-side configuration of the smart gateway with 4-core or more CPU resources, with AI power $\geq$ 128 T.

In terms of system performance, Mode 1 video decoding and analysis need to be completed by the platform side, which is less efficient; Mode 2 video decoding and analysis are completed by each regional platform, which is more efficient; and, in Mode 3, video decoding and analysis are completed by the intelligent gateway at the station end, which is highly efficient.

In terms of system reliability, Mode 1 has low reliability, and production operations may be halted in the event of system downtime, which is not allowed to happen. Mode 2 has the highest reliability, and in the event of system downtime, it can be switched to the regional platform for access. Mode 3 is higher reliability, and in the event of system downtime, the station-side smart gateway cyclical tasks can be carried out normally.

In terms of construction cycle, Mode 1 has a long construction cycle, high pressure on system logic strategy and resource scheduling, and high technical difficulty to realize. Mode 2 has a long construction cycle, and multiple clusters can simplify the scheduling strategy, reduce the pressure of resource scheduling on the platform, and lower the technical difficulty of the platform. Mode 3 has a short construction cycle, and the station ends are dispersed to independently schedule resources.

In terms of economic cost, Mode 1 has a low overall construction cost, but the number of substations that can be accessed is low, and the long-term operational efficiency is low. Mode 2 has a high overall construction cost, high number of substations that can be accessed, and moderate long-term operational efficiency. Mode 3 has a high overall construction cost, meets the number of substations accessed in mega-cities, and has high long-term operational efficiency.

3.3. Solutions and Optimizations

After comparative analysis, this paper integrates the three modes, carries out three-point optimization, integrates the regional layer capacity support module to the platform side, maximizes the application of the AI intelligent gateway at the station side, designs the NVR at the station side to push the video streams to the intelligent gateway at the station side and the master station platform, and proposes a highly concurrent and usable "cloud-side-end collaboration" substation operation support system architecture based on edge computing on the basis of the base of the edge–end collaboration architectural model, which can satisfy the business requirements of mega-city power grids.

Optimization 1: Integrate the regional layer capacity support module to the platform side, adopt cluster deployment to realize the upper and lower levels of mutual master backup, and construct a highly available, highly stable, and highly concurrent system architecture. Under high concurrent access to achieve access load, AI computing, network communication, and other resource balancing are utilized to improve the efficiency of resource use and to ensure that the system has high reliability and compatibility with the gateway. It solves the complexity of the underlying framework of Mode 1 and the extremely high requirements for the carrying capacity of hardware resources such as the communication network, master arithmetic, and storage.

Optimization 2: Combining the current equipment state class identification and equipment defect class identification algorithms with an accuracy rate of more than 85%, configuring intelligent gateways containing AI expansion capabilities at the station side, unifying interface protocols and standards, giving full play to the edge-side computational power, sending universal algorithms from the cloud side, iterative training at the edge side, and feeding back the training results to the cloud side are utilized to continuously perform algorithm tuning. It solves the problem of high operation and maintenance costs of each regional layer in Mode 2.

Optimization 3: The station-side NVR pushes the video stream to the station-side intelligent gateway and the master station platform in one shot, decoupling the core video stream of the business from the intelligent gateway whose technology is not yet mature, and solving the problem that Mode 3 overly relies on the intelligent gateway and cannot be promoted on a large scale.

4. Architecture of "Cloud, Edge, and End Collaboration" Substation Operation Support System

The proposed "cloud, edge, and end collaboration" substation operation support system is based on the "cloud collaborative" substation operation support architecture, with the introduction of edge computing for real-time data collection and processing in substations. With the "cloud, edge, and end collaboration" architecture, real-time monitoring, fault diagnosis, load forecasting, and optimization scheduling of substations can be achieved. Additionally, this architecture can provide remote access and management capabilities, improving the efficiency of maintenance personnel and the reliability of substation operation.

4.1. Architecture of the "Cloud, Edge, and End Collaboration" System

By comparing and analyzing the existing architectures and optimizing them from different aspects, it can also be seen that the proposed "cloud-side-end collaboration" system architecture is technically feasible. Traditional IoT technology is divided into a perception layer, network layer, and application layer, which are responsible for the recognition and perception of objects and data collection, transferring information, and processing data, combining the demand and intelligent application, respectively, but with the requirements of the blowout growth of the data volume and the application scenarios, it puts a great pressure on the network transmission of data real-time, bandwidth loading capacity, and so on [27]. The introduction of edge computing and edge intelligence effectively relieves the burden of communication architecture and data processing, making the operation of the architecture more three-dimensional and efficient. The essence of edge computing is to decentralize some of the functions of the cloud center to the edge side of the network close to the data source in order to achieve local/nearby processing of data and related applications. In the context of massive data generated by the Internet of Everything, edge computing not only reduces the pressure of data traffic in the cloud center, but also improves the data processing efficiency, with low latency, low broadband, and high real-time characteristics, making it one of the key technologies to promote the development of the Internet of Things in electric power [28]. In the "cloud-side-end cooperative" system architecture, the cloud center is responsible for handling non-real-time, high-complexity, global data services; the edge side supports small real-time local data services, which do not incur high equipment costs in computation and storage, and are highly economical; and the user side serves as a source of data, providing diversified and fine-grained information to support the upper-level decision-making and precise analysis.

4.2. Key Technologies for "Cloud, Edge, and End Collaboration" Collaboration

The substation operation support system architecture proposed in this paper utilizes IoT technology, combined with cloud-side collaboration technology, including four major capabilities: data collaboration, application collaboration, intelligent collaboration, and operation and maintenance collaboration [29].

Data consistency, management consistency, and service consistency between the cloud side and the plant and station side are ensured.

(1) Data synergy: It means that the gateway realizes real-time synchronization between the business management data on the cloud side and the real-time production monitoring data on the plant and station side through the unified physical model, API service, data service, and messaging service to ensure that the data above and below the cloud are consistent and the application experience is consistent.

(2) Application synergy: It means that the gateway is docked to the cloud side, and it should have the ability to unify the management of plant-side applications as well as the remote deployment, remote upgrade, and remote monitoring of plant-side applications. The cloud-side application store is the server side, supporting the full life cycle management of application trial, uploading, releasing, subscribing, deploying, and operation and maintenance. The application store on the plant and station side

is the client side, which unifies the operation framework under the cloud and can collaborate and interact with the cloud.

(3) Intelligent collaboration: It means that the gateway is docked to the cloud side with the abilities of unified training, distribution, deployment, and remote upgrade of algorithms, and it is desirable for the gateway to have the ability of algorithm reception and operational reasoning on the plant side. Algorithms trained on the cloud side can be pushed online to the plant-side gateway for inference, and sample data generated on the plant side can be sent back to the cloud side. The downloading time of the model update should be limited to less than 2 h, and the update process should only interrupt the current interface service without affecting the interface services of other algorithms that have not been updated. In order for interface services that have not been updated with algorithms to be able to continue to perform operations, interface services during the update shall be interrupted for no more than ten minutes, and interface services after the update has been completed shall be able to continue to perform the unfinished operations prior to the update.

(4) Operation and maintenance synergy: It means that the gateway, docked on the cloud side, should have the ability to unify operation and maintenance and monitor the equipment, intelligent inspection devices, and intelligent sensor terminals on the plant side; the gateway should have the ability to perform self-tests, alarms, and logging services for the equipment, intelligent inspection devices, intelligent sensor terminals, etc. on the plant side.

4.3. Architecture of "Cloud, Edge, and End Collaboration" Substation Operation Support System

Based on the summary and analysis of the existing architecture, and combined with the actual problems in the field of substations, the "cloud-side-end" cooperative substation operation system architecture is proposed, and the system architecture is shown in Figure 4. Based on the above discussion, the "cloud-side-end" cooperative substation operation system architecture is sorted out.

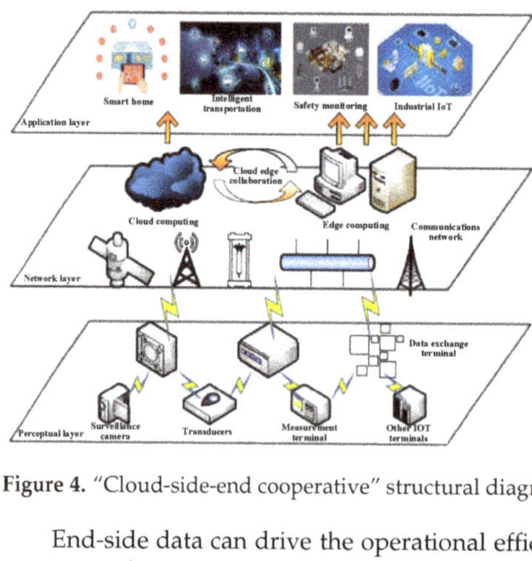

Figure 4. "Cloud-side-end cooperative" structural diagram.

End-side data can drive the operational efficiency of various aspects of substation monitoring from the bottom-up by mining the value of data. Combined with the traditional top-down operation and management mode of a power grid, the idea of cloud–sideend collaboration is utilized to design an application framework that can realize the two-way interaction between end-side data and cloud-side decision-making and utilize the edge computing technology to realize the data application in a distributed and hierarchical manner [30]. The Figure 5 shows the main framework of the system architecture design based on "cloud-side-end collaboration".

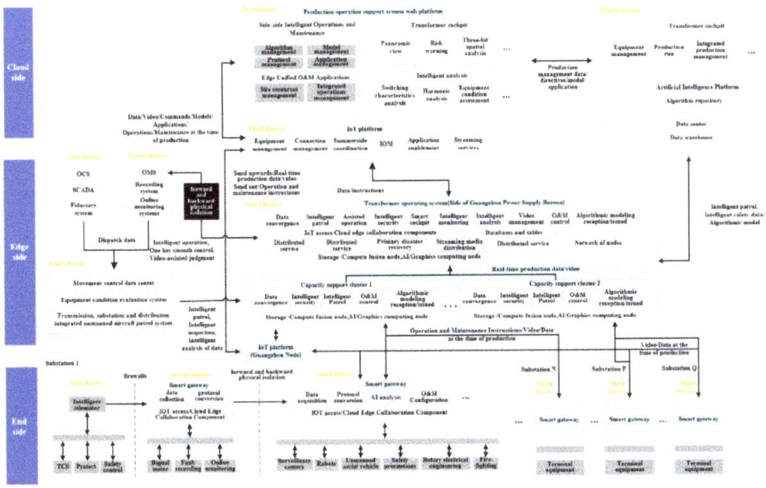

Figure 5. General architecture diagram.

In edge–end collaboration, the end side uploads smart meters, geographic meteorology, and other data to the nearby edge nodes, which use the processed real-time data and stored historical data as inputs to provide a basis for the edge nodes to realize the substation status. Compared with the traditional monitoring system, which can only be dispatched through the cloud center, the edge nodes can be regarded as multiple "monitors" assigned by the cloud center in each region to undertake fine-grained monitoring tasks.

Cloud collaboration is mainly responsible for the data interaction between the cloud center and special equipment. For example, important maintenance equipment should have a direct communication channel with the cloud center to ensure the safe and stable operation of the equipment.

4.4. Deployment Architecture of "Cloud, Edge and End Collaboration" Substation Operation Support System

Figure 6 shows the architecture of the "cloud, edge, and end collaboration" substation operation support system. On the platform side, the cloud platform allocates basic hardware and software resources, including application servers, database servers, data collection servers, streaming media servers, analysis servers, middleware servers, storage servers, load balancing servers, and cluster management servers. The system adopts distributed architecture, the basic application platform is constructed based on Spring Cloud technology route, all servers are deployed in dual-machine hot standby or cluster mode, and the system adopts application-level disaster recovery in off-site server rooms, which ensures the reliability of the system operation while greatly enhancing the system's concurrent access and bearing capacity.

In response to the demand for concurrent access by multiple users and concurrent requests for data from multiple stations, load balancing is used to optimize the connection mechanism for user access and data concurrency, ensuring smooth system access and stability under a large number of concurrent connections. For the business requirements of video stream forwarding, video stream parsing, and image and video analysis, the streaming media server and analysis server are deployed in a cluster mode, which improves the performance of video distribution, AI analysis capability, and operational reliability, and at the same time ensures that the bureau platform has the ability to flexibly expand in the long term [31].

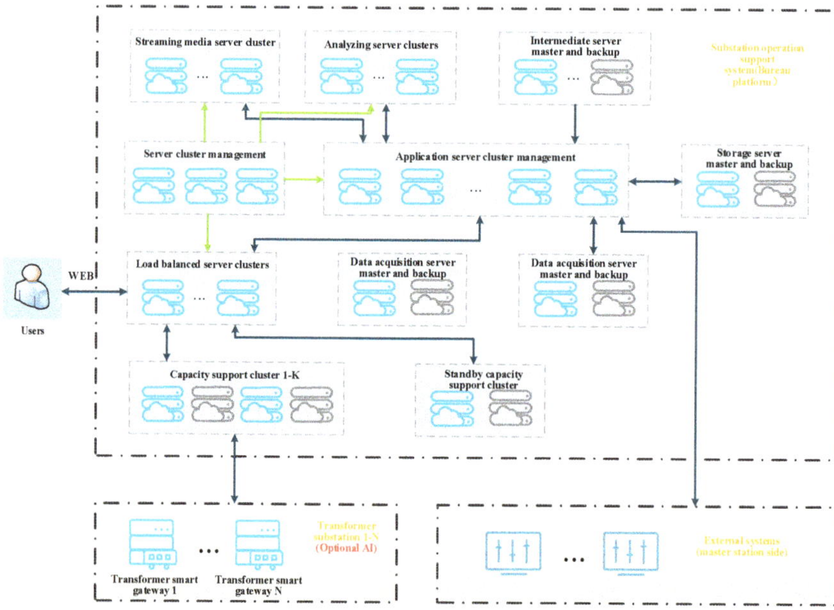

Figure 6. Architecture diagram of "Cloud, Edge, and End Collaboration" substation operation support system.

4.5. Technical Architecture of "Cloud, Edge and End Collaboration" Substation Operation Support System

The technical architecture of the "cloud, edge and end collaboration" substation operation support system is divided into a perception layer, edge layer, network layer, digital platform layer, and application layer [32]. The structure is shown in Figure 7.

The sensing layer collects substation equipment data in an all-round way through digital meters, online monitoring devices, sensors, cameras, and other intelligent terminals.

The edge layer carries out data forwarding, analysis, and other processing through the intelligent gateway (with AI function). The edge layer uses a cloud platform to allocate basic hardware and software resources, including application servers, database servers, data collection servers, etc. [33]. The system can adopt a distributed architecture to ensure the reliability of system operation. Enhancing video distribution performance, AI analysis capability, and operational reliability can ensure that the bureau platform has the ability to flexibly expand capacity in the long term.

The network layer, on the other hand, ensures stable data transmission through communication protocols and communication statutes, ensures high-speed data transmission, and improves the reliability of the system. The substation operation support system utilizes the integrated data network for data transmission.

The digital platform layer realizes the operation and maintenance monitoring of data and ensures the application of equipment infrastructure through IoT platforms, data centers, and cloud platforms.

The application layer, on the other hand, utilizes artificial intelligence, media, etc. to present applications and security management to ensure real-time monitoring.

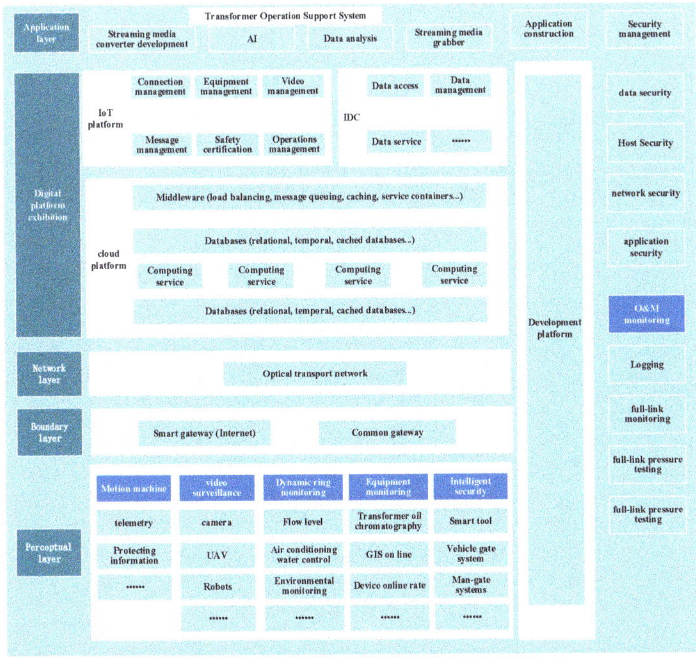

Figure 7. Technical architecture diagram.

5. Engineering Application Examples

The "cloud, edge, and end collaboration" substation operation support system introduces edge computing and edge intelligence, bringing cloud functionalities closer to the edge where data sources are located. This enables on-site data processing and related applications, effectively alleviating the burden on communication architecture and data processing. It reduces the data traffic pressure on the cloud center while improving data processing efficiency. This architecture is characterized by low latency, low bandwidth, and high real-time performance. To validate the effectiveness of the edge–end collaborative structure on the system, the proposed architecture is subjected to practical simulation testing [34].

The hardware environment for testing consists of a Qitian M620 desktop computer, equipped with an Intel Core i3-9100 CPU @3.60 GHz*4 processor, 16 GB of memory, and a 1TB hard drive. As for the software environment, the system used is Windows 10 Enterprise Edition as the test client operating system and LoadRunner 11 as the performance efficiency testing tool. The testing network environment is LAN.

Test content and method: Select the performance points such as environment monitoring, backup monitoring, and environment monitoring inspection result confirmation for download. Simulate 500 concurrent user accesses to the system, and record the response time, transaction success rate, server CPU occupancy rate, memory utilization rate, and other indices.

In the Section 2 of this paper, three mainstream architectures existing in the current power system are analyzed and compared in detail, which are the master connection mode, the hierarchical classification mode, and the side–end collaboration mode. Compared with Mode 1 and Mode 2, Mode 3 not only collects and forwards data, but also realizes the task execution function. In the current mainstream architecture, Mode 3 has relatively high performance and benefits. In order to better validate the feasibility of changing the architecture, this test will also simulate the testing of concurrent access to the system for

500 users in Mode 3 and analyze the comparison through the final test results. The results are shown in Table 3 and Figure 8.

Table 3. Results of 500 concurrent user accesses to the system.

Performance Names	Number of Concurrent Users		Average Response Time (s)		Success Rate	
	New mode	**Mode 3**	**New mode**	**Mode 3**	**New mode**	**Mode 3**
Environment monitoring	500	500	3.166	5.423	99.99%	92.65%
Equipment monitoring	500	500	5.656	8.462	99.99%	93.42%
Intelligent alarm	500	500	11.428	16.452	99.98%	93.36%
Inspection result confirmation download	500	500	0.742	2.389	99.99%	92.33%

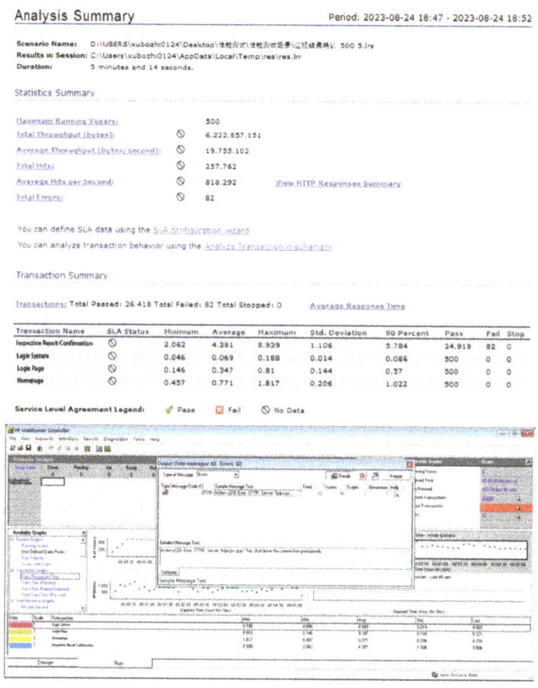

Figure 8. Test result chart.

Test results: In the simulation of 500 users' concurrent access, the average response time of the environment monitoring performance point of the new mode is 3.166 s, with a passing rate of 99.9%, while the average response time of the environment monitoring performance point of the Mode 3 case is 5.423 s, with a passing rate of 92.65%; the average response time of the device monitoring performance point of the new mode is 5.656 s, with a passing rate of 99.99%, while the average corresponding time for the device monitoring performance point in the case of Mode 3 is 8.462 s, with a passing rate of 93.42%; the average response time for the smart alarms of the new mode is 11 s, with a passing rate of 99.99%, while the average response time for the smart alarms of Mode 3 is 16.452 s, with a passing rate of 93.36%. The event passing rate of the inspection results for the new mode is 99.99% and the event passing rate of the inspection results for Mode 3 is 92.33%. Therefore, compared with the model considered, the new system architecture is better at meeting most of the mega-city power grid business requirements.

Compared with the traditional mainstream architecture Mode 3, the new system architecture shortens the response time of environment monitoring by 2.257 s, equipment monitoring by 2.806 s, intelligent alarm by 5.024 s, and inspection result confirmation by 1.647 s. Meanwhile, compared with the traditional mainstream architecture Mode 3, the success rate of key task execution increases by 7.34% for environment monitoring, 6.57% for equipment monitoring, 6.62% for intelligent alarm, and 7.66% for inspection result confirmation download. Meanwhile, compared with the traditional mainstream architecture Mode 3, the success rate of environment monitoring increased by 7.34%, the success rate of equipment monitoring increased by 6.57%, the success rate of intelligent alarm increased by 6.62%, and the success rate of inspection result confirmation download increased by 7.66%. Due to the current immature intelligent gateway technology capability, the side–end system architecture of Mode 3 is still focused on a small range of applications such as inspection and maintenance centers, and fails to scale up. Meanwhile, under the comparison of results, compared with the side–end collaborative architecture of Mode 3, the new system architecture has a greater improvement in real-time processing and accuracy.

Under the proposed architecture, the intelligent inspection result confirmation and defect management are analyzed. The results are shown in Figures 9 and 10. It can be seen that the average time for inspection result confirmation is only 0.3 s, and the average time for defect management is only 0.3 s, and the correct rate is 100%. It can be shown that the architecture proposed in this paper can effectively support the technology in the case of concurrent access by 500 users.

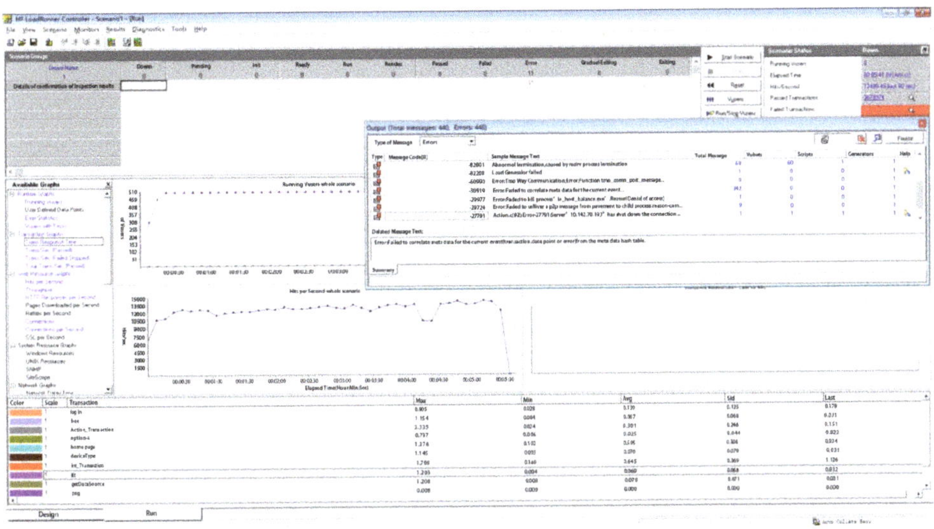

Figure 9. Chart confirming the results of inspections.

Based on the comparison of simulation test and analysis results, the proposed "cloud-side-end coordination" substation operation support system architecture is better in terms of performance. Compared with the traditional cloud coordination system architecture, which is only applicable to small-scale grid systems, the "cloud-side-end coordination" system architecture can better meet the business requirements of most mega-city grids.

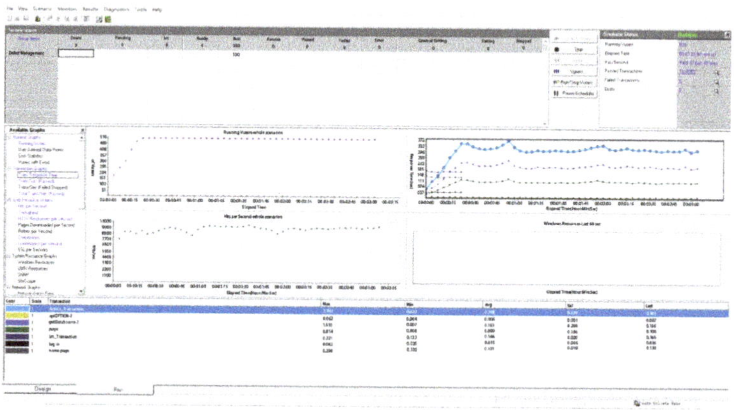

Things	Average value	Error%	Note
Actiong keyholding	0.3	0	

Figure 10. Defect management results chart.

6. Conclusions

By organizing and summarizing the existing technologies related to substation systems, the "cloud-edge-end" collaborative capability is analyzed, and a "high availability, high concurrency, high performance, high security" architecture for the "cloud, edge, and end collaboration" substation operation support system is proposed. In the simulation test section, the system architecture shows better real-time performance and accuracy to better cope with the business requirements of large-scale power grids. Meanwhile, based on simulation test comparisons, the proposed architecture for the substation operation support system achieves precise monitoring of equipment operation, unified management of operational data, and integration of production command decision-making. It can effectively respond to grid operation risks and comprehensively support the digital transformation of substations. Application examples demonstrate that the proposed architectural approach has significant advantages in terms of large-scale substation integration and reasonable distribution of computing power. The passing rate under different performance conditions exceeds 99%, providing effective technical support for concurrent access by 500 users and meeting the requirements for the scaled-up promotion of digital substations. Compared with the traditional system architecture Mode 3, the substation operation architecture proposed in this paper in can greatly improve the real-time performance and accuracy of the monitoring system and can play a better role in large-scale urban power grid operation and management.

In the proposed architecture for the substation operation support system, further analysis and research are required to consider system performance when there is concurrent access by a larger number of users. In subsequent research, based on a unified technical approach, the architecture of cloud–edge collaboration will be implemented by selecting pilot substations and converter stations to complete the coordination and verification of edge and end devices.

Author Contributions: Y.L.: Conceptualization, methodology, writing—original draft preparation; Y.B.: software, data curation, supervision, writing—review and editing; L.Z.: validation, funding acquisition, writing—review and editing. All authors have read and agreed to the published version of the manuscript.

Funding: Education Department of Hunan Province: kq2202213.

Data Availability Statement: Data are contained within the article.

Conflicts of Interest: Authors Yun Long and Yude Bao were employed by the company Guangdong Power Grid Co., Ltd. The remaining authors declare that the research was conducted in the absence of any com-mercial or financial relationships that could be construed as a potential conflict of interest.

References

1. Ma, C.; Deng, T.; Su, Y.; Ding, P.; Guo, H.; Lu, T. How to Combine Standardization Digital Transformation Technology with Power Business Scenarios: Demand Analysis, Technical Route and Application Practice. In Proceedings of the 2022 IEEE 6th Conference on Energy Internet and Energy System Integration (EI2), Chengdu, China, 11–13 November 2022; pp. 1729–1733.
2. Hui, Y. Application of RTU System in Substation Reconstruction Project. *Electr. Eng.* **2023**, 125–127+131.
3. Guo, W.; Meng, J. The design of substation RTU monitoring system based on PLC and GPRS. In Proceedings of the 2011 International Conference on Electrical and Control Engineering, Yichang, China, 16–18 September 2011; pp. 896–899.
4. Si, X.; Li, Z.; Wang, L.; Yang, T.; Li, X. A Protection Scheme Based on Current Phase Change for Active Distribution System. *Shandong Electr. Power* **2019**, *46*, 13–17.
5. Byk, F.; Frolova, Y.; Myshkina, L. The efficiency of distributed and centralized power system integration. *E3S EDP Sci.* **2019**, *114*, 05007. [CrossRef]
6. Hashemi, M.; Zarif, H.M. A novel hierarchical distributed framework for optimal reactive power dispatch based on a system of systems structure. *Comput. Electr. Eng.* **2019**, *78*, 162–183. [CrossRef]
7. Liu, Y.; Liu, Z.; Li, L.; Sun, Y.; Zhang, S. Review on Distributed Control Methods for Active Distribution Network with Distributed Genernation. *Shandong Electr. Power* **2021**, *48*, 13–18+38.
8. Yang, G.; Qian, A. Distributed hierarchical coordinated control of active distribution network with multiple microgrids based on sparse communication network. *Autom. Electr. Power Syst.* **2018**, *42*, 135–141.
9. Li, W.; Weng, X.; Li, H.; Zhang, J.; Xia, T. The research of AIOps of electric transmission and transformation equipment based on big data a nalysis model. *J. Inf. Technol.* **2021**, *7*, 54–58+64.
10. Wang, Y.B.; Wang, Z.; Zhao, D.Y.; Zhang, H.F. Intelligent Operation and Maintenance of Substations Based on Internet of Things (IoT) Technology. *Appl. Mech. Mater.* **2015**, *742*, 708–716.
11. Zhou, K.; Wang, X.; Qiu, X.; Wu, M.; Wu, C.; Li, W. Analysis of Transparent Substation Architecture and Key Technologies Empowered by Digital Technology. *Electr. Meas. Instrum.* **2023**, 1–10.
12. Aiqing, M. Analysis on Technical Management of Power Grid Substation Operation and Maintenance. *Electr. Eng.* **2021**, *50*, 166–167.
13. Fan, Z. Research and Application of Monitoring System for Substation Equipment Based on Edge Computing. *Electr. Eng.* **2021**, 89–91.
14. Chen, C.; Gong, L.; Luo, X. Research on the management model of distribution IoT based on cloud-edge collaboration. *Zhejiang Electr. Power* **2023**, *42*, 48–56.
15. Yan, Z. Design and implementation of power video monitoring network based on cloud computing technology. *Sci. Technol. Innov. Her.* **2018**, *15*, 93–94.
16. Nan, Y.; Jin, Y. Intelligent detection technology for power video monitoring malfunction and state based on cloud computing method. *East China Electr. Power* **2013**, *41*, 2494–2497.
17. Qin, R. Research on the construction method of the "three-chemical" system of substation operation and maintenance. *Lamps Light.* **2021**, 133–134.
18. Divan, D.; Luckjiff, G.; Brumsickle, W.; Freeborg, J.; Bhadkamkar, A. I-grid TM: Infrastructure for nationwide real-time power monitoring. In Proceedings of the Conference Record of the 2002 IEEE Industry Applications Conference. 37th IAS Annual Meeting (Cat. No.02CH37344), Pittsburgh, PA, USA, 13–18 October 2002; Volume 3, pp. 1740–1745.
19. Haba, C.; Breniuc, L. Power Monitoring System Based on ADM1191 Device. In *Buletinul Institutului Politehnic din Iaşi*; Electrotehnică, Energetică, Electronică; Technical University of Iaşi: Iaşi, Romania, 2016; Volume 61.
20. Dhiman, A.; Perveen, R. UPIoT Based Power Monitoring of Transformer. In *Materials Today: Proceedings*; Part 2; Elsevier: Amsterdam, The Netherlands, 2022; Volume 71, pp. 227–231.
21. Jadhav, A.R.; MPR, S.K.; Pachamuthu, R. Development of a Novel IoT-Enabled Power—Monitoring Architecture with Real-Time Data Visualization for Use in Domestic and Industrial Scenarios. *IEEE Trans. Instrum. Meas.* **2021**, *70*, 1002314. [CrossRef]
22. Li, B.; Jia, B.; Chen, S.; Yang, B.; Sun, Y.; Qi, B. Prospect of application of edge computing in the field of supply and demand. *Electr. Power* **2018**, *51*, 154–162.
23. Okay, F.Y.; Ozdemir, S. A fog computing based smart grid model. In Proceedings of the 2016 International Symposium on Networks, Computers and Communications (ISNCC), Hammamet, Tunisia, 11–13 May 2016; pp. 1–6.
24. Chen, S.; Wen, H.; Wu, J.; Lei, W.; Hou, W.; Liu, W.; Xu, A.; Jiang, Y. Internet of Things Based Smart Grids Supported by Intelligent Edge Computing. *IEEE Access* **2019**, *7*, 74089–74102. [CrossRef]
25. Masoud, F.K.; Asghar, F.A. Adaptive distributed scheduling of the resources in smart grids with considering demand response. *Trans. Inst. Meas. Control* **2023**, *45*, 2779–2793.

26. Pan, J.; McElhannon, J. Future Edge Cloud and Edge Computing for Internet of Things Applications. *IEEE Internet Things J.* **2018**, *5*, 439–449. [CrossRef]
27. Alam, S.; Siddiqui, S.T.; Ahmad, A.; Ahmad, R.; Shuaib, M. Internet of things (IoT) enabling technologies, requirements, and security challenges. In *Advances in Data and Information Sciences: Proceedings of ICDIS 2019*; Springer: Singapore, 2020; pp. 119–126.
28. Lu, Z.; Fan, X. Safety issues of power grid operation and equipment maintenance and improvement measures. *China High New Technol.* **2022**, 56–57.
29. Li, B.; Jia, B.; Cao, W.; Tian, S.; Qi, B.; Sun, Y.; Zhu, W.; Zheng, A. Application prospect of edge computing in power demand response business. *Power Syst. Technol.* **2018**, *42*, 79–87.
30. Qi, B.; Xia, Y.; Li, B.; Shi, K.; Xue, M.F. Family energy management system based on edge computing: Architecture, key technology and implementation. *Electr. Power Constr.* **2018**, *39*, 33–41.
31. Gong, G.; Luo, A.; Chen, Z.; Luan, J.Z.; An, X.N.; Wang, X.; Su, C.; Wen, Y. Cyber physical system of active distribution network based on edge computing. *Power Syst. Technol.* **2018**, *42*, 3128–3135.
32. Kumar, N.; Zeadally, S.; Rodrigues, J.J. Vehicular delay-tolerant networks for smart grid data management using mobile edge computing. *IEEE Commun. Mag.* **2016**, *54*, 60–66. [CrossRef]
33. Nauman, A.; Alruwais, N.; Alabdulkreem, E.; Nemri, N.; Aljehane, N.O.; Dutta, A.K.; Assiri, M.; Khan, W.U. Empowering smart cities: High-altitude platforms based Mobile Edge Computing and Wireless Pow-er Transfer for efficient IoT data processing. *Internet Things* **2023**, *24*, 2542–6605. [CrossRef]
34. Ma, Y. *Deepening Research on the Functions of Technical Support System under the Intensification Mode of Substation Operation*; Nanrui Science and Technology Co.: Nanjing, China, 2012.

MDPI AG
Grosspeteranlage 5
4052 Basel
Switzerland
Tel.: +41 61 683 77 34

Energies Editorial Office
E-mail: energies@mdpi.com
www.mdpi.com/journal/energies

www.ingramcontent.com/pod-product-compliance
Lightning Source LLC
Chambersburg PA
CBHW061543220926
51850CB00005B/239